Recent Advancements in Bioremediation of Metal Contaminants

Satarupa Dey
Shyampur Siddheswari Mahavidyalaya, India

Biswaranjan Acharya
School of Computer Engineering, KIIT University (Deemed), India

A volume in the Advances in Environmental Engineering and Green Technologies (AEEGT) Book Series

IGI Global
PUBLISHER of TIMELY KNOWLEDGE

Published in the United States of America by
IGI Global
Engineering Science Reference (an imprint of IGI Global)
701 E. Chocolate Avenue
Hershey PA, USA 17033
Tel: 717-533-8845
Fax: 717-533-8661
E-mail: cust@igi-global.com
Web site: http://www.igi-global.com

Library of Congress Cataloging-in-Publication Data

Names: Dey, Satarupa, 1981- editor. | Acharya, Biswa Ranjan, 1985- editor.
Title: Recent advancements in bioremediation of metal contaminants / Satarupa Dey and Biswa Ranjan Acharya, editors.
Description: Hershey, PA : Engineering Science Reference, an imprint of IGI Global, [2020] | Includes bibliographical references and index. | Summary: "This book explores bioremediation of pollutants from industrial wastes and examines the role of diverse forms of microbes in bioremediation of wastewater. Covering a broad range of topics including microorganism tolerance, phytoremediation, and fungi, the role of different extremophiles and biofilms in bioremediation are also discussed"-- Provided by publisher.
Identifiers: LCCN 2019060095 (print) | LCCN 2019060096 (ebook) | ISBN 9781799848882 (hardcover) | ISBN 9781799852124 (paperback) | ISBN 9781799848899 (ebook)
Subjects: LCSH: Bioremediation.
Classification: LCC TD192.5 .R427 2020 (print) | LCC TD192.5 (ebook) | DDC 628.1/6836--dc23
LC record available at https://lccn.loc.gov/2019060095
LC ebook record available at https://lccn.loc.gov/2019060096

This book is published in the IGI Global book series Advances in Environmental Engineering and Green Technologies (AEEGT) (ISSN: 2326-9162; eISSN: 2326-9170)

British Cataloguing in Publication Data
A Cataloguing in Publication record for this book is available from the British Library.

All work contributed to this book is new, previously-unpublished material. The views expressed in this book are those of the authors, but not necessarily of the publisher.

For electronic access to this publication, please contact: eresources@igi-global.com.

Advances in Environmental Engineering and Green Technologies (AEEGT) Book Series

Sang-Bing Tsai
Zhongshan Institute, University of Electronic Science and Technology of China, China & Wuyi University, China
Ming-Lang Tseng
Lunghwa University of Science and Technology, Taiwan
Yuchi Wang
University of Electronic Science and Technology of China Zhongshan Institute, China

ISSN:2326-9162
EISSN:2326-9170

Mission

Growing awareness and an increased focus on environmental issues such as climate change, energy use, and loss of non-renewable resources have brought about a greater need for research that provides potential solutions to these problems. Research in environmental science and engineering continues to play a vital role in uncovering new opportunities for a "green" future.

The **Advances in Environmental Engineering and Green Technologies (AEEGT)** book series is a mouthpiece for research in all aspects of environmental science, earth science, and green initiatives. This series supports the ongoing research in this field through publishing books that discuss topics within environmental engineering or that deal with the interdisciplinary field of green technologies.

Coverage

- Alternative Power Sources
- Radioactive Waste Treatment
- Contaminated Site Remediation
- Sustainable Communities
- Water Supply and Treatment
- Renewable Energy
- Industrial Waste Management and Minimization
- Cleantech
- Air Quality
- Electric Vehicles

IGI Global is currently accepting manuscripts for publication within this series. To submit a proposal for a volume in this series, please contact our Acquisition Editors at Acquisitions@igi-global.com or visit: http://www.igi-global.com/publish/.

Titles in this Series

For a list of additional titles in this series, please visit:
http://www.igi-global.com/book-series/advances-environmental-engineering-green-technologies/73679

Innovations, Models, and Policies of Green Engineering Technologies
Daniel Dacuma Dasig Jr (De La Salle University Dasmarinas, Philippines) and Mirzi Llego Betasolo (Papua New Guinea University of Technology, Papua New Guinea)
Engineering Science Reference • © 2020 • 300pp • H/C (ISBN: 9781799840336) • US $195.00

Energy-Efficient Underwater Wireless Communications and Networking
Nitin Goyal (Chitkara University Institute of Engineering and Technology, Chitkara University, Punjab, India) Luxmi Sapra (Chitkara University Institute of Engineering and Technology, Chitkara University, Punjab, India) and Jasminder Kaur Sandhu (Chitkara University Institute of Engineering and Technology, Chitkara University, Punjab, India)
Engineering Science Reference • © 2020 • 300pp • H/C (ISBN: 9781799836407) • US $195.00

Energetics Perspective on the Environmental and Human Impact of Buildings
Teodora Melania Şoimoşan (Technical University of Cluj-Napoca, Romania) and Ligia Mihaela Moga (Technical University of Cluj-Napoca, Romania)
Engineering Science Reference • © 2020 • 300pp • H/C (ISBN: 9781799835721) • US $195.00

Molecular Plant Breeding and Genome Editing Tools for Crop Improvement
Pradip Chandra Deka (Sir Padampat Singhania University, India)
Engineering Science Reference • © 2020 • 321pp • H/C (ISBN: 9781799843122) • US $195.00

Population Dynamics and Conservation of Red Deer Emerging Research and Opportunities
Junaid Ahmad Malik (Govt. Degree College, Bijbehara, India)
Engineering Science Reference • © 2020 • 200pp • H/C (ISBN: 9781799828112) • US $175.00

Smart Design and Economic Perspectives of Renewable Energy Systems Emerging Research and Opportunities
Ali Mohamed Eltamaly (Mansoura University, Egypt) and Mohamed Abdelaziz Mohamed (Minia University, Egypt)
Engineering Science Reference • © 2020 • 165pp • H/C (ISBN: 9781799841746) • US $165.00

Analyzing the Physics of Radio Telescopes and Radio Astronomy
Kim Ho Yeap (Universiti Tunku Abdul Rahman, Malaysia) and Kazuhiro Hirasawa (University of Tsukuba, Japan)
Engineering Science Reference • © 2020 • 324pp • H/C (ISBN: 9781799823810) • US $225.00

IGI Global
PUBLISHER of TIMELY KNOWLEDGE

701 East Chocolate Avenue, Hershey, PA 17033, USA
Tel: 717-533-8845 x100 • Fax: 717-533-8661
E-Mail: cust@igi-global.com • www.igi-global.com

Editorial Advisory Board

Table of Contents

Foreword xv

Preface xvi

Acknowledgment xx

Chapter 1
Different Bioremediation Techniques for Management of Waste Water: An Overview 1
Biswajit Patra, Sambalpur University, Odisha, India
Saroj Kumar Deep, Sambalpur University, Odisha, India
Surya Narayan Pradhan, School of Life Sciences, Sambalpur University, Odisha, India

Chapter 2
Tolerance of Microorganisms to Heavy Metals 19
Joan Mwihaki Nyika, Technical University of Kenya, Kenya

Chapter 3
Microbial Enzymes and Their Mechanisms in the Bioremediation of Pollutants 36
Karthika Rajamanickam, Mahendra Arts and Science College (Autonmous), India
Jayanthi Balakrishnan, Mahendra Arts and Science College (Autonomous), India
Selvankumar Thangaswamy, Mahendra Arts and Science College (Autonomous), India
Govarthanan Muthusamy, Kyungpook National University, South Korea

Chapter 4
The Use of Microorganism-Derived Enzymes for Bioremediation of Soil Pollutants 54
Joan Mwihaki Nyika, Technical University of Kenya, Kenya

Chapter 5
Role of Bacterial Chromate Reductase in Bioremediation of Chromium-Containing Wastes 72
Satarupa Dey, Shyampur Siddheswari Mahavidyalaya, India

Chapter 6
Bacterial Remediation of Chromium From Industrial Sludge 97
Dipankar Roy, Calcutta University, India & St. Xavier's College, India
Arup Kumar Mitra, St. Xavier's College, India

Chapter 7
Phytoremediation: A Modern Approach ... 126
Suparna Pal, Lady Brabourne College, Kolkata, India

Chapter 8
Ricinus communis: A Potent Lead (Pb) Accumulator ... 147
Raikamal Pal, Shri Shikshayatan College, India

Chapter 9
Environmental Fluoride: Impact on Humans, Animals and Its Remediation Strategies ... 165
Junaid Ahmad Malik, Government Degree College, Bijbehara, India

Chapter 10
Cadmium- and Lead-Tolerant PGPRs as Proficient Toxicity Alleviators for Agricultural Crops ... 189
Amit Kumar Pal, Department of Botany, University of Kalyani, India
Anjan Hazra, Agricultural and Ecological Research Unit, Indian Statistical Institute, Kolkata, India
Chandan Sengupta, Department of Botany, University of Kalyani, India

Chapter 11
Fungi-Mediated Detoxification of Heavy Metals ... 205
Suchhanda Ghosh, Shri Shikshayatan College, India

Chapter 12
An Eco-Friendly Approach for the Eradication of Heavy Metal Contaminants by Nano-Bioremediation ... 220
Chandana Mohanty, Kalinga Institute of Industrial Technology (Deemed), India
Sneha Shriparna Satpathy, Kalinga Institute of Industrial Technology (Deemed), India
Sweta Mohanty, Kalinga Institute of Industrial Technology (Deemed), India

Chapter 13
Sustainable Treatment of Landfill Leachate Using Constructed Wetlands: An Eco-Friendly Approach ... 237
Vivek Rana, Central Pollution Control Board, Ministry of Environment, Forest, and Climate Change, Government of India, Delhi, India

Chapter 14
Safety and Efficacy of Pseudomonas Exopolymer in Sequestration of Iron From Aqueous Environments ... 256
Moushumi Ghosh, Thapar Institute of Engineering and Technology, India
Divya Sharma, Thapar Institute of Engineering and Technology, India
Taranpreet Kaur, Thapar Institute of Engineering and Technology, India

Chapter 15
Recent Advancements in Bioremediation of Metal Contaminants ... 273
Yuvarajan Ragunathan, Mahendra Arts and Science College (Autonomous), India

Kannan Nallakumar, Mahendra Arts and Science College (Autonomous), India
Thirumalaisamy Rathinavel, Mahendra Arts and Science College (Autonomous), India
Muthusamy Govarthanan, Kyungpook National University, South Korea
Selvankumar Thangaswamy, Mahendra Arts and Science College (Autonomous), India

Compilation of References 288

About the Contributors 356

Index 361

Detailed Table of Contents

Foreword xv

Preface xvi

Acknowledgment xx

Chapter 1
Different Bioremediation Techniques for Management of Waste Water: An Overview 1
Biswajit Patra, Sambalpur University, Odisha, India
Saroj Kumar Deep, Sambalpur University, Odisha, India
Surya Narayan Pradhan, School of Life Sciences, Sambalpur University, Odisha, India

Water contamination remains an issue. A combination of biodegradation and nanotechnology is proposed as a potential proficient, minimal effort, and naturally amiable system to deal with it. Among different mediations, bioremediation procedures can conceivably be utilized to decrease the versatility of materials in the subsurface, reducing the potential for human and ecological exposure. The metabolic diversity of microorganisms ensures an assortment of substrates to be expended. Photosynthetic microorganisms have been found as a compelling and eco-friendly species that can remove carbon, nitrogen, and phosphorous in the manufactured sewage and wastewater. This chapter particularly emphasizes environmentally friendly NMs that give information for removing contaminants from wastewater and effluents. Additionally, various nanocomposites and different natural methods utilized in the wastewater treatment process are also briefly discussed.

Chapter 2
Tolerance of Microorganisms to Heavy Metals 19
Joan Mwihaki Nyika, Technical University of Kenya, Kenya

Heavy metal pollution is a growing environmental concern due to the increase in anthropogenic-based sources. Microorganisms have high adsorptive capacities and surface-area-to-volume ratio that enable the uptake of these contaminants and their conversion to innocuous complexes in the process of bioremediation. This chapter explores the mechanisms and specific microorganisms that are resistant to metal toxicity. A wide range of bacterial, algae, and fungal species used as biosorbents are highlighted. Mechanisms such as reduction of metal cations, their sequestration, and binding on cell barriers are discussed. To optimise the efficacy of microorganisms in bioremediation processes, adoption of genetic and nano-technologies is recommended.

Chapter 3
Microbial Enzymes and Their Mechanisms in the Bioremediation of Pollutants 36
Karthika Rajamanickam, Mahendra Arts and Science College (Autonmous), India
Jayanthi Balakrishnan, Mahendra Arts and Science College (Autonomous), India
Selvankumar Thangaswamy, Mahendra Arts and Science College (Autonomous), India
Govarthanan Muthusamy, Kyungpook National University, South Korea

Bioremediation is the process, with the help of microbes or their enzymes, to remove the pollutants present in the environment and change them into non-toxic compounds. Microbial enzymes have a wide range of metabolic activities and are involved in the transformation of pollutants. Enzymes like oxidoreductase, hydrolases, monooxygenase, dioxygenase, methyltransferases, and lipases are involved in the degradation process. Oxidoreductase catalyzes the transfer of electron and proton from the reduced organic substrate to another chemical compound from donor to acceptor. Monooxygenase and dioxygenases are the transferring oxygen from molecular oxygen (O2) utilizing FAD/NADH/NADPH as a co-substrate in this process. Lyases catalyze the cleavage of the bonds by elimination, leaving double bonds. Peroxidases catalyze the oxidation of lignin and other phenolic compounds at the expense of hydrogen peroxide (H2O2) in the presence of a mediator. Lipases also involve catalyzing the hydrolysis of triacylglycerols to glycerol and free fatty acids.

Chapter 4
The Use of Microorganism-Derived Enzymes for Bioremediation of Soil Pollutants 54
Joan Mwihaki Nyika, Technical University of Kenya, Kenya

Contamination of soils by xenobiotic compounds is a growing concern for environmentalists amidst the rise of anthropogenic activities that encourage such contamination practices. The use of microbial enzymes is a viable alternative to degrade and mineralize these contaminants, which is a growing research interest owing to its eco-friendly nature. This chapter explores the categories of enzymes used in soil bioremediation such as oxidoreductases and hydrolases, their mechanism of action, and their merits and demerits. Furthermore, molecular biology techniques useful in enhancing the production capacity, stability, activity, and shelf life of bioremediation enzymes is discussed. Ultimately, the need to develop bioremediation enzymes in bulk, using cheap technologies while optimising their activity, stability, and shelf life for effective soil decontamination is emphasized.

Chapter 5
Role of Bacterial Chromate Reductase in Bioremediation of Chromium-Containing Wastes 72
Satarupa Dey, Shyampur Siddheswari Mahavidyalaya, India

Chromium toxicity is a major environmental concern as it is the chief environmental pollutant released by paint, stainless steel, and mining industries. In nature, chromium exists in two stable valance states: Cr(VI) and Cr(III). Cr(VI) is highly toxic and soluble at neutral pH, whereas Cr(III) is insoluble at normal pH and is less toxic. Thus, it is essential to draw strategies for mitigation of Cr(VI), and microbial reduction of toxic Cr(VI) has been identified as a bioremediation technique not only to detoxify chromium but also to recover the non-toxic Cr(III) by physical means. Chromate reductase, the central enzyme involved in bioreduction of Cr(VI) to Cr(III) may be both intracellular as well as extracellular in nature. Most of the chromate reductase enzyme belongs to the oxidoreductase group such as nitroreductase, iron reductase, quinone reductase, hydrogenase, flavin reductase, as well as NAD(P)H-dependent reductase. Detailed analysis of the structure of the enzymes will help us in the suitable application of these enzymes in bioremediation of metal-contaminated wastes.

Chapter 6
Bacterial Remediation of Chromium From Industrial Sludge 97
Dipankar Roy, Calcutta University, India & St. Xavier's College, India
Arup Kumar Mitra, St. Xavier's College, India

Chromium-like heavy toxic metals seriously influence the metabolism of living organisms and cause permanent threatening of health. Microorganisms can help to detoxify those hazardous heavy metals in the environment by the process of bioremediation. Two bacterial genera were isolated from industrial sludge designated P1 and P2. From the 16srRNA study, it is revealed that P1 is *Bacillus cereus* and P2 is *Enterobacter* sp. They are deposited in NCMR and NCBI and received the accession no. MCC 3868 for P1 and MCC 3788 for P2. P1 is gram positive, motile, and P2 is gram negative, motile. Eighteen antibiotics have been taken for antibiotic assay; P1 is resistant to 12; P2 is resistant to 8 antibiotics. For growth pattern analysis in chromium, three parameters have been selected, and they are temperature, pH, and biomass. In LD50 and above parameters, total chromium uptake by those bacteria in stressed conditions have been recorded. The two bacteria are not antagonistic to each other so they are used to bioremediate chromium from their contaminated sites and also treated as consortium.

Chapter 7
Phytoremediation: A Modern Approach 126
Suparna Pal, Lady Brabourne College, Kolkata, India

This chapter includes the sources of cadmium and chromium contamination of soil and various detrimental effects on plants and animals. Ecofriendly approach of soil clean up by phytoremediation is the main focus of the author. Heavy metal-induced oxidative stress of plants and their detoxification potentiality has been discussed here to create a wholesome idea about the basic and acute need of phytoremediation. Both enzymatic and non-enzymatic antioxidative defense mechanisms and various other biochemical parameters of metal hyperaccumulator plants are mentioned.

Chapter 8
Ricinus communis: A Potent Lead (Pb) Accumulator 147
Raikamal Pal, Shri Shikshayatan College, India

Contamination of soil and ground water with heavy metals is a great threat to human health, vegetation, and wildlife. Pb is the second most hazardous substance according to ATSDR. The main sources of Pb entering an ecosystem are atmospheric Pb (mainly from automobile emission), paint chips, fertilizers, and pesticides and Pb acid batteries or other industrial Pb products. Phytoremediation could provide sustainable techniques for metal remediation. Roots of *Ricinus communis* were found to accumulate maximum amount of Pb (275.12mg/kg dry wt,). Depending on soil Pb content, the concentration of Pb in shoots of *Ricinus communis* also varied. In most cases only a small part of Pb was translocated in the aerial parts. In 95% of the plant samples collected, the root Pb concentration are much greater than those of the shoot lead content, indicating low mobility of Pb from roots to the shoots. Their ability to accumulate higher amounts of Pb in their roots and considering their rapid growth rate and biomass, this plant has the potential for removal of Pb from contaminated soil.

Chapter 9
Environmental Fluoride: Impact on Humans, Animals and Its Remediation Strategies 165
Junaid Ahmad Malik, Government Degree College, Bijbehara, India

Noteworthy multi-disciplinary undertakings have been taken to investigate the effects of natural fluoride ion (F) contamination since the preceding century. Fluoride is a hazard to the earth and human prosperity. Developed and developing countries are standing up to such enormous extents of issues in light of fluoride in the drinking water. Human use to fluoride has bourgeoned since World War II, on account of fluoridated water and toothpaste just as to the normal defilement by huge ventures, from aluminium to pesticides, where fluoride is an essential mechanical concoction similarly as a waste product. The chapter deals with the proportion of fluoride in nature and its impact on human prosperity, generally on the brain, endocrine system, thyroid, pineal gland, immune system, reproductive system, and organ systems. High assemblies of F in soil may really bargain the life of plants, obliterate soil microbial development, upset the soil environment, and cause soil and water defilement. This chapter further emphasizes various biological approaches for the remediation.

Chapter 10
Cadmium- and Lead-Tolerant PGPRs as Proficient Toxicity Alleviators for Agricultural Crops 189
Amit Kumar Pal, Department of Botany, University of Kalyani, India
Anjan Hazra, Agricultural and Ecological Research Unit, Indian Statistical Institute, Kolkata, India
Chandan Sengupta, Department of Botany, University of Kalyani, India

Agricultural lands are being polluted with different contaminants due to various anthropogenic activities like toxic discharge from Ni-Cd battery industry, tannery industry, alloying of metals like steel, application of agrochemicals, etc. Cadmium and lead contamination in agricultural land are directed towards global food insecurity. Bioremediation, stress alleviation, and phytostimulation by Cd and Pb tolerant PGPR is a promising eco-friendly method to develop sustainable agricultural system. At present, cadmium and lead-tolerant plant growth promoting rhizobacteria (PGPR) can be a sustainable option for heavy metal-contaminated agricultural lands. PGPRs such as *Bacillus*, *Bradyrhizobium*, *Enterobacter*, *Klebsiella*, *Micrococcus*, *Pseudomonas*, *Ralstonia*, etc. can survive the metal stress and stimulate the plant growth under Cd and Pb contaminated condition by direct or indirect plant growth promoting ability. So, these PGPRs could be exploited as biofertilizers and bioremediators under Cd or Pb stressed conditions for futuristic agricultural development.

Chapter 11
Fungi-Mediated Detoxification of Heavy Metals 205
Suchhanda Ghosh, Shri Shikshayatan College, India

Heavy metal pollution is one of the major environmental problems today. Therefore, the elimination of heavy metal ions from wastewater is important to protect public health. The use of biological material in the removal and recovery of toxic metals from industrial wastes has gained important credibility during recent years. Several microorganisms including bacteria, algae, yeast, and fungi have been reported to effectively accumulate or adsorb heavy metals through biosorption. Fungal biomaterial has been proved to be efficient as a biosorbent. High percentage of the cell wall material and availability of fungal biomass as a by-product of various antibiotic and food industries makes it an obvious choice. Thus, the chapter deals with detoxification of heavy metals from contaminated sources using biomaterials with special reference to fungi.

Chapter 12
An Eco-Friendly Approach for the Eradication of Heavy Metal Contaminants by Nano-Bioremediation 220
Chandana Mohanty, Kalinga Institute of Industrial Technology (Deemed), India
Sneha Shriparna Satpathy, Kalinga Institute of Industrial Technology (Deemed), India
Sweta Mohanty, Kalinga Institute of Industrial Technology (Deemed), India

Nanomaterials manifest distinct physical and chemical properties and have received much attention from researchers in different areas of environmental sciences, specifically in bioremediation. However, bioremediation may not always impart contrivable approaches when subjected to high concentrations of contaminants that are harmful to most microorganisms, which include heavy metals and salts. Nanotechnology on the other hand exhibits a number of potential environmental benefits such as treatment and remediation, pollution prevention, and sensing and detection of pollutants. Nanomaterials used towards bioremediation provide less-toxic effects on indigenous microorganisms and improve microbial biodegradation activity. Credibility of nanotechnology to cut down pollution is in its developing stage and could potentially revolutionize the field of environmental sustainability. Nano-bioremediation is a new emerging technique for remediation of pollutants using biosynthetic nanoparticles.

Chapter 13
Sustainable Treatment of Landfill Leachate Using Constructed Wetlands: An Eco-Friendly Approach 237
Vivek Rana, Central Pollution Control Board, Ministry of Environment, Forest, and Climate Change, Government of India, Delhi, India

Sanitary landfilling is the major method of disposal of municipal solid waste (MSW) in developing countries. The disposal of MSW in landfills generates a large amount of highly toxic leachate, which has high potential hazards for the public, flora, fauna health and ecosystems. Advanced leachate treatment systems using biological and chemical treatment methods are recently implemented in developed countries, but high investment and operating costs restricted their application in most of the developing countries. To overcome this problem, an alternative sustainable treatment technology such as phytoremediation could be beneficial. The constructed wetland treatment system is an economical alternative for leachate treatment using local resources and is an energy-efficient technology. These green systems utilize anaerobic and aerobic reactions to break down, immobilize, or incorporate organic substances and other contaminants from polluted effluent. This chapter highlights the recent advances in the treatment of landfill leachates using constructed wetlands.

Chapter 14
Safety and Efficacy of Pseudomonas Exopolymer in Sequestration of Iron From Aqueous Environments 256
Moushumi Ghosh, Thapar Institute of Engineering and Technology, India
Divya Sharma, Thapar Institute of Engineering and Technology, India
Taranpreet Kaur, Thapar Institute of Engineering and Technology, India

The present study reports the iron binding characteristics and safety of an exopolymer (EBP) of an environmental isolate of *Pseudomonas* sp. The EBP was predominantly polysaccharide in composition with pyruvic and uronic acid residues. A prevalence of carboxyl and hydroxyl groups was observed in the Fourier-transform infrared spectroscopy (FTIR) results, while scanning electron microscopy (SEM) revealed a porous structure in a linear fashion with large number of grooves. The purified EBP was stable for over two months and exhibited rapid binding of iron (25mg/L) within 10 minutes at ambient temperature. X-ray diffraction (XRD) and energy-dispersive X-ray spectroscopy (EDAX) analysis of iron challenged EBP suggested the involvement of carboxyl groups in potentiating iron removal. Both Langmuir and Freundlich adsorption isotherms depicted high iron removal capacity in comparison to reported biomasses or biopolymers. Cytotoxic effects were not observed upon challenging various doses of EBP in RAW 264.7 cell lines implying a strong possibility of application of the EBP.

Chapter 15
Recent Advancements in Bioremediation of Metal Contaminants 273
Yuvarajan Ragunathan, Mahendra Arts and Science College (Autonomous), India
Kannan Nallakumar, Mahendra Arts and Science College (Autonomous), India
Thirumalaisamy Rathinavel, Mahendra Arts and Science College (Autonomous), India
Muthusamy Govarthanan, Kyungpook National University, South Korea
Selvankumar Thangaswamy, Mahendra Arts and Science College (Autonomous), India

Biofilms are an accumulation of single or various populations of microorganisms that are present on the surfaces through membrane-bound substances due to the gene expression, which differs from free-floating expression and leads to expressed genes regulating biofilm formation and development. In this regard, recent advances in microbial-based heavy metals have propelled bioremediation as a prospective alternative to conventional techniques. Adsorption and biodegradation of organic contaminants and the immobilization, mobilization, and/or transformation of metals are the main remediation processes that can be mediated by the action of several microorganisms surviving in hostile environments with high concentrations of pollutants. The chapter discussed the formation and regulation of biofilms to degrade the metal contaminant, the importance of gene transfer, and applications of biofilm-mediated bioremediation processes.

Compilation of References 288

About the Contributors 356

Index 361

Foreword

Environmental contamination of heavy metals from both natural and anthropogenic sources has been considered as a major environmental concern. Heavy metals never be degraded, only be remediated, they persists and accumulate over time, which in turn increases human exposure and subsequently results in serious negative environmental consequences. To decontaminate the soils, sediments and waters, the scientists, industrialist and technologists have evolved different strategies over the years. Numerous physicochemical along with some biological technologies have been developed in the last decades to remediate sites contaminated with heavy metals, but most are expensive, not eco-friendly and safe.

Most of these physicochemical technologies are also inefficient at low metal concentrations and large polluted areas. Hence it has become a necessity to develop viable technologies employing microorganism and plants to remediate these metallic residues.

In bioremediation processes, biological organisms are used to remediate contaminated environments. These processes are often considered as highly specific in removal of a particular heavy metal when compared with the traditional physicochemical techniques.

Scientists have also got some notable success in the field of bioremediation as an alternative for treatment of metal contaminated wastes and nowadays several industrial sectors are promoting the sale of plant or microbe-based technologies to deal with specific environmental contamination challenges. Besides, these technologies are cost effective and environment friendly and do not disturb the sites in cleaning process.

In order to give a boost to this technology, I would like to appreciate the sincere efforts of my colleagues, Dr. Satarupa Dey and Biswaranjan Acharya, to publish this book which covers different aspects of bioremediation. I hope this book will be extensively useful for the researchers and academician working in the field of bioremediation. I consider this book to become a precious resource in the field of scientific knowledge on different aspect of bioremediation, which is emerging as a promising technology of today.

Maulin P. Shah
Industrial Waste Water Research Lab, Division of Applied and Environmental Microbiology, Enviro Technology Limited, India

Preface

Heavy metals and metalloids are constantly released from different industrial sources and are considered as a source of concern for most of the developing countries. They are continuously released from different industrial sources, municipal wastes and mining industry and their rate of production and release have increase in manifold level in last few decades. These heavy metals are not degraded and persist in nature and tend to bioaccumulate resulting in several health hazards and negative impact on environment.

Bioremediation processes are multidisciplinary endeavor in which biological organisms in both live as well as dead stage is used to remove metal contamination from environment. However their use is highly specific compared to the traditional chemical metal removal techniques. The advantages of the bioremediation techniques include low energy consumption, low cost as well as less chemical usage. For bioremediation purposes a wide variety of biological materials are usually used which includes bacteria, algae, fungi and plant material and bioremediation can be classified according to the bioremediators used into bacterial bioremediation, mycoremediation, phycoremediation and phytoremediation and so on. These organisms use different methods such as accumulation, adsorbtion and reduction of metal to less toxic form for the process of bioremediation. For reduction of toxic metal to less toxic form involvement of different enzymes becomes eminent. In the present book experts have dealt with different aspects of bioremediation.

This book is suitable for professionals, researchers, and graduate students working in the field of environmental science, applied microbiology, toxicology, environmental chemistry, and soil and water science. This book will also be suitable for courses in environmental sciences, as well as for industrial and government researchers who desire knowledge about bioremediation. It can be used also as a reference book for researcher working in the field of metal bioremediation with microorganisms and will provide insights in the field of phytoremediation and its applications. Use of nanotechnology in the process of removal of metals is also a developing technology which has been covered in the book. Bioremediation of industrial wastes and the present status of bioremediation has also been covered which will be helpful for industries and will help in their policy makings.

In total of 15 chapters covering different aspects of bioremediation has been included in this book. The first part broadly deals with the general introduction to various aspects of bioremediation along with different interactions between microbial community and heavy metals and different mechanisms of heavy metal resistance. The second part concentrates around the different enzymes that play an important role in the process of bioremediation. The third part different strategies for bioremediation are described which includes phytoremediation, biosorption and bioaccumulation using different bacterial and fungal strains. The last part comprises of field applications of bioremediation using wetlands and use of nanotechnology in bioremediation. In these 15 chapters we have tried to configure a comprehensive understanding of this area.

As water contamination with different types of heavy metals remains a great problem. This has increased lately with overall increase of human population as well as industrial development which has resulted in urgent need and implementation of different mitigation program in order to save our nature. It is proposed that combination of biodegradation and nanotechnology has a potential proficiency for bioremediation in a natural amiable system with minimal effort.

The first chapter gives an introduction about different methods in which bioremediation procedures can conceivably be utilized to decrease the metal contaminants, thereby reducing the health hazards in human and will protect the nature. The chapter also briefly describes the role of microorganisms in bioremediation. This chapter emphasizes the role of environment friendly nanomaterials for removing contaminants from wastewater and effluents. Additionally, use of various nanocomposites and different natural methods utilized in wastewater treatment process are also briefly discussed in the chapter.

Microorganisms play an important role in bioremediation of metal contaminants. The microbes growing in metal contaminated environment have naturally developed different mechanism to protect itself from the toxicity of the metal. This includes a wide range of mechanisms such as uptake of metal, reduction to less toxic form, biosorption and so on. The second chapter explores this topic and explains the mechanisms specific to microorganisms that render them resistant to metal toxicity. A wide range of bacterial, algae and fungal species which are used extensively as biosorbents are highlighted in the second chapter. Mechanisms such as reduction of metal cations, their sequestration and binding on cell barriers are discussed in detail in the chapter.

Microbial enzymes play a pivotal role in bioremediation of metal contaminants by converting it into a less toxic form. The third chapter entitled "Microbial Enzymes and Their Mechanisms in the Bioremediation of Pollutants" deals with the role of different enzymes such as oxidoreductase, hydrolases, monooxygenase, dioxygenase, methyltransferases, and lipases are involved in the degradation process. The role of these enzymes and their mode of action were discussed in detail throughout the chapter.

The fourth chapter entitled "The Use of Microorganism-Derived Enzymes for Bioremediation of Soil Pollutants" also deals with the same theme and stresses on the use of different enzymes in removal of contaminants from soil pollutant. The use of microbial enzymes as a viable alternative to degrade and mineralize these contaminants is considered as a growing research interest as it is an eco-friendly method and these aspects are dealt in details in the current chapter. This chapter also deals with the use of molecular biology techniques and explains its role in enhancing the production capacity, stability, activity and shelf life of the bioremediation enzymes. Ultimately, the need to develop bioremediation enzymes in bulk, using different cheap technologies has been emphasized for effective soil decontamination.

Hexavalent chromium is considered as highly toxic, carcinogenic and teratogenic in nature whereas, Cr(III) is insoluble at normal pH and is comparitively less toxic. Cr(VI) is mainly released by different industries such as painting, varnishing and also from mining of chromite ores. Chromium toxicity is a major environmental concern and thus it is essential to draw up strategies for mitigation of Cr(VI) and microbial reduction of toxic Cr(VI) has been identified as a bioremediation technique not only to detoxify chromium but also to recover the non-toxic Cr(III) by physical means. The fifth chapter entitled "Role of Bacterial Chromate Reductase in Bioremediation of Chromium-Containing Wastes" deals with the role of Chromate reductase enzyme which is the central enzyme in bioreduction of Cr(VI) to Cr(III) in bioremediation strategies. The chapter clearly explains the role and mechanisms of different types of chromate reductases in details. Detailed analysis of the structure of different type of chromate reductase enzyme have been discussed which may help us to find out the suitable application of these enzymes in bioremediation of metal-contaminated wastes.

Use of different type of bacterium in consortium has been considered as a suitable alternative in treating metal contaminated waste and sludges. The sixth chapter of the book is a research paper that deals with the use of two bacterial genera *Bacillus cereus* and *Enterobacter* sp in removal of chromium from industrial sludge. The two bacteria were characterized in details in the current chapter and optimization of different growth conditions were carried out to judge the efficiency of both the isolate in chromium removal. The results showed that the two bacteria are not antagonistic to each other so they can be used to bioremediate chromium from their contaminated sites and also can be used as a consortium.

Phytoremediation i.e. use of plant materials is also another cheap and ecofriendly method which is extensively used for removal of contaminants from polluted waste water. The seventh chapter entitled "Phytoremediation: A Modern Approach" includes removal of toxic metals such as cadmium and chromium by this an ecofriendly approach of soil clean up. This chapter discusses heavy metal induced oxidative stress of plants and their detoxification potentiality and the author have tried to create a wholesome idea about the basic and acute need of phytoremediation. Both enzymatic and non-enzymatic antioxidative defense mechanism and various other biochemical parameters of metal hyperaccumulator plants are mentioned in detail. Recent progress of nanoparticle-based soil metal immobilization to reduce metal contamination has also been highlighted in this chapter.

The eighth chapter "*Ricinus communis*: A Potent Lead (Pb) Accumulator" deals with the removal of Pb using *Ricinus communis*. Pb is considered as the second most hazardous substance causing several health hazards. The main sources of Pb in ecosystem are from automobile emission, paint chips, fertilizers and pesticides and Pb acid batteries. Phytoremediation could provide sustainable techniques for metal remediation.

The ninth chapter deals with fluoride which is a hazard on the earth and human prosperity. Developed and developing countries are standing up to such enormous extents of issues due to presence of fluoride in the drinking water. This chapter emphasizes on various biological approaches for the remediation of F-contaminated environment, and exploring their potential applications in environmental clean-up.

Several microbes having bioremediation ability are also known to possess plant growth promoting activity which serves duel purposes of both metal bioremediation and formation of vegetation cover over the waste pollutants. The tenth chapter entitled "Cadmium and Lead-Tolerant PGPRs as Proficient Toxicity Alleviators for Agricultural Crops" deals with this aspect. At present, cadmium and lead tolerant plant growth promoting rhizobacteria (PGPR) are considered as a sustainable option for heavy metal-contaminated lands. PGPRs such as *Bacillus, Bradyrhizobium, Enterobacter, Klebsiella, Micrococcus, Pseudomonas, Ralstonia* etc. are known to survive the metal stress and stimulate the plant growth under Cd and Pb contaminated condition. The chapter deals with different aspects of this plant growth promoting bacteria.

Different groups of fungi also play a vital role in detoxification of metal contaminants from polluted waste lands. The eleventh chapter entitled "Fungi-Mediated Detoxification of Heavy Metals" deals with use of fungal biomaterial for bioremediation of metal contaminants. Fungal biomass is considered as an efficient as biosorbent. High percentage of the cell wall material and availability of fungal biomass as a by-product of various antibiotic and food industries makes it an obvious choice for bioremediation. The chapter deals with detoxification of heavy metals from contaminated sources using biomaterial with special reference to fungi.

Nanomaterials manifest distinct physical and chemical properties and have received much attention from researchers in different areas of environmental sciences, specifically in bioremediation. Bioremediation may not always remain as a contrivable approach when the metals are present in such a high concentration that is detrimental to most microorganisms. The twelfth chapter entitled "An Eco-Friendly Approach for the Eradication of Heavy Metal Contaminants by Nano-Bioremediation" deals with the role of nanotechnology in environmental bioremediation. Nanotechnology exhibits a number of potential environmental benefits such as treatment and remediation, pollution prevention and sensing and detection of pollutants. Nanomaterials used towards bioremediation provide less-toxic effects on indigenous microorganisms and improve microbial biodegradation activity. Credibility of nanotechnology to cut down pollution is in its developing stage and could potentially revolutionize the field of environmental sustainability. Nano-bioremediation is a new emerging technique for remediation of pollutants using biosynthetic nanoparticles.

The thirteenth chapter deals with sustainable treatment of landfill leachate using constructed wetlands Sanitary landfilling is the major method of disposal of municipal solid waste in developing countries and its disposal generates a large amount of highly toxic leachate, which has high potential hazards. Advanced leachate treatment systems using biological and chemical treatment methods are recently implemented in developed countries but have high operating costs which have restricted their application in most of the developing countries. To overcome this problem, an alternative sustainable treatment technology such as phytoremediation could be beneficial. The constructed wetland treatment system is an economical alternative for leachate treatment using local resources and is an energy-efficient in nature and uses anaerobic and aerobic reactions to break down, immobilize or incorporate organic substances and other contaminants from polluted effluent. The chapter highlights the recent advances in the treatment of landfill leachates using constructed wetlands.

The fourteenth chapter deals with use of bacterial exopolymers in removal of metals from contaminated land. Microbial exoproducts, most notably exobiopolymers (EBPs), have found applications in removal of metals from industrial wastes, oil refining, waste water treatment and as thickeners and emulsifying agents. In the present chapter exopolymer from *Pseudomonas* has been used to remove iron from water.

Biofilms are an accumulation of single or various populations of microorganisms on the surfaces which also can be used as a tool for bioremediation. In this regard, recent advances in microbial-based heavy metal have propelled bioremediation as a prospective alternative to conventional techniques. Adsorption and biodegradation of organic contaminants and the immobilization, mobilization, and/or transformation of metals are the main remediation processes that can be mediated by the action of several microorganisms surviving in hostile environments with high concentrations of pollutants. The last chapter discussed the formation and regulation of biofilms to degrade the metal contaminant, the importance of gene transfer and discuss applications of biofilm-mediated bioremediation processes.

Finally, I would like to acknowledge the efforts of all the contributors for their continuous effort in bringing the book to fruition. The continued assistance of the Editorial Department of IGI global is also highly appreciated.

Satarupa Dey
Shyampur Siddheswari Mahavidyalaya, India

Acknowledgment

We are grateful and acknowledge our in- depth gratitude to all the contributors and the reviewers who trusted us and supported immensely in shaping this book. We hope that they will be as proud of this book as we are. We would also like to thank IGI global and all the team members for giving us this opportunity to edit this book and consider this book for publication despite COVID-19 pandemic-associated international restrictions, lockdown, work from home, and several other limiting factors. All chapters included in the book are with proper citation. However, any missed information is purely unintentional.

We would like to express our gratitude to our mentors, teachers, and students who have always motivated us and given us the confidence and strength to accomplish this task.

Also we would like to thank our colleagues, family and friends who provided encouragement and support during the work on this book.

Satarupa Dey
Shyampur Siddheswari Mahavidyalaya, India

Biswaranjan Acharya
School of Computer Engineering, KIIT University (Deemed), India

Chapter 1
Different Bioremediation Techniques for Management of Waste Water: An Overview

Biswajit Patra
https://orcid.org/0000-0002-8248-0487
Sambalpur University, Odisha, India

Saroj Kumar Deep
https://orcid.org/0000-0003-0948-8199
Sambalpur University, Odisha, India

Surya Narayan Pradhan
https://orcid.org/0000-0003-1672-855X
School of Life Sciences, Sambalpur University, Odisha, India

ABSTRACT

Water contamination remains an issue. A combination of biodegradation and nanotechnology is proposed as a potential proficient, minimal effort, and naturally amiable system to deal with it. Among different mediations, bioremediation procedures can conceivably be utilized to decrease the versatility of materials in the subsurface, reducing the potential for human and ecological exposure. The metabolic diversity of microorganisms ensures an assortment of substrates to be expended. Photosynthetic microorganisms have been found as a compelling and eco-friendly species that can remove carbon, nitrogen, and phosphorous in the manufactured sewage and wastewater. This chapter particularly emphasizes environmentally friendly NMs that give information for removing contaminants from wastewater and effluents. Additionally, various nanocomposites and different natural methods utilized in the wastewater treatment process are also briefly discussed.

DOI: 10.4018/978-1-7998-4888-2.ch001

INTRODUCTION

Water on the planet is the most significant natural resource. Presently ecological water sullying has been ascendant in most recent couple of decades as a result of expanded human exercises, risky farming practices and industrialization (Azubuike et al. 2016). Thus, the principle challenge is consistent defilement of water resources by different inorganic and natural pollutants. So, conventional treatment strategies are not proficient to whole expulsion of toxins and accessible methods for waste water have numerous disadvantages (Ferroudi et al. 2013). Presently, contamination in view of overwhelming metals in ground water, waste water and lakes caused problematic wellbeing impacts in living creatures. This section centers the distinctive nanoparticles for contaminant water treatment forms. The procedure of contamination evacuation relies principally upon the idea of the toxin, which may include: agrochemicals, chlorinated mixes, colors, ozone depleting substances, overwhelming metals, hydrocarbons, atomic waste, plastics and sewage (Akhavan and Azimirad 2009). Pollutant nature, profundity and level of contamination, kind of condition, area, cost, and ecological approaches are a portion of the determination criteria that are viewed as while picking any bioremediation strategy (Beaver et al. 2014). Besides, given the idea of exercises prompting raw petroleum contamination, all things considered, contamination of nature with toxins barring hydrocarbons can without much of a stretch be forestalled and controlled. The natural wastewater treatment is generally applied however these are normally moderate, restricted because of the nearness of non-biodegradable contaminant, and at some point makes toxicity to microorganisms because of some poisonous contaminants (Chen et al. 2016). The physical procedures, for example, ultrafiltration could evacuate the pollutant and impurity matters by changing primary stage to next however delivering more concentrate sludge, which is harmful and difficult to dispose. The genuine necessity for most incredible advancements for management of city and mechanical waste and polluted waters (Chong et al. 2010). Nanotechnology in various literature described as one of the best exceptional procedures for contaminate water treatment. Nano-materials have nano structures that people have created, having shape and size of nanometers. Nano-materials created in assortment of structures, for example, nanotubes, nanowires, particles, films, colloids and quantum dots (Douglas et al. 2016). Nano-adsorbent has been delivered using the particles of components which are unnaturally vibrant and have highest adsorption capacity on the outside of nano-material. The used materials for improvement of nano-adsorbents include activated silica, carbon, metal oxides, clay materials and modified compounds as nono-composites. Various sorts of nano-catalysts and electrocatalysts are utilized for degradation of contamination in wastewater. Fenton mediated catalysts for improving oxidation of natural toxins and various catalysts having antimicrobial activities [Eldyasti et al. 2011; Feng et al. 2014]. In nanomembrane classification, the pres-sure determined treatment of contaminated wastewater can demonstrate for improving water quality of desire. Among various sorts of film filtration (FF), and nano-filtration (NF) are mainly applied for management of contaminated water in ventures as an effect of minimal effort, little pore sizes, more efficiency and ease to handle.

Current Techniques for Removal of Contaminants

Some physical treatment strategies like membrane filtration, ion exchange, precipitation, cementation, electrocoagulation, film filtration are utilized for the expulsion of waste water. Reductants like vaporous hydrogen sulfide and dithionites are utilized for reduction process. Chemical washing and chelate flushing techniques are used for the evacuation of metal contaminant (Feng et al. 2011; Gadhe et al.

2015). Bioremediation by natural treatment strategies considered as the eco-accommodating and cost effective method for disposal of toxins from soil and water. But sometimes contaminant themselves become toxic to the micro organisms involved in the process. So, researcher find alternative solution to maintain high rate bioremediation. Bacteria, algae, fungi and various plant species utilized to stabilize and detoxify environmental pollutant into nontoxic state (He et al. 2016; Huang et al. 2014). Biofiltration, biosorption, bioaugmentation, microbes sulfate decrease and biophysiochemical strategies likewise utilized for evacuation process.

Nanotechnology and nano materials (NMs) utilized in squander water treatment process. NMs have created in various structures like nanotubes, particles, colloids, quantum dots, films and nanowires (Idi et al. 2015; Iram et al. 2010; Jiang et al. 2013). As indicated by NMs nature it named Nano-membranes (NMb), Nano- catalysts (NCt) and Nano-adsorbents (NAd).

NAd: Adsorption procedure relies upon recitation dividing of pollutant and adsorption coefficient, for example substantial metals or natural contaminants under harmony conditions. Their chemical, physical and material behavior are likewise recognized with their outward surface variation, range and composition (Kato et al. 2012; Khusnutdinova et al. 2012; Kirkpatrick et al. 2014). Nano-adsorbents are extensively ordered into different gatherings dependent on their role in adsorption system. This incorporates metal nano-materials, nano blended oxides, metallic oxide NPs and magnetic NPs. Oxide nano-materials are inorganic NPs which are generally prearranged by metals and non-metals (Lalley et al. 2016). These nano-materials are extensively used for dangerous pollutant expulsion from contaminant and polluted water. They have a feature of titanium dendrimers composites, zinc oxides, titanium oxides, manganese oxides, ferric oxides and magnesium oxide. Iron coated NMs additionally utilized because of its basic blend process (Lenz et al. 2009). The progressions of these NAd shows high liking for expulsion of various toxins like Cr^{3+}, Ni^{2+}, Co^{2+}, Cd^{2+}, Pb^{2+} and $Cu^{2+}$from waste waters (Maximous et al. 2009). Some manganese based oxide (MnO) NPs give an idea about their most adsorption capacity because of their highest zone of surface and polymorphic arrangements (Mehrizad et al. 2011). This is generally utilized for evacuation of different substantial metalloids, for example, arsenic from polluted water. Zinc Oxide (ZnO) has a permeable smaller scale and nano-arrangement with high morphological surface territory for adsorption of substantial metals (Nasreen et al. 2013). NPs gatherings, microspheres with nano-sheets, nanoplates and progressive Zinc Oxide nano-bars were broadly utilized for nano-adsorbent for the expulsion of overwhelming metals from polluted water. Magnesium based oxide (MgO) are utilized for evacuation of various types of irresistible metalloids from tainted water (Pan et al. 2009; Gu et al. 2015; Gopakumar et al. 2019). MgO micro-spherical shapes are effective structure, which develop the adsorption partiality for the expulsion of substantial particles. Weak dispersion capability sometimes creat problem in division and smallest particles dimensions are barely any issues survive for utilizing Carbon Nano Tubes (CNT) as adsorbant (Pendergast and Hoek 2011; Petrinic et al. 2007; Rashidi et al. 2012). Graphenes are the allotropy of the carbon occurring extraordinary highlights that create it exceptionally constructive for some ecological applications. Graphene oxide (GO) are the carbon based nano-material occurring two-dimensional configuration delivered by graphite layer by means of complex strategy (Ren et al. 2011; Liwarska-Bizukojc 2019). Adsorptions of substantial particles from waste and polluted water by utilizing the NAd are influenced by numerous elements, for example, pH, temperature, adsorbent portion and hatching. Ren et al. (2011) announced that hydrogen ion concentration assumes the fundamental function for adsorption of substantial particles from polluted water (Santhosh et al. 2016).

NCt: The nano-catalysts are inorganic materials, like metal oxides and semiconductors are most extensive contemplation of the analysts in utilization of polluted water management (Tawabini 2015).

Various NCt were used for polluted water management to check antimicrobial activities, electrocatalysts and photocatalysts for improving substance oxidation (Lenz et al. 2009; Gadhe et al. 2015). Among different nano photocatalysts created, TiO_2 broadly useful in photocatalyst because of the highest reactivity below bright beam (k < 390 nm) and substance strength (Varma et al. 2013). Additionally, ZnO having photo-catalytic action and contains wide band hole similar to TiO2. Different doped NCt are ZnS:Mn, ZnO:Co, CdS:Eu, ZnS:Cu, ZnSe:Mn, ZnS:Pb, and CdS:Mn. There are so many dopants, for example, Si, Co, Mg, Cr, Fe, Mn, Al and Ga were utilized for improving the surface zone of metal oxide nano-structure (Visa 2016; Wang et al. 2015). Photocatalysis have demonstrated as potential method for cleaning and management of different sorts of contaminated water. Matsuura et al., (2010) and Tawabini (2015) demonstrated that the capacity of AgNPs on TiO2 films accomplished 6.9 occasions of more antimicrobial action against *E. coli* bacteria contrasted with TiO2 under visible light (Varma et al. 2013; Xiang et al. 2004). So, mesoporous combination of Ag with TiO2 films (Ag/TiO2) demonstrated highest antibacterial activities when contrasted with the viable P-25 TiO2 spinning film (Zhang et al. 2013). Despite the fact that Pt could be utilized as an electro-catalyst, and it has different drawbacks. For example, Pt is a valuable metal and it has constrained accessibility, and significant expense limited the enthusiasm to utilize it as catalyst (Zhang et al. 2013; Wray and Farrington 1957). What's more, during electro-catalysis Pt can limit the response because of harming of transitional compounds. Attractively divisible nano-particles of metal oxide can be utilized as catalysts for evacuation of various kinds of contaminations. The attractive isolated particles having carbonaceous materials, ferrite with Ba, iron oxide phase, Mn, Co and maghemite have highest ability to recuperate from water through attractive techniques (Santhosh et al. 2016).

NMb: For contaminated and polluted water treatment, nano-layer detachment innovation has utilized for victorious emigration of colors, substantial particles and various contaminants. For fluid osmotic partition, zeolite based nano-layers can be utilized (Romuald et al. 2012). Electrospun nanofiber films (ENMs) are the as of late developing layers which bring forth a novel method to treat wastewater. If there should arise an occurrence of expulsion of salts from different water in desalination method, the utilization of nano-fiber films have been demonstrated a powerful route for salty evacuation due lower operational weight, improved motion, and less vitality necessity (Visa 2016; Wei et al. 1997).

Response of $CaCO_3$ Nanoparticles (NPs) in Waste Water Treatment

$CaCO_3$ NPs are high surface polarity, bio-compatible and high hydrophilicity. It is easy to produce, eco friendly and low cost (Wang et al. 2015; Zhang et al. 2013). CaCO3 NPs were synthesized from Sodium carbonate and Calcium nitrate solutions. Sodium carbonate were dissolved in deionised water with Sodium nitrate and Sodium hydroxide. Firstly, Sodium hydroxide aqueous solution was added to support alkaline pH appropriate for Calcium carbonate precipitation while Sodium nitrates were mixed to reduce the solubility of Calcium nitrate by common ion effect. Secondly, calcium nitrate was dissolved in deionised water and the two aqueous solutions were ready for mixing. The calcium nitrate solution was mixed drop wise to the sodium carbonate solution in continuous stirring at standardised speed and incubation periods [Zhang et al. 2013; Romuald et al. 2012]. The outcome solution were poured in a separatory funnel allowing the precipitated particles to settle down.

Nanocomposites in Waste Water Treatment

With reference to the large-scale function of NPs in polluted water management, there are some inherent mechanical and technological chokepoints, like complicated separation, aggregation, leakage into the contact water, probable unfavorable effect forced on human health and ecosystem. Nanocomposite is basically solid material, together with porous media, gels, copolymers and colloids in a broad sense (Agnihotri et al. 1999; Chakraborty and Bhatia 1996). However, nanocomposites were supposed to be the most capable path to promote water nanotechnology from lab experiment to large scale management (Chakraborty et al. 1994). Additionally, nanocomposites of organic and inorganic supports have unique functions for waste water treatment (Chen et al. 1989; Chen et al. 2000; Gao et al. 2007; Haruta et al. 2003).

Response of Lime in Waste Water Treatment

Limes of various forms are cost effective alkaline materials used to purify waste water and sludge. Additionally lime allows waste water to purify, soft and eliminate water cloudiness. Calcium oxide and Calcium hydroxide also used to clean drinking water (Hu and Deng 2003; Hu et al. 2004; Chen et al. 2000). CaO MgO (Calcined dolomite utilized in water treatment plants. High doses of lime kill bacteria and virus from infected water (Hu et al. 2004).

Response of Cow Dung in Waste Water Treatment

Utilization of cow dung and urine (Goumutra) as adsorbent for the elimination of metal and particle ions in batch experiment (Ojedokun and Bello 2016; Vinay et al. 2019). It is a bioorganic waste contains 61% silica, 0.9% mag-nesium oxide, 0.312% calcium sulphate, 20%iron oxide, 20% aluminium oxide and 12.48% calcium oxide (Ojedokun et al. 2016; Kitamura et al. 2002). Benefits of utilizing cow dung as activated carbonis not only spinning around its low cost-effective value, but also can prevent the other ecological and environmental problems of foul odor resulting from its waste. raditionally, Gomutra is beingsprinkled in courtyards and house for its holy purpose. This confers the purity, prosperity, happiness, wealth, positive health etc (Vinay et al. 2019; Mani and Kumar 2014). Cattle dung accelerate the rhizospheric expansion and enhanced the plant root association in copper mines of Globe and Tucson, USA (Mani and Kumar 2014).

Response of Alum in Waste Water Treatment

Alum (Aluminium sulphate) an inorganic salt mostly utilized as coagulant in water treatment process (Jagaba et al. 2018). Coagulation by hydrolyzing salty metals from Fe or Al and these are the chief response stage that drives the elimination of natural organic matter (NOM) and other pollutants in potable water treatment (Koepenick 1976; Mendham et al. 2000; Ogino et al. 1987; Oniyama and Walhbeck 1995; Reddy and Nancollas 1971). Polluted water Treatment Using Alum, the Combinations of Alum-Chitosan, Alum-Zeolite, Alum-Ferric Chloride and *Moringa Oleifera* as Adsorbent and Coagulant was studied by Jagaba et al., 2018 (Jagaba et al. 2018). Aluminium oxide nanoparticles (Al_2O_3 NPs) and microparticles (Al_2O_3 MPs) and their effects on activated sludge and synthetic waste water were studied by Liwarska-Bizukojc (2019) (Liwarska-Bizukojc 2019; Wei et al. 1997). Regarding the chances of NPs in the conventional activated sludge systems, this was considered that nanoparticles usually adsorbed onto activated sludge flocs and entered the interior of the microbial cells (Wray and Farrington 1957; Xiang et al. 2004; Yu et al. 2004).

Palm Kernel Shell in Waste Water Treatment

Agricultural waste like palm kernel shell utilized as effective adsorbent for the removal of toxic metal (Baby et al. 2019). These techniques used as cost effective and greener approach to utilized the agri-waste without any pesticide action (Ahluwalia and Goyal 2007).

Corncobs in Waste Water Treatment

Agri-waste of corncobs also used as bioadsorbent for water and contaminated water treatment (Singh et al. 2017). Nayda et al., 2018 suggested corncob as a effective, inexpensive, ecofriendly biosorbent for the removal of dye from aquous solutions. FTIR and CSLM results confirmed that lignocelluloses and proteins also involved in this process for purification (Beatriz and Juan 2011; Parthasarathy and El-Halwagi 2000).

Response of Aquatic Plants to Absorb Metal NPs

Hydrophytic species have highest capability for elimination of metal NPs from waste water (Liwarska-Bizukojc 2019). The benefits of phyto-remediation are a high remediation intensity that has not compromised to the chemical and physical process, environmental safety, low cost and the possibility of further extraction of contaminants from the plants (Wray and Farrington 1957; Gu et al. 2015). Mostly free-floating hydrophytic species, *Salvinia* and *Pistia* with nanoparticles have the special attention for removal of contaminants from polluted water (Liwarska-Bizukojc 2019). The charges of heavy metals (Cd, V, Ni) into the atmosphere far exceeds their removal by natural processes, as a result leading to the accumulation of heavy metals in the marine ecosystem (Mani and Kumar 2014). Several microalgae, green algae, blue green algae have been used for treatment of contaminated aquatic ecosystem. Some selected micro algae like *Chlorella pyrendoidosa, Aspergillus niger, Spirogyra sp., Oedogonium, Rhizoclonium, Spirullina, Saccharomyces, Cladophora, Ascophyllum* etc. used for removal of heavy metals (Mani and Kumar 2014; Yuan et al. 2011; Yin et al. 2012).

Nano-Cellulose for Waste Water Treatment

The nanocellulose mainly represents a group of particles that contain bacterial nanocellulose (BNC), nanocellulose crystals (NCCs), and nanocellulose fibers (NCFs) (Yu et al. 2004). NCFs were divided micro fibrils from the plant cell wall. Their arrangements occupy infringement the composite fiber matrix by mechanical and chemical treatments. NCCs are strong, lightweight, tough, highly stable, and durable. They also have abrasion stability and high wear (Liwarska-Bizukojc 2019).

Removal of Oil Contaminants from Waste Water

Oil contaminations in natural waterbodies can hazard to hydrophytic ecosystem. Waste water from Industries was released waste into water bodies and during transport petroleum chemicals are connected with normal water which gives serious problems to hydro ecosystem. So, absorption of oil from industrial water are necessary to solved this problem. Evaluated to traditional methods, like flotation and skimming have economically viable. So, Scientists are focusing on sustainable nanomaterials for oil absorption. Newl y developed cellul ose aero gels exhi bited maxim um absor ption capacity of 35 g/g for peanut oil and showed excellent reus ability of less than 15 cycles (Gopakumar et al. 2019). Because of its highly hydrophobic surface and porous nature, the composite illustrated the excellent oil absorption activities (Hu and Deng 2003).

The Molecular Imprinting Materials (MIMs) for Waste Water Treatment

In MIMs, preferred efficient monomers and crosslinker were copolymerized in existence of specific targeted molecules, performing as a molecular template to generate a polymeric material of highest specificity. The functional monomers primarily form a composite with the template particle through covalent or noncovalent connections. Polymerization of the efficient monomers with a bifunctional or-trifunctional crosslinker consequences in the template's docking into the polymeric matrix. The template compounds are consequently eliminated, revealing exact binding position that are harmonizing in size and shape to the template molecule (Kotrotsiou and Costas 2019). During the adsorption progression, for the elimination of a targeted compounds from a fluid medium, the molecules are adsorbed to the binding position via the action of the same covalent interaction forces. So, the main factor to the successful arrangement of MIMs with more specificity and selectivity towards a certain targeted compounds are the selection of the proper functional monomers that effect the covalent interactions between the monomers and template compound (Nasreen et al. 2013). Another choice to covalent imprinting method. By this method, a metal coordination mechanism between the template molecule and the functional monomer recognized as an substitute to covalent bonding method (Reddy and nancollas 1971).

Molecular Imprinting for Analysis of Biological and Environmental Samples

For the detection of waste water samples, various analytical techniques can be employed like High Performance Liquid Chromatography (HPLC), Gas Chromatography and Mass Spectroscopy (GCMS) etc. However, these technologies are sensitive and cannot handle directly complex biological and environmental samples (Kotrotsiou and Costas 2019).

Sensors for Environmental Monitoring

Lab analysis techniques usually time consuming. So, advancements have been made in the improvement of conventional chemical and biological microsensors for environmental observation. Many commercial products are now available due to recent limitations in the detectional sensitivity, accuracy, selectivity, shelf life, reliability, operational strength and cost. So, highly sensitive devices are used for the detection of toxins in waste water. Additionally, molecular imprinting new technologies are useful for the creation of sensing microdevices in water quality monitoring. Surface Plasmon Resonance (SPR), Surface enhanced Ranman Scattering (SERS), Quartz Crystal Microbalance etc are developed microsensors for environmental monitoring applications (Kotrotsiou and Costas 2019; Sawada and Ohtaki 1998).

Electrospinning Techniques

This technique used to develop an electrospun nanofibrous membrane by surface modification. Electrospun nanofibrous membranes had been utilized for the nonstop filtration. Additionally, electrospun polymeric interconnected webs used as porous carry membranes in nanofiltration (NF), thin film composite (TFC), reverse osmosis (RO), ultrafiltration (UF), and membrane distillation (MD) membranes (Ray et al. 2019). Various electrospinning configurations are multispinnerets, coaxial spinneret, and single spinneret. By using a solitary spinneret, particular component nano-fibers could be created by homogeneous polymeric solution and multi-component nano fibers could be electrospun by emulsions and blends (Mushtaq et al. 2015). In coaxial spinneret, two dissimilar solutions flow by two coaxial syringes or capillaries, resulting in a core-shell surface structure. In Multispinnerets, Membranes created by different types of nanofibers and generate electrospinning of different polymeric fluids from various spinnerets (Ramakrishna and Shirazi 2013; Tucker et al. 2012; Rutledge and Fridrikh 2007; Bognitzki et al. 2001; Ngiam et al. 2007; Reneker et al. 2000; Spivak et al. 2000; Lin et al. 2004). However, free surface electrospinning methods introduced to owing possibility of clogging and blockage. According to these techniques, numerous self organized electrically driven jets could be obtained from cylindrical surfaces and planar by applying more voltage by electric fields (Kotrotsiou and Costas 2019; Ray et al. 2019). The needle based electrospinning methods have resulted in the generation of novel nanofibrous membranes with a broad array of applications, such as nanocomposites, fabrication and purification membrane. Mostly, the fabrication charge of the electrospinning technique for nanofibers with little diameters are less. So, the utilize of electrospun nanofibrous materials are restricted to low volume function (Bhardwaj and Kundu 2010, Bandyopadhyay et al. 2006). Another method is melt electrospinning, which utilized to produce nanofibrous microstructures from polymer melts for function in fabric and nanocomposites (Deitzel et al. 2001; Sukigara et al. 2003; Subramanian and Seeram 2013. For characterization of electrospun nanofibrous materials, membrane performance could be scrutinized directly. The results of structure, specifications, morphology and membrane material are crucial for different applications (Ray et al. 2016; Koppel 1982). Main characterization techniques separated by two classes i.e morphological and physical. Physical methodology include membrane permeability, gas and liquid flow test, liquid dislocation method and solute transport techniques. But the morphological method include Scanning Electron Microscopy (SEM), Transmission Electron Microscopy (TEM), Energy Dispersive Spectroscopy (EDS), Atomic Force Microscopy (AFM), Field Emission Scanning Electron Microscopy (FESEM), X-Ray Diffraction (XRD), Nuclear Magnetic Resonance (NMR), Fourier Transform Infrared Spectroscopy (FTIR), Raman spectroscopy, Contact angle measuring device, Confocal microscopy (Ray et al. 2019; Zong et al. 2002; Kang and Cao 2014; Ray et al. 2017; Ray et al. 2014; Aussawasathien et al. 2008). The processes of electrospinning are (i) Polymer solution parameters, (ii) Process parameters and (iii) Ambient parameters. In polymeric solution, some factors are included i.e. concentration, molecular weight, viscosity and conductivity. When the concentration of polymer increases, the combination of bubbles with fibers and beads are obtained and when the concentration achieved optimum level, smooth nano fibers are obtained, where as in higher concentration stage, helix shaped micro ribbons are observed (Ray et al. 2019; Chang and Chen 1984). The molecular weight of the polymer reflects the degree of polymeric chains in the solution. More molecular weight produces flat nano fibres and raising the molecular weight yields micro-ribbon structures (Shao et al. 2004; Homaeigohar and Elbahri 2014). The solution viscocity formulating the nano structure and surface morphology of the fibers. The viscocity of solution changed by concentration of the sample medium. The conductivity of solution estimated by the variety of solvent and polymer. Natural polymers are poly electrolytic and low charge density (Liu et al. 2002; Agrawal et al. 1982). So, the conductivity could be adjusted by adding ionic salts (KH_2PO_4 and NaCl) (Theron et al. 2005). Processing techniques are crucial for conversion of polymeric solutions into fine fibres and smooth surface by electro-spinning. So, it required high voltage and temperature.

CONCLUSION

Every discovery has its individual reimbursement and explicit toxin evacuation productivity. So as to decrease the health risk hazard there are opportunity to get ready such catalysts having least harmfulness to environment (Lenz et al. 2009; Vucak et al. 1998). Further work has requisite to rethink the ecotoxicity potential for every new change in catalysts and for existing particles. In addition, life span appraisal of nano-materials is essentially required to address their general advantages and risks. Nano-innovation is seldom received to mass procedures (Zhang et al. 2013; Zhelev and Bhaw 2000; Kuo and Smith 1998; Statyukha et al. 2008). Further work has required on building up cost effective techniques for synthesizing nano-materials and testing the effectiveness at large scale for fruitful field application. This review also point out that water and wastewater treatment utilizing nanomaterials has a promising field for future research (Varma et al. 2013). Conventional wastewater treatment advancements remain incapable for giving satisfactory safe water, because of expanding request of water coupled with stringent health guidelines and emerging contaminants (Maximous et al. 2009; Vucak et al. 1998). Nanotechnology-based multifunctional and exceptionally effective procedures are giving reasonable answers for wastewater treatments that don't depend on enormous frameworks or centralized system. One capable approach to promote the function of nanomaterials is to expand a nanocomposite materials that take recompense of both the hosts and the impregnated functional nanoparticles (Chen et al. 2000; Sawada and Ohtaki 1998; Sunqing et al. 2000). Hosts like biopolymers, polymers, activated carbons, minerals, or membranes could facilitate the dispersion and stability of the loaded nanoparticles (Hu et al. 2004; Ng et al. 2009). They also promote the transfer of contaminants in the hosts, further enhancing the interfacial interaction. Many researchers have reported numerous naturally occurring materials for the trapping of heavy metal ions. Hence, little efforts have been made to use cow dung as adsorbent for the removal of heavy metals from aqueous solutions (Wahlbeck 1992; Parthasarathy and El-Halwagi 2000). This aspect needs to be investigated further in order to promote large-scale use of the adsorbent. The mixing of Al_2O_3 NPs at specific concentrations improve the quality of separation process in treatment of sludge and contaminated water. Electrospun nanofibrous membranes have attracted much attention from researchers because of its high versatility (Theron et al. 2005). Many scientists also focused on the functionalities of electrospun nanofibers to improve their applicability on an industrial pollutant.

REFERENCES

Agnihotri, R., Mahuli, S. K., Chauk, S. S., & Fan, L. S. (1999). Influenceof Surface Modifiers on the Structure of Precipitated Calcium Carbonate. *Industrial & Engineering Chemistry Research*, *38*(6), 2283–2291. doi:10.1021/ie9900521

Agrawal, P., Lee, C., Lim, H. C., & Ramkrishna, D. (1982). Theoretical Investigations of Dynamic Behavior of Isothermal Continuous Stirred Tank Biological Reactors. *Chemical Engineering Science*, *37*(3), 453–462. doi:10.1016/0009-2509(82)80098-5

Ahluwalia, S. S., & Goyal, D. (2007). Microbial and plant derived biomass for removal of heavy metals from wastewater. *Bioresource Technology*, *98*(12), 2243–2257. doi:10.1016/j.biortech.2005.12.006 PMID:16427277

Akhavan, O., & Azimirad, R. (2009). Photocatalytic property of Fe2O3 nanograin chains coated by TiO_2 nanolayer in visible light irradiation. *Applied Catalysis A, General*, *369*(1-2), 77–82. doi:10.1016/j.apcata.2009.09.001

Aussawasathien, D., Teerawattananon, C., & Vongachariya, A. (2008). Separation of micron to sub-micron particles from water: Electrospun nylon-6 nanofibrous membranes as pre-filters. *Journal of Membrane Science*, *315*(1-2), 11–19. doi:10.1016/j.memsci.2008.01.049

Azubuike, C. C., Chikere, C. B., & Okpokwasili, G. C. (2016). Bioremediation techniques- classification based on site of application: Principles, advantages, limitations and prospects. *World Journal of Microbiology & Biotechnology*, *32*(11), 4–18. doi:10.100711274-016-2137-x PMID:27638318

Baby, R., Saifullah, B., & Hussein, M. Z. (2019). Palm kernel shell as an effective adsorbent for the treatment of heavy metal contaminated water. *Scientific Reports*, *9*(1), 18955. doi:10.103841598-019-55099-6 PMID:31831850

Bandyopadhyay, S., Ghanekar, M. D., & Pillai, H. K. (2006). Process water management. *Industrial & Engineering Chemistry Research*, *45*(15), 5287–5297. doi:10.1021/ie060268k

Beatriz, C. M., & Juan, A. (2011). Combination treatment of corn starch water by sedimentation, microfiltration and reverse osmosis. *Desalination*, *279*(1-3), 285–290. doi:10.1016/j.desal.2011.06.021

Beaver, J. R., Manis, E. E., Loftin, K. A., Graham, J. L., Pollard, A. I., & Mitchell, R. M. (2014). Land use patterns, ecoregion, and microcystin relationships in U.S. lakes and reservoirs: A preliminary evaluation. *Harmful Algae*, *36*, 57–62. doi:10.1016/j.hal.2014.03.005

Bhardwaj, N., & Kundu, S. C. (2010). Electrospinning: A fascinating fiber fabrication technique. *Biotechnology Advances*, *28*(3), 325–347. doi:10.1016/j.biotechadv.2010.01.004 PMID:20100560

Bognitzki, M., Czado, W., Frese, T., Schaper, A., Hellwig, M., Steinhart, M., Greiner, A., & Wendorff, J. H. (2001). Nanostructured fibers via electrospinning. *Advanced Materials*, *13*(1), 70–72. doi:10.1002/1521-4095(200101)13:1<70::AID-ADMA70>3.0.CO;2-H

Chakraborty, D., Agarwal, V. K., Bhatia, S. K., & Bellare, J. (1994). Steadystate Transitions and Polymorph Transformations inContinuous Precipitation of Calcium Carbonate. *Industrial & Engineering Chemistry Research*, *33*(9), 2187–2197. doi:10.1021/ie00033a024

Chakraborty, D., & Bhatia, S. K. (1996). Formation and Aggregation ofPolymorphs in Continuous Precipitation. 2. Kinetics of CaCO3Precipitation. *Industrial & Engineering Chemistry Research*, *35*(6), 1995–2006. doi:10.1021/ie950402t

Chaney, R. L., Angle, J. S., Broadhurst, C. L., Peters, C. A., Tappero, R. V., & Sparks, D. L. (2007). Improved understanding of hyperaccumulation yields commercial phytoextraction and phytomining echnologies. *Journal of Environmental Quality*, *36*(5), 1429–1443. doi:10.2134/jeq2006.0514 PMID:17766822

Chang, H., & Chen, L. (1984). Bifurcation Characteristics of Nonlinear Systems under Conventional PID Control. *Chemical Engineering Science*, *39*(7-8), 1127–1142. doi:10.1016/0009-2509(84)85075-7

Chen, J., Yan, L. G., Yu, H. Q., Li, S., Qin, L. L., Liu, G. Q., Li, Y., & Du, B. (2016). Efficient removal of phosphate by facile prepared magnetic diatomite and illite clay from aqueous solution. *Chemical Engineering Journal*, *287*, 162–172. doi:10.1016/j.cej.2015.11.028

Chen, J. F., Wei, Y. H., & Guo, F. (2000). Synthesis of Nanoparticles With Novel Technology: High-Gravity Reactive Precipitation. *Industrial & Engineering Chemistry Research*, *39*(4), 948–954. doi:10.1021/ie990549a

Chen, L. S., Ma, Y. W., & Cotterell, B. (1989). Impact Fracture Energyof Mineral-Filled Polypropylene. *Polymer Engineering and Science*, *29*(8), 505–512. doi:10.1002/pen.760290804

Chong, M. N., Jin, B., Chow, C. W. K., & Saint, C. (2010). Recent developments in photocatalytic water treatment technology: A review. *Water Research*, *44*(10), 2997–3027. doi:10.1016/j.watres.2010.02.039 PMID:20378145

Das, A. K., Shenoy, U. V., & Bandyopadhyay, S. (2009). Evolution of resource allocation networks. *Industrial & Engineering Chemistry Research*, *48*(15), 7152–7167. doi:10.1021/ie9003392

Deitzel, J., Kleinmeyer, J., Harris, D., & Tan, N. B. (2001). The effect of processing variables on the morphology of electrospun nanofibers and textiles. *Polymer*, *42*(1), 261–272. doi:10.1016/S0032-3861(00)00250-0

Douglas, G. B., Lurling, M., & Spears, B. M. (2016). Assessment of changes in potential nutrient limitation in an impounded river after application of lanthanum-modified bentonite. *Water Research*, *97*, 47–54. doi:10.1016/j.watres.2016.02.005 PMID:26879191

Eldyasti, A., Andalib, M., Hafez, H., Nakhla, G., & Zhu, J. (2011). Comparative modeling of biological nutrient removal from landfill leachate using a circulating fluidized bed bioreactor (CFBBR). *Journal of Hazardous Materials*, *187*(1-3), 140–149. doi:10.1016/j.jhazmat.2010.12.115 PMID:21255923

Essalhi, M., & Khayet, M. (2013). Self-sustained webs of polyvinylidene fluoride electrospun nanofibers at different electrospinning times: 1. Desalination by direct contact membrane distillation. *Journal of Membrane Science*, *433*, 167–179. doi:10.1016/j.memsci.2013.01.023

Feng, Y. Z., Grogan, P., Caporaso, J. G., Zhang, H. Y., Lin, X. G., Knight, R., & Chu, H. (2014). pH is a good predictor of the distribution of anoxygenic purple phototrophic bacteria in Arctic soils. *Soil Biology & Biochemistry*, *74*, 193–200. doi:10.1016/j.soilbio.2014.03.014

Feng, Y. Z., Lin, X. G., Zhang, J., Mao, T. T., & Zhu, J. G. (2011). Soil purple phototrophic bacterial diversity under double cropping (rice-wheat) with free-air CO2 enrichment (FACE). *European Journal of Soil Science*, *62*(4), 533–540. doi:10.1111/j.1365-2389.2011.01357.x

Ferroudi, N., Nzimoto, J., Davidson, A., Talbot, D., Briot, E., Dupuis, V., & Abramson, S. (2013). Maghemite nanoparticles and maghemite/silica nanocomposite microspheres as magnetic fenton catalysts for the removal of water pollutants. *Applied Catalysis B: Environmental*, *136*, 9–18. doi:10.1016/j.apcatb.2013.01.046

Fiset, J. F., Blais, J. F., & Riverso, P. A. (2008). Review on the removal of metal ions from effluents using seaweeds, alginate derivatives and other sorbents. *Revue des Sciences de l'Eau*, *21*(3), 283–308. doi:10.7202/018776ar

Gadhe, A., Sonawane, S. S., & Varma, M. N. (2015). Enhancement effect of hematite and nickel nanoparticles on biohydrogen production from dairy wastewater. *International Journal of Hydrogen Energy*, *40*(13), 4502–4511. doi:10.1016/j.ijhydene.2015.02.046

Gao, C. Z., Dong, Y., Zhang, H. J., & Zhang, J. M. (2007). Utilisation of Distiller Waste and Residual Mother Liquor to Prepare Precipitated Calcium Carbonate. *Journal of Cleaner Production*, *15*(15), 1419–1425. doi:10.1016/j.jclepro.2006.06.024

Gopakumar, D.A., Vishnu, A., Daniel, P., Leu, S.Y., Abdul, K.H.P.S., & Sabu, T. 2019. Nanocellulose based membrane for water purification. *Nanoscale Materials in Water Purification*, *4*, 59-85.

Gu, S., Cai, J., Wang, J., Yuan, Y., Chang, D., & Chikhotkin, V. F. (2015). Compound Method to Disperse $CaCO_3$ Nanoparticles to Nano-Size in Water. *Journal of Nanoscience and Nanotechnology*, *15*(12), 9488–9493. doi:10.1166/jnn.2015.10332 PMID:26682370

Haruta, S., Hanafusa, T., Fukase, H., Miyajima, H., & Oki, T. (2003). An Effective Absorption Behaviour of Insulin for DiabeticTreatment Following Intranasal Delivery using PorousSpherical Calcium Carbonate in Monkey and Healthy HumanVolunteers. *Diabetes Technology & Therapeutics*, *5*(1), 1–9. doi:10.1089/152091503763816409 PMID:12725701

He, S. Y., Feng, Y. Z., Ni, J., Sun, Y. F., Xue, L. H., Feng, Y. F., Yu, Y., Lin, X., & Yang, L. (2016). Different responses of soil microbial metabolic activity to silver and iron oxide nanoparticles. *Chemosphere*, *147*, 195–202. doi:10.1016/j.chemosphere.2015.12.055 PMID:26766356

Homaeigohar, S., & Elbahri, M. (2014). Nanocomposite electrospun nanofiber membranes for environmental remedi-ation. *Materials (Basel)*, *7*(2), 1017–1045. doi:10.3390/ma7021017 PMID:28788497

Hu, Z. S., & Deng, Y. L. (2003). Supersaturation Control in AragoniteSynthesis Using Sparingly Soluble Calcium Sulfate as Reactants. *Journal of Colloid and Interface Science*, *266*(2), 359–365. doi:10.1016/S0021-9797(03)00699-4 PMID:14527459

Hu, Z. S., Deng, Y. L., & Sun, Q. H. (2004). Synthesis of Precipitated Calcium Carbonate Nanoparticles using a Two-MembraneSystem. *Colloid Journal*, *66*(6), 745–750. doi:10.100710595-005-0017-4

Huang, F., Ge, L., Zhang, B., Wang, Y., Tian, H., Zhao, L. P., He, Y., & Zhang, X. (2014). A fullerene colloidal suspension stimulates the growth and denitrification ability of wastewater treatment sludge-derived bacteria. *Chemosphere*, *108*, 411–417. doi:10.1016/j.chemosphere.2014.02.042 PMID:24630252

Idi, A., Nor, M. H. M., Wahab, M. F. A., & Ibrahim, Z. (2015). Photosynthetic bacteria: An eco-friendly and cheap tool for bioremediation. *Reviews in Environmental Science and Biotechnology*, *14*(2), 271–285. doi:10.100711157-014-9355-1

Iram, M., Guo, C., Guan, Y. P., Ishfaq, A., & Liu, H. Z. (2010). Adsorption and magnetic removal of neutral red dye from aqueous solution using Fe3O4 hollow nanospheres. *Journal of Hazardous Materials*, *181*(1-3), 1039–1050. doi:10.1016/j.jhazmat.2010.05.119 PMID:20566240

Jagaba, A. H., Sule, A., Ibrahim, M. L., Ab, A. A. L., & Umaru, I. (2018). Wastewater Treatment Using Alum, the Combinations of Alum-Ferric Chloride, Alum-Chitosan, Alum-Zeolite and Alum- Moringa Oleifera as Adsorbent and Coagulant. *International J of Eng Management*, *3*, 67–75.

Jiang, S., Park, S., Yoon, Y., Lee, J. H., Wu, W. M., Phuoc Dan, N., Sadowsky, M. J., & Hur, H.-G. (2013). Methanogenesis facilitated by geobiochemical iron cycle in a novel syntrophic methanogenic microbial community. *Environmental Science & Technology*, *47*(17), 10078–10084. doi:10.1021/es402412c PMID:23919295

Kang, G. D., & Cao, Y. M. (2014). Application and modification of poly (vinylidene fluoride)(PVDF) membranes –a review. *Journal of Membrane Science*, *463*, 145–165. doi:10.1016/j.memsci.2014.03.055

Kato, S., Hashimoto, K., & Watanabe, K. (2012). Methanogenesis facilitated by electric syntrophy via (semi)conductive iron-oxide minerals. *Environmental Microbiology*, *14*(7), 1646–1654. doi:10.1111/j.1462-2920.2011.02611.x PMID:22004041

Khusnutdinova, A. N., Ovchenkova, E. P., Khristova, A. P., Laurinavichene, T. V., Shastik, E. S., Liu, J., & Tsygankov, A. A. (2012). New tolerant strains of purple nonsulfur bacteria for hydrogen production in a two-stage integrated system. *International Journal of Hydrogen Energy*, *37*(10), 8820–8827. doi:10.1016/j.ijhydene.2012.02.003

Kirkpatrick, B., Kohler, K., Byrne, M., Fleming, L. E., Scheller, K., Reich, A., Hitchcock, G., Kirkpatrick, G., Ullmann, S., & Hoagland, P. (2014). Human responses to Florida red tides: Policy awareness and adherence to local fertilizer ordinances. *The Science of the Total Environment*, *493*, 898–909. doi:10.1016/j.scitotenv.2014.06.083 PMID:25003583

Kitamura, M., Konno, H., Yasui, A., & Masuoka, H. (2002). ControllingFactors and Mechanism of Reactive Crystallisation of CalciumCarbonate Polymorphs from Calcium Hydroxide Suspensions. *Journal of Crystal Growth*, *236*(1-3), 323–332. doi:10.1016/S0022-0248(01)02082-6

Koepenick, M. (1976). *Papeti'ereduQu´ebec*. Groupe d'informationsd'Affaires.

Koppel, L. B. (1982). Input Multiplicities in Nonlinear Multivariable Control Systems. *AIChE Journal. American Institute of Chemical Engineers*, *28*(6), 935–945. doi:10.1002/aic.690280608

Kotrotsiou, O., & Costas, K. (2019). Water treatment by molecularly imprinted materials. *Nanoscale Materials in Water Purification*, *7*, 179-230.

Kuo, W. C. J., & Smith, R. (1998). Designing for the interactions between water use and effluent treatment. *Transactions of the Institution of Chemical Engineers*, *76*(3), 287–288. doi:10.1205/026387698524938

Kvande & Wahbeck. (2006). Theory for the Determination ofVapour Pressures by the Transpiration Method. *Acta Chem.Scand. Ser. A*, *30*, 297–302.

Lalley, J., Han, C., Li, X., Dionysiou, D. D., & Nadagouda, M. N. (2016). Phosphate adsorption using modified iron oxide-based sorbents in lake water: Kinetics, equilibrium, and column tests. *Chemical Engineering Journal*, *284*, 1386–1396. doi:10.1016/j.cej.2015.08.114

Lenz, M., Enright, A. M., O'aherty, V., van Aelst, A. C., & Lens, P. N. L. (2009). Bioaugmentation of UASB reactors with immobilized *Sulfurospirillum barnesii* for simultaneous selenate and nitrate removal. *Applied Microbiology and Biotechnology*, *83*(2), 377–388. doi:10.100700253-009-1915-x PMID:19305993

Lin, T., Wang, H., Wang, H., & Wang, X. (2004). The charge effect of cationic surfactants on the elimination of fibre beads in the electrospinning of polystyrene. *Nanotechnology*, *15*(9), 1375–1381. doi:10.1088/0957-4484/15/9/044

Liu, W., Graham, M., Evans, E. A., & Reneker, D. H. (2002). Poly (meta-phenylene isophthalamide) nanofibers: Coating and post processing. *Journal of Materials Research*, *17*(12), 3206–3212. doi:10.1557/JMR.2002.0464

Liwarska-Bizukojc, E. (2019). Nano- and microparticles-induced effect on activated sludge properties. *International Journal of Environmental Science and Technology*, *16*(12), 8663–8670. doi:10.100713762-019-02496-z

Mani, D., & Kumar, C. (2014). Biotechnological advances in bioremediation of heavy metals contaminated ecosystems: An overview with special reference to phytoremediation. *International Journal of Environmental Science and Technology*, *11*(3), 843–872. doi:10.100713762-013-0299-8

Maximous, N., Nakhla, G., & Wan, W. (2009). Comparative assessment of hydrophobic and hydrophilic membrane fouling in wastewater applications. *Journal of Membrane Science*, *339*(1), 93–99. doi:10.1016/j.memsci.2009.04.034

Mehrizad, A., Zare, K., Dashti Khavidaki, H., Dastmalchi, S., Aghaie, H., & Gharbani, P. (2011). Kinetic and thermodynamic studies of adsorption of 4-chloro-2-nitrophenol on nano-TiO2. *J. Phys. Theor. Chem*, *8*, 33–38.

Mendham, J., Denney, R. C., Barnes, J. D., Thomas, M. J. K., Denney, R. C., & Thomas, M. J. K. (2000). *Vogel's Quantitative ChemicalAnalysis* (6th ed.). Prentice Hall.

Mushtaq, F., Guerrero, M., Sakar, M. S., Hoop, M., Lindo, A. M., Sort, J., Chen, X., Nelson, B. J., Pellicer, E., & Pane, S. (2015). Magnetically driven bi $2O_3$/ BiOCl-based hybrid microrobots for photocatalytic water remediation. *Journal of Materials Chemistry. A, Materials for Energy and Sustainability*, *3*(47), 23670–23676. doi:10.1039/C5TA05825B

Nasreen, S. A. A. N., Sundarrajan, S., Nizar, S. A. S., Balamurugan, R., & Ramakrishna, S. (2013). Advancement in electrospun nanofibrous membranes modification and their application in water treatment. *Membranes*, *3*(4), 266–284. doi:10.3390/membranes3040266 PMID:24957057

Ng, D. K. S., Foo, D. C. Y., & Tan, R. R. (2009). Automated targeting technique for single-impurity resource conservation networks. Part 2: Single-pass and partitioning waste-interception systems. *Industrial & Engineering Chemistry Research*, *48*(16), 7647–7661. doi:10.1021/ie900127r

Ngiam, M., Ramakrishna, S., Raghunath, M., & Chan, C. K. (2007). Nanofiber patent landscape. *Recent Patents on Nanotechnology*, *2*(2), 137–144. doi:10.2174/187221007780859645 PMID:19076028

Ogino, T., Suzuki, T., & Sawada, K. (1987). The Formation andTransformation Mechanism of Calcium-Carbonate in Water. *Geochimica et Cosmochimica Acta*, *51*(10), 2757–2767. doi:10.1016/0016-7037(87)90155-4

Ojedokun, A. T., & Bello, O. S. (2016). Sequestering heavy metals from wastewater using cow dung. *Water Resources and Industry*, *13*, 7–13. doi:10.1016/j.wri.2016.02.002

Oniyama, E., & Walhbeck, P. G. (1995). Application of TranspirationTheory to TGA Data: Calcium Carbonate and Zinc Chloride. *Thermochimica Acta*, *250*(1), 41–53. doi:10.1016/0040-6031(94)01935-A

Pan, B., Zhang, W., Lv, L., Zhang, Q., & Zheng, S. (2009). Development of polymeric and polymer-based hybrid adsorbents for pollutants removal from waters. *Chemical Engineering Journal*, *151*(1-3), 19–29. doi:10.1016/j.cej.2009.02.036

Parthasarathy, G., & El-Halwagi, M. M. (2000). Optimum mass integration strategies for condensation and allocation of multi component VOCs. *Chemical Engineering Science*, *55*, 881–895. doi:10.1016/S0009-2509(99)00382-6

Pendergast, M. M., & Hoek, E. M. (2011). A review of water treatment membrane nanotechnologies. *Energy & Environmental Science*, *4*(6), 1946–1971. doi:10.1039/c0ee00541j

Petrinic, I., Andersen, N. P. R., Sostar-Turk, S., & Le Marechal, A. M. (2007). The removal of reactive dye printing compounds using nanofiltration. *Dyes Pigm.*, *74*(3), 512–518. doi:10.1016/j.dyepig.2006.11.003

Ramakrishna, S., & Shirazi, M. M. A. (2013). Electrospun membranes: Next generation membranes for desalination and water/wastewater treatment. *Desalination*, *308*, 198–208.

Rashidi, H. R., Sulaiman, N. N., & Hashim, N. A. (2012). Batik industry synthetic wastewater treatment using nanofiltration membrane. *Proc. Eng.*, *44*, 2010–2012. doi:10.1016/j.proeng.2012.09.025

Ray, S. S., Chen, S. S., Hsu, H. T., Cao, D. T., Nguyen, H. T., & Nguyen, N. C. (2017). Uniform hydrophobic electrospun nanofibrous layer composed of polysulfone and sodium dodecyl sulfate for improved desalination perfor-mance. *Separation and Purification Technology*, *186*, 352–365. doi:10.1016/j.seppur.2017.06.032

Ray, S. S., Chen, S. S., Li, C. W., Nguyen, N. C., & Nguyen, H. T. (2016). A comprehensive review: Electrospinning technique for fabrication and surface modification of membranes for water treatment application. *RSC Advances*, *6*(88), 85495–85514. doi:10.1039/C6RA14952A

Ray, S.S., Chen, S.S., Nguyen, N.C., & Hau, T.N. (2019). Electrospinning: A versatile fabrication technique for nanofibrous membranes for use in desalination. *Nanoscale Materials in Water Purification, 9*, 247- 273.

Ray, S. S., Chen, S. S., Nguyen, N. C., Hsu, H. T., Nguyen, H. T., & Chang, C. T. (2014). Poly (vinyl alcohol) incorporated with surfactant based electrospun nanofibrous layer onto polypropylene mat for improved desalination by using membrane distillation. *Desalination, 414*, 18–27. doi:10.1016/j.desal.2017.03.032

Reddy, M. M., & Nancollas, G. H. (1971). The Crystallisation ofCalcium Carbonate: Isotopic Exchange and Kinetics. *Journal of Colloid and Interface Science, 37*, 166–172. doi:10.1016/0021-9797(71)90161-5

Ren, X., Chena, C., Nagatsu, M., & Wang, X. (2011). Carbon nanotubes as adsorbents in environmental pollution management: A review. *Chemical Engineering Journal, 170*(2-3), 395–410. doi:10.1016/j.cej.2010.08.045

Reneker, D. H., Yarin, A. L., Fong, H., & Koombhongse, S. (2000). Bending instability of electrically charged liquid jets of polymer solutions in electrospinning. *Journal of Applied Physics, 87*(9), 4531–4547. doi:10.1063/1.373532

Romuald, B. K., Hamoudi, S., Faical, L., & Khaled, B. (2012). Synthesis of $CaCO_3$ nanoparticles by controlled precipitation of saturated carbonate and calcium nitrate aqueous solutions. *Canadian Journal of Chemical Engineering, 90*(1), 26–33. doi:10.1002/cjce.20673

Rutledge, G. C., & Fridrikh, S. V. (2007). Formation of fibers by electrospinning. *Advanced Drug Delivery Reviews, 59*(14), 1384–1391. doi:10.1016/j.addr.2007.04.020 PMID:17889398

Santhosh, C., Velmurugan, V., Jacob, G., Jeong, S. K., Grace, A. N., & Bhatnagar, A. (2016). Role of nanomaterials in water treatment applications: A review. *Chemical Engineering Journal, 306*, 1116–1137. doi:10.1016/j.cej.2016.08.053

Sawada, K., & Ohtaki, H. (1998). *Crystallisation Process*. John Wiley & Sons, Inc.

Shao, C., Guan, H., Liu, Y., Gong, J., Yu, N., & Yang, X. (2004). A novel method for making ZrO_2 nano fibres via an electrospinning technique. *Journal of Crystal Growth, 267*(1-2), 380–384. doi:10.1016/j.jcrysgro.2004.03.065

Singh, A. K., Srivastava, M., Rajneesh, N. K., & Shukla, S. (2017). Corncobs as low cost bio adsorbent for water and waste water treatment. *Int J of Inn Res Sci Eng Tech., 6*, 20487–20491.

Spivak, A., Dzenis, Y., & Reneker, D. (2000). A model of steady state jet in the electrospinning process. *Mechanics Research Communications, 27*(1), 37–42. doi:10.1016/S0093-6413(00)00060-4

Statyukha, G., Kvitka, O., Dzhygyrey, I., & Jezowski, J. (2008). A simple sequential approach for designing industrial wastewater treatment networks. *Journal of Cleaner Production, 16*(2), 215–224. doi:10.1016/j.jclepro.2006.09.002

Subramanian, S., & Seeram, R. (2013). New directions in nanofiltration applications—Are nanofibers the right mate-rials as membranes in desalination. *Desalination, 308*, 198–208. doi:10.1016/j.desal.2012.08.014

Sukigara, S., Gandhi, M., Ayutsede, J., Micklus, M., & Ko, F. (2003). Regeneration of Bombyx mori silk by electrospinning—part 1: Processing parameters and geometric properties. *Polymer*, *44*(19), 5721–5727. doi:10.1016/S0032-3861(03)00532-9

Sunqing, Q., Junxiu, D., & Guoxu, C. (2000). Wear and Friction Behaviour of CaCO3Nanoparticles used as Additives in Lubricating Oils. *Lubrication Science*, *12*(2), 205–212. doi:10.1002/ls.3010120207

Tai, C. Y., & Chen, P. C. (1995). Nucleation, Agglomeration andCrystal Morphpology of Calcium Carbonate. *AIChE Journal. American Institute of Chemical Engineers*, *41*(1), 68–77. doi:10.1002/aic.690410108

Tawabini, B. S. (2015). Removal of methyl tertiary butyl ether (MTBE) from contaminated water using UV-assisted nano composite materials. *Desalination and Water Treatment*, *55*(2), 549–554. doi:10.1080/19443994.2014.924035

Theron, S., Yarin, A., Zussman, E., & Kroll, E. (2005). Multiple jets in electrospinning: Experiment and modeling. *Polymer*, *46*(9), 2889–2899. doi:10.1016/j.polymer.2005.01.054

Theron, S., Zussman, E., & Yarin, A. (2004). Experimental investigation of the governing parameters in the electro-spinning of polymer solutions. *Polymer*, *45*(6), 2017–2030. doi:10.1016/j.polymer.2004.01.024

Tucker, N., Stanger, J., Staiger, M., Razzaq, H., & Hofman, K. (2012). The history of the science and technology of electrospinning from 1600 to 1995. *Journal of Engineered Fibers and Fabrics*, *7*(2_suppl), 63–73. doi:10.1177/155892501200702S10

Varma, S., Sarode, D., Wakale, S., Bhanvase, B. A., & Deosarkar, M. P. (2013). Removal of nickel from waste water using grapheme nanocomposite. *Int. J. Chem. Phys. Sci.*, *2*, 132–139.

Vinay, S.P., Udayabhanu Nagaraju, G., Chandrappa, C.P., & Chandrasekhar, N. (2019). Novel Gomutra (Cow urine) mediated synthesis of silver oxide nanoparticles and their enhanced photocatalytic, photoluminescence and antibacterial studies. *Journal of Science: Advanced Materials and Devices*, *4*, 392-399.

Visa, M. (2016). Synthesis and characterization of new zeolite materials obtained from fly ash for heavy metals removal in advanced wastewater treatment. *Powder Technology*, *294*, 338–347. doi:10.1016/j.powtec.2016.02.019

Vucak, M., Pons, M. N., Peric, J., & Vivier, H. (1998). Effect of Precipitation Conditions on the Morphology of CalciumCarbonate: Quantification of Crystal Shapes using ImageAnalysis. *Powder Technology*, *97*(1), 1–5. doi:10.1016/S0032-5910(97)03375-5

Wahlbeck, P. G. (1992). TGA Experimental Data – Analysis Providedby Transpiration Theory. *Thermochimica Acta*, *197*(2), 469–478. doi:10.1016/0040-6031(92)85048-Z

Wang, H., Yuan, X., Wu, Y., Zeng, G., Chen, X., Leng, L., & Li, H. (2015). Facile synthesis of amino-functionalized titanium metal-organic frameworks and their superior visible-light photocatalytic activity for Cr (VI) reduction. *Journal of Hazardous Materials*, *286*, 187–194. doi:10.1016/j.jhazmat.2014.11.039 PMID:25585267

Wei, S. H., Mahuli, S. K., Agnihotri, R., & Fan, L. S. (1997). HighSurface Area Calcium Carbonate: Pore Structurual Propertiesand Sulfation Characteristics. *Industrial & Engineering Chemistry Research*, *36*(6), 2141–2148. doi:10.1021/ie960768l

Wray, J. L., & Farrington, D. (1957). Precipitation of Calcite and Aragonite. *Journal of the American Chemical Society*, *79*(9), 2031–2034. doi:10.1021/ja01566a001

Xiang, L., Xiang, Y., Wen, Y., & Wei, F. (2004). Formation of CaCO3Nano-Particles in the Presence of Terpineol. *Materials Letters*, *58*(6), 959–965. doi:10.1016/j.matlet.2003.07.034

Yin, X. X., Wang, L. H., Bai, R., Huang, H., & Sun, G. X. (2012). Accumulation and transformation of arsenic in the blue-green alga Syncchocysis sp. PCC6803. *Water, Air, and Soil Pollution*, *223*(3), 1183–1190. doi:10.100711270-011-0936-0

Yu, J., Lei, M., Cheng, B., & Zhao, X. (2004). Effects of PAA Additiveand Temperature on Morphology of Calcium CarbonateParticles. *Journal of Solid State Chemistry*, *177*(3), 681–689. doi:10.1016/j.jssc.2003.08.017

Yuan, X., Huang, H., Zeng, G., Li, H., Wang, J., Zhou, C., Zhu, H., Pei, X., Liu, Z., & Liu, Z. (2011). Total concentrations and chemical speciation of heavy metals in liquefaction residues of sewage sludge. *Bioresource Technology*, *102*(5), 4104–4110. doi:10.1016/j.biortech.2010.12.055 PMID:21211964

Zhang, F., Ge, Z., Grimaud, J., Hurst, J., & He, Z. (2013). Long-term performance of liter-scale microbial fuel cells treating primary effluent installed in a municipal wastewater treatment facility. *Environmental Science & Technology*, *47*(9), 4941–4948. doi:10.1021/es400631r PMID:23517192

Zhang, S., Wang, R., Zhang, S., Li, G., & Zhang, Y. (2013). Development of phosphorylated silica nanotubes (PSNTs)/ polyvinylidene fluoride (PVDF) composite membranes for wastew-ater treatment. *Chemical Engineering Journal*, *230*, 260–271. doi:10.1016/j.cej.2013.06.098

Zhelev, T. K., & Bhaw, N. (2000). Combined water–oxygen pinch analysis for better waste water treatment management. *Waste Management (New York, N.Y.)*, *20*(8), 665–670. doi:10.1016/S0956-053X(00)00050-7

Zhou, F. L., Gong, R. H., & Porat, I. (2009). Mass production of nanofibre assemblies by electrostatic spinning. *Polymer International*, *58*(4), 331–342. doi:10.1002/pi.2521

Zong, X., Kim, K., Fang, D., Ran, S., Hsiao, B. S., & Chu, B. (2002). Structure and process relationship of electrospun bioabsorbable nanofiber membranes. *Polymer*, *43*(16), 4403–4412. doi:10.1016/S0032-3861(02)00275-6

Chapter 2
Tolerance of Microorganisms to Heavy Metals

Joan Mwihaki Nyika
https://orcid.org/0000-0001-8300-6990
Technical University of Kenya, Kenya

ABSTRACT

Heavy metal pollution is a growing environmental concern due to the increase in anthropogenic-based sources. Microorganisms have high adsorptive capacities and surface-area-to-volume ratio that enable the uptake of these contaminants and their conversion to innocuous complexes in the process of bioremediation. This chapter explores the mechanisms and specific microorganisms that are resistant to metal toxicity. A wide range of bacterial, algae, and fungal species used as biosorbents are highlighted. Mechanisms such as reduction of metal cations, their sequestration, and binding on cell barriers are discussed. To optimise the efficacy of microorganisms in bioremediation processes, adoption of genetic and nano-technologies is recommended.

INTRODUCTION

Heavy metals are a variety of elements whose distribution in the environment is ubiquitous. Their distinguishing characteristics are a high atomic weight, density and assorted applications in different areas. Many heavy metals occur naturally in the earth's crust and have essential functions in living organisms. For instance, copper (Cu) is a cofactor in many metabolic reactions and proteins containing this metal enhance iron transport, free radical protection and respiration in living organisms (Singh, Parihar, Singh, Singh & Prasad, 2015). Other metals such as sodium (Na), nickel (Ni), zinc (Zn), manganese (Mn), magnesium (Mg), iron (Fe) and potassium (K) are essential minerals required for growth and functioning of plants and animals. These heavy metals are only required in trace concentrations. Other heavy metals such as cadmium (Cd) and mercury (Hg) do not have biological roles and are harmful even in minute concentrations. The heavy metals however can occur in excess levels in the environment because of anthropogenic activities as illustrated in Figure 1. In such a case, they induce harmful effects to plants, animals, humans and the ecosystems. Heavy metal pollution of the environment through human activi-

DOI: 10.4018/978-1-7998-4888-2.ch002

Figure 1. Anthropogenic sources of named heavy metals

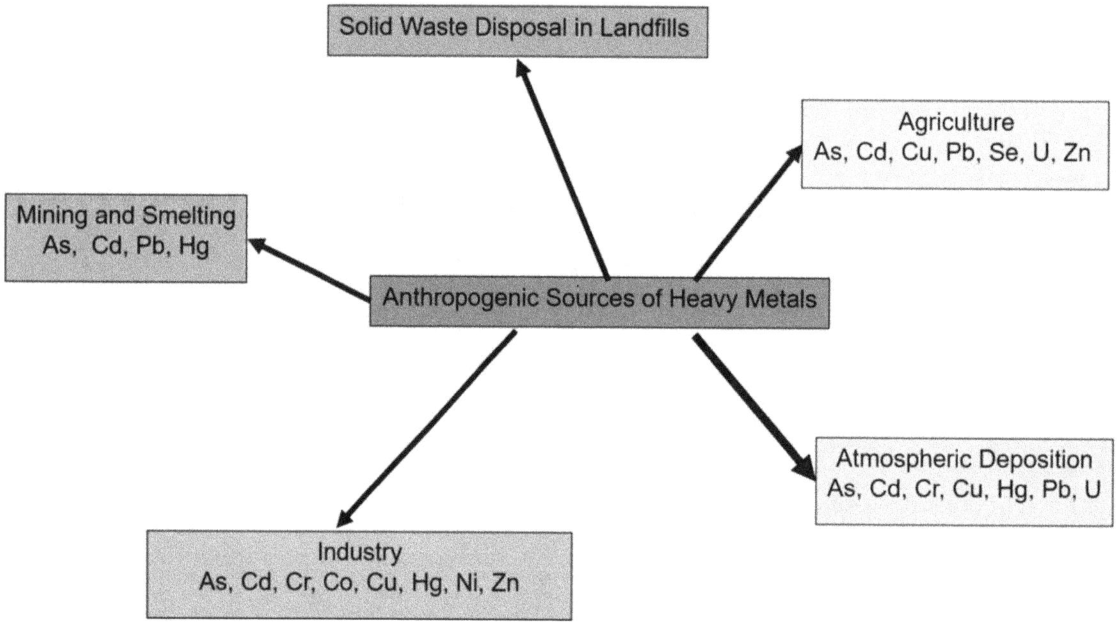

ties is an ecological concern worldwide because metals persist and bio-accumulate in the environment causing disruptions in food chains and ecosystems (De Silva et al. 2012). Additionally, heavy metals do not decompose and cannot be degraded. Once in food chains, they enter in human and animal tissues, which increases the vulnerability of these organisms to diseases and gene transformations/mutations, which are carcinogenic in the long term.

The focus of this book chapter is heavy metal toxicity on microorganisms and approached to bioremediation. Some of the toxic effects of heavy metals on microorganisms documented in literature are summarised in Table 1. To alleviate this pollution problem, a number of biotechnological advances including geno-remediation, cyano-remediation, myco-remediation, rhizo-remediation, bio-stimulation, hyper-accumulation, phyto-stabilization, bio-sorption and bio-mineralization among others have been applied to remove heavy metals from the environment (Samantaray et al., 2014). The toxicity and ubiquitous nature of heavy metals has forced organisms to adapt to their presence. This is a natural process where microbes develop recycling and degradation potential to accumulated heavy metal contaminant resulting to the decrease of their toxicity. Singh et al. (2015) for instance, reported that plants are developing tolerance to heavy metals. Yang, Agouri, Tyrrell and Walsh (2018) associated *enterobacteriaceae* species to heavy metal resistance genes in humans. These bioremediation measures are growing to be effective and economic strategies to detoxify and accumulate metals in microorganisms so that they are no longer harmful to the environment. Apart from intrinsic bioremediation, some chemicals can be added to stimulate microbial bioavailability in the process of bio-stimulation.

Consequently, these microorganisms have developed mechanisms to immobilize and sequester these contaminants. Bacterial that can survive in high levels of heavy metals have been assayed in different sources including tannery effluents, coal mining areas, coastal waters, silver mines, pristine freshwater and sewage (De Silva et al. 2012). This tolerance of microorganisms to heavy metals depends on the availability and concentration of these elements and is influenced by many processes and factors in-

Table 1. Effects of specific heavy metals on microorganisms

Heavy Metal	Effects on Microorganisms	References
Ag	Inhibit growth and cell transduction, cell lysis	Prabhu & Poulose, 2012
As	Enzyme deactivation	Sankarammal, Thatheyus & Ramya, 2014
Cd	Destruction of nucleic acids hindering transcription, cell division and protein denaturation	Fashola, Ngole-Jeme & Babalola, 2016
Cr	Inhibits oxygen uptake, prolongs the lag phase and inhibits growth	Fashola et al., 2016
Cu	Inhibits enzymatic processes and disrupts cellular activities	Prabhu & Poulose, 2012
Hg	Ruptures cell membrane, inhibits enzymatic activities and denatures proteins	Fashola et al., 2016
Ni	Causes oxidative stress and upsets the cell membrane	Fashola et al., 2016
Pb	Destroys proteins and nucleic acids, inhibits transcription	Sankarammal et al., 2014
Se	Inhibits growth	Malik, 2004
Zn	Inhibits growth and can cause death, causes biomass decrease	Malik, 2004

cluding the microbial species, the type of medium and metal species (Naik, Shamim & Dubey, 2012). The tolerance of heavy metals in microorganisms can be used to regulate the ever-increasing pollution through bioremediation. The aim of this chapter is to describe the mechanisms of microbial adaptation to heavy metals and explore ways in which this microbial tolerance could be used in pollution remediation.

MICROBIAL TOXICITY DUE TO HEAVY METALS

Toxicity is the potential of trace and heavy metals to induce harmful effects on microorganisms based on a number of factors such physicochemical features and binding components of media, pH and redox potential. The factors will be discussed in details in the following sections of this book chapter. The toxicity effects could range from complications such as retarded growth and inability to reproduce and to even mortality. The mechanisms of toxicity include the interference and inhibition of enzymatic functions, interference with the metabolism of genetic material (DNA and RNA) or some important metabolic proteins, the damage of ion regulation pathways and the production of reactive oxygen species (ROS) where heavy metals serve as redox catalysts (Hildebrandt et al., 2007; Gauthier et al., 2014). The presence of heavy metals often modifies the biochemical and physiological features of microorganisms. These modifications have been summarized in Table 1 and are dependent on the specificity of the metal contaminant. For instance, Cr and Cd promote the destruction of microorganisms through oxidative damage and weaken the capacity of microbes as tools for bioremediation (Igiri et al., 2018). Chromium (III) inflicts a number of changes in the anatomy and physiology of microbial enzymes by binding on their functional thiol and carboxyl groups (Cervantes et al., 2001). The heavy metal also interferes with replication and transcription through electrostatic reaction with phosphate groups of DNA resulting to mutagenesis (Cervantes et al., 2001).

Through the Haber-Weis and Fenton reactions, heavy metals such as Cu (II) and Cu (I) speed up reactions that produce ROS, which are soluble electron transporters. The reactions therefore cause severe damage to proteins, lipids, DNA and cytoplasmic molecules of microbes as Osman and Cavet (2008) highlighted. Aluminium (Al) facilitates the destruction of DNA by promoting the stability of superoxide radicals. The heavy metal also results to the imbalance of ions by adsorbing on the cell surface and consequently finding its ways to transmembrane carriers through the ionic channels (Chen et al., 2014; Booth et al., 2015). Lead (Pb) and Cd relocate microbes from their ligand interaction regions and the usual binding sites, which results to the damage of DNA structure and cell membrane causing microbial fatalities (Olaniran et al., 2013). Other general effects of heavy metals on microbes include the inhibition of essential enzyme functions through non-competitive and competitive reactions, the disturbance of microbial functioning through oxidative phosphorylation, modifications of their genetic material formation process and retardation of growth (Fashola et al., 2016).

Factors Influencing Metal Toxicity

The toxic effects and physiology of a metal on a biological system is a function of its speciation and bioavailability rather than its total metal concentration (Olaniran, Balgobind & Pillay, 2013). The latter describes the portion on total amount of metal in a given environment and at a given time, which is available for microbial uptake from the environment. Speciation on the other hand, is the quantification and identification of different metal elements and their occurrence. Traditionally, the hazard potential of environmental samples is assayed by quantifying their inherent metal content without distinguishing hazardous from non-hazardous portions (Olaniran et al., 2013). This is despite the fact that a number of factors govern the fate and characteristics of metals in the environment. In the water phase for instance, biological and chemical activity of metals depends on their chemical form. Effects of metals on microbial composition of the environment depends the physicochemical characteristics and binding capacity of involved medium, redox potential and pH of some buffers (Olaniran et al., 2013). The influence of these factors on metal toxicity is discussed in the following sub-sections.

Physicochemical Features and Binding Components in Media

Some pH buffers precipitate and complex metals affecting their bioavailability and speciation especially if they are in high concentrations (Olaniran et al., 2013). Phosphate for example precipitates metals to reduce their bioavailability. The buffer sequesters metals to insoluble phosphate complexes at mildly acidic and neutral pH levels. Different metal species have varied sensitivity levels to these buffers. Formation of metal complexes with zwitterionic[1] buffers in microbial environments is another factor that affects metal toxicity (Mash, Chin, Sigg, Hari & Xue, 2003). Zwitterionic buffers promote bioavailability of metals unlike phosphate. The presence of trisaminomethane (tris)[2] buffers is associated with limited bioavailability of metals (Olaniran et al., 2013). Physicochemical properties and composition of soil influence the speciation of metals and hence, its toxicity and mobility (Olaniran et al., 2013). Soil solid characteristics including their mineralogy, complexation, organic matter content, ion exchange, adsorptive and precipitation properties influence metal distribution. Additionally, soil moisture and time effects influence the toxicity of metals in environments as they control the leaching of these elements. These factors interfere with the buffering capacity of environmental microorganisms to these contaminants.

Redox Potential and pH

Metal speciation is a function of redox potential (Eh) of an environment. Although oxido-reductive processes are slow in environments such as soils, their rate can fluctuate significantly in the presence of microorganisms. Negative redox potential in anaerobic conditions leads to metal precipitation where toxic metal ions such as $Pb^{2+,}$ Cd^{2+} and Fe^{2+} are complexed with sulphides to form nontoxic and insoluble deposits (Olaniran et al., 2013). The increase of sulphide in environments also corresponds to increased activity of *Desulfovibrio* bacteria species that reduce sulphates to sulphides. Positive redox potential promotes bioavailability of metals and under these oxidising conditions, pH decreases. Acid mine drainage is a typical example where *Thiobacillus thiooxidans* converts sulphur and sulphide and then to sulphuric acid (Olaniran et al., 2013).

At basic pH, metals precipitate to insoluble carbonates and phosphates while at acidic pH, they form soluble and free ionic organometals that are bioavailable (Sandrin & Maier, 2002). Under basic conditions, metal ions substitute protons to form hydroxo-metal complexes while in acidic conditions protons block metal binding sites. A fluctuation in pH has great influence on metal bioavailability. At acidic pH, heavy metals are skewed to dissolve to their free ions because many protons are circulating at their binding sites. At this state, there is a high concentration of hydrogen ions and a subsequent reduction in the cohesion of metal and adsorbent cations since the adsorbent surfaces are highly positive, which results to enhanced toxicity (Igiri et al., 2018). This factor mediates metal toxicity in microorganisms so that under mild basic pH, many metals are not bioavailable for microorganisms and form complexes with their ligands, as is the case with nickel (Ni). Increasing the pH however, leads to more toxicity by metal such as Cu, Zn and U to microorganisms (Sandrin & Maier, 2002). Under acidic conditions, electrostatic attraction of metal ions to microbial membranes is limited, which promotes their bioavailability. In enhancing the tolerance to heavy metals by microorganisms, this factor can be optimised and regulated towards bioremediation.

Environmental Factors

The susceptibility of heavy metal pollutants to induce inhibitory or stimulatory effects to microbes is influenced by a number of environmental factors. These include the presence of humic or organic acids of low molecular weight (which alter pH) and temperature. The reactions influenced by these factors include the heavy metals' valence, the bioavailability of metal pollutants to microbes, processes of transportation and transformation (Igiri et al., 2018). Temperature influences heavy metal adsorption and hence its toxicity to microorganisms. High temperatures enhance the adsorbate diffusion rate at the contaminants' outside boundary layer. This phenomenon results to their increased solubility at high temperature and hence, enhanced heavy metal bioavailability (Bandowe et al., 2014). On a positive response, higher temperatures promote higher activity of enzymes in microbes, which is stimulatory to bioremediation. However, the temperature must be optimal as temperatures beyond this limit could reverse the activity following its ability to denature microbial enzymes and interfere with their physiological processes. The stability of metal and microorganism complexes is also influenced by chemical moiety ionization of a microorganism's cell wall, the configuration of the cell wall and available sorption sites. A range of environmental factors including the substrate influence the outcome of degradation processes of the heavy metal-microbe complex as shown in Table 2. In the following sub-topics, the following factors and their interplay in the microorganism-heavy metal complex is discussed with the focus being to demonstrate how bioremediation processes occur and are sustained by microorganisms.

MECHANISMS OF MICROBIAL TOLERANCE TO HEAVY METALS

A number of microorganisms have evolved to become metal resistancc following exposure to heavy metal toxins for short or long-term periods. These organisms determine the fate of these metals in the environment. Selective pressures at metal containing vicinities and habitats prompt the phenomenon of bioremediation (Bruins et al., 2000). Most bioremediation systems are plasmid-mediated and are common in eubacterial groups. However, some of the systems can be a result of chromosomes. Several microorganisms are confirmed to be resistance to certain metal pollutants. Out of these, bacterial species such as *Bacillus, Pseudomonas* and *Staphylococcus* species are widely researched (Bruins et al., 2000). Fungal species such as *Chlorella* and *Aspergillus* were found to possess metal resistance capability (Indu & Shaili, 2011) Resistance to heavy metal toxicity could have developed at the early stages of prokaryotic life of microorganisms since for a long time; their habitats are located in metal containing environs (Olaniran et al., 2013).

Table 2. The factors influencing the bioremediation of heavy metals by microorganisms and the processes affected

Factors Influencing Bioremediation Processes	Affected Activities
Aerobic and anaerobic biological reactions	• Redox potential • The number of microbes available at a target site • The number of electron acceptors available for reactions
Co-metabolism and growth substrate	• The presence of alternative sources of carbon • Concentration • Microbial reaction processes of predation, succession and competition
Limitations in mass transfer	• Nutrient diffusion • Solubility and diffusion of oxygen for ATP production • Solubility of the complex in and with water
Environmental	• Nutrient deficiency • Exhaustion of favourable substrates • Introduction of negative environmental conditions
Microbial	• The capacity to produce toxic substances in response to metal toxicity • Production of enzymes to inhibit toxicity • Horizontal gene transfer resulting to mutation • Modifications of microbial populations that can enhance bioremediation
Substrate/Heavy metal	• Solubility of heavy metals • Potency of contaminants • Concentration levels of heavy metals • Chemical structure of contaminants in relation to the functional groups of microbes/ microbial enzymes

Source: Boopathy (2000)

Bioremediation processes are influenced by mechanisms, which control membrane transport of proteins. During normal conditions, when heavy metal concentration levels are low and non-toxic, non-specific uptake controls the transport of ions in microorganisms (Geva et al., 2016). However, when the metal contaminants are in excess, ion efflux is specific and excludes nonessential metals in favour of essential ones (Dunbar, 2017). This phenomenon also facilitates the up-regulation of genes responsible for triggering detoxifying enzyme production in microorganisms (Dien et al., 2018). Gene-mediated heavy metal detoxification occurs using either plasmid or chromosomal systems. The latter controls detoxification of essential metals while the former, trace metal contaminants. Usually microorganisms develop resistance to heavy metals to protect their sensitive cell organisms and maintain the integrity of both their physiology and anatomy. These protective measures are crucial for survival of microbes. The following sub-sections evaluate mechanisms in which microorganisms detoxify and resist toxic heavy metals.

Formation of Extracellular Barriers

The capsule, plasma membrane or cell wall prevent the entry of metal ions in cells. These carbohydrate-based slime layers, often comprising of fatty acids and nucleic acids prevent cells from parasitism, phagocytosis and desiccation. Microorganisms such as bacterial species adsorb heavy metals using their hydroxyl, phosphate, amino or carboxyl groups found on the capsule or cell wall. The nature of adsorption is passive since dead cells possess these characteristics. Dead bacteria species including *Bacillus, Brevibacterium* and *Pseudomonas* species were reported to have high adsorptive capacities (Pardo, Herguedas, Barrado & Vega, 2003). In living microorganisms, accumulation of heavy metals occurs in two steps namely: -1) swift non-specific adsorption by the cell wall and 2) sluggish active transport to the cell cytoplasm (McEldowney, 2000). Bacterial capsules endowed with polysaccharide carboxyl groups adsorb heavy metals. This is evident in *Klebsiella aerogenes*, *Enterobacter chloaceae*, *Marinobacter* and *Acinetobacter* species whose extracellular biopolymers adsorb such contaminants (Bhaskar & Bhosle, 2006). These microorganisms have extracellular polymeric substances (EPSs) also known as exopolymers that strongly bind on metals such as U, Cd and Pb resulting to their detoxification as observed in *Micrococcus luteus*, *Staphyloccus aureus* and *Azobacter* species (Gomathy & Sabarinathan, 2010).

Apart from EPSs, microorganisms produce siderophores[3], which are complexes of low molecular weight that facilitate iron transport into the cell. These extracellular molecules bind on ferric and other trivalent metal ions to form complexes to detoxify the metals upon uptake by receptors. A case example is the decreased copper toxicity in *cyanobacterium Anabaena* species following the production of a siderophore (Gomathy & Sabarinathan, 2010). Microorganisms also produce biosurfactants[4] that form complexes with heavy metals such as Zn, Pb and Cd. These complexes are innocuous to cells (Gomathy & Sabarinathan, 2010).

Efflux of Heavy Metals

Most microorganisms resist heavy metal toxicity using the efflux or active transport mechanism. Using this approach, metal ions are exported out of cells. Genes that enable efflux are located on plasmids and/or chromosomes of the microorganisms. Metal ions get into bacterial cell systems during nutrient uptake. For example, in *Rastonia metallidurans*, heavy metals such as Mn, Ni, Co, Zn and Cd enter in the system during Mg transport while Cr enters during the transport of sulphate (Nies, 2000). The transport of metal ions from cells requires an electrochemical gradient and involves ATP hydrolysis. Proteins used during this efflux system are of three categories: - 1) P-type ATPases, 2) CDF (cation diffusion facilitator) and 3) RND (resistance, nodulation, cell division) (Nies, 2003). The first category transfers metals such as Cu^{2+}, Ag^{2+}, Zn^{2+}, Cd^{2+} and Pb^{2+}, which have high sulfhydryl group affinity while the second category transports divalent metals such as Co^{2+}, Ni^{2+} and Fe^{2+}. These two categories of proteins transport metals through the plasma membrane into the periplasm while the last category of RND proteins transports these cations from the periplasm across the plasma membrane.

A case example of a microorganism that uses the efflux system to resist metal toxicity is *R. metallidurans* CH34, which has a *Czc* operon that uses an electrochemical gradient efflux system. The operon[5] has three subunits: - *Czc A-C*. The last two subunits control the efflux system while *Czc A* is the cation-proton antiporter that functions as the RND protein (Nies, 2000). CPx-type ATPases are examples of P-type ATPases that export monovalent cations from *Streptococcus mutans*, *S. aureus*, *P. putida* and *Enterococcus hirae* (Nies, 2003). In the case of *P. putida* KT2440 strain, Cu and Cd efflux was influenced by a chemiosmotic efflux pump similar to the *Czc* system with *cadA*, which is a P-type ATPase conferring the tolerance to these metals (Canovas, Cases & de Lorenzo, 2003). Some bacteria employ both the EPSs and efflux systems for heavy metal resistance, as is the case with the S4 strain of *P. putida* and the arsenic resistant (ars) system found in both gram positive and negative bacteria (Saxena, Joshi & Srivastava, 2002).

Intracellular Sequestration of Heavy Metals

Complexation of heavy metals to various products in the cell cytoplasm is another mechanism that microorganisms use to resist toxicity associated with these cations. The process is propagated by metal binding peptides such as phytochelatins and metallothioneins, which have cysteine residues and can bind on metals using sulfhydryl groups. The process occurs in the intracellular space whereby sequestered contaminants are stored in peptide ligands (Mishra & Malik, 2013; Diep et al., 2018). Cyanobacterium such as *Synechococcus* species can synthesize metallothionein under the influence of *smtA* and *smtB* genes in the presence of Zn and Cd ions (Ybarra & Webb, 1999). Intracellular sequestration of Zn, Cd and Cu is also reported in *P. putida*, *P. diminuta* (Ibrahim, Ahmad & Baba, 2001) and *Rhizobium leguminosarum* cells (Lima, Corticeiro & Figueira, 2006).

Reductive Reactions of Microorganisms on Heavy Metals

Microorganisms such as bacteria, which are found in different ecological niches can reduce heavy metals such as vanadium (V), molybdenum (Mo) and Cr. In this case, microorganisms use metalloids and metals as electron acceptors/donors during energy production processes. Resultant metals that are in their oxidised form are less toxic and serve as electron acceptors in anaerobic respiration of host microorganisms. Table 3 shows examples of microorganisms and the metals they transform from more toxic to less toxic forms by reduction.

Extracellular Sequestration

This mechanism involves accumulation of heavy metals in the cellular components of the outer membrane or periplasm to complex them to insoluble products. A number of microorganisms use this mechanism

Table 3. Microorganisms with capacity to detoxify metals through reduction

Microorganism	Metal Ions Reduced	References
Bacillus thermoamylovorans, S. oneidensis	Te^{4+}/Te^{0}	Klonowska, Heulin & Vermeqlio, 2005
***Strenotrophomonas* species**	Au^{3+}/Au^{0}	Song et al., 2008
Thiobacillus ferrooxidans	Mo^{6+}/Mo^{5+}	Song et al., 2008
Shewanella putrefaciens, Geobacter sulphurreducens	Tc^{7+}/Tc^{4+}	Wildung, Gorby & Krupka, 2000
S. oneidensis, G. metallireducens	V^{5+}/V^{4+}	Ortiz-Bernad, Anderson, Vrionis & Lovley, 2004
Shewanella oneidensis, B. thermoamylovorans, R. metallidurans	$Se^{6+}/Se^{4+}/Se^{0}$ Se^{4+}/Se^{0}	Slobodkina, Bonch-Osmolovskaia & Slobodkin, 2007
Thermoterrabacterium ferrireducens, S. putrefaciens, D. desulfuricans	U^{6+}/U^{4+}	Ianieva, 2009
S. aureus	As^{5+}/As^{3+}	Ianieva, 2009
B. cereus, K. pneumonia, P. stutzeri	Hg^{2+}/Hg^{0}	Ianieva, 2009
Geobacter species, B. thermoamylovorans, G. metallireducens	Fe^{3+}/Fe^{2+}	Ianieva, 2009
Pseudomonas species, Brevibacterium species, Microbacterium species, Desulfomicrobium norvegicum	Cr^{6+}/Cr^{3+}	Ianieva, 2009

to resist heavy metal toxicity. An example is *Pesudomonas syringae* that synthesis periplasmic (Cop A and B) and outer membrane proteins (Cop C) that bind in copper making it innocuous (Slobodkina et al., 2007). Similar behaviour was observed in *Pseudomonas pickettii* US321 strain where copper was complexed on the outer membrane and transported into the cytoplasm (Gilotra & Srivastava, 1997). Some microorganisms expel toxic cations from the cytoplasm to the periplasm. A similar tendency is evident in *Synechocystis* PCC 6803 strain that exports Zn, *Salmonella* species and *P. putida* S4 strain that transport Ag (Saxena et al., 2002).

Extracellular sequestration occurs through metal precipitation to insoluble substances. Sulphate reducing bacteria such as *Kelbsiella planticola* strain that transforms thiosulfate to hydrogen sulphide in anaerobic conditions precipitates Cd^{2+} to insoluble sulphides is an example (Ianieva, 2009). The same mechanism is used to precipitate Cd ions in *P. aeruginosa* and *Nostac muscorum* (Moskvina, Brekhovskikh & Nikandrov, 2003). Lead was complexed to lead phosphate in *Vibrio harveyi* while copper was precipitated under carbon-limited conditions and in the presence of phosphate and hydroxyl residues in *P. putida* S4 strain (Ianieva, 2009).

Genetically Controlled Metal Resistance

Genes that control plasmids or extra-chromosomal resistance factors influence resistance to heavy metals in some microorganisms. Cd-resistance, which is plasma-mediated has been reported to occur in S. aureus and the resistance is dependent on decreased uptake of the cation by tolerant strains (Manasi, Rajesh & Rajesh, 2016). Plasmid-mediated resistance to Hg is evident in Pseudomonas species and S. aureus. In this case, Hg is transformed to harmless organomercurial [RHg(I)] (Manasi et al., 2016). The genetic controlled resistance to heavy metals by microorganisms is also associated with antimicrobial resistance. This mechanism is ecologically imperative and is transferrable to sensitive microorganisms as has been done on Escherichia coli and Aerobacter aerogenes where Hg-resistance genes have been introduced to reduce the metal's toxicity (Alavi et al., 2011).

The advancement of genetic engineering has seen the synthesis of microbes with the capacity to express and sometimes overexpress heavy metal resistance genes. Case examples are Chlamydomonas reinhardtii that tolerates Cd toxicity (Frederick et al., 2013), E. coli ArsR and S. cerevisiae CP2 HP3 that detoxify As, Cd and Zn ions through gene overexpression (Igiri et al., 2018). Biofilms and nano-techniques are also being used to enhance bioremediation of heavy metals particularly the ability of microorganisms to target specific metals even at lethal concentrations. According to Grujic, Vasic, Radojevic, Comic and Ostojic (2017), these technologies have been used to enhance the efficacy of Rhodotorula mucilaginosa in clearing heavy metals from planktonic cells to more than 95% from 10% without the modifications. Gupta et al. (2016) also noted that nanoparticles have been used to bioccumulate Cu, Ni, Pb and Zn ions in some bacteria for immobilisation and elimination from polluted environs. The study emphasised the need to incorporate genetic and nano-technologies in microbial resistance research to optimise their metal toxicity alleviation abilities.

MICROORGANISMS AS TOOLS FOR REMEDIATION OF POLLUTION BY HEAVY METALS

The rise of pollution due to anthropogenic activities has resulted to discharge of solid waste and sewage to the environment leading to exposure to heavy metals. These elements are difficult to remove from these environs and are inhibitors of biodegradation pathways. Additionally, they are very toxic. Through bioaccumulation, many microorganisms take up heavy metals through active processes such as adsorption or passive processes (Gupta et al., 2016). The high surface area compared to volume ratio of these organisms facilitates this uptake. It is for this reason that a number of microorganisms have been used in cleaning up heavy metal polluted environs as summarised in Table 4. Application of metal resistant strains of algae, fungi and bacteria in consortium, immobilised and/or single forms to remediate heavy metals have been documented successfully according to Igiri et al. (2018).

REMEDIATION OF HEAVY METALS BY ALGAE, FUNGI AND BACTERIA

Unlike other microbial biosorbents, algae are autotrophic, produce high quantities of biomass but require low nutrients. Through integration into cells or via adsorption, algae detoxify and immobilise heavy metals. They have chemical moieties such as amide, phosphate, carboxyl and hydroxyl groups that are binding sites to these elements (Irigi et al., 2018). They can also be combined with cyanobacteria to degrade toxic metals in a process known as phycoremediation (Chabukdhara, Gupta & Gogoi, 2017). Examples of algae species used in heavy metal bioremediation are shown in Table 4.

The excellent metal uptake and recovery capacity of fungi has been optimised to use them as biosorbents. Fungal cells used in bioremediation of heavy metals are either passive (lifeless) or active (Javaid, Bajwa & Manzoor, 2011). Some of these species act as surfactants to complex metal ions and detach from their environs. As such, the resultant products are biodegradable and of low toxicity. Case examples of fungi species and the metal ions they adsorb and immobilise are shown in Table 4.

Table 4. Examples of microorganisms used in heavy metal bioremediation and their percentage sorption efficacies

Microbial Group	Bioremediation Microorganism used	Metals	Adsorption Efficiency (%)	References
Protozoa	*Tetrahymena rostrata*	Hg	40	Muneer, Iqbal, Shakoori & Shakoori, 2013
Algae	*Nostoc species*	Fe	98	Kumaran, Sundaramanicam & Bragadeeswaran, 2011
Algae	*Nostoc species* *Chlorella vulgaris*	Ni	88 41	Kumaran et al., 2011 Wong, Li, Zhang, Qi & Min, 2002
Algae	*Nostoc species* *Chlorella vulgaris*	Cd Zn Cd	50 95 96	Kumaran et al., 2011 Goher et al., 2016
Algae	*Chlorella vulgaris*	Pb	99	Goher et al., 2016
Algae	*Spirulina species* *Spirogyna species*	Cr	98	Mane & Bhosle, 2012
Fungi	*Asper niger* *Aspergillus versicolor* *Aspergillus species*	Ni	98 30 90	Magyarosy et al., 2002 Tastan, Ertugrul & Donmez, 2010 Tastan et al., 2010
Fungi	*Candida parapsilosis*	Hg	80	Muneer et al., 2013
Fungi	*Aspergillus Niger* *Aspergillus lentulus* *Sphaerotilus natans* *Aspergillus versicolor*	Cu	50 99 58 29	Tastan et al., 2010 Jha, Dikshit & Pandy, 2011 Ashokkumar, Loashini & Bhavya, 2017 Tastan et al. 2010
Fungi	*Phanerochaete chrysosporium* *Sphaerotilus natans* *Mutant S. cerevisiae* *Aspergillus niger* *Gleophyllum sepiarium* *Aspergillus species* *Aspergillus versicolor*	Cr	98 82 99 65 95 92 99	Igiri et al., 2018
Consortium organisms	*Arthribacter and Acinetobacter species* *Bacillus subtilis and P. aeruginosa* *B. subtilis and S. cerevisiae* *B. subtilis and P. aeruginosa*	Hg, Cr	78 99 99 97	De, Ramaiah & Vardanyan, 2008
Bacteria	*B. firmus* *Pseudomonas species*	Zn	62 50	Muneer et al., 2013
Bacteria	*Vibrio fluvialis* *B. licheniformis* *V. parahaemolyticus* *P. aeruginosa* *K. pneumoniae*	Hg	60 73 80 30 29	Jafari, Cheraghi, Mirbakhsh, Mirza & Maryamabadi, 2015
Bacteria	*Enterobacter cloacae*	Co	8	Jafari et al., 2015
Bacteria	*Acinetobacter species* *Pseudomonas species* *Micrococcus species* *Desulfovibrio desulfuricans*	Ni	69 53 55 90	Igiri et al., 2018
Bacteria	*Bacillus species* *D. desulfuricans* *Staphylococcus species* *Acinetobacter species* *Strenotrophomonas species* *Sporosarcina saromensis* *Immobilised B. subtilis* *Immobilised P. aeruginosa* *E. coli*	Cr	99 56 45 87 81 83 99 99 45	Indu & Shaili, 2011
Bacteria	*B. Iodinium* *Streptomycetes species* *Staphylococcus species* *Pseudomonsa species* *B. firmus* *Micrococcus species* *Gemella species* *Methylobacterium organophilum*	Pb	87 33 83 88 98 37 55 18	Kumar, Bhatia, Singh, Rani & Bishnoi, 2011

A number of bacterial species have different metal biosorptive capacities as shown in Table 4. This capacity depends on the environmental conditions and pre-treatment measures conducted on these microbes. Bacteria are the most commonly used biosorbents because of their ubiquitous nature, resilience to harsh environmental conditions and their adaptability to conditions that allow optimised growth (Gupta et al., 2016).

CONCLUSION

This chapter affirms that microbial resistance to heavy metals is a promising tool that can be optimised to enhance detoxification and biosorption of these environmental contaminants. Sanitisation of environments off these pollutants is crucial for sound ecological functioning. The preference to bioremediation method is due to its green nature in that it uses microorganisms that are ubiquitous in nature, have high affinity to metals and can withstand extreme temperature and pH conditions. Through mechanisms such as reduction of metal cations, efflux of metal ions, creation of extracellular barriers using EPSs, biosurfactants and siderophores, intracellular and extracellular sequestration, these microbes immobilise and transform heavy metals to innocuous products. To optimise the outcome of these biosorbents further research is necessary through application of genetic transfers using biofilms and the use of nano-technologies for better precision of target heavy metals and detoxification efficacies.

REFERENCES

Alavi, M., Antonic, V., Ravizee, A., Weina, P., Izadjoo, M., & Stojadinovic, A. (2011). An Enterobacter plasmid as a new genetic background for the transposon Tn1331. *Infection and Drug Resistance*, *4*, 209–213. doi:10.2147/IDR.S25408 PMID:22259249

Ashokkumar, P., Loashini, V. M., & Bhavya, V. (2017). Effect of pH, Temperature and biomass on biosorption of heavy metals by *Sphaerotilus natans. International Journal of Microbiology and Mycology*, *6*(1), 32–38.

Bandowe, M., Bigalke, M., Boamah, L., Nyarko, E., Saalia, F., & Wilcke, W. (2014). Polycyclic aromatic compounds (PAHs and oxygenated PAHs) and trace metals in fish species from Ghana (West Africa): Bioaccumulation and health risk assessment. *Environment International*, *65*, 135–146. doi:10.1016/j.envint.2013.12.018 PMID:24486971

Bhaskar, P. V., & Bhosle, N. B. (2006). Bacterial extracellular polymeric substance (EPS): A carrier of heavy metals in the marine food chain. *Environment International*, *32*(2), 191–198. doi:10.1016/j.envint.2005.08.010 PMID:16256198

Boopathy, R. (2000). Factors limiting bioremediation technologies. *Bioresource Technology*, *74*(1), 63–67. doi:10.1016/S0960-8524(99)00144-3

Booth, S., Weljie, A., & Turner, J. (2015). Metabolomics reveals diferences of metal toxicity in cultures of Pseudomonas pseudoalcaligenes KF707 grown on different carbon sources. *Frontiers in Microbiology*, *6*, 827. doi:10.3389/fmicb.2015.00827 PMID:26347721

Bruins, R., Kapil, S., & Oehme, F. W. (2000). Microbial Resistance to Metals in the Environment. *Ecotoxicology and Environmental Safety*, *45*(3), 198–207. doi:10.1006/eesa.1999.1860 PMID:10702338

Canovas, D., Cases, I., & de Lorenzo, V. (2003). Heavy metal tolerance and metal homeostasis in *Pseudomonas putida* as revealed by complete genome analysis. *Environmental Microbiology*, *5*(12), 1242–1256. doi:10.1111/j.1462-2920.2003.00463.x PMID:14641571

Cervantes, C., Campos-Garc'ıa, J., Devars, S., Gutierrez-Corona, F., Loza-Tavera, H., Torres-Guzman, C., & Sanchez, R. (2001). Interactions of chromium with microorganisms and plants. *FEMS Microbiology Reviews*, *25*(3), 335–347. doi:10.1111/j.1574-6976.2001.tb00581.x PMID:11348688

Chabukdhara, M., Gupta, S. K., & Gogoi, M. (2017). Phycoremediation of heavy metals coupled with generation of bioenergy. In *Algal Biofuels* (pp. 163–188). Springer. doi:10.1007/978-3-319-51010-1_9

Chen, S., Yin, H., Ye, J., Peng, H., Liu, Z., Dang, Z., & Chang, J. (2014). Influence of co-existed benzo[a] pyrene and copper on the cellular characteristics of *Stenotrophomonas maltophilia* during biodegradation and transformation. *Bioresource Technology*, *158*, 181–187. doi:10.1016/j.biortech.2014.02.020 PMID:24603491

De, J., Ramaiah, N., & Vardanyan, L. (2008). Detoxification of toxic heavy metals by marine bacteria highly resistant to mercury. *Marine Biotechnology (New York, N.Y.)*, *10*(4), 471–477. doi:10.100710126-008-9083-z PMID:18288535

De Silva, A., Carvalho, M., De Souza, S., Dias, P., Filho, R., Saramago, C., Bento, C., & Hofer, E. (2012). Heavy metal tolerance (Cr, Ag and Hg) in bacteria isolated from sewage. *Brazilian Journal of Microbiology*, *43*(4), 1620–1631. doi:10.1590/S1517-83822012000400047 PMID:24031994

Dien, V. T., Morris, S. E., Karadeema, R. J., & Romesberg, F. E. (2018). Expansion of the genetic code via expansion of the genetic alphabet. *Current Opinion in Chemical Biology*, *46*, 196–202. doi:10.1016/j.cbpa.2018.08.009 PMID:30205312

Diep, P., Mahadevan, R., & Yakunin, A. (2018). Heavy metal removal by bioaccumulation using genetically engineered microorganisms. *Frontiers in Bioengineering and Biotechnology*, *6*, 1–20. doi:10.3389/fbioe.2018.00157 PMID:30420950

Dunbar, W. S. (2017). Biotechnology and the Mine of Tomorrow. *Trends in Biotechnology*, *35*(1), 79–89. doi:10.1016/j.tibtech.2016.07.004 PMID:27612568

Fashola, M. O., Ngole-Jeme, V. M., & Babalola, O. O. (2016). Heavy metal pollution from gold mines: Environmental effects and bacterial strategies for resistance. *International Journal of Environmental Research and Public Health*, *13*(11), 1047. doi:10.3390/ijerph13111047 PMID:27792205

Frederick, T. M., Taylor, E. A., Willis, J. L., Shultz, M. S., & Woodruff, P. J. (2013). Chromate reduction is expedited by bacteria engineered to produce the compatible solute trehalose. *Biotechnology Letters*, *35*(8), 1291–1296. doi:10.100710529-013-1200-z PMID:23563698

Gauthier, T., Norwood, P., Prepas, E., & Pyle, G. (2014). Metal-PAH mixtures in the aquatic environment: A review of co-toxic mechanisms leading to more-than-additive outcomes. *Aquatic Toxicology (Amsterdam, Netherlands)*, *154*, 253–269. doi:10.1016/j.aquatox.2014.05.026 PMID:24929353

Geva, P., Kahta, R., Nakonechny, F., Aronov, S., & Nisnevitch, M. (2016). Increased copper bioremediation ability of new transgenic and adapted Saccharomyces cerevisiae strains. *Environmental Science and Pollution Research International*, *23*(19), 19613–19625. doi:10.100711356-016-7157-4 PMID:27392627

Gilotra, U., & Srivastava, S. (1997). Plasmid-encoded sequestration of copper by *Pseudomonas pickettii* strain US321. *Current Microbiology*, *34*(6), 378–381. doi:10.1007002849900199 PMID:9142746

Goher, M. E., El-Monem, A. M. A A., Abdel-Satar, M., Ali, M. H., Hussian, A.-E. M., & Napiórkowska-Krzebietke, A. (2016). Biosorption of some toxic metals from aqueous solution using non-living algal cells of *Chlorella vulgaris*. *Journal of Elementology*, *21*(3), 703–714.

Gomathy, M., & Sbarinathan, K. (2010). Microbial mechanisms of heavy metal tolerance- A review. *Agricultural Research Communication Centre*, *31*(2), 133–138.

Gruji, C. S., Vasic, S., Radojevic, I., Comic, L., & Ostojic, A. (2017). Comparison of the *Rhodotorula mucilaginosa* biofilm and planktonic culture on heavy metal susceptibility and removal potential. *Water, Air, and Soil Pollution*, *228*(2), 73. doi:10.100711270-017-3259-y

Gupta, A., Joia, J., Sood, A., Sood, R., & Sidhu, C. (2016). Microbes as Potential Tool for Remediation of Heavy Metals: A Review. *Journal of Microbial & Biochemical Technology*, *8*(4), 364–372. doi:10.4172/1948-5948.1000310

Hildebrandt, U., Regvar, M., & Bothe, H. (2007). Arbuscular mycorrhiza and heavy metal tolerance. *Phytochemistry*, *68*(1), 139–146. doi:10.1016/j.phytochem.2006.09.023 PMID:17078985

Ianieva, O. (2009). *Heavy metal resistance mechanisms in bacteria*. Retrieved from: https://www.ncbi.nlm.nih.gov/pubmed/20455433

Ibrahim, Z., Ahmad, W. A., & Baba, A. B. (2001). Bioaccumulation of silver and the isolation of metal-binding protein from *P. diminuta*. *Brazilian Archives of Biology and Technology*, *44*(3), 223–225. doi:10.1590/S1516-89132001000300001

Igiri, B., Okoduwa, S., Idoko, G., Akabuogu, E., Adeyi, A., & Ejiogu, I. (2018). Toxicity and bioremediation of heavy metals contaminated ecosystem from tannery wastewater: A review. *Journal of Toxicology*, *2018*, 1–16. doi:10.1155/2018/2568038 PMID:30363677

Indu, T. S., & Shaili, S. (2011). Bioremediation and bioconversion of chromium and pentachlorophenol in tannery effluent by microorganisms. *International Journal of Technology*, *3*, 224–233.

Jafari, S. A., Cheraghi, S., Mirbakhsh, M., Mirza, R., & Maryamabadi, A. (2015). Employing response surface methodology for optimization of mercury bioremediation by *Vibrio parahaemolyticus* PG02 in coastal sediments of Bushehr, Iran. *Clean - Soil, Air. Water (Basel)*, *43*(1), 118–126.

Javaid, N. M. A., Bajwa, K. U. R., & Manzoor, R. A. (2011). Biosorption of heavy metals by pre-treatment of biomass of *Aspergillus niger*. *Pakistan Journal of Botany*, *43*(1), 419–425.

Jha, S., Dikshit, S., & Pandy, G. (2011). Comparative study of agitation rate and stationary phase for the removal of Cu^{2+} by *A. lentulus*. *International Journal of Pharma and Bio Sciences*, *2*, 208–211.

Klonowska, A., Heulin, T., & Vermeqlio, A. (2005). Selenite and tellurite reduction by *Shewanella oneidensis*. *Applied and Environmental Microbiology*, *71*(9), 5607–5609. doi:10.1128/AEM.71.9.5607-5609.2005 PMID:16151159

Kumar, R., Bhatia, D., Singh, R., Rani, S., & Bishnoi, N. R. (2011). Sorption of heavy metals from electroplating effluent using immobilized biomass *Trichoderma viride* in a continuous packed-bed column. *International Biodeterioration & Biodegradation*, *65*(8), 1133–1139. doi:10.1016/j.ibiod.2011.09.003

Kumaran, N. S., Sundaramanicam, A., & Bragadeeswaran, S. (n.d.). Adsorption studies on heavy metals by isolated cyanobacterial strain (nostoc sp.) from uppanar estuarine water, southeast coast of India. *Journal of Applied Sciences Research*, *7*(11), 1609–1615.

Lima, A. I. G., Corticeiro, S. C., & Figueira, E. M. A. P. (2006). Glutathione-mediated cadmium sequestration in *Rhizobium leguminosarum*. *Enzyme and Microbial Technology*, *39*(4), 763–769. doi:10.1016/j.enzmictec.2005.12.009

Magyarosy, A., Laidlaw, R., Kilaas, R., Echer, C., Clark, D., & Keasling, J. (2002). Nickel accumulation and nickel oxalate precipitation by *Aspergillus niger*. *Applied Microbiology and Biotechnology*, *59*(2-3), 382–388. doi:10.100700253-002-1020-x PMID:12111174

Malik, A. (2004). Metal bioremediation through growing cells. *Environment International*, *30*(2), 261–278. doi:10.1016/j.envint.2003.08.001 PMID:14749114

Manasi, M., Rajesh, N., & Rajesh, V. (2016). Evaluation of the genetic basis of heavy metal resistance in an isolate from electronic industry effluent. *Journal of Genetic Engineering and Biotechnology*, *14*(1), 177–180. doi:10.1016/j.jgeb.2016.02.002 PMID:30647612

Mane, P. C., & Bhosle, A. B. (2012). Bioremoval of some metals by living algae *spirogyra* sp. and *spirullina* sp. from aqueous solution. *International Journal of Environmental of Research*, *6*(2), 571–576.

Mash, H. E., Chin, Y. P., Sigg, L., Hari, R., & Xue, H. B. (2003). Complexation of copper by zwitterionic aminosulfonic (good) buffers. *Analytical Chemistry*, *75*(3), 671–677. doi:10.1021/ac0261101 PMID:12585500

McEldowney, S. (2000). The impact of surface attachment on cadmium accumulation by *Pseudomonas fluorescens* H2. *FEMS Microbiology Ecology*, *33*(2), 121–128. doi:10.1111/j.1574-6941.2000.tb00734.x PMID:10967211

Mishra, A., & Malik, A. (2013). Recent advances in microbial metal bioaccumulation. *Critical Reviews in Environmental Science and Technology*, *43*(11), 1162–1222. doi:10.1080/10934529.2011.627044

Moskvina, M. I., Brekhovskikh, A. A., & Nikandrov, V. V. (2003). Role of the heterotrophic bacteria associated with the cyanobacterium *Nostoc muscorum* in cadmium sulfide synthesis. [In Russian]. *Mikrobiologiia*, *72*(2), 284–285. PMID:12751257

Muneer, B., Iqbal, M. J., Shakoori, F. R., & Shakoori, A. R. (2013). Tolerance and biosorption of mercury by microbial consortia: Potential use in bioremediation of wastewater. *Pakistan Journal of Zoology*, *45*(1), 247–254.

Naik, M., Shamim, K., & Dubey, S. (2012). Biological characterization of lead resistant bacteria to explore role of bacterial metallothionein in lead resistance. *Current Science*, *103*, 426–429.

Nies, D. H. (2000). Heavy metal-resistant bacteria as extremophiles: Molecular physiology and biotechnological use of *Ralstonia* sp. CH34. *Extremophiles*, *4*(2), 77–82. doi:10.1007007920050140 PMID:10805561

Nies, H. D. (2003). Efflux-mediated heavy metal resistance in prokaryotes. *FEMS Microbiology Reviews*, *27*(2-3), 313–339. doi:10.1016/S0168-6445(03)00048-2 PMID:12829273

Olaniran, A., Balgobind, A., & Pillay, B. (2013). Bioavailability of heavy metals in soil: Impact on microbial biodegradation of organic compounds and possible improvement strategies. *International Journal of Molecular Sciences*, *14*(5), 10197–10228. doi:10.3390/ijms140510197 PMID:23676353

Ortiz-Bernad, I., Anderson, R. T., Vrionis, H. A., & Lovley, D. R. (2004). Vanadium respiration by *Geobacter metallireducens*: Novel strategy for in situ removal of vanadium from groundwater. *Applied and Environmental Microbiology*, *70*(5), 3091–3095. doi:10.1128/AEM.70.5.3091-3095.2004 PMID:15128571

Osman, D., & Cavet, J. (2008). Copper Homeostasis in Bacteria. *Advances in Applied Microbiology*, *65*, 217–247. doi:10.1016/S0065-2164(08)00608-4 PMID:19026867

Pardo, R., Herguedas, M., Barrado, E., & Vega, M. (2003). Biosorption of cadmium, copper, lead and zinc by inactive biomass of *Pseudomonas putida*. *Analytical and Bioanalytical Chemistry*, *376*(1), 26–32. doi:10.100700216-003-1843-z PMID:12734614

Prabhu, S., & Poulose, E. K. (2012). Silver nanoparticles: Mechanism of antimicrobial action, synthesis, medical applications, and toxicity effects. *International Nano Letters*, *2*(32), 1–10. doi:10.1186/2228-5326-2-32

Samantaray, D., Mohapatra, S., & Mishra, B. (2014). Bioremediation of industrial effluents. In *Microbial Biodegradation and Bioremediation* (pp. 325–339). Elsevier. doi:10.1016/B978-0-12-800021-2.00014-5

Sandrin, T., & Maier, R. M. (2002). Effect of pH on cadmium toxicity, speciation, and accumulation during naphthalene biodegradation. *Environmental Toxicology and Chemistry*, *21*(10), 2075–2079. doi:10.1002/etc.5620211010 PMID:12371483

Sankarammal, M., Thatheyus, A., & Ramya, D. (2014). Bioremoval of Cadmium Using *Pseudomonas fluorescens*. *Open Journal of Water Pollution and Treatment*, *2*(2), 92–100. doi:10.15764/WPT.2014.02010

Saxena, D., Joshi, N., & Srivastava, S. (2002). Mechanism of copper resistance in a copper mine isolate *Pseudomonas putida* strain S4. *Current Microbiology*, *45*(6), 410–414. doi:10.100700284-002-3787-5 PMID:12402081

Singh, S., Parihar, P., Singh, R., Singh, V., & Prasad, S. (2015). Heavy metal tolerance in plants: Role of transcriptomics, proteomics, metabolomics and lonomics. *Frontiers in Plant Science*, *6*, 1143. PMID:26904030

Slobodkina, G. B., Bonch-Osmolovskaia, E. A., & Slobodkin, A. I. (2007). Reduction of chromate, selenite, tellurite, and iron (III) by the moderately thermophilic bacterium *Bacillus thermoamylovorans* SKC1. [in Russian]. *Mikrobiologiia*, *76*(5), 602–607. PMID:18069319

Song, H. P., Li, X. G., Sun, J. S., Xu, S. M., & Han, X. (2008). Application of magnetotactic bacterium, *Strenotrophomonas* sp. to the removal of Au (III) from contaminated wastewater with a magnetic separator. *Chemosphere*, *72*(4), 616–621. doi:10.1016/j.chemosphere.2008.02.064 PMID:18439649

Tastan, B. E., Ertugrul, S., & Dönmez, G. (2010). Effective bioremoval of reactive dye and heavy metals by *Aspergillus versicolor. Bioresource Technology*, *10*(3), 870–876. doi:10.1016/j.biortech.2009.08.099 PMID:19773159

Wildung, R. E., Gorby, Y. A., Krupka, K. M., Hess, N. J., Li, S. W., Plymale, A. E., McKinley, J. P., & Fredrickson, J. K. (2000). Effect of electron donor and solution chemistry on products of dissimilatory reduction of technetium by *Shewanella putrefaciens. Applied and Environmental Microbiology*, *66*(6), 2451–2460. doi:10.1128/AEM.66.6.2451-2460.2000 PMID:10831424

Wong, S. C., Li, X. D., Zhang, G., Qi, S. H., & Min, Y. S. (2002). Heavy metals in agricultural soils of the Pearl River Delta, South China. *Environmental Pollution*, *119*(1), 33–44. doi:10.1016/S0269-7491(01)00325-6 PMID:12125727

Yang, Q., Agouri, S., Tyrrell, J., & Walsh, T. (2018). Heavy metal resistance genes are associated with blaNDM-1 and blaCTX-M15 carrying enterobecteriaceae. *Antimicrobial Agents and Chemotherapy*, *62*(5), e02614–e02642. doi:10.1128/AAC.02642-17 PMID:29507071

Ybarra, G. R., & Webb, R. (1999). Effects of divalent metal cations and resistance mechanisms of the cyanobacterium *Synechococcus* sp. strain PCC 7942. *Journal of Hazardous Substance Research*, *2*(1), 1–9. doi:10.4148/1090-7025.1011

ENDNOTES

1 A dipolar ion that has two or more functional groups, which are either cation or anion though their net charge is zero.

2 Organic compound used to buffer solutions.

3 They are high-affinity-Fe-chelating molecules secreted by microorganisms to transport ion in between membranes of cells.

4 Molecules with amphiphilic properties and are secreted by hydrophilic and hydrophobic moieties of microorganisms to give them the ability to accumulate and react with other substances in fluids.

5 A functional DNA unit with a cluster of controlled by a common promoter.

Chapter 3
Microbial Enzymes and Their Mechanisms in the Bioremediation of Pollutants

Karthika Rajamanickam
Mahendra Arts and Science College (Autonmous), India

Jayanthi Balakrishnan
Mahendra Arts and Science College (Autonomous), India

Selvankumar Thangaswamy
https://orcid.org/0000-0002-3500-8681
Mahendra Arts and Science College (Autonomous), India

Govarthanan Muthusamy
Kyungpook National University, South Korea

ABSTRACT

Bioremediation is the process, with the help of microbes or their enzymes, to remove the pollutants present in the environment and change them into non-toxic compounds. Microbial enzymes have a wide range of metabolic activities and are involved in the transformation of pollutants. Enzymes like oxidoreductase, hydrolases, monooxygenase, dioxygenase, methyltransferases, and lipases are involved in the degradation process. Oxidoreductase catalyzes the transfer of electron and proton from the reduced organic substrate to another chemical compound from donor to acceptor. Monooxygenase and dioxygenases are the transferring oxygen from molecular oxygen (O_2) utilizing FAD/NADH/NADPH as a co-substrate in this process. Lyases catalyze the cleavage of the bonds by elimination, leaving double bonds. Peroxidases catalyze the oxidation of lignin and other phenolic compounds at the expense of hydrogen peroxide (H_2O_2) in the presence of a mediator. Lipases also involve catalyzing the hydrolysis of triacylglycerols to glycerol and free fatty acids.

DOI: 10.4018/978-1-7998-4888-2.ch003

INTRODUCTION

Over the last three decades, there is increasing environmental pollution that leads to public health issues due to global concern (Kimani, 2007). Mainly the anthropogenic pollution causes a serious problem for the environment. The pollution made by human activity due to the expansion of industries, the use of chemicals, consumption of massive quantity of Petro based raw materials, and a huge amount of chemical fertilizers in developed as well as developing countries (Fereidoun et al, 2007). Air, water, and land pollution are considered as major environmental pollution all over the world. Earth contains two-third of water but freshwater has only less than one percentage consumed by the living organisms. There is a chance of freshwater reached the condemning state and leads to scarcity of water. Human beings are responsible to take care of freshwater and benefited sufficiently. On a global scale, people live with a serious shortage of water it is estimated that nearly 1.3 billion people and children die with drinking dirty water 1.9 million which cause serious disease to the children (UNICEF, 2017). By 2025 it is predicted that nearly 60% of people significantly affected due to the water scarcity. Serious poverty for food may also occur because of water shortage or irrigation by polluted water. Water pollution is another source of water scarcity. If the emission of harmful pollutants to the environment it causes serious health problems to the living organisms. Unsafe disposal of wastewater is the most challenging in the ambient environment (Kanu et al, 2011). The freshwater reservoirs affected mainly by these wastewaters and it has emerged as a major challenge in developing countries (Fakayode, 2005).

Soil is a potent and renewable living system that is wellbeing in the food production and fibre for global balance, ecosystem function that balance the physical, chemical and biological factors (Karlen et al, 2003). Soil contains the biological elements to sustain biological activity, promote the air quality, water environments and maintain the health of the plant, animal and humans. Soil is a living resource for all living organisms, but huge contamination may affect the soil and the soil may be considered as functionally dead (Doran and Zeiss, 2000). Pollution of soil caused due to improper industrial and agricultural processes and deforestation. Nearly one-third of topsoil in the world gets depleted and most of the topsoil and its nutrients could be out off within the next 60 years. In most developing countries agriculture considered as one of the greatest economies and soil as a great resource of that country. According to the Earth Institute use of fertilizer in heavy amount plays an important role in air and land pollution. Land pollution is wealth and health of land would be destructed due to the misuse of land resources by human activities. When the humans apply chemicals for the agricultural land such as pesticides, herbicides, chemical fertilizers to the soil, disposal of wastes improperly, the introduction of a large quantity of heavy metals to the land, hydrocarbons and chlorinated hydrocarbons, radioactive materials, disease-causing agents, mining, pharmaceuticals and domestic sources like plastic bags, bottles.

Development of urbanization and industrialization economically have led to a rise in energy consumption and discharges of large amount of hazardous waste to the environment. The global environmental pollution includes emission of greenhouse gases, acid deposition, wastewater management which should be looking into multiple prospects including social, economic and engineering systems (Loux et al, 2011). This type of environmental pollution can cause adverse effects to humans, plants and animals like prenatal disorders, neurobehavioral disorders, cardiovascular problems, infant mortality, mental disorders, asthma, premature death, reduced energy levels of the organism, endothelial dysfunction and can cause various serious health problems (Kelishadi and Poursafa, 2010). Because of these prospective, there is a necessity to take effort to control the pollutants present in the environment otherwise due to inhalation, consumption, mining, transportation, manufacturing and other human activities will destroy the natural ecosystem of the environment.

There is an urgent need to recover the whole environment from this type of pollutants to avoid all of the biological complications. The main intention to retrieve the environment to its original position is to maintain soil health and fertility, purification of groundwater, reuse of wastewater and protect the air from the harmful pollutants. Consequently, the researchers have been developed a new strategy to recover the whole ecosystem from the pollutants. From that various technologies have been developed by the researchers to remediate the polluted sites and hence they achieve successful results. They employ two basic methods for remediating the pollutants: i) engineering ii) biological. The engineering approach deals with physical and chemical methods. The biological approach leads to the implication of biological factors. (Gianfreda and Rao, 2008). The use of microorganism for bioremediation has some disadvantages to overcome this most of the researchers concentrate on enzymatic proteins. (Gianfreda and Rao, 2004). Enzymes possess various advantages over the other factors and it acts as a catalyst for transforming the pollutants into non-toxic ones. Over the traditional technologies, the enzyme technology has more advantages that are not inhibited by several other inhibitors of microbial metabolism. The enzymes act as potent in the given substrate that limiting the microbial activity and it can easily penetrate wherever because of its small size. (Gianfreda and Bollag, 2002).

Microbes can flourish in adverse environmental conditions such as alkaline and acid pH, elevated temperature, and also in a higher concentration of toxic chemicals and can develop resistance against particular toxic chemicals in the environment occasionally, because of their maximum level of genetic mutation. Few microbial species can transform the hazardous toxic compounds into simpler nontoxic chemicals under favourable conditions. Hence, the use of a microbial mechanism for cleaning up the heavy metals and toxic chemicals contaminated lands and water resources are now common practice. The microbial mechanism that affects the bioremediation process of metals is as follows;

1. Metal ions are effluxed outside of the cell.
2. Cations (positively charges heavy metal ions) are biosorped to the anionic (negatively charged) microbial cell membrane and transferred to inside the cell environment through specific transporter and successive bioaccumulation.
3. Microbial enzymes immobilize the metal ions and biotransform them to less toxic form.

A large group of gram-negative and gram-positive bacteria have *ars* genetic system use arsenite and arsenate in their metabolism. These types of bacteria can constitute prospective candidates for As removal existing in industrial areas or other arsenic-loaded environments. The *arsM* gene expression in these bacteria considerably enhances the efficacy of bacteria to convert arsenic in water and soil environment into volatile methylated arsenicals. Thus, genetically engineered bacteria having *ars* gene could be a cost-effective and well-organized approach for the bioremediation of arsenic-contaminated environment (Liu et al, 2011).

Bioremediation

Due to development of the vast expansion of various industries, food, health care, vehicles and its continual discovery of numerous products over recent years has led accumulation of hazardous compounds in the environment and need to international efforts to remediate these environments, in response to enable the affected area to be redeveloped or restored for future generations usage. The quality of life of living biota is directly linked with the quality of the whole environment, at the universal level, it must pay attention to restore the quality of the environment for maintaining the diversity of the earth. But it is certainly mysterious to maintain the quality of the ecological community with these types of advanced development of new technologies, which is to be considered as an unfavourable condition to the environment if proper remediation not yet to be applied. There is various techniques have been evolved to remove the pollutants present in the various fields with naturally occurring elements.

By nature, there are various organisms like fungi, algae, bacteria involved constantly to break down complex organic compounds. Microorganisms are abundantly distributed in terrain because their existence ability is very majestic and it also easily grown on a wide range of environmental conditions (Tang et al, 2007). Certain microorganisms can convert or utilize the toxic pollutants to obtain the pollutants like the carbon source and biomass production.

The process or technique which is used of living biological systems for the removal of pollution from the air, soil, and water termed as "Bioremediation". It employs the biological systems inborn in microbes and plants to eliminate hazardous compounds and retain the environment to its original condition (Ayangbenro and Babalola, 2017). The main principle of bioremediation involves reducing the solubility by changing pH, redox reactions and adsorption of contaminants. Mainly bioremediation is the process based on natural process, most studied reviewed that bioremediation is the most adequate technology than the other technologies. Most bioremediation processes carried out under aerobic conditions because it encourages proliferating by the addition of electron donors or electron acceptors in the presence of air (Colberg and Young, 1995). The main advantage of bioremediation is the destruction of a wide variety of contaminants and legally considered as the complete transformation of hazardous compounds to non-toxic products. It is also cost-effective and less time-consuming.

Figure 1. Strategies involved in the bioremediation process

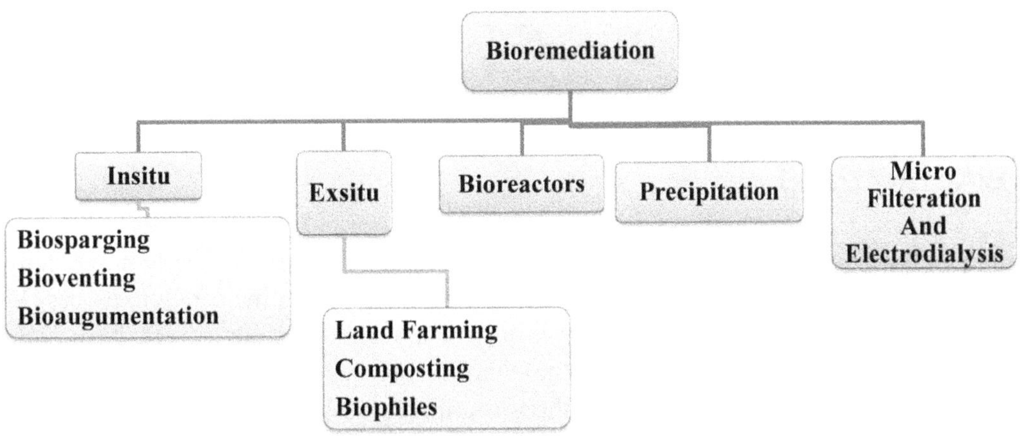

Strategies Involved in the Bioremediation Process

Table 1. Various methods involved in the bioremediation process

S.No	Techniques	Types	Interpretation	Applications
1.	*Exsitu*	Biosparging Bioventing Bio augmentation	The air is injected to increase the concentrations of groundwater and stimulate the rate of degradation. Through wells by supplying oxygen and nutrients to contaminated soil to enhance degradation. Introduce a new group of natural microbial strains or a genetically engineered variant to degrade contaminated soil or water.	To degrade the metals, inorganic compounds based on chemical solubility, geological factors, and Distribution of pollutants
2.	*Insitu*	Land farming Composting Biopiles	The contaminated soil is scrape and disperse over a prepared bed and periodically tilled until pollutants are degraded. Combining contaminated soil with nonhazardous organic amendments such as manure or agricultural wastes. The hybrid of land farming and composting.	Surface application, aerobic process, application of organic materials
3.	Bioreactors	Slurry reactors Aqueous reactors	Both reactors used to conceive a three-phase (solid, liquid, and gas) mixing condition to enhance the bioremediation rate of soil-bound and water-soluble pollutants as a water slurry of the contaminated soil and biomass.	Rapid degradation kinetics under optimized environmental parameters.
4.	Precipitation		Non-directed Physico-chemical complexation reaction between dissolved contaminants and charged cellular components	Removal of Heavy Metals
5.	Microfiltration		Filters membranes are used at a constant pressure	Wastewater treatment
6.	Electrodialysis		Uses cation and anion exchange membrane pairs	Removal of dissolved solids.

MICROBIAL ENZYMES

Enzymes play a crucial role in all the biological transformations occurring in the living systems. An enzyme may be protein or glycoproteins that are precisely convoluted in the catalytic process known as active sites. Enzymes involved in metamorphosis of structural and toxicological effects of pollutants and may involve in the conversion of harmful to non-harmful substances. For the degradation process, the enzymes considered as a good alternative rather than microbes because microbes have several disadvantages (Gianfreda and Rao, 2004). Enzymes have more advantages than any other technologies like microbial remediation. The enzymes can be used under extreme conditions (thermophilic and psychrophilic) and active at low pollutant concentrations and are very effective in the presence of microbial preys (Gianfreda and Bollag, 2002). All enzymes are eco-friendly and it has the capability of remediation of

Table 2. Various Microbial enzymes involved in the Bioremediation process and their soures and their applications

S.No	Enzyme	Source	Pollutant To Be Degraded	Application
1.	**Oxidoreductase** -Mono oxygenase	*Pseudomonas putida* *B. cepacia*	Halogen compounds, nitrates	Protein engineering, bioremediation, synthetic chemistry
2.	Dioxygenase	*Pseudomonas putida* *P. mendocina*	Aromatic compounds	Synthetic chemistry, pharmaceutical industry, bioremediation,
3.	Laccases	*Pycnoporussanguineis* *Trametes hispida* *Pyricularia oryzae* *Trametes versicolor* *Cerrena unicolor* *Pycnoporus cinnabarinus* *Plant materials*	Azo dyes, Bleach plant effluents	Textile effluent degradation, chlorophenols and urea derivatives degradation
4.	**Peroxidase** - lignin Peroxidase	*White-rot fungi*	Biopolymers	Food industry, paper and pulp industry, textile industry, pharmaceutical industry
5.	Maganese Peroxidase	*Phanerochaete chrysopsorium* *White-rot fungi*	Degradation of phenols, lignin, pentachlorophenol, and dyes	Food industry, Paper and pulp industry, textile industry, pharmaceutical industry, bioremediation
6.	Versatile Peroxidase	*Agrobacterium, Pseudomonas sp., Flavobacterium sp., Nocardia sp., Bacillus cereus*	Methoxybenzenes and phenolic aromatic	Industrial biocatalyst, bioremediation,
7.	**Hydrolase** - lipase	*Candida rugosa* *Rhizopus delemar* *Comamonas acidovorans* *White-rot fungi*	Ortho and paradiphenols, aminophenols, polyphenols, polyamines, lignins, and aryldiamines	Food industry, paper and pulp industry, textile industry, nanotechnology, synthetic chemistry, bioremediation, cosmetics
8.	**Methyl transferases**	*Bacillus licheniformis Alicyclobacillus tengchogenesis, Brevibacillus sp., Alicyclobacillus tengchogenesis*	methylcobalamin, S-adenosylmethionine	Industrial biocatalyst, bioremediation,
9.	**Proteases**	*C. keratinophilum, Amycolatopsis, Bacillus sp. P. chrysosporium and Trametes versicolor*	Enzymes that hydrolyze peptide bonds in aqueous environment.	Leather, laundry, biocatalyst, bioremediation, and so forth.
10.	**Cellulases**	*Trichoderma spp, Penicillium camemberti, Azospirilium lipoferum, Bacillus subtilis.*	Enzymes hydrolyze cellulose	Textile industry,paper, pulp, laundry, agricultue and medicine.

many pollutants that are very toxic to the living organisms (Alcalde et al, 2006). At different denaturing agents, the specialized enzymatic proteins are stable with elevated catalytic activity.

The enzymes act as both intracellularly and extracellularly as well as in free cell form and immobilized form. Intracellular enzymes represent the presence of enzymes inside their originating cells while extracellular represents the absence of their originating cells.

TYPES OF ENZYMES

Oxidoreductase

Microorganisms like bacteria and fungi are the producers of oxidoreductase for the deterioration of contaminants. By biochemical reactions, the microbes extract vitality with the help of enzymes to cleave chemical bonds and transmit of electrons from the reduced organic substrate to the other chemical compound. The pollutants finally oxidized to non-toxic compounds by the oxidation-reduction reaction (Park et al, 2006). Xenobiotics such as phenolic or anilinic compounds, asphaltenes and PCBs, polychlorophenols, PAHs and other toxic pollutants detoxify by oxidoreductase enzyme. Oxidoreductase detoxifies the organic pollutants through polymerization, copolymerization and by humification of various phenolic substances (Williams, 1977). It also involved in the decolourization and degradation of azo dyes (Rubilar et al, 2008). Several types of bacteria involved to detoxification of radioactive hazardous compounds from an oxidized soluble form to reduced insoluble form. Mainly filamentous fungal can reach the soil pollutants than bacteria. The activity of fungi due to presence of extracellular oxidoreductases enzymes that are released by the mycelium of fungi into the pollutants and it can degrade the contaminants (Duran and Esposito, 2000) apart from microbes, plants can release the oxidoreductase enzyme to degrade the pollutants and it is considered as the phytoremediation. The plant families like Fabaceae, Solanaceae and Gramineae are found to release the oxidoreductase enzyme (Newman et al, 1998). The presence of reactive diffusible redox mediators (RMs) in enzyme-based systems can dramatically enhance the reaction rate and increase the range of substrates that can be degraded by these oxides- reductase enzymes.

ENZYMATIC

Oxygenases

Oxygenases are the group of oxidoreductase enzymes and play an important role in the metabolism of contaminants in the cleavage of the aromatic ring by increasing the water solubility. Oxygenases act as broad substrate and active upon chlorinated aliphatics. The mechanism behind the cleavage is the establishment of oxygen atoms into the organic molecules utilizing cosubstrate of FAD/NADH/NADPH. The enzyme oxygenases classified into two types - monooxygenases and dioxygenases based on the incorporation of oxygen molecules.

Monooxygenase

Monoxygenases are the group of an enzyme that incorporates one atom of oxygen molecule into the substrate. Based on cofactor monooxygenase further classified into two subclasses i) flavin-dependent monooxygenase ii) P450 monooxygenases.

Flavin dependent monooxygenases are widespread and consist of flavin as a prosthetic group that requires NADP or NADPH as a coenzyme. These enzymes act on many substrates by activation of molecular oxygen. Due to enzymatic activation of molecular oxygen C4a- peroxyflavin is formed which can act as an electrophile. An oxygen atom is inserted into the substrate and the other ie reduced to water (Chaiyen et al, 2012). P450 monooxygenases are heme-containing oxygenase that can catalyze a variety of enzymatic reactions to transform most harmful chemicals into detoxified derivatives (Isin and Guengerich, 2007). Most of the monooxygenase enzyme having the cofactor but certain monooxygenase contains independent cofactor. For activation, the monooxygenase enzyme requires molecular oxygen and utilizes the substrate as a reducing factor. The enzyme catalyzes various reactions like desulfurization, dehalogenation, denitrification, ammonification, hydroxylation, biotransformation, and biodegradation of various aromatic and aliphatic compounds. The monooxygenase enzyme also involved in the detoxification of hydrocarbons such as methanes, alkanes, cycloalkanes, alkenes, haloalkenes, ethers, and aromatic and heterocyclic hydrocarbons (Fox, et al 1990, Grosse et al 1999).

Microbial Dioxygenases

Dioxygenases are the multi-component enzymes that absorb two oxygen molecules to the substrate. Mainly Dioxygenases enzyme oxidizes the aromatic compounds into aliphatic compounds. Based on the detoxification ramification the phenomenon can be classified into two types i) convergent mode and ii) divergent mode. In convergent mode, aromatic compounds get cleavage by substrates like catechol, gentsate, protocatechuate and their derivatives. In divergent mode metal-dependent dioxygenase carrier formed by two possible ways i) Meta cleavage and ii) ortho cleavage pathway(Takami, 1997).

The dioxygenases classified into two classes 1. Extradiol dioxygenases 2.intradiol dioxygenases. In their active site Extradiol dioxygenases contain non-heme iron (II), catalyzes cleavage of ring in the carbon-carbon (C-C) bond contiguous to the vicinal hydroxyl groups (meta-cleavage) whereas in their active site Intra dioldioxygenases contain non-heme iron (III), catalyzes ring cleavage at the C-C bond between the vicinal hydroxyl groups (ortho-cleavage).

Laccases

Laccases are the group of oxidoreductase enzymes containing multicopper oxidases mainly produced by certain types of fungi, bacteria, plants, and insects. The laccase enzyme catalyzes the one-electron oxidation of phenols, anilines and aromatic compounds with the reduction of water from molecular oxygen. The radicals are produced by polymeric products by cross-coupling or self-coupling with other molecules. During coupling and polymerization of differently substituted substrates like demethoxylation dechlorination, and decarboxylation may also occur. For the transformation of phenolic substrates to phenoxy radicals, it requires oxygen (Gianfreda et al, 1999). Laccases can also able to oxidize the non-phenolic aromatics compounds when 2,2'-azino-bis(3-ethylbenzthiazoline-6-sulfonic acid, (ABTS),

3- hydroxyanthranilate, 1-hydroxy benzotriazole (HTB) act as mediators and others are present in the reaction mixture (Bourbonnais et al, 1995 and Pointing, Vrijmoed, 2000).

Laccase enzyme not only oxidizes phenolic and methoxyphenolic acids but attack their methoxy groups and it also involves in lignin depolymerization. Also besides, these compounds can be utilized by microorganisms as nutrients with the help of laccase enzymes (Xu, 1996). With the change of pH, the specificity of the substrate and laccase affinity can be varied. The enzyme laccase can be inhibited by various reagents such as azide, cyanide, halides (excluding iodide) and hydroxide(Clarissa et al, 2018). High nitrogen levels are usually required to obtain greater amounts of laccase. Among the other biological enzymes, laccases possess great potential for biotechnological and bioremediation applications. Laccases also widely used for the degradation of pharmaceuticals and personal care products (PPCPs) and antibiotics. These two products are the most commonly used medicinal products and leftover in the environment with any treatment of degradation. Conventional wastewater treating process cannot be efficiently used against these sediments present over the soil, land and water resources. These products lead to the origin of the antibiotic-resistant bacteria with antibiotic resistance genes in the environment which will affect the present level antibiotics in the medical treatment process (Larsson 2014). Sulfonamides and tetracyclines are more easily attacked by laccase compared with quinolones. This is presumably due to the strong electron-donating aromatic amine group in sulfonamides and the phenol group in tetracyclines, which are not found in quinolones (Ding et al, 2016).

Peroxidases

Peroxidases are the class of enzymes that reduce the hydrogen peroxide and oxidation of extensive range of non-phenolic substrates, organic and inorganic compounds. It is widely found in nature and produced by various sources like plants, microbes and animals. Based on sources and activity of peroxidase enzyme mainly it can be grouped into three types. lignin peroxidase (LiP), manganese-dependant peroxidase (MnP), and versatile peroxidase (VP) are mainly used for bioremediation process.

Lignin Peroxidase

Lignin peroxidases are the group of enzymes produced by the white-rot fungus during secondary metabolism in the existence of hydrogen peroxide as co-substrate for the degradation cell wall constituent of lignin. During this reaction hydrogen peroxide reduced to water by yielding of an electron from lipase. By adding an electron the LiP gets oxidized from veratryl alcohol in a reduced state, and veratryl aldehyde is formed. Veratryl aldehyde then again reduced to veratryl alcohol by gaining an electron from the substrate. Finally, aromatic compounds get oxidized in the non-enzymatic reactions (Yoshida, 1998, Ten Have and Teunissen 2001). LiP can oxidize aromatic compounds with redox potentials higher than 1.4 V (NHE) by single-electron abstraction, but the exact redox mechanism is still failing to understood (Piontek et al, 2001).

Manganese Peroxidases (MnP).

MnP is an extracellular heme enzyme produced by the lignin-degrading basidiomycetes fungus, which involved in multistep reactions followed that oxidation of Mn2+ to the oxidant Mn3+. Mn2+ triggers the MnP production and act as a substrate for degradation process. The Mn3+, get provoked by MnP, acts as a negotiator for the oxidation of various phenolic compounds. At the end of the reaction, Mn3+ chelate oxalate is small enough to disseminated into areas inaccessible even to the enzyme, as in the case of lignin or analogous structures such as xenobiotic pollutants covered wide within the soil, which is not necessarily feasible to the enzymes (Ruiz-Duenas et al, 2007).

Versatile Peroxidases (VP).

VP enzymes can precisely oxidized to Mn2+, methoxy benzenes, phenolic aromatic compounds along with substrates like MnP, LiP, and horseradish peroxidase. VP has incredible immense substrate specificity and tendency to oxidize the substrates when compared to other peroxidases while in the absence of manganese. It has also been manifested that VP enzyme can oxidize both phenolic and non-phenolic lignin model dimers (Tsukihara et al, 2006). Overproduction of VP is highly eminent for biotechnological applications and bioremediation of pollutants.

Methyltransferases

Methyl-transferase is one of the vital enzymes that catalyze the transferring the methyl group from donor substance like methylcobalamin, S-adenosylmethionine to the acceptor compounds. e.g Arsenic methyltransferase, here the acceptor group is Arsenic metal. The methylation process in Arsenic metal is a cascade reaction in which hazardous inorganic arsenic methylates are converted into less toxic pentavalent mono, di and tri-methylated arsenicals. Highly toxic intermediates of arsenic methylation reaction are trivalent methyl arsonous acid [MAs(III)] and dimethyl arsinous acid [DMAs(III)] which are easily converted by oxidation into less toxic methyl arsenate [MAs(V)], dimethyl arsenate [DMAs(V)] and trimethylarsine oxide [TMAs(V)O]. Microbial metal detoxification includes many mechanisms, among them methylation of arsenic is the most common which utilizes the arseniteS-adenosyl methionine methyltransferase (Ars M) enzyme. Arsenite S-adenosyl methyltransferase (Ars M) is a microbial homolog of As(III)MT present in eukaryotes, cloned from *Rhodopseudomonas palustris*. This gene is expressed in an arsenic-hypersensitive strain of *Escherichia coli*. As(III)-resistance cells in *E. coli* having recombinant *arsM* associated with the transformation of medium arsenic to the methylated pentavalent species such as DMA(V) and TMAO and the final product trimethylarsine [TMA(III)] gas. The experiment exhibited that methylation of environmental arsenic through ArsM by conversion to soluble and gaseous methylated species is a detoxifying process that may contribute to the global cycling of arsenic (Jie Qin et al, 2006).

*B. subtilis*168 expressing the arseniteS-adenosyl methionine methyltransferase gene (CmarsM) from the thermophilic alga *Cyanidio schyzonmerolae* converted the inorganic As in the medium into dimethylarsenate and trimethylarsine oxide within 48 h and volatized considerable amounts of dimethylarsine and trimethylarsine. This genetic engineering converted the As contaminated organic waste during composting through enhanced methylation and volatilization (Huang et al, 2016 and Huang et al 2015). Similarly, ArsM from *Chlamydomonas reinhardtii* and *Spirulina platensis* expressed into *Rhizobium leguminosarumbv. Trifolii* and *E.coli* AW3110, respectively. The legume-rhizobia symbionts can be used as the host for bioremediation (43.44). Purified fungal methyltransferase (*WaarsM)* isolated from *Westerdykella aurantiaca* catalyzes the synthesis of methylated arsenicals from both AsIII and AsV, that also establish AsV reductase activity (Verma et al, 2016).

Marinomonas communis a naturally occurring marine bacterium, not genetically engineered confirmed to have the efficiency to eliminate arsenic from the culture medium amended with arsenate (Takeuchi et al, 2007). Consequently, the bacteria may be a potential species for bioremediation of arsenic-laden aquatic environment. The *Lactobacillus acidophilus* have the capability to fix and eliminate arsenic from water contaminated with a higher concentration of arsenic (50–1000 ppb) and maximum removal was observed after 4 h of exposure (Singh and Sharma, 2020).

Lipases

Lipases are the group of enzymes that can able to degrade the lipids produced from a wide variety of organisms. Current works have shown that lipase is closely related to the structure of organic pollutants present in the soil. Crude oil or petroleum-contaminated soil-based researches has reported for the enhanced lipase activity and this enzyme is responsible for the reduction of contaminants from contaminated soil (Riffaldi et al, 2006). The enzyme lipases extracted from a variety of sources such as bacteria, plants, actinomycetes, and animal cells. Microbial lipases are multifaceted enzymes and catalyze a wide range of reactions such as hydrolysis, inter-esterification, esterification, alcoholysis, and aminolysis. Due to this versatile property, microbial lipases have potent applications in industries than the other sources.

Lipases are universal enzymes that have peculiar catalytic nature i.e., hydrolysis that converts from triacylglycerols to free fatty acids and glycerol. Hydrolytic reactions through lipase enzyme occur at the interface between lipid-water, where lipolytic/hydrolytic substrates usually form stability between different states such as monomeric, micellar, and emulsified. Generally, lipases have been divided into two types based on the criteria such as (a) Increased enzyme activity when the triglycerides form an emulsion and (b) lipases having a loop of protein (lid) wrapping on the active site (Sharma et al, 2011).

Triglyceride is the key constituent of natural oil or fat. Hydrolysis of triglyceride substrate consecutively yields diacylglycerol, monoacylglycerol, glycerol, and fatty acids. Hydrolyzed products such as glycerol and fatty acids are commonly used as raw materials for various industrial purposes, for example, monoacylglycerol is used as an emulsifying agent in the food, cosmetic, and pharmaceutical industries. The lipase enzyme adsorbed on to the oil-water interface in the bulk of the water phase. Then it breaks the ester bonds of triolein to yield consecutively diolein, monoolein, and glycerol. During the catalysis oleic acid is formed at each consecutive reaction stage. The glycerol formed is hydrophilic and thus dissolves into the water phase. Lipases produced from the various wastes are used to treat cooking oil through bioremediation, oil spills, and hydrocarbon degradation (Clarissa et al,2018).

Proteases

Proteases that produced by microbes that break down the proteins and also hydrolyse the peptide bonds. Proteases have a broad range of application in detergent, leather, pharmaceutical and food industry. [Rao et al 1998]. Based on the cleavage of the polypeptide chain proteases divided into two types: endopeptidases and exopeptidases. The exopeptidases cleave the bonds near the terminal end of amino or carboxylic group of chan. The proteases acts on free carboxyl terminals and free amino group are known as carboxypeptidase and aminopeptidase respectively. The endopeptidase cleaves the bond at the inner side of the peptide region. It is a negative impact on enzyme activity in the existence of free amino and carboxyl-terminal region.

Mainly proteases pre-owned in the cheese and detergent industry for protein denaturation. Mainly proteases used in the leather industry for removing the parts that attached to the skin and hairs from the skin. In the pharmaceutical industry, proteases can be used to evolving potential restorative agents. Clostridial collagenase or subtilisin is the group of proteases enzyme which is used in combination with broad-spectrum antibiotics in the treatment of burns and wounds (Rao et al 1998).

Cellulases

Cellulase is the group of enzymes mainly produced by fungi, bacteria and actinomycetes. These group of microbes secrete cellulases in the cell-free surface, cell-bound and always diverse from each other. The microbial cellulase categorized into (1) endoglucanase which cleaves the regions of the cellulose fibre, low crystallinity, creating free chain ends; (2) exoglucanase or cellobiohydrolase which break down the cellulose molecule while removing cellobiose units from the free chain ends; (3) βglucosidase which hydrolyzes cellobiose to glucose units. (Rixon et al 1992). During hydrolysis, the cellulases degrade the cellulose to reducing sugars. The cellulase enzyme eliminates the cellulose microfibrils which are fabricated during washing of cotton-based clothes. Its also used in the textile industry for colour brightening and softening of the material. The bacillus strains produced the alkaline cellulases and neutral and acidic cellulases produced by Trichoderma and Humicola fungi which are used for several purposes. The enzyme mainly used for removal of ink in the paper during recycling of paper in paper and pulp industry. It also used in the brewing industry to increase the juice liberation from the fruit pulp and for the production of alcohol from cellulosic biomass (Leisola et al 2006).

SOME OTHER ENZYMES

Nitrilases

Nitrilases are the group of enzymes which cleave the non-peptide carbon and nitrogen bonds and have the capability of hydrolyzing the nitriles stereospecifically and cyanides. These enzymes directly involved in the transformation of nitriles to carboxylic acids and ammonia. Nitrilases are produced by some bacterial species like *Nocardia* sp. and *Rhodococcus* sp., fungi like *Aspergillus niger, Fusarium solani* and produced in a lesser amount from filamentous fungi. (Banerjee et al,2002). Many researchers scrutinized the enzyme generated from fungi have greater potential when compared to bacterial enzymes. The enzyme produced by filamentous fungi act upon aromatic, aliphatic and alicyclic nitriles. It also shows the highest activity towards benzonitile and 3,4 -cyanopyidine under mild conditions with excellent regio and enantioselective and exhibit high activity and thermostability (Martinkova et al, 2009). The mechanism behind the transformation process the nitrile compound have the carbon atom bearing positive charge subjected to nucleophilic attack which results in imines group hydrolyzed into ketone and ammonia is formed as by-product. With the addition of water molecule, the acyl-enzyme is hydrolyzed into a carboxylic acid.

Xylanases

Xylanase is the group of enzymes which involve in the breakdown of xylan. Mainly these enzymes present in the cell wall of the plant and obtained from some kinds of bacteria, protozoa, fungi and yeasts. Xylan is the type of hemicellulose polysaccharide which contain xylopyranosyl units with 1,4 glycosidic linkages. The xylanase enzyme can break down the cell wall of the plants (Gomez et al, 2008). Xylanases able to hydrolyse of xylans and it is a heterogeneous polymer which needs high specificity action to degrade the backbone and side chains of xylan. Some other hydrolytic enzyme may involve in the conversion of xylan derivatives (Dodd and Cann, 2010).). The xylanase divided into five types: i) Endo -1-4-β-xylanases which cleaves the glycosidic linkages at the backbone of xylan and it undergoes polymerization of the substrate. ii) β- Xylosidases able to hydrolyze p-nitrophenyl and o-nitrophenyl-β-D-xylopyranoside iii) α- Glucuronidase able to cleaves at the backbone of glucuronoxylan in α-1, 2 bonds between the glucuronic acid residues and β-Dxylopyranosyl. iv) Arabinofuranosidases hydrolyze L-arabinose residues at 2,3 β-D-xylopyranosyl. V) Acetylxylan esterase cleaves the O-acetyl substituents at 2, 3 positions of xylose residues in acetylated xylansdehalogenases. (Juturu and Wu, 2011)

Esterases

Esterases can able to hydrolyze the ester bonds in fatty acids, the carboxyl ester act as a substrate to form alcohol and carboxylic acids. This enzyme plays an important role in the process of esterification, transesterification, aminolysis, alcoholysis which act as biocatalysts as well as this enzyme act crucial tool in the bioremediation process. (Sharma et al, 2017)

CONCLUSION

Herewith this chapter concluded that the applications of several enzymes in the remediation of contaminants is truly effective and worldwide attention. It has been recognized as an effective tool for the treatment of various types of pollutants present in the environment. Environmental pollution is rapidly expanding with increased development in the entire industrial sector of the developing world. Owing to this heavily in practice human-made industrialization and synthetic manufacturing processes, the continuous disposal/discharge of toxic compounds with/without partial or insufficient treatments is among a major a cause of environmental contamination, both terrestrial and aquatic systems. With the help of microbial enzymes remediate the environment is an attractive topic of interest and the application of these enzymes in a wide range of industries. The advances in the emerging fields like environmental biotechnology, microbiology and molecular biology would be open up for magnifying the great potentials of these enzymes and the researchers able to know knowledge about the mechanisms behind the activity of these enzymes must be important.

REFERENCES

Alcalde, M., Ferrer, M., Plou, F. J., & Ballesteros, A. (2006). Environmental biocatalysis: From remediation with enzymes to novel green processes. *Trends in Biotechnology*, *24*(6), 281–287. doi:10.1016/j.tibtech.2006.04.002 PMID:16647150

Ayangbenro, A. S., & Babalola, O. O. (2017). A new strategy for heavy metal polluted environments: A review of microbial biosorbents. *International Journal of Environmental Research and Public Health*, *14*(1), 94. doi:10.3390/ijerph14010094 PMID:28106848

Banerjee, A., Sharma, R., & Banerjee, U. C. (2002). The nitrile-degrading enzymes: Current status and future prospects. *Applied Microbiology and Biotechnology*, *60*(1-2), 33–44. doi:10.100700253-002-1062-0 PMID:12382040

Bourbonnais, R., Paice, M. G., Reid, I. D., Lanthier, P., & Yaguchi, M. (1995). Lignin oxidation by laccase isozymes from *Trametes versicolor* and role of the mediator 2,2¢-azinobis(3-ethylbenzthiazoline-6-sulfonate) in kraft lignin depolymerization. *Applied and Environmental Microbiology*, *61*(5), 1876–1880. doi:10.1128/AEM.61.5.1876-1880.1995 PMID:7646025

Chaiyen, P., Fraaije, M. W., & Mattevi, A. (2012). The enigmatic reaction of flavins with oxygen. *Trends in Biochemical Sciences*, *37*(9), 373–380. doi:10.1016/j.tibs.2012.06.005 PMID:22819837

Clarissa, H. (2018). Brazilian fruit processing, wastes as a source of lipase and other biotechnological products: A review. *Annals of the Brazilian Academy of Sciences*, *90*(3), 2927–2943. doi:10.1590/0001-3765201820180095 PMID:30304225

Colberg, P. J. S., & Young, L. Y. (1995).Anaerobic Degradation of Nonhalogenated Homocyclic Aromatic Compounds Coupled with Nitrate, Iron, or Sulfate Reduction. Microbial Transformation and Degradation of Toxic Organic Chemicals, 307–330.

Ding, H., Wu, Y., Zou, B., Lou, Q., Zhang, W., Zhong, J., Lu, L., & Dai, G. (2016). Simultaneous removal and degradation characteristics of sulfonamide, tetracycline, and quinolone antibiotics by laccase-mediated oxidation coupled with soil adsorption. *Journal of Hazardous Materials*, *307*, 350–358. doi:10.1016/j.jhazmat.2015.12.062 PMID:26826938

Dodd, D., & Cann, I. O. (2010). Enzymatic deconstruction of xylan for biofuel production. *Biology and Bioenergy*, *1*(1), 2–17. PMID:20431716

Doran, J. W., & Zeiss, M. R. (2000). Soil health and sustainability: Managing the biotic component of soil quality. *Applied Soil Ecology*, *15*(1), 3–11. doi:10.1016/S0929-1393(00)00067-6

Duran, N., & Esposito, E. (2000). Potential applications of oxidative ´ enzymes and phenoloxidase-like compounds in wastewater and soil treatment: A review. *Applied Catalysis B: Environmental*, *28*(2), 83–99. doi:10.1016/S0926-3373(00)00168-5

Fakayode, S. O. (2005). Impact assessment of industrial effluent on water quality of the receiving Alaro river in Ibadan Nigeria. *AJEAM-RAGEE*, 1-13

Fereidoun, H., Nourddin, M. S., Rreza, N. A., Mohsen, A., Ahmad, R., & Pouria, H. (2007). The Effect of Long-Term Exposure to Particulate Pollution on the Lung Function of Teheranian and Zanjanian Students, Pakistan. *The Journal of Physiology*, *3*(2), 1–5.

Fox, B. G., Borneman, J. G., Wackett, L. P., & Lipscomb, J. D. (1990). Haloalkene oxidation by the soluble methane monooxygenase from *Methylosinus trichosporium* OB3b: Mechanistic and environmental implications. *Biochemistry*, *29*(27), 6419–6427. doi:10.1021/bi00479a013 PMID:2207083

Gianfreda, L., & Bollag, J.-M. (2002). Isolated enzymes for the transformation and detoxification of organic pollutants. In R. G. Burns & R. Dick (Eds.), *Enzymes in the Environment: Activity, Ecology and Applications* (pp. 491–538). Marcel Dekker. doi:10.1201/9780203904039.ch19

11. Gianfreda, L., & Rao, M. A. (2004). Potential of extra cellular enzymes in remediation of polluted soils: A review. *Enzyme and Microbial Technology*, *35*(4), 339–354. doi:10.1016/j.enzmictec.2004.05.006

Gianfreda, L., & Rao, M. A. (2008). Bonifica di suoli contaminati e depurazione dell'acqua. In M. Gennari & M. Trevisan (Eds.), *Agrofarmaci. Conoscenze per un Uso Sostenibile* (pp. 521–564). Oasi Alberto Perdisa.

Gianfreda, L., Xu, F., & Bollag, J. M. (1999). *Laccases: a useful group of deligniðcation and possible industrial applications*. In A. Messerschmidt (Ed.), *Multi-Copper Oxidases* (pp. 201–224). World Scientiðc.

Gomez, L. D., Steele-King, C. G., & McQueen-Mason, S. J. (2008). Sustainable liquid biofuels from biomass: The writing's on the walls. *The New Phytologist*, *178*(3), 473–485. doi:10.1111/j.1469-8137.2008.02422.x PMID:18373653

Grosse, S., Laramee, L., Wendlandt, K.-D., McDonald, I. R., Miguez, C. B., & Kleber, H.-P. (1999). Purification and characterization of the soluble methane monooxygenase of the type II methanotrophic bacterium Methylocystis sp. strain WI 14. *Applied and Environmental Microbiology*, *65*(9), 3929–3935. doi:10.1128/AEM.65.9.3929-3935.1999 PMID:10473397

Guo, Y. (2016). Arsenic methylation by an arsenite S-adenosylmethionine methyltransferase fromSpir-ulinaplatensis. *Journal of Environmental Sciences (China)*, 34–35.

Huang, K., Chen, C., Shen, Q., Rosen, B. P., & Zhao, F.-J. (2015). Genetically engineering *Bacillus subtilis*with a heat-resistantarsenitemethyltransferase for bioremediation of arsenic contaminated organic waste. *Applied and Environmental Microbiology*, *81*(19), 6718–6724. doi:10.1128/AEM.01535-15 PMID:26187966

Huang, K., Chen, C., Zhang, J., Tang, Z., Shen, Q., Rosen, B. P., & Zhao, F.-J. (2016). Efficient Arsenic Methylation and Volatilization Mediated by a Novel Bacterium From an Arsenic-Contaminated Paddy Soil. *Environmental Science & Technology*, *50*(12), 6389–6396. doi:10.1021/acs.est.6b01974 PMID:27258163

Isin, E. M., & Guengerich, F. P. (2007). Complex reactions catalyzed by cytochrome P450 enzymes. *Biochimica et Biophysica Acta*, *1770*(3), 314–329. doi:10.1016/j.bbagen.2006.07.003 PMID:17239540

Juturu, V., & Wu, J. C. (2011). Microbial xylanases: Engineering, production and industrial appli- cat-ions. *Biotechnology Advances*. PMID:22138412

Kanu, I., & Achi, O.K. (2011). Industrial Effluents and their Impact on Water Quality of Receiving Rivers in Nigeria. *Journal of Applied Technology in Environmental Sanitation, 1*(1), 75-86.

Karlen, D. L., Doran, J. W., Weinhold, B. J., & Andrews, S. S. (2003). Soil quality Humankind's foundation for survival. *Journal of Soil and Water Conservation*, 58.

Kelishadi, R., & Poursafa, P. (2010). Air pollution and non-respiratory health hazards for children. *Archives of Medical Science*, *6*(4), 483–495. doi:10.5114/aoms.2010.14458 PMID:22371790

Kimani, N. G. (2007). *Environmental Pollution and Impacts on Public Health: Implications of the Dandora Dumping Site Municipal in Nairobi*. United Nations Environment Programme.

Larsson, D. G. (2014). Pollution from drug manufacturing: Review and perspectives. *Philosophical Transactions of the Royal Society of London. Series B, Biological Sciences*, *369*(1656), 369. doi:10.1098/rstb.2013.0571 PMID:25405961

Leisola, M., Jokela, J., Pastinen, O., & Turunen, O. (2006). *Industrial use of Enzymes—Essay*. Laboratory of Bioprocess Engineering, Helsinki University of Technology.

Liu, S., Zhang, F., Chen, J., & Sun, G. (2011). Arsenic removal from contaminated soil via biovolatilization by genetically engineered bacteria under laboratory conditions. *Journal of Environmental Sciences (China)*, *23*(9), 1544–1550. doi:10.1016/S1001-0742(10)60570-0 PMID:22432292

Loux, N. T., Su, Y. S., & Hassan, S. M. (2011). Issues in assessing environmental exposures to manufactured nanomaterials. *International Journal of Environmental Research and Public Health*, *8*(9), 3562–3578. doi:10.3390/ijerph8093562 PMID:22016703

Martinkova, L., Vejvoda, V., Kaplan, O., Kubac, D., Malandra, A., Cantarella, M., Bezouska, K., & Kfen, V. (2009). Fungal nitrilases as biocatalysts: Recent developments. *Biotechnology Advances*, *27*(6), 661–670. doi:10.1016/j.biotechadv.2009.04.027 PMID:19427375

Newman, L. A., Doty, S. L., Gery, K. L., Heilman, P. E., Muiznieks, I., Shang, T. Q., Siemieniec, S. T., Strand, S. E., Wang, X., Wilson, A. M., & Gordon, M. P. (1998). Phytoremediation of organic contaminants: A review of phytoremediation research at the University of Washington. *Soil & Sediment Contamination*, *7*(4), 531–542. doi:10.1080/10588339891334366

Park, J.-W., Park, B.-K., & Kim, J.-E. (2006). Remediation of soil contaminated with 2,4-dichlorophenol by treatment of minced shepherd's purse roots. *Archives of Environmental Contamination and Toxicology*, *50*(2), 191–195. doi:10.100700244-004-0119-8 PMID:16392021

Piontek, K., Smith, A. T., & Blodig, W. (2001). Lignin peroxidase structure and function. *Biochemical Society Transactions*, *29*(2), 111–116. doi:10.1042/bst0290111 PMID:11356137

Pointing, S. B., & Vrijmoed, L. L. P. (2000). Decolorization of azo and triphenylmethane dyes by Pycnoporus sanguineus producing laccase as the sole phenoloxidase. *World Journal of Microbiology & Biotechnology*, *16*(3), 317–318. doi:10.1023/A:1008959600680

Qin, J., Rosen, B. P., Zhang, Y., Wang, G., Franke, S., & Rensing, C. (2006). Arsenic detoxification and evolution of trimethylarsine gas by a microbial arsenite S-adenosylmethionine methyltransferase. *Proceedings of the National Academy of Sciences of the United States of America*, *103*(7), 2075–2080. doi:10.1073/pnas.0506836103 PMID:16452170

Rao, M. B., Tanksale, A. M., Ghatge, M. S., & Deshpande, V. V. (1998). Molecular and biotechnological aspects of microbial proteases. *Microbiology and Molecular Biology Reviews*, *62*(3), 597–635. doi:10.1128/MMBR.62.3.597-635.1998 PMID:9729602

Reports of UNICEF. (2017) *Thirsting for a Future: Water and children in a changing climate*. UNICEF.

Riffaldi, R., Levi-Minzi, R., Cardelli, R., Palumbo, S., & Saviozzi, A. (2006). Soil biological activities in monitoring the bioremediation of diesel oil-contaminated soil. *Water, Air, and Soil Pollution*, *170*(1–4), 3–15. doi:10.100711270-006-6328-1

Rixon, E., Ferreira, L. M. A., Durrant, A. J., Laurie, J. I., Hazlewood, G. P., & Gilbert, H. J. (1992). Characterization of the gene celD and its encoded product 1,4-β-D-glucan glucohydrolase D from Pseudomonas fluorescens subsp. cellulosa. *The Biochemical Journal*, *285*(3), 947–955. doi:10.1042/bj2850947 PMID:1497631

Rubilar, O., Diez, M. C., & Gianfreda, L. (2008). Transformation of chlorinated phenolic compounds by white rot fungi. *Critical Reviews in Environmental Science and Technology*, *38*(4), 227–268. doi:10.1080/10643380701413351

Ruiz-Duenas, F. J., Morales, M., Pérez-Boada, M., Choinowski, T., Martínez, M. J., Piontek, K., & Martínez, Á. T. (2007). Manganese oxidation site in Pleurotus eryngii versatile peroxidase: A site-directed mutagenesis, kinetic, and crystallographic study. *Biochemistry*, *46*(1), 66–77. doi:10.1021/bi061542h PMID:17198376

Sharma, D., Sharma, B., & Shukla, A. K. (2011). Biotechnological approach of microbial lipase: A review. *Biotechnology (Faisalabad)*, *10*(1), 23–40. doi:10.3923/biotech.2011.23.40

Sharma, T., Sharma, A., & Kanwar, S. S. (2017). An overview on esterases: structure, classification, sources and their application. In Recent advances in biotechnology (vol. 2, pp. 216-229). Shree Publishers.

Singh, A. L., & Sarma, P. N. (2010). Removal of arsenic(III) from waste water using *Lactobacillus acidophilus*. *Bioremediation Journal*, *14*(2), 92–97. doi:10.1080/10889861003767050

Takami, H., Kudo, T., & Horikoshi, K. (1997). Isolation of extradiol dioxygenasegenes that is phylogenetically distant from other meta cleavage dioxygenase genes. *Bioscience, Biotechnology, and Biochemistry*, *61*(3), 530–532. doi:10.1271/bbb.61.530 PMID:9095556

Takeuchi, M., Kawahata, H., Gupta, L. P., Kita, N., Morishita, Y., Yoshiro, O., & Komai, T. (2007). Arsenic resistance and removal by marine and non-marine bacteria. *Journal of Biotechnology*, *127*(3), 434–442. doi:10.1016/j.jbiotec.2006.07.018 PMID:16934903

Tang, C. Y., Criddle, Q. S., Fu, C. S., & Leckie, J. O. (2007). Effect of flux (trans membrane pressure) and membranes properties on fouling and rejection of reverse osmosis and nano filtration membranes treating perfluoro octane sulfonate containing waste water. *Journal of Environmental Science and Technology*, *41*(6), 2008–2014. doi:10.1021/es062052f PMID:17410798

Ten Have, R., & Teunissen, P. J. M. (2001). Oxidative mechanisms involved in lignin degradation by white-rot fungi. *Chemical Reviews*, *101*(11), 3397–3413. doi:10.1021/cr000115l PMID:11749405

Tsukihara, T., Honda, Y., Sakai, R., Watanabe, T., & Watanabe, T. (2006). Exclusive overproduction of recombinant versatile peroxidase MnP2 by genetically modified white rot fungus, Pleurotus ostreatus. *Journal of Biotechnology*, *126*(4), 431–439. doi:10.1016/j.jbiotec.2006.05.013 PMID:16820241

Verma, S., Verma, P. K., Meher, A. K., Dwivedi, S., Bansiwal, A. K., Pande, V., Srivastava, P. K., Verma, P. C., Tripathi, R. D., & Chakrabarty, D. (2016). A novel arsenic methyltransferase gene of *Westerdykell aaurantiaca* isolated 1 from arsenic contaminated soil: Phylogenetic, physiological, biochemical studies and its role in arsenic bioremediation. *Metallomics*, *8*(3), 344–353. doi:10.1039/C5MT00277J PMID:26776948

Williams, P. P. (1977). Metabolism of synthetic organic pesticides by anaerobic microorganisms. *Residue Reviews*, *66*, 63–135. doi:10.1007/978-1-4612-6352-4_3 PMID:335464

Xu, F. (1996). Catalysis of novel enzymatic iodide oxidation by fungal laccase. *Applied Biochemistry and Biotechnology*, *59*(3), 221–230. doi:10.1007/BF02783566

Yoshida, S., Chatani, A., Honda, Y., Watanabe, T., & Kuwahara, M. (1998). Reaction of manganese peroxidase of Bjerkandera adusta with synthetic lignin in acetone solution. *Journal of Wood Science*, *44*(6), 486–490. doi:10.1007/BF00833415

Zhang, Xu, Cao, Chen, Rosen, & Zhao. (2017). Arsenic methylation by a genetically engineered Rhizobium-legume symbiont.Plant. *World Journal of Microbiology & Biotechnology*, *16*(3), 317–318.

Chapter 4
The Use of Microorganism-Derived Enzymes for Bioremediation of Soil Pollutants

Joan Mwihaki Nyika
https://orcid.org/0000-0001-8300-6990
Technical University of Kenya, Kenya

ABSTRACT

Contamination of soils by xenobiotic compounds is a growing concern for environmentalists amidst the rise of anthropogenic activities that encourage such contamination practices. The use of microbial enzymes is a viable alternative to degrade and mineralize these contaminants, which is a growing research interest owing to its eco-friendly nature. This chapter explores the categories of enzymes used in soil bioremediation such as oxidoreductases and hydrolases, their mechanism of action, and their merits and demerits. Furthermore, molecular biology techniques useful in enhancing the production capacity, stability, activity, and shelf life of bioremediation enzymes is discussed. Ultimately, the need to develop bioremediation enzymes in bulk, using cheap technologies while optimising their activity, stability, and shelf life for effective soil decontamination is emphasized.

INTRODUCTION

Environmental contamination by hazardous compounds that bioaccumulate and resist degradation is one of the world's contemporary ecological problems. Contamination emanates mainly from agricultural, mining and industrial activities. Production of contaminants is promoted by advances in technologies used in the chemical industry leading to ease in production of dyes, solvents, explosives, pesticides, plastics and fuels. Contamination by these chemicals is long-term and has negative effects to human health and the environment due to their high toxicity and persistence (Godheja, Sk, Siddiqui, & Dr, 2016). Additionally, pollution interferes with nutrient cycles and decomposition capacities of soils. A

DOI: 10.4018/978-1-7998-4888-2.ch004

case example is the excess use of agrochemicals that is reported to pollute land and water resources that affect humans and other animals once they are incorporated in the food chain.

Cleaning up polluted areas such as soils by removing these pollutants is a growing challenge that requires the use of various technologies so that their levels can be acceptable and cause no harm. Some of the physico-chemical methods used apply biological agents such as microorganisms and plants. Of focus in recent research, is the microbial remediation of xenobiotics, which unlike physicochemical methods, which is cost effective in cleaning polluted soils (Piotrowska-Długosz, 2017). This microbial bioremediation may occur as either bio-augmentation or bio-stimulation. The former is the process of introducing non-native microorganisms such as pollutant- degrading bacteria in contaminated environments to enhance remediation reactions (Sharma, Dangi, & Shukla, 2018). Bio-stimulation on the other hand, entails modifying the environment to enhance the reaction of native bacteria that help in bioremediation processes. These modifications include the addition of electron acceptors and reaction controlling nutrients such as nitrogen, carbon, oxygen and phosphorous (Karigar & Rao, 2011). Specific plant or animal-origin and microbiological enzymes facilitate these microbiological reactions. In particular, cell-free enzymes instead of whole organisms are used in these reactions because they grow independent of nutrients (Rao, Scelza, Scotti, & Gianfreda, 2010). This chapter explores the potential of microbial enzymes in remediating xenobiotics in soils.

THE CONCEPT OF ENZYMOLOGY IN SOIL POLLUTANT BIOREMEDIATION

Enzymes are proteins that speed up biochemical reactions without being changed. To enhance the conversion of reactants to products, enzymes lower activation energy used during the process (Piotrowska-Długosz, 2017). Regions of enzymes involved in catalytic processes are known as active sites whose association with the rest of the protein occur through covalent or non-covalent bonding. An enzyme may have one or more essential catalytic groups or active sites. These are known as apoenzymes if they are made of protein or prosthetic group if they are nonproteins or holoenzyme if it is a combination of the two (Karigar & Rao, 2011). The nomenclature of enzymes is related to their catalytic group, their function and/or the reactions that they catalyse. Identification and classification of enzymes is dependent on their enzyme commission (EC) number, which is defined by the international union of biochemistry and molecular biology[1] (Karigar & Rao, 2011). All enzymes are categorised in six groups: - synthetases also known as ligases, isomerases, lyases, hydrolases, transferases and oxidoreductases.

The use of enzymes in soil decontamination from xenobiotics is advantageous unlike the use of classic microorganisms since the proteins have some unique aspects. Enzyme activity in the soil is associated with different locations and sources. Sources are mainly plant or microbial origin. Their location maybe extracellular or intracellular environments (Piotrowska-Długosz, 2017). In the former, enzymes are in the aqueous phase and occur as temporarily distinct from the substrate or as a complex with organic colloids and clay minerals. In the latter, enzymes occur in non-multiplicative or multiplicative cells and/or cell debris of in dead cells. The locations and sources of these proteins make their associations with soil biological and physicochemical features close. It is because of this relationship that environmental pollution, agricultural practices and human activities influence enzymatic activities in soils (Cele & Maboeta, 2016). Enzymes can therefore be used to optimise short and long-term modifications in soils. According to Piotrowska-Dlugosz (2017), many environmental and human-activity based factors that modify soil quality are attributable to changes in its physicochemical aspects but concurrently, they modify the enzyme pool in soils, which is a bio-indicator. Burns et al. (2013) noted that enzymes in soils play crucial roles in organic matter putrefaction processes. They catalyse reactions necessary for survival of microorganisms in the soil matrix, stabilize its structure and influence decomposition of wastes from a variety of sources, nutrient cycling and formation of organic matter. Reviews on the classification, structure, origin, function and behaviour of enzymes in soils have been documented previously (Rao et al., 2010; Karigar & Rao, 2011; Burns et al., 2013; Sharma et al., 2018).

Enzymatic systems and their use in soil bioremediation is a widely accepted concept globally. However, its success is influenced by the specificity and efficiency of these proteins as Cele and Maboeta (2016) noted. Some enzymes have narrow (stereo-, chemo, regional- specificity) while other have wide-range specificity and can be applied to different compounds. Irrespective of specificity, enzymes result to extensive toxicological and structural transformation of pollutants where in some cases the products become harmless inorganic compounds or their toxicity is rFeduced significantly. Rao et al. (2010) emphasized the potency of these proteins and claimed that some of them can decontaminate soils without any chemical reactions occurring. Enzymes are also used in extreme situations where microbial activity is limited. As such, they have no negative effects in the environment and their application in bioremediation activities is considered eco-friendly (Godheja et al., 2016). The following subsections discuss the specific groups of enzymes that have been used in soil bioremediation purposes and their functioning principles.

Bioremediation of Heavy Metals

Microorganisms use toxic heavy metals as sources of food during bioremediation that can occur ex-situ or in-situ. The latter describes the decontamination of heavy metals at their pollution site while in the former, transformation of these pollutants occurs away from their original contamination site (Raghunandan et al., 2018). Microbes are able to transform heavy metals to non-toxic forms, as they are ubiquitous in nature. They use two mechanisms to maintain bioremediation of heavy metals namely: 1) production of degradation enzymes that target the pollutants and 2) developing resistant to heavy metals. The microbes transform, adsorb, volatilise, immobilise through physico-bio-chemical processes, oxidise and bind on heavy metals through sorption making them less or non-toxic in catalytic reactions (Alvarez et al., 2017). A number of microorganisms and the metals they remediate is summarised in Table 1 (Verma & Kuila, 2019).

TYPES OF ENZYMES USED IN SOIL BIOREMEDIATION

Oxidoreductases

Oxidoreductases describe a group of enzymes that speed up reactions through the transfer of electron transfer from a reductant (electron donor) to an oxidant (electron acceptor). Various higher plants, fungi and bacteria species secret and produce these enzymes. During microbial remediation, oxidoreductases undergo oxidative coupling of contaminants in soils to release methanol, carbon dioxide and chloride ions (Sharma et al., 2018). The reactions result to production of energy and heat, which is used by microorganisms to power their metabolic activities. The energy is also used to break chemical bonds of substrates and transfer electrons to recipient acceptors resulting to innocuous compounds (Karigar & Rao, 2011).

Table 1. Microbes used in bioremediation of heavy metals

Microbes	Metals Remediated
Sphingomonas desiccabilis, Bacillus subtilis BR151	Arsenic (As)
B. idriensis, Methyloccocus capsulatus	Cadmium (Cd), Chromium (VI) (Cr)
Caulobacter crescentus	Cd (II)
Pseudomonas strain K-62, Achromobacter sp. AO22	Mercury (Hg)
Escherichia coli	As
P. fluorescens OS8, E. coli MC1061, B. subtilis BR151,	Cd,Hg, lead (Pb), Nickel (Ni), Zinc (Zn)
E. coli JM 109, Staphylococcus aureus RN4220	Cd (II), Hg, As
P. putida 06909	
Pseudomonas K-62	Hg
E. coli SE5000	Ni
Acidithiobacillus ferrooxidans	Hg
Maroxella sp.	Cd (II), Hg
Mesorhizobium huakuii B3	Cd (II)
Deinococcus radiodurans	Hg

Source: (Verma & Kuila, 2019)

Immobilisation of contaminants using oxidoreductases occurs in five mechanisms namely: adsorption, covalent binding on a carrier, encapsulation, entrapment and enzyme cross-linking (Zdarta et al., 2018). Adsorption involves hydrophobic and ionic interactions that are non-specific using enzymes while covalent binding uses enzymes with –OH, -SH and NH_2 functional groups. Encapsulation and entrapping involves the physical enclosure of a pollutant molecule without altering its structure. Cross-linking is the binding of bifunctional groups such as carbodiimides and glutaraldehyde on enzymes to stabilize pollutants even without using supporting agents (Zdarta et al., 2018). Enzymes from different microbes could use one or more of these immobilisation techniques based on its molecular structure (Table 2).

Oxidoreductases are widely used in bioremediation of manmade and natural contaminants in soils. Some oxidoreductase enzymes sourced from bacterial species reduce radioactive metals found in soils. For instance, *Bacillus safensis* CFA-06, a gram-positive bacterium produces oxidoreductases with the capacity to decontaminate soil from petroleum compounds. Laccases and peroxidases, which are subcategories of oxidoreductases, have been used in removal of coloured compounds from textile industry waste in soils (Novotný et al., 2004). A study by Newman et al. (1998) reported the ability of plants belonging to the families of *Solonaceae, Gramineae* and *Fabaceae* to secret extracellular proteins that degrade petroleum-based hydrocarbons and chlorinated compounds in soils. Other categories of oxidoreductases including tyrosinase and dehydrogenases have been used in soil contaminant bioremediation (Piotrowska-Długosz, 2017). These enzymes modify lignin, which is an irregular, highly complex and stable polymer due to their high redox and nonspecific oxidation capacity (Sharma et al., 2018). Table 2 summarises some of the microbes used as sources of oxidoreductases, their target soil pollutants and the mechanism used to in the bioremediation process (Zdarta et al., 2018). The categories of oxidoreductases used in decontaminating soils are discussed in details in the following subheadings.

Oxygenases

This category of enzymes is mainly used in aerobic degradation of xenobiotic contaminants with aromatic compounds and catalyse the breaking of the aromatic ring by adding oxygen molecules during decontamination (Piotrowska-Długosz, 2017). Based on the number of oxygen molecules used in the reaction, oxygenases are either mono-oxygenases or di-oxygenases that add one and two oxygen molecules, respectively. Mono-oxygenases are useful in hydroxylation, denitrification, desulphurisation and dehalogenation of aromatic compounds in soil contaminants. (Arora, 2010). They occur in two categories based on the cofactor in use, 1) p450 and 2) Flavin-dependent mono-oxygenases. The latter are useful in degradation of chlorine-based pesticides such as endosulfan (Bajaj, Pathak, Mudiam, Mayilraj, & Manickam, 2010) while the former, are isolated from prokaryotes such as the bacterium *Bacillus megaterium*, BM3 and degrade aromatic and fatty acid containing contaminants in soils (Roccatano, 2015). Contaminants such as halide-based aromatic, aliphatic and heavy metal compounds are co-degraded by methane mono-oxygenase, which is isolated from the bacterium *Methylocella palustris* (Bajaj et al., 2010). Arora (2010) also documented the isolation of quinol and tetranomycin F1, which are enzymes of this category that do not require cofactor activation from *Escherichia coli* and *Streptomyces glauscens*, respectively.

Mono-oxygenases of the p450 family occur in many eukaryotic and prokaryotic organisms and oxidize many industrial waste xenobiotic compounds to metabolise them. These enzymes recycle their redox group, which is non-covalently bound to the cofactor and have a porphyrin[2] group that contains iron. For instance, the white-rot fungus (*Phanerochaete chrysosporium*) contains about 12 families of cytochrome p450 enzymes in its genetic component and has been used in metabolism of herbicides such as chlortouluron, norflurazon and atrazine in polluted soils (Tuomela & Hatakka, 2011). Using the protein, *Pesudomonas syringae*, a cytochrome p450 mono-oxygenase that uses nicotinamide adenine dinucleotide phosphate (NADPH)[3] in its reduced form and has been developed synthetically. It is isolated on the surface of *E.coli* and is useful in bioremediation of pharmaceuticals and selective synthesis of soil foreign chemicals according to Yim et al. (2006).

Di-oxygenases are classified into two, 1) aromatic ring cleaving and 2) aromatic ring hydroxylation enzymes. The former break up aromatic rings while the latter adds oxygen molecules to metabolize xenobiotic compounds (Sharma et al., 2018). Degradation of toluene for instance is done by toluene di-oxygenase, which is isolate from *Pseudomonas putida* F1. By behaving as a mono-and di-oxygenase, the enzyme can co-metabolise aliphatic olefins and aromatic contaminants to decontaminate soils (Mukherjee, 2012). Bioremediation of oxoquinoline and quinaldine carbon double bonds to form carbon monoxide has been done using ring opening di-oxygenases while bacterial isolated catechol di-oygenases that occur in soils transform aromatic to aliphatic compounds (Muthukamalam, Sivagangavathi, Dhrishya, & Sudha Rani, 2017). In soils polluted by industrial aromatic chemicals such as dyes and pharmaceuticals, dioxygenases cleave their aromatic rings at the first and second positions to decontaminate them. A case example is the use of *P. putida* derived naphthalene di-oxygenase to degrade naphthalene in polluted soils (Guzik, Hupert-Kocurek, & Wojcieszysk, 2013). Demarche et al. (2012) documented the use of the di-oxygenase tyrosinase to catalyse hydroxylation of phenols to produce catechols and additional oxidation of catechols to innocuous o-quinones in a sequential reaction.

Laccases

These are a family of oxidases that have multiple copper molecules and are produced by a variety of bacteria, insects, fungi and plants. They catalyse the reduction of aromatic and phenolic compounds to water and oxygen and occur in isomeric forms (Piotrowska-Długosz, 2017). Unique properties of these oxidases that prompt their application in bioremediation include their stability at pH levels near 7 and the broad range of specificity to various substrates that translates to wide application (Giardina et al., 2010). Additionally, laccases use harmless di-oxygen instead of oxygen peroxide as a co-substrate and have a high redox potential (Karigar & Rao, 2011). Microorganisms can produce intra-and-extra-cellular laccases that catalyse the oxidation of aryl diamines, polyamines, polyphenols, aminophenols, para- and ortho-diphenols to non-pollutants (Rodríguez Couto & Toca Herrera, 2006). The enzymes decontaminate soils through demethylation and decarboxylation of their phenols and methoxy-phenol compounds. Additionally, they nitrify microorganisms by catalysing humification and depolymerise lignin, which prevents formation of more phenols (Kim, Park, Lee, & Kim, 2002).

Specific examples of laccases that are sourced from microorganisms and have been used in bioremediation of soil contaminants are provided in literature. For instance, two isomeric laccases that were purified from the fungus *Trametes hispida* oxidised the decolouration of azo dyes and transformed them to harmless products (Legerská, Chmelová, & Ondrejovič, 2016). Oxidases produced by *Rhizoctonia practicola* metabolised phenolic compounds (Strong & Claus, 2011). A study by Chakroun et al. (2010) using a fungus derived laccase, *Trichoderma atroviride* reported its application to degrade a wide range of contaminants including heterocyclic compounds, phenolic and amid-based contaminants of soils. Using technologies such as immobilisation on glass beads, the half-life, activity and stability of these enzymes can be enhanced making them resistant to protease degradation (Sharma et al., 2018).

Peroxidases

Peroxidases are ubiquitously distributed oxidoreductases that metabolise phenol-containing pollutants and lignin using a mediator and hydrogen peroxide. Phenols during the reaction are oxidised to become insoluble and precipitate (Piotrowska-Długosz, 2017). Although these enzymes have a common structure in that they are glycosylated, the catalytic feature containing heme is a differentiating characteristic where some enzymes have it and others do not. Heme containing peroxidases occur in two classes based on their sources, 1) those sourced from animals and 2) those occurring in plants, bacteria and fungi (Sharma et al., 2018). The latter category is further sub-divided into three classes. The first consists of ascorbate peroxidases produced by plants and have cytochrome c peroxidase from yeast. A second class has manganese and lignin peroxidases from fungi species and a last category of horseradish peroxidases from plants. Non-heme enzymes occur as NADH, halogen, alkylhydrogen and thiol containing peroxidases (Karigar & Rao, 2011).

Of all these sub-categories, manganese and lignin peroxidases are the most studied in soil microbial remediation studies due to their high potential. Lignin peroxidases are produced from fungi such as *T. versicolor* and *Phanerochaete chrysosporium* as well as bacteria species. An example is iron (III) lignin peroxidase that oxidises toxic pollutants such as heavy metals with the help of a co-substrate and a mediator such as veratryl ethanol to degrade them (Xu et al., 2014). Tuomela and Hatakka (2011) also documented the application of these peroxidases in soil and wastewater bioremediation. The thermostability and specificity of peroxidases from bacteria is more effective than those from fungal sources as Sharma et al. (2018) noted. Manganese peroxidases are produced by lignin metabolising fungi and oxidise Mn (II) to Mn (III). The enzymes have manganese at their various binding sites, which helps to reduce substrates and chelate xenobiotic compounds indirectly by degrading their amine, aromatic and phenolic groups (Karigar & Rao, 2011). *Trametes* fungi species for example contain a Mn peroxidase (MnP-Tra-48424), which decolorizes dyes and degrades organic solvents and heavy metals of contaminated soils (Xu et al., 2014). Peroxidases are also reported to degrade poly-cyclic aromatic hydrocarbons (Zhang et al., 2016) and decolorize soils contaminated by textile industry dyes (Lee et al., 2016). In the two studies, the enzymes used were sourced from *Trametes versicolor* and *Peniophora incarnate* fungi species, respectively.

Table 2. Examples of microbes used to degrade soil pollutants catalytically via redox reactions and the mechanisms they use

Microorganism	Oxidoreductase category	Target Soil Pollutants	Immobilisation mechanism	Reference
Fungi				Zdarta et al., 2018
Coriolus versicolor	Laccase	Chlorophenols	Covalent immobilisation	
Trametes pubescens	Laccase	Dyes	Entrapment	
Coriolus versicolor	Manganese and lignin peroxidase	Molasses	Crosslinking and covalent immobilisation	
Schicophyllum commune IBL-06	Lignin peroxidase	Textile dyes		
Ganoderma lucidum BL05	Manganese peroxidase	Dyes	Entrapment	
Agaricus bisporus	Tyrosinase	Bisphenol A, phenol	Encapsulation/ entrapment	
Trametes villosa	Laccase	Dyes	Covalent immobilisation	
Phanerochaete chrysosporium	Lignin Peroxidase	Azo dyes		
Cerrena unicolor	Laccase	Triclosan, 4-nonylphenol		
Mycelia sterilia IBR 35219/2	Oxidase	Phenols		
Trametes hirsuta	Laccase	Dyes		
Anthracophyllum discolor	Manganese peroxidase	Phenanthrene, fluoranthene, anthracene, pyrene	Adsorption	
Bacteria				Mahmood et al., 2015
Bacillus sp. ADR	Reductases, laccases, oxidase	Dyes found in soils	Adsorption	
Bacillus subtilis WD23	Laccases			
Enterococcus faecalis, Escherichia coli, Exiguobacterium sp. RD3, Staphyloccocus aureus, Xenophilus azovorans	Reductases		Covalent immobilisation	
Sphingobacterium sp. AT	oxidase		Entrapment	

Hydrolases

Hydrolytic enzymes break various chemical bonds through reduction to convert xenobiotic contaminants to harmless products. The chemical bonds where these enzymes act include carbon-halide, peptide and ester bonds (Karigar & Rao, 2011). Examples of this category of enzymes include phosphatases, phytases, esterases, proteases, lipases and carbohydratases and are produced by various bacteria and fungi species. The preference to these enzymes in soil bioremediation is because of their low specificity to substrates, ready availability, ability to tolerate solvents and their limited stereoselectivity to cofactors as Sharma et al. (2018) noted. In soils, this class of enzymes is known to metabolise organochlorine insecticides such as heptachlor and DDT[4](Karigar & Rao, 2011), carbamate and organophosphate based pesticides, carbendazim based fungicides (Guzik et al., 2013). A number of microorganisms produce hydrolases that are used to degrade soil pollutants as summarised in Table 3. Examples of hydrolytic enzyme sub-classes used in soil bioremediation are discussed in the following subheadings.

Lipases

These are ubiquitous enzymes isolated from animal cells, actinomycetes, plants and bacteria and catalyse the transformation of glycerol and free fatty acids from triacylglycerols hydrocarbons through processes such as aminolysis, alcoholysis, esterification and hydrolysis (Cele & Maboeta, 2016). Lipases occur either as having a protective cover for the active sites or as proteins whose activation is triggered when hydrocarbons form an emulsion (Karigar & Rao, 2011). A lipase from the fungus *P. aeruginosa* SL-72 was used in bioremediation of crude oil polluted soils (Verma, Saxena, Prasanna, Sharma, & Nain, 2012). *Candida rugosa* was also used to hydrolyse ester bonds of triolein, which is found in many food, pharmaceutical and cosmetic wastes in soils (Karigar & Rao, 2011). According to (Margesin, Zimmerbauer, & Schinner, 1999) lipase activity in contaminated soils is a useful indicator of its hydrocarbon degradation capacity.

Proteases

Proteases occur as either proteinases or peptidases and catalyse the breakdown of peptide bonds. Some sub-categories of these enzymes include pepsinases, collagenases and keratinases that catalyse the breakdown of pepsin, collagen and keratin associated xenobiotic compounds, respectively. Of the three categories, keratinases are widely used in soil bioremediation. Keratinases are produced by *rhizopus, neurospora, cephalosporium* and *penicillum* fungi species as well as bacteria species such as *B. subtilis, B. megaterium*, *B. cereus* and *B. amyloloquefaciens* (Piotrowska-Długosz, 2017). Mazotto et al. (2011) documented the use of bacillus species that have keratinases in degrading feather and leather wastes found in soils. Sharma et al. (2018) reported the use of the collagenases in collagen-containing waste degradation such as antibiotics and pharmaceuticals.

Phosphotriesterases

This group of hydrolases that catalyses the breakdown of organophosphate triesters. Organophosphoric compounds are found in agrochemicals and pesticides and accumulate in soils consequently polluting ground-and surface-water resources. Bacterial phosphotriesterases such as *Pseudomonas diminuta* and

Agrobacterium radiobacter have been used to detoxify organophosphotriesterases in contaminated soils (Scott et al., 2011). Sirotkina et al. (2012) also documented the use of these hydrolytic enzymes in metabolising pesticides in polluted soils.

Cellulases

Cellulases are enzymes that degrade cellulose, which is the most abundant component found in soil wastes and the environment. Categories used in bioremediation practices include endoglucanases, exoglucanases and β-glucosidases based on the areas they act on the substrate (Karigar & Rao, 2011) to reduce cellulose to free sugars. Soils polluted by paper, textile and pulp waste have been degraded using bacteria (bacillus species) derived alkaline cellulases and fungi derived acidic cellulases (Sharma et al., 2018). Aslam, Hussain and Qazi (2019) reported the use of *B. amyloliquefacience* ASK11 derived cellulose to degrade leather wastes.

Table 3. Hydrolytic enzymes produced by microorganisms and their bioremediation uses

Microorganisms	Hydrolytic enzyme	Pollutants degraded	Reference
Bacteria: ***Staphyloccocus sp., Chromobacterium sp., Pseudomonas sp., Arthrobacter sp., Alcaligenes sp.*** **Fungi:** ***Penicilium cyclopium, P. verrucosum, P. crustosum, Mucor griseocyanus, M. hiemalis, M. lipolyticus, Rhizopus japanicus, R. delamar, R. nigricans, R. microspores, R. arrhizus.***	Lipases	Organic pollutants such as oils	Sharma et al., 2011
Bacteria: ***Bacillus sp.* AC-1, LFC-15, NZ, *Bacillus thuringiensis, B. licheniformis, B. cereus, Cellulomonas cellulans* MTCC 23, *Clostridium thermocellum, Anoxybacillus flavithermus sp.* EHP2, *Anoxybacillus sp.* 527, *Streptomyces sp.* BRC 1, *Microbacterium sp.* MTCC 10047.**	Cellulase	Cellulose containing soil pollutants such as agricultural residues, paper pulp, straws, oils and spices.	Sadhu & Maiti, 2013
***B. licheniformis* NCIM-2042**	Protease	Waste feather in feed supplements of domestic animals.	Bhunia et al., 2013
B. weihenstephanensis			Sahoo et al., 2012
***B. subtilis* LFB-FIOCRUZ 1270, 1273, 1274**			Mazotto et al., 2011
***B. altitudinis* GVC11**		Hides and skins	Vijay Kumar et al., 2011

Other Enzymes Used in Soil Bioremediation

Apart from the well-documented oxidoreductases and hydrolases used in soil bioremediation of xenobiotics and discussed above, other sub-classes are also being adopted for use in this process. Carboxylesterases that hydrolyse carboxylic-esters in fungicides, insecticides and pesticides have been documented. A *Pseudomonas aeruginosa* PA1 derived carboxylesterase was used in decontaminating mercury-contaminated soils (Yin et al., 2016). Haloalkane dehydrogenases, which breakdown carbon halogens in various halide and alcohol containing chemicals are also used in soil bioremediation. These hydrolases are derived from bacteria species such as *P. pavonaceae* and *Sphingomonas paucimobilis* (Sharma et al., 2018) and are used in decontaminating pesticide polluted soils. Piotrowska-Długosz (2017) also noted the use nitrilases that metabolise the C-N triple bond in soils polluted with buctril, bromoxynil and dichlobenil, which are used in the manufacture of pesticides and extractants. This sub-class of hydrolases is isolated from bacteria, fungi and some plant and insect species. Other enzyme categories including cyanases, hydratases, cyanidases and synthases, which are microorganism derivatives are being developed for the bioremediation of xenobiotics in soils (Scott et al., 2011; Piotrowska-Długosz, 2017; Sharma et al., 2018).

MERITS AND DEMERITS OF USING ENZYMES FOR SOIL BIOREMEDIATION

In many soil bioremediation processes, enzymes facilitate the ultimate action of microorganisms in metabolising pollutants. Rao et al. (2010) agreed with this suggestion claiming that enzymes are the ultimate effectors of bioremediation transformations occurring in the environment and biota. These proteins have broad or narrow specificity that allows their use in xenobiotic remediation to be wide. Additionally, they produce assorted transformations to toxicological and structural characteristics of their target contaminants leading to their conversion to innocuous inorganic compounds. Some of the enzymatic processes have no known chemical transformations to substitute them (Demarche et al., 2012). Unlike traditional forms of remediation, enzymatic activities are not inhibited by the metabolism of microorganisms hence their use can occur even in extreme conditions such as in the presence of microbial antagonists and predators as well as at high or low contaminant levels (Rao et al., 2010). Piotrowska-Długosz (2017) disagreed with these suggestions claiming that enzymes have low efficacy at high contaminant concentration as they poison microorganisms while at low concentration, enzymes use soluble carbon sources of substrates, which reduces their catalytic effectiveness.

Enzymes are small sized hence highly mobile and their specificity to given substitutes makes bioremediation activities precise to the target pollutant. The mobility and specificity allow the quick action of these proteins and in their synthetic development, it helps to predict behaviour and develop cheap and optimised dosing regimens for specific soil bioremediation applications (Karigar & Rao, 2011). The use of enzymes in bioremediation is advantageous because their acclimatisation is not a prerequisite and thus, they are applicable in varied environmental conditions such as extreme moisture, temperature and pH conditions. Unlike exclusive microorganism bioremediation, the use of enzymes enables easy bioavailability modifications using surfactants and co-solvents to stabilise such systems and lengthen their survival in soil systems as Burns et al. (2013) noted. Enzymatic systems in microorganisms undergo in situ digestion after bioremediation processes. This ease in controlling their fate is crucial in sustaining ecological balance particularly other uses of microorganisms such as agriculture in addition to xenobiotic bioremediation (Piotrowska-Długosz, 2017). It is from these advantages that the use of enzymes according to Alcade et al. (2006) is referred to us white biotechnology that is part of green chemistry and whose environmental sustainability is widely documented and confirmed.

Despite the many documented merits on the use of enzymes in microorganism bioremediation of soil contaminants, a number of limitations have been identified (Torres, Bustos-Jaimes, & Le Borgne, 2003; Gianfreda & Rao, 2004). Some products of these reactions for instance have been found to be more toxic compared to their parent substrates. This observation necessitates a prior to and after remediation assessment of substrate and product, respectively. Processes of isolating and purifying enzymes from their sources are exorbitant and time ineffective. In recognition of this limitation, genetic modification and molecular biology techniques and tools are being used to improve the production protocols for such enzymes and optimize their stability and activity (Gianfreda & Rao, 2004; Burns et al., 2013). Most enzymatic systems require cofactors to mineralize contaminants completely. Mineralization, which is the complete transformation of toxic to non-toxic compounds however is realized with application of many enzymes for simultaneous transformation and degradation of xenobiotic compounds (Piotrowska-Długosz, 2017). These preconditions make the process expensive. Some enzyme forms used in bioremediation have a short half-life and are unstable in hostile environments such as in soil colloids, extreme alkaline and acid soils where they can be denatured easily. The use of proteases and non-biological denaturation processes can destabilize and immobilise enzyme systems, which reduces their effectiveness in soil bioremediation processes (Torres et al., 2003). With these limitations, future researches should focus on improving the stability, efficiency and kinetic properties of enzymes during soil remediation to make their use viable and realistic.

TOOLS AND TECHNIQUES TO IMPROVE ENZYMATIC BIOREMEDIATION OF SOIL POLLUTANTS

Apart from the aforementioned limitations of enzymatic systems used in soil bioremediation, the macro-structure of these proteins is complex and any physicochemical alteration may result to their denaturation (Tyagi, da Fonseca, & de Carvalho, 2011). The situation occurs in soil matrix, which has many ongoing physicochemical processes that have natural or anthropogenic-based triggers. Furthermore, many of the discussed microbes that are sources of enzymes do not produce them in adequacy, which necessitates novel approaches to produce enzymes that mineralize soil contaminants completely (Baweja, Nain, Kawarabayasi, & Shukla, 2016). Some of these technologies documented in literature are discussed in the following subheadings.

Genetic Engineering

Genetic engineering is the deliberate manipulation of genetic material of an organism to alter its natural characteristics. The technology is used to enhance production of enzymes by isolating their genetic code from original sources and transferring it to a different host where it can be expressed (Alcalde et al., 2006). According to Gupta and Shukla (2016), the enzymes can be overexpressed at different hosts increasing their stability and activity cheaply using recombinant DNA[5] technology. The resultant recombinant enzymes are easier to purify compared to natural ones and can be produced in bulk quantities. Baweja et al. (2016) also noted that through genetic engineering, the substrate range, pH and shelf life of bioremediation enzymes can be optimised. Sharma et al. (2018) gave examples of enzymes such as tetrahydrofuran, laccase CUeO, dye decolorizing peroxidase, polyphenol oxidase and carboxyesterase that have been produced through genetic engineering from their native microorganisms. These include *Pseudono cardia, E. coli, Geotrichum candidum, Marinomonas mediterranea* and human liver in respective order. The coupling of techniques such as error-prone PCR[6], site-specific mutagenesis and shuffling of DNA in recombinant genetic engineering enhances the selectivity of enzymes to a wide range of substrates during bioremediation (Gupta & Shukla, 2016).

Nanoenzymology

The technique produces artificial enzyme mimics as nanoparticles, which speed up bioremediation reactions using similar kinetics and mechanisms as native enzymes under analogous physiological conditions (Xu et al., 2012). The high stability and low cost of production associated with these nanoparticles justifies their growing research and application in bioremediation. Although they do not have active sites, nanoenzymes bind on substrates to trigger chemical reactions. They have been documented to metabolise organic compounds, lignin-based waste and dye waste found in soils (Gao & Yan, 2016). Nanoparticles mimicking laccases have been produced from nucleotide- synchronised copper (II) ions (Gao & Yan, 2016) while those mimicking manganese peroxidases have been produced from iron (III) oxide particles and are used to degrade rhodamine, phenol and methylene blue containing xenobiotic contaminants (Xu et al., 2012). Another study documented the use of cerium and iron oxide mimicked phosphatases to degrade chemicals used in warfare, parathion methyl 1 and organophosphorous based pesticides (Shin, Park, & Kim, 2015). The use of carbon nanomaterials such as graphene quantum dots and graphene oxide that mimic peroxidases in bioremediation of soil chemicals has been documented (Ma et al., 2017).

Enzyme Immobilisation Technologies

Immobilisation technique uses soluble or non-bound enzyme forms to enhance their stability and activity through reduced movement (Meryam Sardar, 2015). Methods used for enzyme immobilisation include their covalent bonding to insoluble compounds such as silica gel, adsorbing them on alginate or glass beads and binding them to affinity tags (Sirisha, Jain, & Jain, 2016). Materials used in the technique do not react with products, have limited substrate diffusion from their large surface area and are cheap. Immobilisation enhances catalytic efficacy of enzymes since their shelf life is lengthened. Their use in bioremediation of xenobiotic compounds is deemed economical (Meryam Sardar, 2015, 2015; Sirisha et al., 2016). Immobilisation of laccases for instance, was shown to protect them from protease denaturation and retain their activity even in extreme pH and temperature conditions (Giardina et al., 2010; Strong & Claus, 2011).

Enzyme Engineering

The procedure involves the altering of basic amino acids of an enzyme to improve its resilience to extreme pH, temperature, stress and promote its activity (Sharma et al., 2018). Just like genetic engineering, this technique uses recombinant DNA to optimise desired outputs of amino acid modifications particularly, in their sequencing. These enzymology advances have borne fruits as proteins with selective ability to degrade soil contaminants such as radionuclides and heavy metals have been produced. For example, modification of the sequence of the dioxygenase, nitrobenzene-1,2 at position 293 adjacent to the binding site increased its oxidation capacity to mineralise 2,6 dinitrotoulene (Ju & Parales, 2006). In another study, modification of 2- nitrotoulene dioxygenase at the amino acid position 258 led to enhanced transformation of nitrotoulene to nitrite and 3-methyl catechol, which are less harmful. (Singh, Kang, Mulchandani, & Chen, 2008).

CONCLUSIONS AND RECOMMENDATIONS

With the increasing concerns on soil pollution and its potential to mediate groundwater resource pollution amidst reducing capacity of the matrix to naturally degrade contaminants, there is need to explore alternatives such as microbial enzyme bioremediation. This chapter proposes the use of this approach in soil bioremediation using a number of enzymes under the oxidoreductases and hydrolases classes. Using documented examples in literature, these groups of enzymes are reported as effective in soil contaminant bioremediation. The chapter also explores the advantages such as specificity to a wide range of substrates, fast action and easy availability, which are associated with enzyme use in bioremediation. However, it is noted that under natural conditions adequate production of stable enzymes is impossible. It is from this recognition that the use of molecular biology, biochemistry and nanoscience in processes such as enzyme engineering, their genetic modification, immobilisation and nano-enzyme mimicking is increasingly gaining attention. These technologies have the capacity to produce enzyme in bulk and modify their characteristics to increase their bioavailability and shelf life, enhance their activity and stability even under stressful conditions.

Although this chapter reports notable progress in research on xenobiotic bioremediation in soils, it reports inadequacies in cost effective and bulk enzyme production and optimised enzyme activity to result to produce innocuous products. Additionally, the need to lengthen the shelf life and promote the stability and activity of enzymes for more prompt bioremediation results is highlighted. Therefore, the need to explore microflora that produces enzymes using appropriate cultivation techniques is imperative. Metaproteomics, metagenomics and metatranscriptomics are such techniques whose role will be identifying and qualifying new and natural enzymes that promote soil contaminant bioremediation processes with greater effectiveness than existent techniques.

Understanding the intricate behaviour of microorganisms in soils is also recommendable. This suggestion comes with the recognition that microbes just as soil matrix have complex communication networks that influence their reaction to contaminants. This understanding is crucial in devising degradation pathways of soil contaminant that promote effective mineralisation, survival of microflora and promote production of enzymes useful in speeding up the process. Adopting techniques using recombinant DNA and nanoscience to produce enzymes in bulk and supplement existent natural occurring ones should be a priority. Overall, caution should be taken to implement these recommendations while at the same time, cutting cost and maintaining ecological integrity for sustainable development.

REFERENCES

Alcalde, M., Ferrer, M., Plou, F. J., & Ballesteros, A. (2006). Environmental biocatalysis: From remediation with enzymes to novel green processes. *Trends in Biotechnology*, *24*(6), 281–287. doi:10.1016/j.tibtech.2006.04.002 PMID:16647150

Alvarez, A., Saez, J. M., Costa, J. S. D., Colin, V. L., Fuentes, M. S., Cuozzo, S. A., Benimeli, C. S., Polti, M. A., & Amoroso, M. J. (2017). Actinobacteria: Current research and perspectives for bioremediation of pesticides and heavy metals. *Chemosphere*, *166*, 41–62. doi:10.1016/j.chemosphere.2016.09.070 PMID:27684437

Arora, P. K. (2010). Application of Monooxygenases in Dehalogenation, Desulphurization, Denitrification and Hydroxylation of Aromatic Compounds. *Journal of Bioremediation & Biodegradation, 01*(03). Advance online publication. doi:10.4172/2155-6199.1000112

Bajaj, A., Pathak, A., Mudiam, M. R., Mayilraj, S., & Manickam, N. (2010). Isolation and characterization of a Pseudomonas sp. strain IITR01 capable of degrading α-endosulfan and endosulfan sulfate: Endosulfan degrading Pseudomonas sp. *Journal of Applied Microbiology, 109*(6), 2135–2143. doi:10.1111/j.1365-2672.2010.04845.x PMID:20825519

Baweja, M., Nain, L., Kawarabayasi, Y., & Shukla, P. (2016). Current Technological Improvements in Enzymes toward Their Biotechnological Applications. *Frontiers in Microbiology, 7*. Advance online publication. doi:10.3389/fmicb.2016.00965 PMID:27379087

Bhunia, B., Basak, B., Mandal, T., Bhattacharya, P., & Dey, A. (2013). Effect of pH and temperature on stability and kinetics of novel extracellular serine alkaline protease (70 kDa). *International Journal of Biological Macromolecules, 54*, 1–8. doi:10.1016/j.ijbiomac.2012.11.024 PMID:23219732

Burns, R. G., DeForest, J. L., Marxsen, J., Sinsabaugh, R. L., Stromberger, M. E., Wallenstein, M. D., Weintraub, M. N., & Zoppini, A. (2013). Soil enzymes in a changing environment: Current knowledge and future directions. *Soil Biology & Biochemistry, 58*, 216–234. doi:10.1016/j.soilbio.2012.11.009

Cele, E. N., & Maboeta, M. (2016). Response of soil enzyme activities to synergistic effects of biosolids and plants in iron ore mine soils. *International Journal of Environmental Science and Technology, 13*(9), 2117–2126. doi:10.100713762-016-1043-y

Chakroun, H., Mechichi, T., Martinez, M. J., Dhouib, A., & Sayadi, S. (2010). Purification and characterization of a novel laccase from the ascomycete Trichoderma atroviride: Application on bioremediation of phenolic compounds. *Process Biochemistry, 45*(4), 507–513. doi:10.1016/j.procbio.2009.11.009

Demarche, P., Junghanns, C., Nair, R. R., & Agathos, S. N. (2012). Harnessing the power of enzymes for environmental stewardship. *Biotechnology Advances, 30*(5), 933–953. doi:10.1016/j.biotechadv.2011.05.013 PMID:21624453

Gao, L., & Yan, X. (2016). Nanozymes: An emerging field bridging nanotechnology and biology. *Science China. Life Sciences, 59*(4), 400–402. doi:10.100711427-016-5044-3 PMID:27002958

Gianfreda, L., & Rao, M. A. (2004). Potential of extra cellular enzymes in remediation of polluted soils: A review. *Enzyme and Microbial Technology, 35*(4), 339–354. doi:10.1016/j.enzmictec.2004.05.006

Giardina, P., Faraco, V., Pezzella, C., Piscitelli, A., Vanhulle, S., & Sannia, G. (2010). Laccases: A never-ending story. *Cellular and Molecular Life Sciences, 67*(3), 369–385. doi:10.100700018-009-0169-1 PMID:19844659

Godheja, J., Sk, S., Siddiqui, S. A., & Dr, M. (2016). Xenobiotic Compounds Present in Soil and Water: A Review on Remediation Strategies. *Journal of Environmental & Analytical Toxicology, 6*(5). Advance online publication. doi:10.4172/2161-0525.1000392

Gupta, S. K., & Shukla, P. (2016). Advanced technologies for improved expression of recombinant proteins in bacteria: Perspectives and applications. *Critical Reviews in Biotechnology, 36*(6), 1089–1098. doi:10.3109/07388551.2015.1084264 PMID:26384140

Guzik, U., Hupert-Kocurek, K., & Wojcieszysk, D. (2013). Intradiol Dioxygenases — The Key Enzymes in Xenobiotics Degradation. In R. Chamy (Ed.), *Biodegradation of Hazardous and Special Products*. doi:10.5772/56205

Ju, K.-S., & Parales, R. E. (2006). Control of Substrate Specificity by Active-Site Residues in Nitrobenzene Dioxygenase. *Applied and Environmental Microbiology, 72*(3), 1817–1824. doi:10.1128/AEM.72.3.1817-1824.2006 PMID:16517627

Karigar, C. S., & Rao, S. S. (2011). Role of Microbial Enzymes in the Bioremediation of Pollutants: A Review. *Enzyme Research, 2011*, 1–11. doi:10.4061/2011/805187 PMID:21912739

Kim, J.-S., Park, J.-W., Lee, S.-E., & Kim, J.-E. (2002). Formation of Bound Residues of 8-Hydroxybentazon by Oxidoreductive Catalysts in Soil. *Journal of Agricultural and Food Chemistry, 50*(12), 3507–3511. doi:10.1021/jf011504z PMID:12033819

Lee, A. H., Kang, C.-M., Lee, Y. M., Lee, H., Yun, C.-W., Kim, G.-H., & Kim, J.-J. (2016). Heterologous expression of a new manganese-dependent peroxidase gene from Peniophora incarnata KUC8836 and its ability to remove anthracene in Saccharomyces cerevisiae. *Journal of Bioscience and Bioengineering, 122*(6), 716–721. doi:10.1016/j.jbiosc.2016.06.006 PMID:27353859

Legerská, B., Chmelová, D., & Ondrejovič, M. (2016). Degradation of Synthetic Dyes by Laccases – A Mini-Review. *Nova Biotechnologica et Chimica, 15*(1), 90–106. doi:10.1515/nbec-2016-0010

Ma, X., Zhang, L., Xia, M., Li, S., Zhang, X., & Zhang, Y. (2017). Mimicking the Active Sites of Organophosphorus Hydrolase on the Backbone of Graphene Oxide to Destroy Nerve Agent Simulants. *ACS Applied Materials & Interfaces, 9*(25), 21089–21093. doi:10.1021/acsami.7b07770 PMID:28621926

Mahmood, S., Khalid, A., Arshad, M., Mahmood, T., & Crowley, D. E. (2015). Detoxification of azo dyes by bacterial oxidoreductase enzymes. *Critical Reviews in Biotechnology*, •••, 1–13. doi:10.3109/07388551.2015.1004518 PMID:25665634

Margesin, R., Zimmerbauer, A., & Schinner, F. (1999). Soil lipase activity – a useful indicator of oil biodegradation. *Biotechnology Techniques, 13*(12), 859–863. doi:10.1023/A:1008928308695

Mazotto, A. M., Coelho, R. R. R., Cedrola, S. M. L., de Lima, M. F., Couri, S., Paraguai de Souza, E., & Vermelho, A. B. (2011). Keratinase Production by Three *Bacillus* spp. Using Feather Meal and Whole Feather as Substrate in a Submerged Fermentation. *Enzyme Research, 2011*, 1–7. doi:10.4061/2011/523780 PMID:21822479

Meryam Sardar, R. A. (2015). Enzyme Immobilization: An Overview on Nanoparticles as Immobilization Matrix. *Biochemistry and Analytical Biochemistry, 04*(02). Advance online publication. doi:10.4172/2161-1009.1000178

Mukherjee, P. (2012). Cloning, Sequencing and Expression of Novel Trichloroethylene Degradation Genes from Stenotrophomonas maltophilia PM102: A Case of Gene Duplication. *Journal of Bioremediation & Biodegradation*, *04*(02). Advance online publication. doi:10.4172/2155-6199.1000177

Muthukamalam, S., Sivagangavathi, S., Dhrishya, D., & Sudha Rani, S. (2017). Characterization of dioxygenases and biosurfactants produced by crude oil degrading soil bacteria. *Brazilian Journal of Microbiology*, *48*(4), 637–647. doi:10.1016/j.bjm.2017.02.007 PMID:28629972

Newman, L. A., Doty, S. L., Gery, K. L., Heilman, P. E., Muiznieks, I., Shang, T. Q., Siemieniec, S. T., Strand, S. E., Wang, X., Wilson, A. M., & Gordon, M. P. (1998). Phytoremediation of Organic Contaminants: A Review of Phytoremediation Research at the University of Washington. *Journal of Soil Contamination*, *7*(4), 531–542. doi:10.1080/10588339891334366

Novotný, Č., Svobodová, K., Erbanová, P., Cajthaml, T., Kasinath, A., Lang, E., & Šašek, V. (2004). Ligninolytic fungi in bioremediation: Extracellular enzyme production and degradation rate. *Soil Biology & Biochemistry*, *36*(10), 1545–1551. doi:10.1016/j.soilbio.2004.07.019

Piotrowska-Długosz, A. (2017). The Use of Enzymes in Bioremediation of Soil Xenobiotics. *Xenobiotics in the Soil Environment, 49, 243–265.*, doi:10.1007/978-3-319-47744-2_17

Raghunandan, K., Kumar, A., Kumar, S., Permaul, K., & Singh, S. (2018). Production of gellan gum, an exopolysaccharide, from biodiesel-derived waste glycerol by sphingomonas spp. *Biotech*, *8*(71), 1–10. doi:10.100713205-018-1096-3 PMID:29354382

Rao, M. A., Scelza, R., Scotti, R., & Gianfreda, L. (2010). Role of enzymes in the remediation of polluted environments. *Journal of Soil Science and Plant Nutrition*, *10*(3). Advance online publication. doi:10.4067/S0718-95162010000100008

Roccatano, D. (2015). Structure, dynamics, and function of the monooxygenase P450 BM-3: Insights from computer simulations studies. *Journal of Physics Condensed Matter*, *27*(27), 273102. doi:10.1088/0953-8984/27/27/273102 PMID:26061496

Rodríguez Couto, S., & Toca Herrera, J. L. (2006). Industrial and biotechnological applications of laccases: A review. *Biotechnology Advances*, *24*(5), 500–513. doi:10.1016/j.biotechadv.2006.04.003 PMID:16716556

Sadhu, S., & Maiti, K. (2013). Cellulase production by bacteria: A review. *British Microbiology Research Journal*, *3*(3), 235–258. doi:10.9734/BMRJ/2013/2367

Sahoo, D. K., Das, A., Thatoi, H., Mondal, K. C., & Mohapatra, P. K. (2012). Keratinase production and biodegradation of whole chicken feather keratin by a newly isolated bacterium under submerged fermentation. *Applied Biochemistry and Biotechnology*, *167*(5), 1040–1051. doi:10.100712010-011-9527-1 PMID:22246728

Scott, C., Begley, C., Taylor, M. J., Pandey, G., Momiroski, V., & French, N. … Russell, R. J. (2011). Free-Enzyme Bioremediation of Pesticides: A Case Study for the Enzymatic Remediation of Organophosphorous Insecticide Residues. In K. S. Goh, B. L. Bret, T. L. Potter, & J. Gan (Eds.), Pesticide Mitigation Strategies for Surface Water Quality (Vol. 1075, pp. 155–174). doi:10.1021/bk-2011-1075.ch011

Sharma, B., Dangi, A. K., & Shukla, P. (2018). Contemporary enzyme based technologies for bioremediation: A review. *Journal of Environmental Management*, *210*, 10–22. doi:10.1016/j.jenvman.2017.12.075 PMID:29329004

Sharma, D., Sharma, B., & Shukla, A. (2011). Biotechnological approach of microbial lipase: A review. *Biotechnology (Faisalabad)*, *10*(1), 23–40. doi:10.3923/biotech.2011.23.40

Shin, H. Y., Park, T. J., & Kim, M. I. (2015). Recent Research Trends and Future Prospects in Nanozymes. *Journal of Nanomaterials*, *2015*, 1–11. doi:10.1155/2015/756278

Singh, S., Kang, S. H., Mulchandani, A., & Chen, W. (2008). Bioremediation: Environmental clean-up through pathway engineering. *Current Opinion in Biotechnology*, *19*(5), 437–444. doi:10.1016/j.copbio.2008.07.012 PMID:18760355

Sirisha, V. L., Jain, A., & Jain, A. (2016). Enzyme Immobilization. Advances in Food and Nutrition Research, 79, 179–211. doi:10.1016/bs.afnr.2016.07.004

Sirotkina, M., Lyagin, I., & Efremenko, E. (2012). Hydrolysis of organophosphorus pesticides in soil: New opportunities with ecocompatible immobilized His6-OPH. *International Biodeterioration & Biodegradation*, *68*, 18–23. doi:10.1016/j.ibiod.2011.12.004

Strong, P. J., & Claus, H. (2011). Laccase: A Review of Its Past and Its Future in Bioremediation. *Critical Reviews in Environmental Science and Technology*, *41*(4), 373–434. doi:10.1080/10643380902945706

Torres, E., Bustos-Jaimes, I., & Le Borgne, S. (2003). Potential use of oxidative enzymes for the detoxification of organic pollutants. *Applied Catalysis B: Environmental*, *46*(1), 1–15. doi:10.1016/S0926-3373(03)00228-5

Tuomela, M., & Hatakka, A. (2011). Oxidative Fungal Enzymes for Bioremediation. Comprehensive Biotechnology, 183–196. doi:10.1016/B978-0-08-088504-9.00370-6

Tyagi, M., da Fonseca, M. M. R., & de Carvalho, C. C. C. R. (2011). Bioaugmentation and biostimulation strategies to improve the effectiveness of bioremediation processes. *Biodegradation*, *22*(2), 231–241. doi:10.100710532-010-9394-4 PMID:20680666

Verma, S., & Kuila, A. (2019). Bioremediation of heavy metals by microbial process. *Environmental Technology and Innovation*, *14*, 100369. doi:10.1016/j.eti.2019.100369

Verma, S., Saxena, J., Prasanna, R., Sharma, V., & Nain, L. (2012). Medium optimization for a novel crude-oil degrading lipase from Pseudomonas aeruginosa SL-72 using statistical approaches for bioremediation of crude-oil. *Biocatalysis and Agricultural Biotechnology*, *1*(4), 321–329. doi:10.1016/j.bcab.2012.07.002

Vijay Kumar, E., Srijana, M., Kiran Kumar, K., Harikrishna, N., & Reddy, G. (2011). A novel serine alkaline protease from *Bacillus altitudinis* GVC11 and its application as a dehairing agent. *Bioprocess and Biosystems Engineering*, *34*(4), 403–409. doi:10.100700449-010-0483-x PMID:21080200

Xu, P., Liu, L., Zeng, G., Huang, D., Lai, C., Zhao, M., Huang, C., Li, N., Wei, Z., Wu, H., Zhang, C., Lai, M., & He, Y. (2014). Heavy metal-induced glutathione accumulation and its role in heavy metal detoxification in Phanerochaete chrysosporium. *Applied Microbiology and Biotechnology*, *98*(14), 6409–6418. doi:10.100700253-014-5667-x PMID:24723291

Xu, P., Zeng, G. M., Huang, D. L., Feng, C. L., Hu, S., Zhao, M. H., Lai, C., Wei, Z., Huang, C., Xie, G. X., & Liu, Z. F. (2012). Use of iron oxide nanomaterials in wastewater treatment: A review. *The Science of the Total Environment*, *424*, 1–10. doi:10.1016/j.scitotenv.2012.02.023 PMID:22391097

Yim, S.-K., Jung, H.-C., Pan, J.-G., Kang, H.-S., Ahn, T., & Yun, C.-H. (2006). Functional expression of mammalian NADPH–cytochrome P450 oxidoreductase on the cell surface of Escherichia coli. *Protein Expression and Purification*, *49*(2), 292–298. doi:10.1016/j.pep.2006.05.013 PMID:16829136

Yin, K., Lv, M., Wang, Q., Wu, Y., Liao, C., Zhang, W., & Chen, L. (2016). Simultaneous bioremediation and biodetection of mercury ion through surface display of carboxylesterase E2 from Pseudomonas aeruginosa PA1. *Water Research*, *103*, 383–390. doi:10.1016/j.watres.2016.07.053 PMID:27486950

Zdarta, J., Meyer, A. S., Jesionowski, T., & Pinelo, M. (2018). Developments in support materials for immobilization of oxidoreductases: A comprehensive review. *Advances in Colloid and Interface Science*, *258*, 1–20. doi:10.1016/j.cis.2018.07.004 PMID:30075852

Zhang, H., Zhang, S., He, F., Qin, X., Zhang, X., & Yang, Y. (2016). Characterization of a manganese peroxidase from white-rot fungus Trametes sp.48424 with strong ability of degrading different types of dyes and polycyclic aromatic hydrocarbons. *Journal of Hazardous Materials*, *320*, 265–277. doi:10.1016/j.jhazmat.2016.07.065 PMID:27551986

ENDNOTES

1. A non-governmental organization founded in 1955 to promote molecular biology and biochemistry research in subjects that influence the world though they are at early developmental stages.
2. A category of pigments that have four heterocyclin groups ringed to form a flat ring and have a metal atom, which is centrally located.
3. This is a cofactor used in anabolic reactions as a reductant.
4. Dichlorodiphenyltrichloroethane is an insecticide whose use has negative environmental impacts.
5. Deoxyribonucleic acid is a molecule, which contains genetic material of an organism that controls its metabolic activities.
6. Polymerase chain reaction is a molecular biology technique used to make copies of some DNA segments.

Chapter 5
Role of Bacterial Chromate Reductase in Bioremediation of Chromium-Containing Wastes

Satarupa Dey
https://orcid.org/0000-0002-0681-8992
Shyampur Siddheswari Mahavidyalaya, India

ABSTRACT

Chromium toxicity is a major environmental concern as it is the chief environmental pollutant released by paint, stainless steel, and mining industries. In nature, chromium exists in two stable valance states: Cr(VI) and Cr(III). Cr(VI) is highly toxic and soluble at neutral pH, whereas Cr(III) is insoluble at normal pH and is less toxic. Thus, it is essential to draw strategies for mitigation of Cr(VI), and microbial reduction of toxic Cr(VI) has been identified as a bioremediation technique not only to detoxify chromium but also to recover the non-toxic Cr(III) by physical means. Chromate reductase, the central enzyme involved in bioreduction of Cr(VI) to Cr(III) may be both intracellular as well as extracellular in nature. Most of the chromate reductase enzyme belongs to the oxidoreductase group such as nitroreductase, iron reductase, quinone reductase, hydrogenase, flavin reductase, as well as NAD(P)H-dependent reductase. Detailed analysis of the structure of the enzymes will help us in the suitable application of these enzymes in bioremediation of metal-contaminated wastes.

INTRODUCTION

Contamination of heavy metal is considered as a serious environmental problem causing serious health hazards all over the world. They are released mainly due to anthropogenic activities, mineral processing and mining activities and their release has increased enormously in the past few decades. Their mitigation has become very necessary as well as a challenging task for mankind and it has received a lot of attention. Consequently more strict legislation for the protection of the environment has gradually become indispensible to reduce the release of heavy metal containing waste in the water bodies.

DOI: 10.4018/978-1-7998-4888-2.ch005

Chromium is one of the frequently used metal and is released from different industrial sources such as electroplating, stainless steel, dye and leather industries. It is emerging as one of the top 20 contaminants and is considered as a hazardous substance for the past 15 years (United States Environmental Protection Agency, 1998; Ryan et al. 2002). In India a huge amount of elemental Cr are released annually through waste water by different industries and concentration of Cr in these water ranges between 2 to 5 g/L much higher than the maximum permissible limit (2 mg/L).

Chromium mainly exists in nature in different valence states ranging from -4 to +6, with the hexavalent species and trivalent species considered as the most stable state. Cr(VI) is predominant in natural aquifers and mobile, can penetrate the biological membrane and is considered as mutagenic, carcinogenic and teratogenic in nature. Hexavalent chromium mostly exists as chromate ions (CrO^4) which has structural similarities with sulphate ions (SO^4) and therefore can enter the cell via the sulphate transport system present in the cell membrane. Once inside the cell chromate ion being an oxyanion readily forms DNA adducts and causes mutations. On the other hand trivalent counterpart Cr(III) prevails in the municipal wastewater rich in organic substances (Fukai, 1967; Jan and Young, 1978), it is thermostable in nature, less mobile and is therefore much less harmful.

Several chemical treatments such as chemical reduction followed by precipitation, and adsorption and processes such as reverse osmosis, electrodialysis and ion exchange processes are used to remove and reduce hexavalent chromium and precipitate them. However, they are expensive with high energy and chemical consumption, which also generates a huge amount of toxic sludge and secondary wastes. Moreover, most of these treatments are applicable if the concentration of chromium ranges between 1 to 100 mg/L. So the use of bioremediation technology is nowadays considered as a more viable option in removal of Cr(VI) from waste waters and contaminated lands. Several bacteria exists which in both living as well as non living condition have the capacity to reduce, accumulate and absorb Cr(VI) ions. Several bacterial strains isolated from different environment are reported to be capable of reducing toxic and mutagenic Cr(VI) to less toxic Cr(III) as well as precipitate it at neutral pH. The enzyme playing a central role in the reduction process is the chromate reductase which is either NADH or NADPH dependent. Mostly these enzymes belong to the hydrolases, dehalogenases, transferases and oxidoreductases class and can either be intracellular or extracellular in nature. Most of the chromium reductases reported till now works in aerobic conditions and is normally associated with soluble protein fractions and the reduction process is carried out either internal or external to the plasma membrane. Whereas, those chromate reductase which works under anaerobic condition, uses Cr(VI) as terminal electron acceptor and is membrane bound.

The present study mainly focuses on the production, mode of action and application of these enzymes under both free and immobilized conditions for bioremediation of hexavalent chromium.

CHROMIUM TOXICITY

As described in the previous section chromium is released from different sources such as industrial effluent and mining effluent which contaminates the soils and water bodies. Chromium exists in nature in different valance form, however, hexavalent and trivalent states are found to be most stable in nature. Hexavalent chromium can also be generated from trivalent chromium due to environmental oxidation. Cr(VI) is highly toxic and can penetrate the cell membrane through the sulphate transporter pathway. When within the cell they are reduced to the trivalent species in presence of ascorbate and glutathione and in this process they generates intermediates like Cr(V) and Cr(IV) (Kanmani et al. 2012). Cr(V) is once again oxidized to Cr(VI) within the cytoplasm generating reactive oxygen species (ROS) which reacts with proteins and nucleic acid causing mutation. On the other hand, Cr(IV) is reported to bind with cellular materials and enzymes which alters their normal physiological functions (Cervantes et al. 2001; Pesti et al. 2000). Cr(III) is however much less toxic, less bioavailable in nature as it cannot penetrate cell membrane and forms insoluble hydroxide or oxide at neutral pH.

Prokaryote cells have been reported to be more resistant to Cr(VI) than eukaryotes (Kamaludeen et al. 2003). Both Cr(VI) and Cr(III) have found to have both positive and negative effects on human being (Bielicka et al. 2005). Cr(VI) has been found to cause lung carcinoma (Browning and Wise, 2017) by causing structural changes in chromosomes and inhalation of chromium was found to be responsible for nasal perforation and respiratory tract diseases. Kornhauser et al. (2002) reported excessive accumulation of Cr in the body may adversely effects of iron metabolism in humans. Cr(VI) also changes the subcellular localization of protein RAD 51 and interfere the process of homologous recombination (HR) DNA repair. Hexavalent chromium known to have adverse effect on plants as it decreases the nutrient uptake, photosynthesis and alters several biochemical processes leading to decrease in growth rate of plants as well as causing chlorosis and necrosis in plants (Shahid et al. 2017).

CHROMIUM RESISTANCE MECHANISM

Resistance was appropriately defined by Gadd et al. 1992, as "the ability of a microorganism to survive toxic effects of metal exposure by means of a detoxification mechanism produced in direct response to the metal species concerned". Bacterial species which are able to withstand high concentration of Cr(VI) have developed several strategies such as, ion transport, reduction of the metal to a less toxic one, DNA repair and metal efflux to evade the toxic effect of metal.

Cr(VI) Reduction

Reduction of toxic metal to its non toxic counterpart is considered as an important alternative strategy for chromium detoxification, however, in most cases such mechanism is not plasmid mediated (Cervantes et al. 2001). Several Cr-resistant bacteria with high Cr(VI)-reducing potential have been reported including *Achromobacter* (Ma et al. 2007), *Arthrobacter* (Dey and Paul, 2013), *Bacillus* (Das et al. 2014; He et al. 2010), *Desulfovibrio* (Goulhen et al. 2006), *Enterobacter* (Wang et al. 1989), *Escherichia* (Shen and Wang, 1994), *Pseudomonas*, *Shewanella* (Myer et al. 2001), *Thermus* (Opperman et al. 2008) and other species. The process of reduction in these isolates can be both enzymatic as well as non-enzymatic in nature. Non enzymatic reduction process is mediated by various chemical compound such as Fe(II) and HS-, which are the end products of iron and sulphate-reducing bacteria. Apart from them ascorbic acid, glutathione (GSH), cysteine, hydrogen peroxide (H_2O_2) which are by product of microbial cells (Poljsak et al., 2010) and are known to reduce Cr(VI) to Cr(III). Apart from these, intracellular components like amino acids, nucleotides, sugars, vitamins, organic acids or glutathione are also capable of reducing Cr(VI). However in most of the cases no relationship exist between Cr(VI) tolerance and Cr(VI) reduction ability and they are found to be independent properties in a bacteria (Bopp and Ehrlich, 1988; Silver, 1997).

Enzymatic reduction of Cr(VI) by bacteria is mediated by several types of chromate reductase enzymes and most of these enzymes belongs to the hydrolases, dehalogenases, transferases and oxidoreductases group. Enzymatic reduction can occur in both aerobic as well as anaerobic processes. In aerobic process bacteria requires oxygen as electron acceptor and use carbon as a substrate. In anaerobic reduction oxygen is replaced by sulphate, nitrate, carbon dioxide, oxidized materials, or organic compounds and includes several processes such as methanogenesis, sulphate and nitrate reduction. Enzymatic reduction of Cr(VI) can also be intracellular in nature or extracellular in nature. In extracellular enzymatic reduction the reducing enzymes are exported in the medium to reduce Cr(VI). Apart from enzymatic reduction there are several other processes which renders Cr(VI) resistance to the cell. They are the following:

Transmembrane Efflux of Chromate

Plasmid is mainly responsible for encoding membrane transporters and main protein behind this resistance mechanism is ChrA, a hydrophobic protein which is coded by plasmids pUM 505 of *Psuedomonas aeruginosa* and pMOL 28 from *Cupriavidus metallidurans* (Cerventes et al. 1990 and Nies et al. 1990). ChrA protein functions as a chemiosmotic pump and plays a vital role in efflux of chromate from the cytoplasm using the proton motive force which is driven by the membrane potential (Pimentel et al. 2002) and prevents the accumulation of Cr(VI) inside the cell. However efflux of chromate is inhibited by sulphate. Efflux of chromate is also seen in isolates *Ochrobactrum tritici* 5vil (Branco et al. 2004) and *Shewanella* sp. ANA-3 (Anguilar-Barajas et al. 2008). There is also a CHR super family of transporters classified as TC# 2.A.51 (Saier, 2003) which is also involved in chromate or sulphate transport.

Protection Against Oxidative Stress

Different bacterial proteins participate as a defense mechanism against oxidative stress induced by chromate and act as an additional mechanism for chromate resistance. This can be encoded by both chromosomal genes as well as plasmid.

Ackerley et al. (2006), have pointed out that *E. coli* protects itself by the activation of enzymes such as superoxide dismutase (SOD) and catalase. Also it was observed that in *E. coli* presence of chromate leads to depletion of pools of glutathione and other thiols, suggesting that these compounds play vital role of detoxification of Cr(VI). *Caulobacter crescentus* (Hu et al. 2005) is also known to produce SOD, glutathione S-transferase, thioredoxin and gluteredoxin to counteract oxidative stress caused by Cr(VI). Similar responses were also evident in *Shewanella oneidensis* MR-1 (Chourey et al. 2006).

Plasmids are also known to encode system devoted to protect bacterial cells from oxidative stress. Plasmid pMOL 28 from *Cupriavidus metallidurans* encoding ChrA protein also codes for ChrC and ChrE protein which are involved in chromate resistance (Juhnke et al. 2002). ChrC is responsible for detoxification of super oxide radicals whereas ChrE causes cleavage of chromium-glutathione complexes.

DNA Repair

Another method of Cr resistance occurs via protection of bacterial cells from DNA damage, which is mainly by the activation of chromosomal genes. In *E. coli* activation of SOS repair system (Llagostera et al. 1986) occurs which protects the DNA from oxidative damage caused due to chromium exposure. The component of recombinational DNA repair system such as DNA helicases, Rec G and Rur B participate to repair the damage in DNA structure caused by Cr(VI) in *Pseudomonas aeruginosa* (Miranda et al. 2005). *Caulobacter crescentus* (Hu et al. 2005) also showed activation of endonuclease and RecA proteins in response to Cr(VI) exposure.

Sulfate Uptake Pathway and Sulfur and Iron Homeostasis

Additional protective mechanism includes reduced uptake of Cr(VI) by sulfate uptake pathway and with sulfur and iron homeostasis. In *Cupriavidus crescentus* sulfate transport system probably reduces chromate uptake (Hu et al. 2005).

Plasmid pANL of *Synechococcus* sp. encodes SrpC and SrpA protein, which increases sulfate uptake and participate in detoxification of hydrogen peroxide which in turn reduces Cr(VI) mediated oxidative damage. *Shewanella oneidensis* shows enhanced expression of genes encoding proteins involved in iron binding (ferritin) and transport (siderophore production) (Brown et al. 2006) when exposed to chromium.

Cr(VI) REDUCTION MECHANISM

Microbial reduction of Cr(VI) to Cr(III) is an additional chromate resistance mechanism which is mainly chromosome encoded trait. Three different types of Cr(VI) reduction mechanisms have been enumerated (Cervantes and Campos-Garcia, 2007).

i. In aerobic condition, where reduction is mainly associated with soluble chromate reductases in the cytosol and it is aided by the presence of NADH/NADPH as electron donor.
ii. In anaerobic condition, where chromate reductases are mainly membrane bound and Cr(VI) serves as electron acceptor in electron transport chain.
iii. Indirect reduction of chromate by microbes generally involved a biotic-abiotic coupling system and is mediated by metabolic by products. Fe(III) or S^{-2} produced by a variety of micro organisms through dissimilatory reduction pathways, catalyze the reduction of chromate.

Cr(VI) Reducing Aerobes

Since the discovery of the first microbe capable of reducing Cr(VI) in 1970,s (Romnenko and Karenkov, 1977), the search for Cr(VI) reducing micro organism is pursued and a huge number of strains has been isolated. Different Cr(VI) reducing microbe including halophilic, alkaliphilic and psychrophilic organisms have been isolated from diverse areas and capable of tolerating extreme conditions (Horton et al. 2006, Amoozegar et al. 2007, Wani et al. 2007 and Ibrahim et al. 2011). This include members like, *Bacillus cereus*, *Planococcus mcmeekinni*, *Micrococcus luteus* (Camargo et al. 2005), *Burkholderia cepacia* (Wani et al. 2007), *Orchrobactrum* sp CSCr-3 (He et al. 2009), *Staphylococcus* sp (Mistry et al. 2010), *Stenotrophomonas maltophilia* ZA-6 (Alam and Ahmed, 2011), *Corynebacterium paurometabolum* (Divyasree et al. 2016), and *Cellulosimicrobium* sp. KX710177 (Bharagava and Mishra, 2018). Most of these Cr(VI) reducing aerobes have been found to reduce Cr(VI) utilizing reductases soluble in cytosol, however, in *Pseudomonas maltophilia* O-2 and *Bacillus megaterium* TKW3 Cr(VI) reduction was mainly associated with membrane cell fraction (Blakell et al. 1993). A list of aerobic chromate reducing bacteria and their reduction efficiency has been presented in Table 1.

Physiology and Biochemistry of Aerobic Cr-reduction

In most cases, aerobic Cr(VI) reduction was observed during growth of the microbes but in some cases it was found that multiplication of cell is not necessary for the process of reduction. Resting cells of different micro-organisms such as *Ochrobactrum* (Sultan and Hasnain, 2007), *Acinetobacter heamolyticus* (Zakaria et al. 2007) and *Pannonibacter phragmitetus* (Chai et al. 2009) were capable of reducing chromate, but in these case high cell densities were required for significant Cr(VI) reduction.

The rate of Cr(VI) reduction increased with increase in initial Cr(VI) concentration in several occasion (Sultan and Hasnain, 2007), although the reverse was true for *Brucella* sp. (Thacker et al. 2007), *Burkholderia cepacia* MCMB-821 (Wani et al. 2007) and *Acinetobacter heamolyticus* (Zakaria et al. 2007). However, in certain cases the chromate reduction rate remains unaffected by initial Cr(VI) concentration (Amoozgar et al. 2007 and Zhu et al. 2008).

Chromate reducing bacteria have been found to utilize a variety of organic compounds, low molecular weight carbohydrates, amino acids and fatty acids as source of electron donor. The rate of chromate reduction in most case is found to be influenced by sulfate concentration (Ahmed et al. 2009). Presence of different metal ions was found to be inhibitory of which mercury is found to inhibit the process noncompetitively. In some cases metal like Cu(II) and Fe(III) was found to be promotive in nature (Camargo et al. 2003).

Metabolic inhibitors like carbonyl cyanide-m-chloro phenyle hydrazone, 2, 4 di nitrophenol, sodium cyanide and sodium flouride were inhibitory in nature, which suggests the involvement of multicomponent of electron transport chain in the process of reduction. However in *Burkholderia cepacia* MCMB-821 (Wani et al. 2007) and *Stenotrophomonas maltophilia* ZA-6 (Alam and Ahmed, 2011), DNP is known to accelerate the respiratory chain linked electron transport mechanism, similar to the stimulation of aerobic respiration where electron flow towards the terminal electron accepter, such as molecular oxygen is enhanced. The enhanced chromate reduction in presence of DNP also suggest that Cr(VI) may act as a terminal electron acceptor in bacteria.

Cr(VI) Reducing Anaerobes

Cr(VI) reduction in anaerobic condition was first reported by Romaneko and Korenkov (1975) using *Pseudomonas dechromaticans*, which was a facultative bacterium and used Cr(VI) as a terminal electron acceptor during the process. In anaerobic Cr(VI) reduction membrane bound reductases belonging to the group flavin reductases, cytochromes and hydrogenases plays a important part and chromate is used as the terminal electron acceptor and their activity is inhibited by the presence of Oxygen. Later Wang et al. 1989, 1991 reported anaerobic Cr(VI) reduction by facultative bacterium, *Enterobacter cloacae*, isolated from industrial wastewater which uses chromate as terminal electron acceptor by membrane bound hydrogenase or by reduced cytochrome. Similar reduction was also noticed with *Shewanella putrefaciens* MR-1 (Myers et al. 2000) where Cr(VI) reduction was carried out anarobically using formate and NADH as electron donor. Here the chromate reductase was associated with the cytoplasmic membrane.

Apart from these several other Cr(VI) reducing bacteria such as *Microbacterium* sp. MP 30 (Pattanapipitpaisal et al. 2001) and a corsortium capable of reducing Cr(VI) and degrading benzoate (Shen et al. 1996) have been reported. Several extremophiles have also been found capable of reducing Cr(VI) in anaerobic condition, such as radiation-resistant *Deinococcus radiodurans* R1 (Fredrickson et al. 2000),

Pyrobaculum islandicum (Kashefi and Loveley, 2000) capable of Cr(VI) reduction at high temperature and *Thermoanaerobacter ethanolicus* isolated from deep subsurface sediments (Roh et al. 2002).

Physiology and Biochemistry of Anaerobic Cr-reduction

Chromate reduction by isolates like *Enterobacter cloacae* HO1 was increased when ascorbate PMS was supplied as electron donor, whereas with isolates like *Desulfovibrio vulgaris* and *Achromobacter* sp. lactate was found to be most efficient. Hydrogen also served as the electron donor in *Desulfovibrio vulgaris* (Lovley and Philips, 1994). The suitable pH ranged from 6.5 to 8.5 in case of *Enterobacter cloacae* HO1, whereas it was 9.0 for *Achromobacter* sp. Reduction rate in case of *Achromobacter* sp. was found to be independent of cell density and initial Cr(VI) concentration. The rate of Cr(VI) reduction decreased with increasing initial Cr(VI) concentration in *Enterobacter cloacae* (Komori et al. 1989). Metabolic inhibitors like carbonyl cyanide-m-chloro phenyle hydrazone, 2, 4 di nitrophenol, sodium cyanide and formaldehyde were also inhibitory to Cr(VI) reduction.

However, there also exists several microbes which are able to reduce Cr(VI) in both aerobic as well as anaerobic conditions but in such strains reduction of Cr(VI) was better under anaerobic condition than in aerobic phase. Facultative anaerobes from tannery effluents showed > 90% Cr(VI) reduction under anaerobic condition, while only 10-50% reduction was achieved under aerobic condition (Srinath et al., 2001). Chromate reduction was repressed by dissolved oxygen in *E. coli* ATCC 33456, where an apparent uncompetitive inhibition of oxygen was noted (Shen and Wang, 1994).

Indirect or Non-Enzymatic Cr(VI) Reduction

Indirect or non enzymatic reduction of chromate by microbes is mediated via biotic abiotic coupling system involving by-products of anaerobic metabolism. Under anaerobic environment sulfate and iron reducing bacteria carry out such process. Fe(II) or S^{-2} produced by these organisms through dissimilatory reduction pathways helps in the reduction of chromate. In *Shewanella alga* ATCC 51181, dissimilatory Fe (III) reduction provided a primary pathway for Cr(VI) reduction by microbial induced ferrous ions (Wielinga *et al.*, 2001). H_2S produced by sulfate reducing bacteria have been also seen to reduce Cr(VI). Chemoautotrophs such as *Thiobacillus ferroxidans* produced sulfate and thiosulfate which catalyzes the reduction of Cr(VI).

ENZYME MEDIATED HEXAVALENT CHROMIUM REDUCTION

Chromate reductase enzyme may be associated with the cell membrane or with the soluble fraction, which usually occur under aerobic or anaerobic condition respectively. The first Cr(VI) reductase was described from the chromate resistant *Enterobacter cloacae* HO1 (Ohtake *et al.*, 1990), which was a membrane associated enzyme that transfers electron to Cr(VI) by NADH-dependent cytochrome (Wang *et al.*, 1991). Researchers have reported chromate reductase from different groups of bacteria and it can be classified broadly into 3 groups: Chromate reductase from aerobes, chromate reductase from anaerobes which were mainly intracellular in origin and extracellular chromate reductase enzyme. A metadata analysis of the available publications from public database (www.ncbi.nlm.nih.gov/pubmed) with relevant keywords revealed that a total of 358000 publications are available on chromate reductase enzyme present in microbes till 14.05.2020. Out of which 58% and 32% of the publication belongs to aerobic and anaerobic chromate reductase respectively (Figure 1). Chromate reduction by extracellular enzyme is however least covered area sharing only 10% of the total publication.

Chromate Reductase of Aerobes

Chromate reductase from an aerobic bacterium was initially isolated in a partially purified form from *Pseudomonas putida* PRS2000 (Ishibashi *et al.,* 1990). Since then, several bacterial Cr(VI) reductases have been characterized. Ishibashi *et al.,* (1990) also suggested that the ability to reduce chromate may be a secondary function for Cr(VI) reductases, which have a different primary role other than Cr(VI) reduction. The nitro reductases NfsA/NfsB from *Vibrio harvaji* possesses a nitrofurazone nitro reductase as primary activity and have developed Cr(VI) reductase activity as secondary function related to bacterial enzymatic adaptation.

ChrR from *Pseudomonas putida* is currently the best studied Cr(VI) reductase. It exhibit a high nucleotide sequence homology of nearly 58% to a nitro reductases of *Vibrio harvaji* KCTC2720, also having Cr(VI) reducing activity. ChrR is a soluble flavin mononucleotide-binding protein (Park *et al.,* 2000) showing a NADH-dependent reductase activity, is also found to reduce ferricyanide (Ackerley *et al.,* 2004) and quinines but have no activity with nitro compounds. This enzyme was classified as class I enzyme. ChrR is found to protect the cell from chromium toxicity by minimizing ROS generation (Ackerley *et al.,* 2004). During reduction, ChrR exhibits a quinone reductase activity that produces a flavin semiquinone. The enzyme through this reaction transfers more than 25% of the NADH electrons to superoxide anions and produced Cr(VI) species transiently this provide a shielding against ROS.

YieF Cr(VI) reductase from *E. coli* shares sequence homology with *Pseudomonas putida* is also a class I enzyme. YieF has a broad substrate range and can also reduce substrates like ferricyanide, vanadium(V), molybdenum(VI), quinines, 2.6, dichloro indophenol (Ackerley *et al.,* 2004). The mechanism of the action of this enzyme involves an obligatory 4-electron reduction of Cr(VI), in which 3 electron from Cr(VI) is simultaneously transferred to generate Cr(III) and one electron to molecular oxygen generating ROS.

Apart from this chromate reductase, a second group of reductase is also designated as class II enzymes, such as NfsA of *E. coli*. This enzyme have been previously been well characterized in nitro reduction context (Whiteway *et al.,* 1998). Despite of broad distribution and high degree of conservation of these enzymes, the biological role of even best studied NfsA of *E. coli* remains unclear as it is induced by chromate. NfsA is found to produce greater amount of ROS than class I enzyme.

The South African gold mines isolate, *Thermus scotocductus* SA-01 is capable of reducing variety of heavy metals. Its chromate reductase was purified, characterized and sequenced and was found to be related to OYE homodimer enzyme with a monomer molecule mass of nearly 36 KDa containing non covalent FMN co-factor and Ca(II) ions for reductase activity (Opperman *et al.,* 2008).

The source and some features of chromate reductases from different aerobic chromate reducing bacteria have been compiled and presented in Table 2.

Chromate Reductase of Anaerobes

Both soluble and membrane associated enzymes are found mediating Cr(VI) reduction under anaerobic conditions. The Cr(VI) reducing activities of anaerobes were associated with electron transfer system initiating the electron shuttle along the respiratory chains (Wang and Shen, 1995).

Under anaerobic conditions, Cr(VI) serves as a terminal electron acceptor through the respiratory chains of *Enterobacter cloacae* (Wang *et al.,* 1991), *E. coli* (Shen and Wang, 1993) and *D. vulgaris* (Lovley and Phillips, 1994). Recent studies have also implicated the involvement of membrane-bound respiratory chain in the transfer of reducing equivalents to Cr(VI) through cytochrome c in *Enterobacter cloacae* and cytochromes b and d in *E. coli*. In the absence of oxygen, the soluble reductase activity mediate electron transport to Cr(VI) in *E. coli* and *D. vulgaris* in which the cytochrome c_3 in the soluble protein fraction of *D. vulgaris* was needed for Cr(VI) reduction. Terminal electron acceptor chromium has been replaced by sulphate and nitrate in *Shewanella*, *Enterobacter* and other sulphate reducing bacteria. (Chardin *et al.,*2003; Myers *et al.,* 2000).

Cr(VI) reduction under anaerobic conditions by cell free extracts of different bacteria are highlighted in Table 3.

Extracellular Chromate Reduction

In some cases the bacterial chromate reductase produced within the cell is exported into the medium to reduce Cr(VI). The extracellular chromate reductases are advantageous for the cells as they protect the cells from toxic Cr(VI), prevent the entry of insoluble Cr(III) into the cells and damages to DNA. These enzymes may be of different types such as flavin reductases, nitrate reductases, flavin protein and ferrireductases which are mostly of soluble type (Cheung and Gu, 2007). Priester et al., 2006 first reported the presence of extracellular Cr(VI) reduction in *P. putida*, where the enzymes where mainly produced intracellularly and released outside the cell to remove Cr(VI). Membrane bound Cr(VI) reductase enzymes have also been reported to reduce Cr(VI) extracellularly by Wang et al., 1991. Similar extracellular enzymes was also reported later by Rath et al., 2019, Dey and Paul, 2016, and Mala et al., 2015 in *Bacillus amyloliquefaciens* (CSB 9), *Arthrobacter* sp. SUK 1205, and *Bacillus methylotrophicus* respectively. In most of the cases these enzymes are reported to be inducible in nature and are controlled by regulatory elements which promote the production of enzyme under unfavorable condition.

MECHANISM OF ACTION OF CHROMATE REDUCTASES

Chromate reductases are of different types depending on the basis of the oxidative stress generated by these enzymes. They have two distinct types of mechanism depending on the number of electron it reduces, it can reduce either one or two electrons during the process of Cr(VI) reduction.

One Electron Reducers

In this process Cr(VI) is reduced to form highly unstable Cr(V) which can again reoxidise to Cr(VI) giving its electrons to molecular oxygen and thus resulting in the generation of more ROS. One electron reducers are mostly flavin dependent enzymes where the flavin nucleotide is tightly and covalently bound. The oxidized flavin nucleotide accepts one electron and generates a stable free radical semiquinone. Chromate reductases like glutathione reductase, cytochrome C and ferridoxin –NAD are reported to be one electron reducers (Ackerley et al., 2004b).

Two Electron Reducers

Two electron reducers are characterized by reduction of Cr(VI) without formation of the intermediate Cr(V). As a result this process is characterized by production of fewer amounts of ROS as compared to the previous type. The enzymes belonging to this class mostly belong to NAD(P) dependent enzymes and the most common types of two electron reducers include ChrR from *P. putida*, YieF and NfsA from *E. coli* (Barak et al., 2006).

Further the enzymes are classified into two types i.e. Class I and Class II depending on the protein sequence homologies (Ackerley et al., 2004b).

Class I Chromate Reductase

All of the Class I chromate reductase enzymes so far described are known to be efficient chromate and quinine reducers, but they have no activity against any nitro compounds. The most common type I enzymes are ChrR and YieF amongst which YieF is a dimeric protein which reduces Cr(VI) to Cr(III) without redox cycling and generates minimum amount of ROS. On the other hand ChrR generates more ROS as they reduce Cr(VI) by a combination of one and two electron reduction steps. Both ChrR and YieF are obtained from *E. coli* (Ackerley et al., 2004) belongs to the NADH-dh[2] family of protein, which is similar to bacterial and eukaryotic NAD(P)H oxidoreductases. Later on He et al., 2010 isolated ChrA protein from *Bacillus cereus* and *Bacillus thuringiensis* which has a capacity to reduce Cr(VI) to Cr(III) and is also a class 1 family enzyme. Similarly, NemA isolated from *E. coli* (Robins et al. 2013) is also a class 1 enzyme belonging to the family of flavoproteins which can catalyse Cr(VI) reduction through addition of one or two electrons from NADH or NADPH.

Class II Chromate Reductase

These are all chromate as well as nitroreductases and bears no homology in protein structure with class I chromate reductase. However, they have a structural homology with the chromate reductase found in *Pseudomonas ambigua* (Park et al., 2002). Other class II chromate reductase includes NfsA (Ackerley et al., 2004), NfsB (Kwak et al., 2003) and ChfN (Park et al., 2002) protein isolated from *E. coli*, *Vibrio harveyi* and *B. subtillis* respectively. NfsA is a 50 kDa protein and semi tight chromate reducer which can reduce nitrocompounds and quinone by two electron transfer method (Ackerley et al., 2004). This protein has been reported to possess therapeutic property and can activate prodrugs for cancer chemotherapy (Carroll et al., 2002).

STRUCTURAL ANALYSIS OF SOME CHROMATE REDUCTASES

A metadata analysis of the available publications from public database (www.ncbi.nlm.nih.gov/pubmed) with relevant keywords revealed that of the total publications made on structural analysis of chromate reductase, about 62%, 33% and 5% of the publication are on C type cytochromes, oxidoreductase and nitroreductase respectively (Figure 2).

C Type Cytochrome

Periplasmic C type cytochrome is used for Cr(VI) reduction by different sulfate-reducing bacteria (SRB). They cannot grow using chromate as terminal electron donor. A wide range of cytochrome enzymes have been reported by many researchers including Cytochrome c3 from *Desulfovibrio vulgaris* (Lovley and Phillips, 1994) and cytochrome c7 from *Desulfuromonas acetoxidans* (Michel et al.,2001) which are capable of reducing chromate. Similarly c type cytochrome ApcA of *Acidiphilium cryptum* JF-5 is periplasmic in nature and is induced by the presence of chromium which triggers Cr(VI) reducing activity by reduced ApcA (Magnuson et al., 2010). On the other hand, extracellular c type cytochrome such as MtrC and OmcA are reported from *Shewanella oneidensis* MR-1(Reardon et al., 2010) which shows high chromate reducing ability (Belchik et al., 2011). The c type cytochrome reduces chromate mainly by using the redox potencial of heme proteins.

Oxidoreductases

Oxidoreductases are a group of enzymes consisting of either oxidases or dehydrogenases that catalyze oxidoreduction reactions. Oxidases are involved when molecular oxygen is an acceptor of hydrogen or electrons, on the other hand, dehydrogenases oxidizes substrate by transferring hydrogen to either NAD+/NADP+ or flavin enzymes. Other oxidoreductases include peroxidases, hydroxylases, oxygenases, and reductases which vary in their location and mechanism of action. Peroxidases are are oxido reductase which are localized in peroxisomes and catalyze the reduction of hydrogen peroxide. Similarly the mode of action of hydroxylases and Oxygenases is by addition hydroxyl and oxygen groups to its substrates respectively. Different oxidoreductases with different metabolic functions have been isolated which catalyze Cr(VI) reduction in bacteria. They include chromate reductase, nitroreductase (Kwak et al., 2003), iron reductase, quinone reductases (Gonzalez et al., 2005), hydrogenases (Chardin et al, 2003), flavin reductases (Ackerley et al., 2004a) as well as NAD(P)H dependant reductases (Puzon et al., 2002).

Nitroreductases

Nitroreductase (Nfs) from *Shewanella oneidensis* MR-1 is capable of reducing chromate under both aerobic and anaerobic conditions are induced by nitrate. However presence of chromate cannot induce its activity (Brown et al., 2006). NfsA obtained from other sources such as E. coli showed Vmax as 250 nmol /min/mg protein, Km of 36 mM (Ackerley et al., 2004), whereas nitroreductase purified from *Vibrio harveyi* reduces chromate showed Kmax as 5.4 μM, Vmax at 10.7 nmol/min/mg protein (Kwak et al., 2003). The mechanism of action of NfsA is mediated by two-step reactions and during reduction process generates more ROS (Ackerley et al., 2004). The main electron donor is NAD(P)H and subsequently reduces chromate similar to ChrR (Kwak et al., 2003).

APPLICATION OF CHROMATE REDUCING ENZYME IN Cr(VI) REMOVAL

An alternative strategy for ex situ removal of chromium from waste water is use of immobilized chromate reductase enzymes. This system has its own advantages as well as disadvantages. The main disadvantages includes stability problem and they also do not self replicate and multiply like bacterial system. Another

Table 1. List of aerobic chromate reducing bacteria and their Cr(VI) reduction efficiency

Bacterial isolates	Source	Cr tolerance	Cr(VI) reduction efficiency	Reference
Acinetobacter haemolyticus	Cr(VI) contaminated water system	70 mg/l	50 mg/l Cr(VI) in 50 h	Zakaria *et al.*, 2007
Acinetobacter sp. S3	Effluents of tanneries	160 mg/l	83.1% of 80 mg/l Cr (VI) in 72 h	Farag and Zaki, 2010
Acinetobacter guillouiae SFC 500-1A	Tannery sludge		62% of 10mg/l in 72 h	Ontañon *et al.*, 2015
Acinetobacter baumannii L2	Crude oil samples	1000 mg/l	99.58% of 1000 mg/l in 24 h	Sathish kumar *et al.*, 2016
Aeromonas sp.	Agricultural soil irrigated with tannery water	4.7 mM	125 µM Cr(VI) in 56 h	Alam and Ahmed, 2011
Amphibacillus sp. KSUCr3	Hypersaline soda lakes	75 mM	5 mM of Cr(VI) in 24 h.	Ibrahim *et al.*, 2011
Arthobacter sp.	Tannery waste contaminated soil		20 mg/l in 46 h	Megharaj *et al.*, 2003
Arthobacter sp. SUK1201	Chromite mine overburden	13 mM	67% of 2 mM in 7days	Dey and Paul, 2012a
Arthobacter sp. SUK1205	Chromite mine overburden	11.8 mM	100 µM in 48 h	Dey and Paul, 2012b
Arthrobacter sp. LLW01	Contaminated soil		50%of 50 µM in 144h	Field *et al.*, 2018
Bacillus sp.	Tannery waste contaminated soil		10 mg/l in 72 h	Megharaj *et al.*, 2003
B. sphaericus	Contaminated soil		62% of 20 mg/l in 48 h	Pal and Paul, 2004
Bacillus sp.	Chromate contaminated soil		95% of 10 mg/l in 24 h	Elangovan *et al.*, 2006
Bacillus sp.ev3	Industrial areas	4800 µg/ml	91% of 100 µg/ml Cr(VI) in 96 h	Rehman *et al.*, 2008
Bacillus cereus, *B. fusiformis* and *B. sphaericus*	Cr(VI) polluted landfill	NR	85% of 1000 µM Cr(VI) in 30 h	Desai *et al.*, 2008
Bacillus sp.	Heavy metal contaminated soil	100 ppm	87.8% of 25 ppm Cr(VI) in 36 h	Parameswari *et al.*, 2009
Bacillus sp.	Soil of iron mineral area	2500 mg/l	10 mg/l of Cr (VI) in 24 h	Cheng and Li, 2009
Bacillus sp. CSB-4	Chromite mine soil	2000 mg/l	90% of 100 mg/l Cr(VI) in 144 h	Dhal *et al.*, 2010
Bacillus cereus SJ1	Wastewater of electroplating factory	30 mM	1 mM Cr(VI) in 57 h	
Bacillus atrophaeus MM20	Tannery waste contaminated soil		94% of 10 mg/l of Cr (VI) in 50 h	Patra et al., 2010
Bacillus sp. strain KSUCr5	Hypersaline Soda lakes	75 mM	40 mg/l of Cr(VI) in 24 h	Ibrahim et al., 2011
Bacillus sp. FM1	Tannery effluent	1,000 mg/l	100 mg/l Cr(VI) in 48 h	Masood and Malik, 2011
Bacterial isolates	**Source**	**Cr tolerance**	**Cr(VI) reduction efficiency**	**Reference**
Bacillus subtilis MNU16	Soil from coal mining site		75% of 50 mg/l in 72 h	Upadhyay et al., 2017
Brucella sp.	Contaminated sites of chemical industry	1,000 mg/l	100% reduction of 50 mg/l Cr(VI)	Thacker et al., 2007
Burkholderia cepacia	Alkaline crater lake	1,000 ppm	98% of 75 ppm Cr(VI) in 36 h	Wani et al., 2007
Cellulosimicrobium sp.MWM81	Contaminated soil		45% of 10 mM in 48 h	Rehman et al., 2013
Cellulosimicrobium sp. KX710177	Tannery wastewater		62% of 300 mg/l in 96 h	Bharagava & Mishra 2018
Corynebacterium paurometabolum	Procured,from culture collection bank	4 mg/l	55% of 4 mg/l in 2 hrs	Divyasree et al., 2016
Clostridium sp. SS1	Activated sludge	50 mg/l	5 mg/l Cr(VI) reduced in 24 h	Nguema and Luo, 2011
E. coli ASU 7	Wastewater canal	7.69 mM	54.62% of 1 ppm Cr(VI) in 48 h	Abskharon et al., 2009
Exiguobacterium sp. GS1	Grass land soil	NR	57.06% of 100 µg/ml in 6 h	Okeke et al., 2009
Leucobacter sp. CRB1	Soil of chromite ore processing residue disposal site	4,000 mg/l	34.5% of 4000 mg/l	Zhu et al., 2008
Nesterenkonia sp. MF2	Tannery effluent	600 mM	0.2 mM Cr(VI) in 24 h	Amoozegar et al., 2007
Ochrobactrum intermedium SDCr-5	Tannery effluent	15 mg/ml	200 µg/ml of Cr(VI) in 72 h	Sultan and Hasnain, 2007
Ochrobactrum sp. CSCr-3	Chromium landfill	800 mg/l	40% of 400mg/l Cr(VI) in 30 h	He et al., 2009
Pannonibacter phragmitetus	Chromium-containing slag heap	500 mg/l	500 mg/l Cr (VI) in 24 h	Chai et al., 2009
Pantoea sp.	Agricultural soil irrigated with tannery Effluents	4.7 mM	125 µM Cr(VI) in 64 h	Alam and Ahmed, 2011
Pediococcus pentosaceus IFR-3	Tannery effluent	2000 mg/l	20 mg/l Cr(VI) in 24 h	Ilias et al., 2011
Providencia sp.	Contaminated soil		100% of 100 mg/L in 30 h	Thacker et al., 2006
Bacterial isolates	**Source**	**Cr tolerance**	**Cr(VI) reduction efficiency**	**Reference**
Pseudomonas sp. C-171	Domestic sewage	2000 ppm	1200 ppm Cr(VI) in 288 h	Rahman et al., 2007
Pseudomonas fluorescens	Heavy metal contaminated soil	100 ppm	93% of 25 ppm Cr(VI) in 36 h	Parameswari et al., 2009
Pseudomonas aeruginosa	NR	40 mg/l	40 mg/l to 18 mg /l in 72 h	Wei-hua et al., 2009
Pseudomonas olevorans	Contaminated soil	15 mM	42.4 µM of 50 µM in 6 h	
Pseudomonas sp. S4	Tannery effluent	200 mg/l	92.5% of 80 mg/l in 72 h	Farag and Zaki, 2010
Pseudomonas sp. G1DM21	Chromium industrial landfill		99.7% of 500 µ M in 48 h	Desai et al., (2008)
Pseudomonas sp. JF122	Contaminated soil		100% of 2 mg/l in 72 h	Zhou & Chen (2010)
Pseudomonas stutzeri L1	Crude oil samples	1000 mg/l	97% of 1000 mg/l in 24 h	Sathish kumar et al., 2016
Serratia sp. Cr-10	Chromium contaminated soil		100% of 10 mg/l after 12 h	Zhang and Li (2011)
Sphaerotilus natans	Activated sludge	NR	75% of 80 mg/l in 240 h	Caravelli et al., 2008
Staphylococcus gallinarum W-61	Agricultural soil irrigated with tannery Effluents	12.4 mM	125 µM Cr(VI) in 24 h	Alam and Ahmed, 2011
Staphylococcus aureus IFR-2	Tannery effluent	2000 mg/l	20 mg/l Cr(VI) in 6 h	Ilias et al., 2011
Stenotrophomonas maltophilia ZA-6	Agricultural soil irrigated with tannery Effluents	16.5 mM	125 µM Cr(VI) within 16 h	Alam and Ahmed, 2011
Streptomyces sp. MS-2	Marine habitat	NR	75 mg/l Cr(VI) within 72 h	Mabrouk, 2008
Thermus scotocductus	Mine groundwater	0.3 mM	0.5 mM Cr(VI) in 10 h	Oppermen and Heerden, 2007

[NR- not reported]

Table 2. Source and some features of chromate reductases of aerobic chromate reducing organisms

Bacterial source	Enzyme	Enzyme type	Electron donor	Nature	K_m, µM Cr(VI)	References
Acidiphilium cryptum JF-5	ApcA	Chromate reductase	Heme	M		Magnuson et al. 2010
Arthrobacter crystallopoites ES 32			NADH	S	1.78	Camargo et al. 2004
Arthrobacter rhombi			NADH	S	170.5	Elangovan et al. 2010
Bacillus sp. QCI-2			Glucose		1.25	Campose et al. 1995
Bacillus sp. ES-29			NADH	S	7.09	Camargo et al. 2003
Bacillus sp. RE	Chromate reductase	NAD(P)H dependent	NADH and NAD(P)H	S	14	Elangovan et al. 2006
Bacillus sp.ev3			NR	S	NR	Rehman et al. 2008
Bacillus fusiformis			NADH	S	200	Desai et al. 2008
Bacillus sphaericus			NADH	S	158.12	Pal et al. 2005
Bacillus sphaericus			NADH	S	NR	Desai et al. 2008
Bacillus cereus			NADH	S	NR	Desai et al. 2008
Bacillus sp.			NADH	S	45	Sarangi and Krishnan, 2008
B. subtilis	YcnD	FMN reductase	FMN			Morokutti et al. 2005
Brucella sp.			NADH	S	NR	Thacker et al. 2007
E. coli	Chr	Chromate reductase	NAD(P)H	S	47.5	Bae et al. 2005
E. coli	YieF	Chromate & quinone reductase	FMN /NAD(P) H	S	NR	Ackerley et al. 2004 a
E. coli	ChrA	Chromate & quinone reductase	FMN /NAD(P) H	S	NR	Barak et al. 2006
E. coli	NfsA	Chromate & nitro reductase	FMN/ NADH	S	NR	Ackerley et al. 2004 b
E.coli (Strain K12)	ChrR	Quinone Reductase	FMN /NADH			Eswaramoorthy et al. 2012
Exiguobacterium sp.			NAD(P)H	S	200	Sarangi and Krishnan, 2008
Gluconobacter hansenii	Gh-ChrR	Chromate reductase	FMN /NADH			Jin et al. 2012
Leucobacter sp.			NAD(P)H	S	55	Sarangi and Krishnan, 2008
Orchrobactrum sp.			NADH/ NAD(P)H	S	NR	Thacker and Madamwar, 2005
Pseudomonas ambigua	Chr	Chromate & nitroreductase	NADH/ NAD(P)H	S	13	Suzuki et al. 1992
Pseudomonas fluorescence			Acetate	M	NR	Bopp and Ehrlich, 1983
Pseudomonas putida	Chr	Chromate & quinone reductase	NADH and NAD(P)H	S	374	Park et al. 2000
Pseudomonas sp.			NADH	S	175	Desai et al. 2008
Pseudomonas olevorans			NR	S	NR	Mistry et al. 2009
Streptomyces sp.			NADH	M	NR	Das and Chandra, 1990
S. thermocarboxydus			Glucose	S	NR	Desjardin et al. 2002
Streptomyces sp. MC1			NADH	S	NR	Polti et al. 2010
Vibrio harveyi	NfsB	Nitro reductase	FMN /(NADP) H	NR	11.8	Kwak et al. 2003

[S-soluble, M- membrane bound, NR-not reported]

Table 3. Anaerobic chromate reduction by cell-free extracts of some selected bacteria

Bacterial source	Enzyme	Enzyme type	Electron donor	Nature	K_m, µM Cr(VI)	References
Desulfovibrio vulgaris	Cytochrome c3	Periplasmic c type cytochrome	H_2	S	NR	Lovely and Philip, 1994
Desulfuromonas acetoxidans	Cytochrome c7	Periplasmic c type cytochrome	Heme	M		Michel et al. 2001
Desulfovibrio desulfurican G20		Thioredoxin oxidoreductase	NADP H	M		Li and Krumholz, 2009
Desulfovibrio alaskensis		Thioredoxin oxidoreductase	NADP H	M		Hauser et al. 2011
Enterobacter cloacae			Acetate	M	NR	Wang *et al.*, 1991
Escherichia coli			NADH	M	NR	Shen and Wang, 1994
Shewanella putrefaciens MR-1	chrBAC	Efflux transporter	Formate	M	NR	Myers *et al.*, 2000
Paracoccus denitrificans[a]	FerB	Ferric reductase B	NADH/ NAD(P)H	S	70	Sedla´cˇek and Kucˇera, 2010
Rhodobacter sphaeroides[a]	Chr	Chromate reductase	NADH	S	15	Nepple *et al.*, 2000
Thermus scotocductus[a]	Chr	Chromate reductase	NAD(P)H	M	55.5	Opperman *et al.*, 2008

[[a]I Isolates capable of reducing hexavalent chromium under aerobic condition.
S-soluble, M-membrane bound, NR-not reported]

main problem associated with use of enzyme in bioremediation includes the ability of the enzyme in using a cost effective electron source for their activity and also easy availability of the enzyme.

Several efficient chromate reducing enzymes has been described by different research group which has been reported to be flavoprotein in nature using expensive NAD(P)H co factor to felicitate Cr(VI) reduction, therefore it is required to find a suitable electron donor which is economically more viable for large scale. Robins et al., 2013 developed an enzyme based system for *ex situ* detoxification of Cr(VI) contaminated waste water. In this effective cell free system they included a cofactor regenerating enzyme partner for inexpensive supply of reduced NAD(P)H, which is vital for Cr(VI) reductase enzyme.

CONCLUSION

The enzymes belonging to different classes such as nitroreductase, iron reductase, quinone reductase, hydrogenase, flavin reductase as well as NAD(P)H dependent reductase have found to play vital role in reducing Cr(VI). Chromate reductases of different types have been reported from different microorganisms which may both soluble or membrane bound in nature and they vary in their ability and mechanisms. The soluble chromate reductases uses NAD(P)H as electron donors to reduce Cr(VI) and are found to be suitable for development of biocatalyst for bioremediational purpose. The membrane bound reductases are mainly belongs to the class flavin reductase, cytochromes and hydrogenases system and use chromate as the terminal electron acceptor. However, till now only a few chromate reductase enzymes have been purified, characterized and genes responsible for the activity have been identified. Cloning techniques can be used to produce large amount of enzymes which can be immobilized to exploit them for bioremediational practice.

Figure 1. Status of research based on pubmed (www.ncbi.nlm.nih.gov/pubmed) searched with different key words related to different types of chromate reductase enzyme

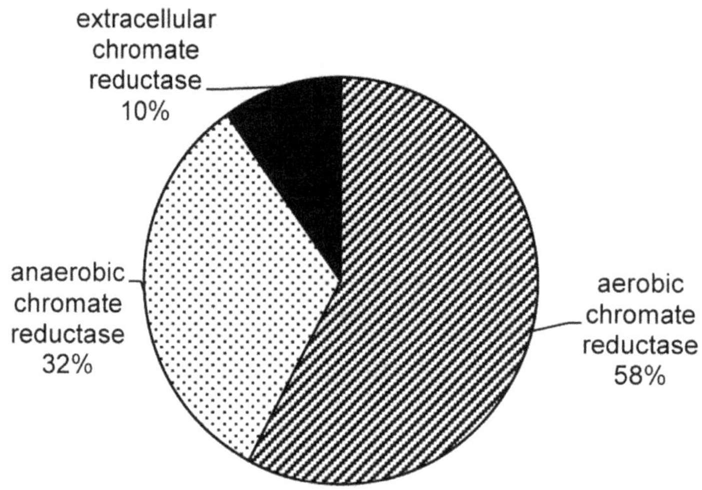

Figure 2. Status of research based on pubmed (www.ncbi.nlm.nih.gov/pubmed) searched with different key words related to structural analysis of chromate reductase enzyme and its mode of action

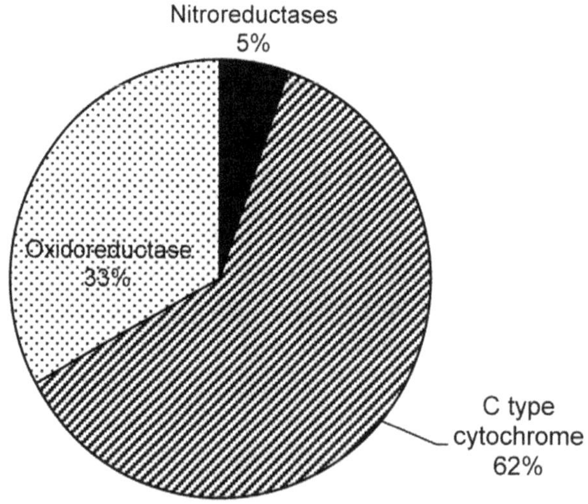

REFERENCES

Ackerley, D. F., Gonzalez, C. F., Keyhan, M., Blake, I. I. R., & Matin, A. (2004b). Mechanism of chromate reduction by the *E. coli* protein, NfsA, and the role of different chromate reductases in minimizing oxidative stress during chromate reduction. *Environmental Microbiology*, *6*(8), 851–860. doi:10.1111/j.1462-2920.2004.00639.x PMID:15250887

Ackerley, D. F., Gonzalez, C. F., Park, C. H., Blake, R. II, Keyhan, M., & Matin, A. (2004a). Chromate reducing properties of soluble flavoproteins from *Pseudomonas putida* and *E. Coli*. *Applied and Environmental Microbiology*, *70*(2), 873–882. doi:10.1128/AEM.70.2.873-882.2004 PMID:14766567

Amoozegar, M. A., Ghasemi, A., Razavi, M. R., & Naddaf, S. (2007). Short Communication: Evaluation of hexavalent chromium reduction by chromate-resistant moderately halophile, *Nesterenkonia* sp. strain MF2. *Process Biochemistry*, *42*(10), 1475–1479. doi:10.1016/j.procbio.2007.07.001

Bae, W. C., Lee, H. K., Choe, Y. C., Jahng, D. J., Lee, S. H., & Kim, S. J. (2005). Purification and characterization of NADPH dependent Cr(VI) reductase from *Escherichia coli* ATCC 33456. *Journal of Microbiology (Seoul, Korea)*, *43*, 21–27. PMID:15765053

Barak, Y., Ackerley, D. F., Dodge, C. J., Banwari, L., Alex, C., Francis, A. J., & Matin, A. (2006). Analysis of novel soluble chromate and uranyl reductases and generation of an improved enzyme by directed evolution. *Applied and Environmental Microbiology*, *72*(11), 7074–7082. doi:10.1128/AEM.01334-06 PMID:17088379

Belchik, S. M., Kennedy, D. W., Dohnalkova, A. C., Wang, Y., Sevinc, P. C., Wu, H., Lin, Y., Lu, P. H., Fredrickson, J. K., & Shi, L. (2011). Extracellular reduction of hexavalent chromium by cytochromes MtrC and OmcA of *Shewanella oneidensis* MR-1. *Applied and Environmental Microbiology*, *77*(12), 4035–4041. doi:10.1128/AEM.02463-10 PMID:21498755

Bharagava, R. N., & Mishra, S. (2018). Hexavalent chromium reduction potencial of *Cellulosimicrobium* sp. isolated from common effluent treatment plant of tannery industries. *Ecotoxicology and Environmental Safety*, *147*, 102–109. doi:10.1016/j.ecoenv.2017.08.040 PMID:28841524

Bielicka, A., Bojanowska, I., & Wisniewski, A. (2005). Two faces of chromium pollutant and bioelement. *Polish Journal of Environmental Studies*, *14*(1), 5–10.

Blake, R. C. II, Choate, D. M., Bardhan, S., Revis, N., Barton, L. L., & Zocco, T. C. (1993). Chemical transformation of toxic metals by *Pseudomonas* strain from a toxic waste site. *Environmental Toxicology and Chemistry*, *12*(8), 1365–1376. doi:10.1002/etc.5620120806

Bopp, L. H., & Ehrlich, H. L. (1988). Chromate resistance and reduction in *Pseudomonas fluorescens* strain LB300. *Archives of Microbiology*, *150*(5), 426–431. doi:10.1007/BF00422281

Branco, R., Chung, A. P., Johnston, T., Gurel, V., Morais, P., & Zhitkovich, A. (2008). The chromate inducible chr BACF operon from the transposable element TnOtChr confers resistance to chromium(VI) and superoxide. *Journal of Bacteriology*, *190*(21), 6996–7003. doi:10.1128/JB.00289-08 PMID:18776016

Brown, S., Thompson, M., VerBerkmoes, N., Chourey, K., Shah, M., Zhou, J., Hettich, R., & Thompson, D. (2006). Molecular dynamics of the *Shewanella oneidensis* response to chromate stress. *Molecular & Cellular Proteomics*, *5*(6), 1054–1071. doi:10.1074/mcp.M500394-MCP200 PMID:16524964

Browning, C. L., & Wise, J. P. Sr. (2017). Prolonged exposure to particulate chromate inhibits RAD51 nuclear import mediator proteins. *Toxicology and Applied Pharmacology*, *331*, 101–107. doi:10.1016/j.taap.2017.05.030 PMID:28554658

Camargo, F. A. O., Okeke, B. C., Bento, F. M., & Frankenberger, W. T. (2003). In vitro reduction of hexavalent chromium by a cell-free extract of *Bacillus* sp. ES 29 stimulated by Cu^{2+}. *Applied Microbiology and Biotechnology*, *62*(5-6), 569–573. doi:10.100700253-003-1291-x PMID:12679851

Camargo, F. A. O., Okeke, B. C., Bento, F. M., & Frankenberger, W. T. (2004). Hexavalent chromium reduction by immobilized cells and cell-free extract of *Bacillus sp.* ES29. *Bioremediation Journal*, *8*(1-2), 23–30. doi:10.1080/10889860490453140

Camargo, F. A. O., Okeke, B. C., Bento, F. M., & Frankenberger, W. T. (2005). Diversity of chromium-resistant bacteria isolated from soils contaminated with dichromate. *Applied Soil Ecology*, *29*(2), 193–202. doi:10.1016/j.apsoil.2004.10.006

Campos, J., Martinez-Pacheco, M., & Cervantes, C. (1995). Hexavalent-chromate reduction by a chromate resistant *Bacillus* sp. *Antonie van Leeuwenhoek*, *68*(3), 203–208. doi:10.1007/BF00871816 PMID:8572677

Carroll, C. C., Warnakulasuriyarachchi, D., Nokhbeh, M. R., & Lambert, I. B. (2002). *Salmonella typhimurium* mutagenicity tester strains that over express oxygen insensitive nitroreductases nfsA and nfsb. *Mutation Research*, *501*(1-2), 79–98. doi:10.1016/S0027-5107(02)00018-0 PMID:11934440

Cervantes, C., & Campos-Garcia, J. (2007). *Reduction and efflux of chromate by bacteria. Molecular Microbiology of Heavy Metals*. Springer-Verlag.

Cervantes, C., Campos-Garcia, J., Devars, S., Corona, F. G., Loza-Tavera, H., Guzman, J., & Moreno-Sanchez, R. (2001). Interactions of chromium with microorganisms and plants. *FEMS Microbiology Reviews*, *25*(3), 335–347. doi:10.1111/j.1574-6976.2001.tb00581.x PMID:11348688

Cervantes, C., Ohtake, H., Chu, L., Misra, T. K., & Silver, S. (1990). Cloning, nucleotide sequence and expression of the chromate resistance determinant of *Pseudomonas aeruginosa* plasmid pUM505. *Journal of Bacteriology*, *172*(1), 287–291. doi:10.1128/JB.172.1.287-291.1990 PMID:2152903

Chai, L. Y., Huang, S. H., Yang, Z. H., Peng, B., Huang, Y., & Chen, Y. H. (2009). Hexavalent chromium reduction by *Pannonibacter phragmitetus* BB isolated from soil under chromium-containing slag heap. *Journal of Environmental Science and Health. Part A, Toxic/Hazardous Substances & Environmental Engineering*, *44*(6), 615–622. doi:10.1080/10934520902784690 PMID:19337925

Chardin, B., Giudici-Orticoni, M. T., De Luca, G., Guigliarelli, B., & Bruschi, M. (2003). Hydrogenases in sulfate-reducing bacteria function as chromium reductase. *Applied Microbiology and Biotechnology*, *63*(3), 315–321. doi:10.100700253-003-1390-8 PMID:12861426

Cheng, G., & Li, X. (2009). Bioreduction of chromium (VI) by *Bacillus* sp. isolated from soils of iron mineral area. *European Journal of Soil Biology*, *45*(5-6), 483–487. doi:10.1016/j.ejsobi.2009.06.009

Cheung, K. H., & Gu, J. D. (2007). Mechanism of hexavalent chromium detoxification by microorganisms and bioremediation application potential: A review. *International Biodeterioration & Biodegradation, 59*(1), 8–15. doi:10.1016/j.ibiod.2006.05.002

Chourey, K., Thompson, M. R., Morrrell-Falvey, J., Verberkmoes, N. C., Brown, S. D., Shah, M., Zhou, J., Doktycz, M., Heitich, R. L., & Thompson, D. K. (2006). Global molecular and morphological effects of 24-hour chromium (VI) exposure on *Shewanella oneidensis* MR-1. *Applied and Environmental Microbiology, 72*(9), 6331–6344. doi:10.1128/AEM.00813-06 PMID:16957260

Das, S., & Chandra, A. L. (1990). Chromate reduction in *Streptomyces. Experientia, 46*(7), 731–733. doi:10.1007/BF01939949 PMID:2164957

Das, S., Mishra, J., Das, S. K., Pandey, S., Rao, D. S., Chakraborty, A., Sudarshan, M., Das, N. N., & Thatoi, H. N. (2014). Investigation on mechanism of Cr(VI) reduction and removal by *Bacillus amyloliquefaciens*, a novel chromate tolerant bacterium isolated from chromite mine soil. *Chemosphere, 96*, 112–121. doi:10.1016/j.chemosphere.2013.08.080 PMID:24091247

Desai, C., Jain, K., & Madamwar, D. (2008). Evaluation of in vitro Cr(VI) reduction potential in cytosolic extracts. *Bioresource Technology, 99*(14), 6059–6069. doi:10.1016/j.biortech.2007.12.046 PMID:18255287

Desjardin, V., Bayard, R., Huck, N., Manceau, A., & Gourdon, R. (2002). Effect of microbial activity on the mobility of chromium in soils. *Waste Management (New York, N.Y.), 22*(2), 195–200. doi:10.1016/S0956-053X(01)00069-1 PMID:12003148

Dey, S., & Paul, A. K. (2012). Optimization of cultural conditions for growth associated chromate reduction by *Arthrobacter* sp. SUK 1201 isolated from chromite mine overburden. *Journal of Hazardous Materials, 213–214*, 200–206. doi:10.1016/j.jhazmat.2012.01.078 PMID:22361630

Dey, S., & Paul, A. K. (2013). Hexavalent chromium reduction by aerobic heterotrophic bacteria indigenous to chromite mine overburden. *Brazilian Journal of Microbiology, 44*(1), 307–315. doi:10.1590/S1517-83822013000100045 PMID:24159321

Dey, S., & Paul, A. K. (2016). Evaluation of chromate reductase activity in the cell-free culture filtrate of *Arthrobacter* sp. SUK 1201 isolated from chromite mine overburden. *Chemosphere, 156*, 69–75. doi:10.1016/j.chemosphere.2016.04.101 PMID:27176938

Dhal, B., Thatoi, H., Das, N., & Pandey, B. D. (2010). Reduction of hexavalent chromium by *Bacillus* sp. isolated from chromite mine soils and characterization of reduced product. *Journal of Chemical Technology and Biotechnology (Oxford, Oxfordshire), 85*, 1471–1479. doi:10.1002/jctb.2451

Divyasree, C., & Prabhakaran Subramanian, S. (2016). Studies on the bioremediation of hromium from aqueous solutions using *Corynebacterium paurometabolum. Transactions of the Indian Institute of Metals, 70*, 497–509.

Elangovan, R., Abhipsa, S., Rohit, B., Ligy, P., & Chandraraj, K. (2006). Reduction of Cr(VI) by a *Bacillus* sp. *Biotechnology Letters, 28*(4), 247–252. doi:10.100710529-005-5526-z PMID:16555008

Elangovan, R., Philip, L., & Chandraraj, K. (2010). Hexavalent chromium reduction by free and immobilized cell-free extract of *Arthrobacter rhombi*-RE. *Applied Biochemistry and Biotechnology, 160*(1), 81–97. doi:10.100712010-008-8515-6 PMID:19165627

Farag, S., & Zaki, S. (2010). Identification of bacterial strains from tannery effluent and reduction of hexavalent chromium. *Journal of Environmental Biology, 31*, 877–882.

Field, E. K., Blaskovich, J. P., Peyton, B. M., & Gerlach, R. (2018). Carbon-dependent chromate toxicity mechanism in an environmental *Arthrobacter* isolate. *Journal of Hazardous Materials, 355*, 162–169. doi:10.1016/j.jhazmat.2018.05.020 PMID:29800910

Fredrickson, J. K., Kostandarithes, H. M., Li, S. W., Plymale, A. E., & Daly, M. J. (2000). Reduction of Fe (III), Cr (VI), U (VI), and Tc (VII) by *Deinococcus radiodurans* R1. *Applied and Environmental Microbiology, 66*(5), 2006–2011. doi:10.1128/AEM.66.5.2006-2011.2000 PMID:10788374

Gadd, G. M. (1992). Metals and microorganisms: A problem of definition. *FEMS Microbiology Letters, 100*(1-3), 197–203. doi:10.1111/j.1574-6968.1992.tb05703.x PMID:1478456

Gonzalez, C. F., Ackerley, D. F., Lynch, S. V., & Matin, A. (2005). ChrR a soluble quinine reductase of *Pseudomonas putida* that defends against H2O2. *The Journal of Biological Chemistry, 280*(24), 22590–22595. doi:10.1074/jbc.M501654200 PMID:15840577

Goulhen, F., Gloter, A., Guyot, F., & Bruschi, M. (2006). Cr(VI) detoxification by Desulfovibrio vulgaris strain Hildenborough: Microbe-metal interactions studies. *Applied and Environmental Microbiology, 71*(6), 892–897. PMID:16896506

Hauser, L. J., Miriam, L., Brown, S. D., Larimer, F., Keller, K. L., Rapp-Giles, B. J., & … . (2011). Complete genome sequence and updated annotation of *Desulfovibrio alaskensis* G20. *Journal of Bacteriology, 193*(16), 4268–4269. doi:10.1128/JB.05400-11 PMID:21685289

He, M., Li, X., Guo, L., Miller, S. J., Rensing, C., & Wang, G. (2010). Characterization and genomic analysis of chromate resistant and reducing *Bacillus cereus* Strain SJ1. *BMC Microbiology, 10*(221), 1–10. doi:10.1186/1471-2180-10-221 PMID:20723231

He, Z., Gao, F., Sha, T., Hu, Y., & He, C. (2009). Isolation and characterization of a Cr(VI)-reduction *Ochrobactrum* sp. strain CSCr-3 from chromium landfill. *Journal of Hazardous Materials, 163*(2-3), 869–873. doi:10.1016/j.jhazmat.2008.07.041 PMID:18722054

Horton, R. N., Apel, W. A., Thompson, V. S., & Sheridan, P. P. (2006). Low temperature reduction of hexavalent chromium by a microbial enrichment consortium and a novel strain of *Arthrobacter aurescens*. *BMC Microbiology*, 6. PMID:16436214

Hu, P., Brodie, E. L., Suzuki, Y., McAdams, H. H., & Andersen, G. L. (2005). Whole-genome transcriptional analysis of heavy metal stresses in *Caulobacter crescentus*. *Journal of Bacteriology, 187*(24), 8437–8449. doi:10.1128/JB.187.24.8437-8449.2005 PMID:16321948

Ibrahim, A. S. S., El-Tayeb, M. A., Elbadawi, Y. B., & Al-Salamah, A. A. (2011). Isolation and characterization of novel potent Cr(VI) reducing alkaliphilic *Amphibacillus* sp. KSUCr3 from hypersaline soda lakes. *Electronic Journal of Biotechnology*, *14*(4). Advance online publication. doi:10.2225/vol14-issue4-fulltext-4

Ishibashi, Y., Cervantes, C., & Silver, S. (1990). Chromium reduction in *Pseudomonas putida*. *Applied and Environmental Microbiology*, *56*(7), 2268–2270. doi:10.1128/AEM.56.7.2268-2270.1990 PMID:2389940

Juhnke, S., Peitzsch, N., Hubener, N., Große, C., & Nies, D. H. (2002). New genes involved in chromate resistance in *Ralstonia metallidurans* strain CH34. *Archives of Microbiology*, *179*(1), 15–25. doi:10.100700203-002-0492-5 PMID:12471500

Kamaludeen, S. P. B., Megharaj, M., Juhasz, A. L., Sethunathan, N., & Naidu, R. (2003). Chromium microorganism interactions in soils: Remediation implications. *Reviews of Environmental Contamination and Toxicology*, *178*, 93–164. doi:10.1007/0-387-21728-2_4 PMID:12868782

Kanmani, P., Aravind, J., & Preston, D. (2012). Remediation of chromium contaminants using bacteria. *International Journal of Environmental Science and Technology*, *9*(1), 183–193. doi:10.100713762-011-0013-7

Kashefi, K., & Loveley, D. R. (2000). Reduction of Fe(III), Mn(IV), and toxic metals at 100 °C by *Pyrobaculum islandicum*. *Applied and Environmental Microbiology*, *66*(3), 1050–1056. doi:10.1128/AEM.66.3.1050-1056.2000 PMID:10698770

Kobori, H., Harrison-Bernard, L. M., & Navar, L. G. (2001). Expression of angiotensinogen mRNA and protein in angiotensin II-dependent hypertension. *Journal of the American Society of Nephrology*, *12*, 431–439. PMID:11181790

Komori, K., Wang, P. C., Toda, K., & Ohtake, H. (1989). Factors affecting chromate reduction in *Enterobacter cloacae* strain HO1. *Applied Microbiology and Biotechnology*, *31*(31), 567–570. doi:10.1007/BF00270796

Kornhauser, C., Wrobel, K., Wrobel, K., Malacara, J. M., Nava, L. E., Gómez, L., & González, R. (2002). Possible adverse effect of chromium in occupational exposure of tannery workers. *Industrial Health*, *40*(2), 207–213. doi:10.2486/indhealth.40.207 PMID:12064563

Kwak, Y. H., Lee, D. S., & Kim, H. B. (2003). *Vibrio harveyi* nitroreductase is also a chromate reductase. *Applied Microbiology and Biotechnology*, *69*, 4390–4395. PMID:12902220

Llagostera, M., Garido, S., Guerrero, R., & Barbe, J. (1986). Induction of SOS genes of *Escherichia coli* by chromium compounds. *Environmental Mutagenesis*, *8*(4), 571–577. doi:10.1002/em.2860080408 PMID:3525136

Lovley, D. R., & Phillips, E. J. P. (1994). Reduction of Chromate by *Desulfovibrio vulgaris* and its C_3 Cytochrome. *Applied and Environmental Microbiology*, *60*(2), 726–728. doi:10.1128/AEM.60.2.726-728.1994 PMID:16349200

Ma, Z., Zhu, W., Long, H., Chai, L., & Wang, Q. (2007). Chromate reduction by resting cells of *Achromobacter* sp. Ch-1 under aerobic conditions. *Process Biochemistry*, *42*(6), 1028–1032. doi:10.1016/j.procbio.2007.03.007

Magnuson, T. S., Swenson, M. W., Paszczynski, A. J., Deobald, L. A., Kerk, D., & Cummings, D. E. (2010). Proteogenomic and functional analysis of chromate reduction in *Acidiphilium cryptum* JF-5, an Fe(III)-respiring acidophile. *Biometals*, *23*(6), 1129–1138. doi:10.100710534-010-9360-y PMID:20593301

Masood, F., & Malik, A. (2011). Hexavalent chromium reduction by *Bacillus* sp. Strain FM1 isolated from heavy-metal contaminated soil. *Bulletin of Environmental Contamination and Toxicology*, *86*(1), 114–119. doi:10.100700128-010-0181-z PMID:21181113

Megharaj, M., Avudainayagam, S., & Naidu, R. (2003). Toxicity of hexavalent chromium and its reduction by bacteria isolated from soil contaminated with tannery waste. *Current Microbiology*, *47*(1), 51–54. doi:10.100700284-002-3889-0 PMID:12783193

Michel, C., Brugna, M., Aubert, C., Bernadac, A., & Bruschi, M. (2001). Enzymatic reduction of chromate: Comparative studies using sulphate reducing bacteria. *Applied Microbiology and Biotechnology*, *55*(1), 95–100. doi:10.1007002530000467 PMID:11234966

Miranda, A. T., González, M. V., González, G., Vargas, E., Campos-García, J., & Cervantes, C. (2005). Involvement of DNA helicases in chromate resistance by *Pseudomonas aeruginosa* PAO1. *Mutation Research*, *578*(1-2), 202–209. doi:10.1016/j.mrfmmm.2005.05.018 PMID:16105671

Mistry, K., Desai, C., La, S., Patel, K., & Patel, B. (2010). Hexavalent chromium reduction by *Staphylococcus sp.* isolated from Cr (VI) contaminated land fill. International. *Journal of Biotechnology and Biochemistry*, *6*, 117–129.

Mistry, K., Desai, C., & Patel, K. (2009). Reduction of chromium (VI) by bacterial strain KK15 isolated from contaminated soil. *Journal of Cell and Tissue Research*, *9*, 1821–1826.

Myers, C. R., Carstens, B. P., Antholine, W. E., & Myers, J. M. (2001). Chromium(VI) reductase activity is associated with the cytoplasmic membrane of anaerobically grown *Shewanella putrefaciens* MR01. *Journal of Applied Microbiology*, *88*(1), 98–106. doi:10.1046/j.1365-2672.2000.00910.x PMID:10735248

Nepple, B. B., Kessi, J., & Buchofen, R. (2000). Chromate reduction by *Rhodobacter sphaeroides*. *Journal of Industrial Microbiology & Biotechnology*, *25*(4), 198–203. doi:10.1038j.jim.7000049

Nies, A., Nies, D. H., & Silver, S. (1990). Nucleotide sequence and expression of a plasmid encoded chromate resistance determinant from *Alcaligenes eutrophus*. *The Journal of Biological Chemistry*, *265*, 5648–5653. PMID:2180932

Ohtake, H., Fujii, E., & Toda, K. (1990). Reduction of toxic chromate in an industrial effluent by use of a chromate- reducing srain of *Enterobacter cloacae*. *Environmental Technology*, *11*(7), 663–668. doi:10.1080/09593339009384909

Ontañon, O. M., González, P. S., & Agostini, E. (2015). Biochemical and molecular mechanisms involved in simultaneous phenol and Cr(VI) removal by *Acinetobacter guillouiae* SFC 500-1A. *Environmental Science and Pollution Research International*, *22*(17), 13014–13023. doi:10.100711356-015-4571-y PMID:25916475

Opperman, D. J., Piater, L. A., & van Heerden, E. (2008). A novel chromate reductase from *Thermus scotoductus* SA- 01 related to old yellow enzyme. *Journal of Bacteriology*, *190*(8), 3076–3082. doi:10.1128/JB.01766-07 PMID:18263719

Pal, A., Dutta, S., & Paul, A. K. (2005). Reduction of hexavalent chromium by cell free extract of *Bacillus sphaericus* AND 303 isolated from serpentine soil. *Current Microbiology*, *51*(5), 327–330. doi:10.100700284-005-0048-4 PMID:16163455

Pal, A., & Paul, A. K. (2004). Aerobic chromate reduction by chromium-resistant bacteria isolated from serpentine soil. *Microbiological Research*, *159*(4), 347–354. doi:10.1016/j.micres.2004.08.001 PMID:15646381

Parameswari, E., Lakshmanan, A., & Thilagavathi, T. (2009). Chromate resistance and reduction by bacterial isolates. *Australian Journal of Basic and Applied Sciences*, *3*, 1363–1368.

Park, C. H., Gonzalez, C. F., Ackerley, D. F., Keyhan, M., & Matin, A. (2002). Molecular engineering of soluble bacterial proteins with chromate reductase activity. *Proceedings of the 1st International Conference on Remediation of Contaminated Sediments.*

Park, C. H., Keyhan, B., Wielinga, B., Fendorf, S., & Matin, A. (2000). Purification to homogeneity and characterization of a novel *Pseudomonas putida* chromate reductase. *Applied and Environmental Microbiology*, *66*(5), 1788–1795. doi:10.1128/AEM.66.5.1788-1795.2000 PMID:10788340

Patra, R. C., Malik, S., Beer, M., Meghraj, M., & Naidu, R. (2010). Molecular characterization of chromium (VI) reducing potential in Gram positive isolated from contaminated sites. *Soil Biology & Biochemistry*, *42*(10), 1857–1863. doi:10.1016/j.soilbio.2010.07.005

Pattanapipitpaisal, P., Brown, N. L., & Macaskie, L. E. (2001). Chromate reduction by *Microbacterium liquefaciens* immobilized in polyvinyl alcohol. *Biotechnology Letters*, *23*(1), 61–65. doi:10.1023/A:1026750810580

Pesti, M., Gazdag, Z., Emri, T., Farkas, N., Koosz, Z., Belagy, J., & Pocsi, I. (2002). Chromate sensitivity in fission yeast is caused by increased glutathione reductase activity and peroxide overproduction. *Journal of Basic Microbiology*, *42*(6), 408–419. doi:10.1002/1521-4028(200212)42:6<408::AID-JOBM408>3.0.CO;2-8 PMID:12442303

Poljsak, B., Pocsi, I., Raspor, P., & Pesti, M. (2010). Interference of chromium with biological systems in yeasts and fungi: A review. *Journal of Basic Microbiology*, *50*(1), 21–36. doi:10.1002/jobm.200900170 PMID:19810050

Polti, M. A., Amoroso, M. J., & Abate, C. M. (2010). Chromate reductase activity in *Streptomyces* sp. MC1. *The Journal of General and Applied Microbiology*, *56*, 11–18. doi:10.2323/jgam.56.11 PMID:20339215

Priester, J. H., Olson, S. G., Webb, S. M., Neu, M. P., Hersman, L. E., & Holden, P. A. (2006). Enhanced exopolymer production and chromium stabilization in *Pseudomonas putida* unsaturated biofilms. *Applied and Environmental Microbiology, 72*(3), 1988–1996. doi:10.1128/AEM.72.3.1988-1996.2006 PMID:16517647

Puzon, G. J., Petersen, J. N., Roberts, A. G., Kramer, D. B., & Xun, L. (2002). A bacterial flavin reductase system reduces chromate to a soluble chromium (III)–NAD(+) complex. *Journal of Biochemical and Biophysical Research Communication, 294*(1), 76–81. doi:10.1016/S0006-291X(02)00438-2 PMID:12054743

Reardon, C. L., Dohnalkova, A. C., Nachimuthu, P., Kennedy, D. W., Saffarini, D. A., Arey, B. W., Shi, L., Wang, Z., Moore, D., McLean, J. S., Moyles, D., Marshall, M. J., Zachara, J. M., Fredrickson, J. K., & Beliaev, A. S. (2010). Role of outer-membrane cytochromes MtrC and OmcA in the biomineralization of ferrihydrite by *Shewanella oneidensis* MR-1. *Geobiology, 8*(1), 56–68. doi:10.1111/j.1472-4669.2009.00226.x PMID:20002197

Rehman, A., Zahoor, A., Muneer, B., & Hasnain, S. (2008). Chromium tolerance and reduction potential of a *Bacillus* sp.ev3 isolated from metal contaminated wastewater. *Bulletin of Environmental Contamination and Toxicology, 81*(1), 25–29. doi:10.100700128-008-9442-5 PMID:18498008

Rehman, Y., Rizvi, F. Z., Faisal, M., & Hasnain, S. (2013). Arsenic and chromium reduction in co-cultures of bacteria isolated from industrial sites of Pakistan. *Microbiology, 82*(4), 458–433. doi:10.1134/S0026261713040188

Robins, K. J., Hooks, D. O., Rehm, B. H. A., & Ackerley, D. F. (2013). Escherichia coli NemA is an efficient chromate reductase that can be biologically immobilized to provide a cell free system for remediation of hexavalent chromium. *PLoS One, 8*(3), 1–8. doi:10.1371/journal.pone.0059200 PMID:23527133

Roh, Y., Liu, S. V., Li, G., Huang, H., Phelps, T. J., & Zhou, J. (2002). Isolation and characterization of metal-reducing thermoanaerobacter strains from deep subsurface environment of the Piceane basin, Colorado. *Applied and Environmental Microbiology, 68*(12), 6013–6020. doi:10.1128/AEM.68.12.6013-6020.2002 PMID:12450823

Romanenko, V. I., & Korenkov, V. N. (1977). A pure culture of bacteria utilising chromates and bichromates as hydrogen acceptors in growth under anaerobic conditions. *Microbiology, 46*, 329–332.

Saier, M. H. Jr. (2003). Tracing pathways of transport protein evolution. *Molecular Microbiology, 48*(5), 1145–1156. doi:10.1046/j.1365-2958.2003.03499.x PMID:12787345

Sarangi, A., & Krishnan, C. (2008). Comparison of in vitro Cr(VI) reduction by CFEs of chromate resistant bacteria isolated from chromate contaminated soil. *Bioresource Technology, 99*(10), 4130–4137. doi:10.1016/j.biortech.2007.08.059 PMID:17920879

Sathish Kumar, K., Murugan, K., Benelli, G., Higuchi, A., & Rajasekar, A. (2016). Bioreduction of hexavalent chromium by *Pseudomonas stutzeri* L1 and *Acinetobacter baumannii* L2. *Annals of Microbiology, 67*, 91-98.

Sedláček, V., & Kučera, I. (2010). Chromate reductase activity of the *Paracoccus denitrificans* ferric reductase B (FerB) protein and its physiological relevance. *Archives of Microbiology*, *192*(11), 919–926. doi:10.100700203-010-0622-4 PMID:20821194

Shahid, M., Shamshad, S., Rafiq, M., Khalid, S., Bibib, I., & Khan, N. (2017). Chromium speciation, bioavailability, uptake, toxicity and detoxification in soil-plant system: A review. *Chemosphere*, *178*, 513–533. doi:10.1016/j.chemosphere.2017.03.074 PMID:28347915

Shen, H., & Wang, Y. T. (1993). Characterization of enzymatic reduction of hexavalent chromium by *Escherichia coli* ATCC 33456. *Applied and Environmental Microbiology*, *59*(11), 3771–3777. doi:10.1128/AEM.59.11.3771-3777.1993 PMID:8285683

Shen, H., & Wang, Y. T. (1994). Biological reduction of chromium by *E. coli*. *Journal of Environmental Engineering*, *120*(3), 560–572. doi:10.1061/(ASCE)0733-9372(1994)120:3(560)

Silver, S. (1997). The bacterial view of the periodic table: specific functions for all elements. In J. F. Banfield & K. H. Nealson (Eds.), *Geomicrobiology: Integration between Microbes and Minerals. Mineralogical society of America* (pp. 345–360). doi:10.1515/9781501509247-012

Srinath, T., Khare, S., & Ramteke, P. W. (2001). Isolation of hexavalent chromium-reducing Cr-tolerant facultative anaerobes from tannery effluent. *The Journal of General and Applied Microbiology*, *47*(6), 307–312. doi:10.2323/jgam.47.307 PMID:12483605

Sultan, S., & Hasnain, S. (2007). Reduction of toxic hexavalent chromium by *Ochrobactrum intermedium* strain SDCr-5 stimulated by heavy metals. *Bioresource Technology*, *98*(2), 340–344. doi:10.1016/j.biortech.2005.12.025 PMID:16488604

Suzuki, T., Miyata, N., Horitsu, H., Kawai, K., Takamizawa, K., Tai, Y., & Okazaki, M. (1992). NAD(P)H-Dependent Chromium(VI) Reductase of *Pseudomonas ambigua* G-1: A Cr(V) intermediate is formed during the reduction of Cr(VI) to Cr(III). *Journal of Bacteriology*, *174*(16), 5340–5345. doi:10.1128/JB.174.16.5340-5345.1992 PMID:1322884

Thacker, U., Parikh, R., Shouche, Y., & Madamwar, D. (2007). Reduction of chromate by cell-free Extract of *Brucella* sp. isolated from Cr(VI) contaminated sites. *Bioresource Technology*, *98*(8), 1541–1547. doi:10.1016/j.biortech.2006.06.011 PMID:16931000

Upadhyay, N., Vishwakarma, K., Singh, J., Mishra, M., Kumar, V., Rani, R., Mishra, R. K., Chauhan, D. K., Tripathi, D. K., & Sharma, S. (2017). Tolerance and reduction of chromium(VI) by *Bacillus* sp. MNU16 isolated from contaminated coal mining soil. *Frontiers in Plant Sciences*, *8*, 778. doi:10.3389/fpls.2017.00778 PMID:28588589

Wang, P. C., Mori, T., Komori, K., Sasatsu, M., Toda, K., & Ohtake, H. (1989). Isolation and characterization of an *Enterobacter cloacae* strain that reduces hexavalent chromium under anaerobic conditions. *Applied and Environmental Microbiology*, *55*(7), 1665–1669. doi:10.1128/AEM.55.7.1665-1669.1989 PMID:16347962

Wang, P. C., Toda, K., Ohtake, H., Kusaka, I., & Yale, I. (1991). Membrane bound respiratory system of *Enterobacter cloacae* strain HO1 grown anaerobically with chromate. *FEMS Microbiology Letters*, *76*(1), 11–16. doi:10.1111/j.1574-6968.1991.tb04408.x PMID:1851711

Wang, Y. T., & Shen, H. (1995). Bacterial reduction of hexavalent chromium. *Journal of Industrial Microbiology*, *14*(2), 159–163. doi:10.1007/BF01569898 PMID:7766208

Wani, R., Kodam, K. M., Gawai, K. R., & Dhakephalkar, P. K. (2007). Chromate reduction by *Burkholderia cepacia* MCMB-821 isolated from the pristine habitat of alkaline crater lake. *Applied Microbiology and Biotechnology*, *75*(3), 627–632. doi:10.100700253-007-0862-7 PMID:17361433

Whiteway, J., Koziarz, P., Veall, J., Sandhu, N., Kumar, P., Hoecher, B., & Lambert, I. B. (1998). Oxygen-insensitive nitroreductases: Analysis of the roles of *nfsA* and *nfsB* in development of resistance to 5-nitrofuran derivatives in *Escherichia coli*. *Journal of Bacteriology*, *180*(21), 5529–5539. doi:10.1128/JB.180.21.5529-5539.1998 PMID:9791100

Wielinga, B., Mizuba, M. M., Hansel, C. M., & Fendorf, S. (2001). Iron promoted reduction of chromate by dissimilatory iron-reducing bacteria. *Environmental Science & Technology*, *35*(3), 522–527. doi:10.1021/es001457b PMID:11351723

Zakaria, Z. A., Zakaria, Z., Surif, S., & Ahmad, W. A. (2007). Hexavalent chromium reduction by *Acinetobacter haemolyticus* isolated from heavy metal-contaminated wastewater. *Journal of Hazardous Materials*, *146*(1-2), 30–38. doi:10.1016/j.jhazmat.2006.11.052 PMID:17188812

Chapter 6
Bacterial Remediation of Chromium From Industrial Sludge

Dipankar Roy
https://orcid.org/0000-0001-9144-3226
Calcutta University, India & St. Xavier's College, India

Arup Kumar Mitra
St. Xavier's College, India

ABSTRACT

Chromium-like heavy toxic metals seriously influence the metabolism of living organisms and cause permanent threatening of health. Microorganisms can help to detoxify those hazardous heavy metals in the environment by the process of bioremediation. Two bacterial genera were isolated from industrial sludge designated P1 and P2. From the 16srRNA study, it is revealed that P1 is Bacillus cereus *and P2 is* Enterobacter *sp. They are deposited in NCMR and NCBI and received the accession no. MCC 3868 for P1 and MCC 3788 for P2. P1 is gram positive, motile, and P2 is gram negative, motile. Eighteen antibiotics have been taken for antibiotic assay; P1 is resistant to 12; P2 is resistant to 8 antibiotics. For growth pattern analysis in chromium, three parameters have been selected, and they are temperature, pH, and biomass. In LD50 and above parameters, total chromium uptake by those bacteria in stressed conditions have been recorded. The two bacteria are not antagonistic to each other so they are used to bioremediate chromium from their contaminated sites and also treated as consortium.*

INTRODUCTION

Toxic heavy metals are mixing regularly in the environment as a result of industrial activities indicate potential hazard to ecosystem. Toxicants may be inorganic cationic metallic ions of mercury, cadmium, chromium, lead, nickel, Zinc, cobalt etc. Some of the heavy metal are essential and are required by the organism as micronutrient, as known as 'trace elements' [Bruins MR et al., 2000] such as Copper, Zinc

DOI: 10.4018/978-1-7998-4888-2.ch006

and Magnesium etc. Heavy toxic metals also seriously influence the metabolism of living organisms and cause permanent threatening of health such as Chromium, Cobalt and Cadmium etc.

Microorganisms can help to detoxify chromium in the environment. By the process of bioremediation those microorganisms are used to remove and recover chromium ions from polluted areas. The process can function naturally or can be improved through the addition of electron acceptors, nutrients, or other factors.

The cost effective and eco-friendly newer biotechnological processes viz. bioremediation through microbial metal reabsorption have been widely accepted.

To overview the mechanisms by which microorganisms interact with heavy metals, and to highlight recent advances in the application of these processes to bioremediation of metal contamination. In order to address the aims of this study, the following objectives were taken into consideration.

ü Isolation of Metal resistant bacteria from industrial sludge.
ü Determination of Chromium content into industrial sludge.
ü Chromium tolerance Study & LD_{50} analysis.
ü 16s rRNA and FAME study for identification of Samples.
ü Study on Morphology, various biochemical character and antibiotic sensitivity of isolated strains.
ü Study of growth patterns and bioaccumulation of samples in different stressed condition.
ü Antagonistic assay of the samples for developing consortium.

BACKGROUND

Among 94 naturally occurring elements only 17 heavy metals have importance for organisms and ecosystems (Weast, 1984). Among these elements, Al, Co, Se, and Si play a role in promoting plant growth and may be essential for particular taxa. Cobalt is an essential trace element that is an integral part of vitamin B12, which is essential in the metabolism of folic acid and fatty acids. Tungsten (W), on the other hand, appears to be essential only in hyperthermophilic bacteria, *Pyrococcus furiosus*, found in hydrothermal vent. However, elevated concentration of both essential and nonessential metals can result in growth inhibition and toxicity symptoms. For Example: Lead interferes with haemoglobin formation and causes anaemia due to deficiency of haemoglobin. Lack of haemoglobin may further cause kidney and brain damage. Cobalt may cause nausea and vomiting, deafness, nerve problems, ringing in the ears (tinnitus), thickening of the blood, thyroid problems. Enzymes like catalase, peroxidase and cytochrome oxidase with iron as their component are affected by chromium toxicity (Nath *et al.*, 2008).

There are a variety of natural and anthropogenic sources of heavy metals in the environment. On a worldwide basis, the disposal of commercial products that contain chromium may be the largest contributor, accounting for 51% of the total chromium released to soil (Nriagu and Pacyna 1988). Other significant sources of chromium release into soil include the disposal of coal fly ash and bottom fly ash from electric utilities and other industries (33.1%), agricultural and food wastes (5.3%), animal wastes (3.9%), and atmospheric fallout (2.4%) (Nriagu and Pacyna 1988). Solid wastes from metal manufacturing constituted >0.2% to the overall chromium release in soil. It is estimate that the total emission of chromium to the atmosphere is about 1.92×10^5t.

From polluted environment several techniques are there to remove or recover chromium such as adsorption processes, electrochemical techniques, ion exchange, reverse osmosis, chemical precipitation, chemical oxidation or reduction reaction, evaporative recovery & sludge filtration are among them. These techniques are not so much effective because of high reagent requirement (effective when metal concentration in solution is greater than 100mg/L) and high energy requirement, generation of toxic waste and the unpredictable nature of metal ion removal.

The microbial processes for bioremediation of chromium from waste employ living cells, non-living biomass, or biopolymers as biosorbents. A wide variety of fungi, algae and bacteria are now under study or are already in use as biosorbents (Gadd, 1992; Volesky and Holan, 1995). Recent work (Johnson and Hallberg, 2004; Hall et al., 2005; Hall and Puhlmann, 2004) indicated that most of the transition between metal speciation forms was controlled by microbial behaviour.

In this case some metals are reduced and converted from more soluble form to less soluble metal forms including sulfidic and phosphidic metal salts by bacteria. It is seen that *Desulfovibrio* spp.a sulfate reducing bacteria, can cause metal precipitation under anaerobic condition (Roane and Pepper, 2000). According to P. Kanmani • J. Aravind • D. Preston, 2012 Cr(VI), which is potent, extremely toxic carcinogenic when enters the cell by membrane sulphate channels interacts with some reductive enzymes like chromate reductase, glutathione, ascorbate etc forms Cr(III). In this reaction Cr(IV) and/or Cr(V), free oxygen radical can be formed. Cr(III) is an essential nutrient in this diet, aiding in the metabolism of glucose and lipids (Anderson 1997). Cr(III) is impermeable to the cell. Cr(V) and Cr(IV) also convert to the Cr(VI) by transferring the electron to oxygen. This process results in the formation of reactive oxygen species (ROS) capable of combining with DNA–protein complexes.

Figure 1. Transport of Cr(VI) into the cell
Source: P. Kanmani • J. Aravind • D. Preston, 2012

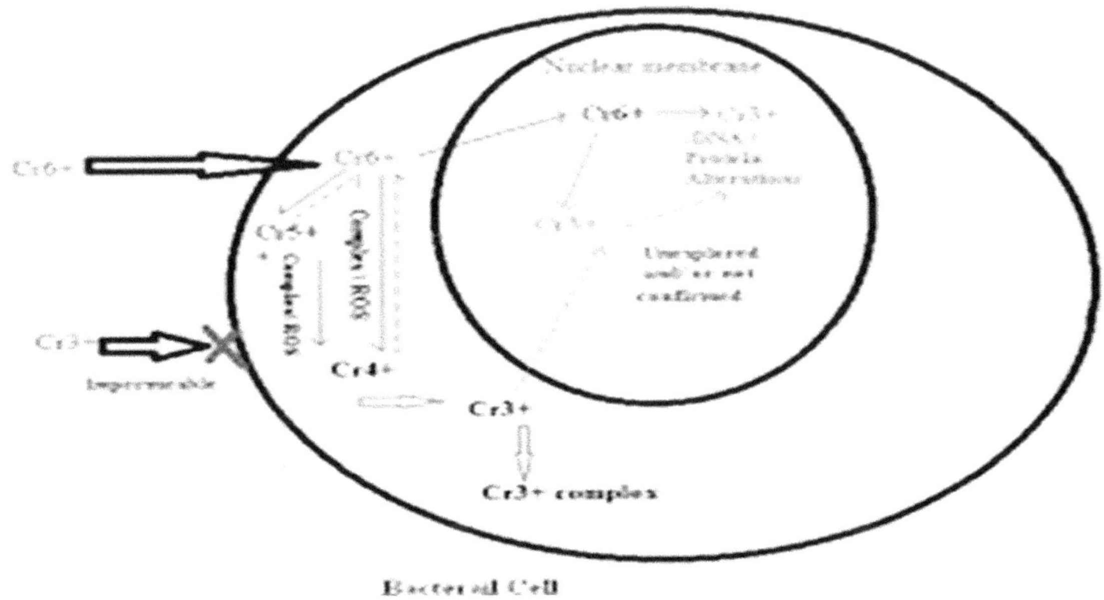

MAIN FOCUS OF THE CHAPTER

Issues, Controversies, Problems

Heavy metals are toxic to plants, animals and also for microorganisms. For that they are environmental pollutants especially in the site of anthropogenic sources (Imran et al. 2008; Mukesh et al.2008). In human body various types of life threatening problems occurred in presence of heavy metals because human body needs very trace amounts of these metals in metabolic pathways. For example GI hemorrhage, hemolysis, acute renal failure (Cr^{6+} ingestion), pulmonary fibrosis, lung cancer (inhalation) are occurred by Chromium. Mercury, Manganese and Zinc cause damage of central nervous system, corrosive effect on skin etc. Cadmium and Lead damage kidney and lung, causes cancer, gastrointestinal disorder and mental disorder. Bronchitis and dermatitis are caused by arsenic (Donald,2003).

There are several problems occurred in plants also, such as leaf chlorosis in hyacinth, toxicity in seed germination and growth, effect on root and shoot growth rate, effect on the cortical cells on meristematic zone etc and effect is increased with increasing the concentration of chromium (Muhammad Faisal and Shahida Hasnain, 2006).

Microorganisms also affected by toxic heavy metals in several ways such as breaking of fatal enzymatic functions, by producing reactive oxygen species in the reaction where they act as redox catalyst, affecting of DNA production and protein synthesis directly, destruction of ion regulation and also several biochemical and physiological properties can be altered in presence of toxic heavy metals. Mercury can affect the transcription process (Maier *et al.* 2000). Protein synthesis is inhibited by mercury, lead and cadmium (Maier *et al.* 2000). Copper, nickel, zinc, mercury, lead and cadmium can disrupt the cellular membrane as well as inhibition of enzymatic activities (Maier *et al.* 2000). Arsenic can cause the DNA damage (Maier *et al.* 2000). Several heavy metals can inhibit cell division also. Chromium can denature the microorganisms by oxidative damage (Igiri, Okoduwa 2018). Chromium Cr (III) can react with carboxyl and thiol groups of enzymes and also reacts electrostatically with phosphate groups of DNA (Igiri, Okoduwa 2018). For that DNA mutagenesis is occurred and effect on transcription, Replication etc. Heavy metal cobalt when applied on E.coli cells it inactivates the [Fe-S] enzymes, aconitase enzyme in sufC mutant. In general it has direct effect on scaffold proteins involved in [Fe-S] cluster biosynthesis. It is indicated that cobalt is toxic for iron metabolism (Caroline Ranquet et al, 2007). Another study revealed that it effects on various metabolic process such as Fe–S cluster assembly, sulfur assimilation, production of free radicals and reduction of free thiol pool. It also creates cobalt- protoporphyrin IX which inhibits their electron transport capacity and resulted in a substantially decreased respiration (Tomas Majtan et al, 2011).

SOLUTIONS AND RECOMMENDATIONS

These wastes are very dangerous for human health and all living beings. For that the safe management is very necessary to protect the environment so their safe management has received considerable attention worldwide.

Treatment is an important phase in the management of heavy metal wastes, it aims to reduce the volume of generated wastes to enhance the safety and reduce the cost of further management phases.

Microorganisms utilize the heavy metals and toxic materials in their Bio geochemical cycling for detoxifying them. Almost all metal–microbe interactions have been examined as a means for removal, recovery or detoxification of inorganic and organic metal pollutants (Lovley and Coates, 1997; Francis, 1998; Stephen and Macnaughton, 1999; Eccles, 1999).

The microbial processes for bioremediation of toxic metals from waste streams employ living cells, non-living biomass, or biopolymers as biosorbents. A wide variety of fungi, algae and bacteria are now under study or are already in use as biosorbents for heavy metal remediation (Gadd, 1992; Volesky and Holan, 1995). Recent work (Johnson and Hallberg, 2004; Hall et al., 2005; Hall and Puhlmann, 2004) indicated that most of the transition between metal speciation forms was controlled by microbial behaviour.

The cost effective and ecofriendly newer biotechnological processes viz. bioremediation and biobeneficiation through microbial metal reabsorption have been widely accepted.

Bioleaching/biosolubilization of metal sulphide ores is an ideal alternative for the mitigation of pollution even at mining sites. It has been found that maximum rates and yields of metal extraction can be enhanced at elevated temperatures (Norris, 1990). Ultimately there is a need to search such metal tolerant, metal absorbent organisms for biogeotechnological applications.

From polluted environment several techniques are there to remove or recover heavy metals adsorption processes, electrochemical techniques, ion exchange, reverse osmosis, chemical precipitation, chemical oxidation or reduction reaction, evaporative recovery & sludge filtration are among them. These techniques are not so much effective because of high reagent requirement (effective when metal concentration in solution is greater than 100mg/L) and high energy requirement, generation of toxic waste and the unpredictable nature of metal ion removal. The heavy metals are highly soluble, for that separation of heavy metals by physical and chemical techniques is problematic and also incomplete. So, there is a need to evaluate alternative techniques which are suitable, appropriate and applicable to all conditions.

The use of microorganisms to remediate polluted environments is sustainable and helps to restore the natural state of the polluted environment with long term environmental benefits and cost effectiveness.

These wastes are very dangerous for human health and all living beings. For that the safe management is very necessary to protect the environment so their safe management has received considerable attention worldwide.

Many organisms have the natural capacity to biosorb toxic metals. Following figure shows the main types of microorganisms associated in bioremediation of toxic heavy metals. It shows that bacteria are the main microorganism that use for this purpose. Second is fungi and algae and yeast are also quite responsible for this job.

In this study bacteria is selected as the metal bioremediation because:

- Small sized bacteria are most abundant and flexible.
- They are capabile to grow under controlled conditions
- Their resistance against a wide range of varying environmental conditions.
- The metal uptake capacities of bacteria generally range between 568 to 0.70 mg g-1
- A great deal of heterogeneity exists among different bacterial species in relation to their number of surface binding sites, binding strength for different ions and the binding mechanisms.
- Cell walls are composed of peptidoglycans which consist of linear chains of the N-acetyl glucosamine & β-1,4-N-acetyl muramic acid with peptide chains.
- Gram positive cell walls and surfaces have a negative charge density due to peptidoglycan network, a macromolecule consisting of strands of alternating glucosamine and muramic acid resi-

dues, which are often N-acetylated. Carboxylate groups at the carboxyl terminus of individual strands provide anionic character to the cell wall.

- The phosphodiesters of teichoic acid and the carboxyl groups of teichuronic acid contribute to the ion exchange capacity of cell walls

Treatment is an important phase in the management of heavy metal wastes, it aims to reduce the volume of generated wastes to enhance the safety and reduce the cost of further management phases.

Microorganisms utilize the heavy metals and toxic materials in their Bio geochemical cycling for detoxifying them. Almost all metal–microbe interactions have been examined as a means for removal, recovery or detoxification of inorganic and organic metal pollutants (Lovley and Coates, 1997; Francis, 1998; Stephen and Macnaughton, 1999; Eccles, 1999).

The microbial processes for bioremediation of toxic metals from waste streams employ living cells, non-living biomass, or biopolymers as biosorbents. A wide variety of fungi, algae and bacteria are now under study or are already in use as biosorbents for heavy metal remediation (Gadd, 1992; Volesky and Holan, 1995). Recent work (Johnson and Hallberg, 2004; Hall et al., 2005; Hall and Puhlmann, 2004) indicated that most of the transition between metal speciation forms was controlled by microbial behaviour.

BACKGROUND RESEARCH

Materials and Methods:

Ø Sampling:

In Howrah district of West Bengal there are belts of small scale galvanizing industries. Among them three most big are selected where more labours working. From there soil samples are collected. We all know that cobalt and chromium are mixed with zinc as impurities. When galvanization is done then the impurities are mixed with soil. For that reason this industry is selected. Mainly surface soils are collected with few inches depth. These soil samples are collected in sterile containers.

Ø Isolation:

Soil samples were serially diluted and plated on nutrient agar plates. Plates were incubated in 37^{0}C for 24-48 hours. According to colony morphology, character and colour eight different isolates were identified (Given in table-1).

Ø Study of Chromium content in samples:

Samples were pulverized and homogenized first and then 50% HNO_3 was added to a digestion vessel. After mixing well this solution was heated at 95°C±5 on a hot plate and refluxed for 10-15 minutes without boiling. After cooling the same step is repeated until no brown fumes are released. Ultrapure water and 30% H_2O_2 were added to the sample when it is cool. The solutions were again heated until the effervescence reduced; aliquots of 1 ml of 30% H_2O_2 were added until the effervescence was minimal or the sample's appearance suffered no further changes. The heating was repeated and continue until the volume of acid-peroxide digestate reduced. After cooling, this was dilute to 100 ml with water. Particulates in the digestate then be removed by filtration by Whatman No. 41 filter paper. Samples are now ready for USEPA 3050B study in mg/kg. (Table-2)

Ø Chromium Tolerance Test:

According to table-2 three concentrations of chromium were selected. These respective concentrations are made in nutrient broths and pure bacterial cultures of these eight bacterial strains are inoculated. Also the strains were inoculated in nutrient broth without metal concentrations for control. After 24 hours of incubation the growth of every tube is measured in Spectrophotometer and compare with the control (Table-3). The highest percentage of chromium tolerant strains (two) ware selected.

Ø LD_{50} study:

From metal tolerance test a clear reference is created about concentration of chromium and percent of metal tolerance. According to this table several nutrient broths were prepared with a series of chromium concentrations within this range. In the same process of metal tolerance assay OD reading was recorded and with comparison of Nutrient broth culture without chromium metal concentration was selected for each strain where half of the bacterial population remains. More than one value was recorded for each strain and Standard error was calculated (Table-4, Table-5 and Fig-2) with the following calculation:

$$\text{Standard Error(SE)} = \frac{Standard\ deviation(\sigma)}{\sqrt{Sample\ Size(n)}}$$

$$\text{SE} = \frac{(\sigma)}{\sqrt{(n)}}$$

Ø 16s rRNA study:

DNA was isolated from the culture P1 and P2. Quality was evaluated on 1.2% Agarose Gel. Isolated DNA was amplified with *16S rRNA* Specific Primer (8Fand 1492R) using Veriti® 99 well Thermal Cycler (Model No. 9902). The PCR amplicon was enzymatically purified and further subjected to Sanger Sequencing. Bi-directional DNA sequencing reaction of PCR amplicon was carried out with 8F and 1492R primers using BDT v3.1 Cycle sequencing kit on ABI 3730xl Genetic Analyzer. Consensus sequence of 1471 bp for P1 and 1461bp for P2 *16S rDNA* were generated from forward and reverse sequence data using aligner software.The *16S rDNA* sequence was used to carry out BLAST alignment search tool of NCBI Genbank database. Based on maximum identity score first Fifteen sequences were selected and aligned using multiple alignment software program ClustalW. Distance matrix was generated using RDP database and the Phylogenetic tree was constructed using MEGA5 (Fig-3, Fig-4, Fig-5 & Fig-6)

Ø FAME analysis: This analysis needs two steps:

i) Fatty acid Saponification, methylation and extraction: The selected samples were treated for saponification, methylation and extraction by the method described in Sharmili AS, 2016 and Myron Sasser, 2006. After extraction the samples were kept for evaporation. Then the evaporated, dried samples were reconstituted by organic solvent. Then the samples were stored at -20°C until further use.

ii) FAME analysis by Gas chromatography: After preparing these samples were analyzed in Gas chromatograph attached with MIDI Sherlock Microbial Identification System and PLFA Software is being used for identification of bacterial strains based on gas chromatographic (GC) analysis of extracted microbial fatty acid methyl esters (FAMEs) and FAME profile compared with profile stored with MIDI's Sherlock bacterial library and PLFA in environmental soil samples quantified by using Sherlock's PLFA analysis software.

Ø Cell morphology and Motility study:

For cell morphology analysis Gram Staining is done and gram character and cell morphology is observed. For motility hanging drop method is used and observed under phase contrast microscope (Table-8).

Ø Biochemical study:

- **Indole Test:** Selected strains were grown in 1%Tryptone broth and incubate for 24 hours. Next day the Kovac's reagent is mixed with these strains with control (Table-9).
- **Methyl Red and Voges Proskaur Test:** Two sets of selected stains were first grown on MRVP broth and allowed to incubate for 24 hours. Next day in one set few drops of Methyl red indicator is added and in other set 10-12 drops of V-P reagent 1 and 2-3 drops of V-P reagent 2 was added sequentially (Table-9).
- **Citrate utilization Test:** Simmon's Citrate Agar slants were made and then two strains streaked on the slants with a comparative control and kept them for incubation at 37°C for 24 hours (Table-9).
- **Catalase test:** Selected strains were grown on the trypticase soy broth slants at pH 7.3 with an uninoculated broth tube. Next H_2O_2 was added drop wise in selected cultures and in uninoculated tube (Table-9)
- **TSI Agar test:** Using sterile condition and technique each experimental organism was inoculated into its appropriately labelled tubes by means of a stab-and–streak inoculation. Control was also maintained. All the tubes were incubated at 37°C for 24 hours (Table-9).
- **Urease Test:** Urea agar medium was prepared using 20% aqueous urea solution and 0.2% phenol red solution. Selected strains were streaked on the respective tubes and kept for incubator for 24 hours at 37°C (Table-9).
- **Carbohydrate Fermentation test:** In this experiment Glucose, Sucrose, Xylose, Lactose and Mannitol were used as carbohydrate source. The fermentation medium was prepared with adding the carbohydrate source in their respective medium. Here durham's tube was used for gas production and phenol red is used as indicator which is red in neutral pH and yellow in acidic pH. After inoculating the selected strains these were kept in incubator for 24 hours at 37°C (Table-9).
- Antibiotic Sensitivity test:

For P1 and P2 18 antibiotics were used. The P1 and P2 were spreaded on the nutrient agar plates and then antibiotic discs were placed. The result was given below. (Table-10)

Ø Chromium Uptake in LD_{50} concentration of the respective strains:

This was done by two processes:

i) Process 1: Measurement of chromium in nutrient broth for each strain

The chromium concentration was analyzed by standard Diphenyl Carbazide (DPC) method for hexavalent chromium where chromogenic reaction is occurred which was measured by spectrophotometer (Table-8 & 9).

ii) Process 2: Determination of uptake chromium concentration

After determining the chromium concentration it was deducted from the original chromium concentration in which they grow which is the concentration of chromium what the strain uptake (Table-8 & 9).

Ø Study of growth patterns and bioaccumulation of samples in different stressed condition.

Three parameters were selected for this analysis pH. Temperature and Biomass of selected strains. In all cases chromium concentration was in its LD_{50} value.

§ **pH:** Chromium uptake was monitored in three pH ranges-pH5, pH7 and pH10 (Sengupta et al, 2014,Tarangini,2009). For preparation of different pHs Bicarbonate-carbonate buffer and Citrate buffer was used. HCl and NaOH were used for pH regulation. After growth the solutions were centrifuged at 10,000 rpm for 10mins. The supernatant was collected for chromium concentration assessment by DPC method stated previously.

§ **Temperature:** According to Sengupta et al, 2014, Tarangini, 2009 and R.J.Pathak et.al, 2012 three temperature ranges 21°C, 30°C and 37°C was selected for metal uptake. For incubation BOD incubator was used where different temperature was achieved. Same amount of cultures was given and pH 7 is maintained in those cultures. Same amount of culture was prepared i0n three temperatures and growth pattern was monitored upto stationary phase. Cultures were taken from stationary phase to centrifuge at 10,000 rpm for 10 mins. The supernatant was collected to analyse chromium by DPC method as previously stated.

§ **Biomass:** According to Sengupta et al. 2014, and Tarangini, 2009 three biomass 400µl, 700µl and 1000µl were selected. For P1 1000µl contains $47x10^6$ cells, 700µl contains $329x10^5$ cells and 400µl contains $141x10^5$ cells. For P2 1000µl contains $49x10^6$ cells, 700µl contains $343x10^5$ cells and 400µl contains $147x10^5$ cells. Same pH (pH 7) and temperature (37°C) was maintained in this experiment. In P1 Cr 160 1000µl flask 0.0239 gm of dry weight biomass of P1 was inoculated. In P2 Cr 180 1000µl flask 0.0293 gm dry weight of P2 was inoculated. In P1 Cr 160 700µl and P2 Cr 180 700µl flasks 0.0189 gm and 0.0249 gm dry weight were inoculated respectively in their respective flasks. 0.014gm dry weight of P1 and 0.0196gm dry weight of P2 were inoculated in P1 Cr 160 400µl and P2 Cr 180 400µl flasks respectively. After growth upto stationary phase the cultures were centrifuged at 1000 rpm for 10 mins and collected supernatant for assessment of chromium content into the supernatant by the same DPC method and calculated the uptake of chromium as indicated before.

Ø Antagonism Assay:

Both P1 and P2 are cross streaked on the agar plates.

Results:

❖ Isolation:

Eight samples named S1, S2, S3, S4, S5, S6, S8 and S9 were isolated from the industrial sludge samples. Their colony colour and Nature of colony are indicated according the Table-1.

❖ Study of metal content in sample:

Table 1. Colony colour and nature of colony of isolated strains

Strain No.	Colony Feature	
	Colour Of Colony	Nature Of Colony
S1	White	Irregular, boil like.
S2	Creamy	Round Shaped, Slimy growth
S3	White	Irregular, boil like.
S4	Creamy	Round Shaped, Slimy growth
S5	White	Round Shaped, Shiny
S6	White	Round Shaped, Shiny
S8	Creamy	Round Shaped, Slimy growth
S9	White	Irregular, boil like.

By acid digestion and ICP-MS study chromium content of sludge sample was determined. This is given in Table-2.

❖ Metal Tolerance Test:

Table 2. Chromium content in sample sludge

Analysis	Method	Result	Unit
Chromium	USEPA3050B/3051A/3052	58.57	mg/kg

According to Table-2 the metal content in sludge samples were 58.57mg/kg or 58.57 ppm. So we have selected the near about double, six times and 10 times concentrations of Chromium. Table-3 of following depicts the % of metal tolerance. From this table we have seen that S1 and S5 have the highest tolerance among others and this more than 50%. So, these strains were selected. Isolated strains were reexamined in chrome agar for differentiate the cells in colony formation type and Color.

❖ LD_{50} study:

Table 3. Percentage of chromium tolerant of isolated strains

		Values are on the basis of O.D reading			
Sample No.	Growth in NB	100 ppm	% of Metal Tolerence	300 ppm	600 ppm
S1	0.526	0.832	58.17	0.269	0.25
S2	0.751	0.413	-45	0.203	0.173
S3	0.54	0.4	-25.92	0.194	0.101
S4	0.55	0.14	-75.54	0.089	0.083
S5	0.6	0.98	63.33	0.16	0.143
S6	0.48	0.369	-23.25	0.173	0.064
S8	0.24	0.277	15.42	0.129	0.116
S9	0.401	0.409	2	0.211	0.19

The optical density values of P1 and P2 are given in the table-4. By analyzing these values the LD_{50} of P1 and P2 are determined and these are 180ppm and 160ppm chromium for P1 and P2 respectively. These results are given in Table-4 & 5.

❖ 16s rRNA study:

Table 4. Optical density values of growth of P1 and P2 in nutrient broth

Sample	**Growth in Nutrient broth**
P1	0.540 O.D value
P2	0.6 O.D value

- **For P1:**

The culture, which was labeled as **P1** showed similarity with ***Bacillus cereus*** based on nucleotide homology and Phylogenetic analysis.

Table 5. LD50 Values of Chromium for P1 and P2

Sample	O.D values in 180 ppm	Mean	O.D values in 160 ppm	Mean	Std. Error
P1			0.283		
			0.281	0.274	0.013317
			0.259		
P2	0.285				0.0252080
	0.27	0.302			
	0.352				

Figure 2. Standard error graph for comparison of Chromium LD50 Values of P1 (160ppm) and P2 (180ppm)

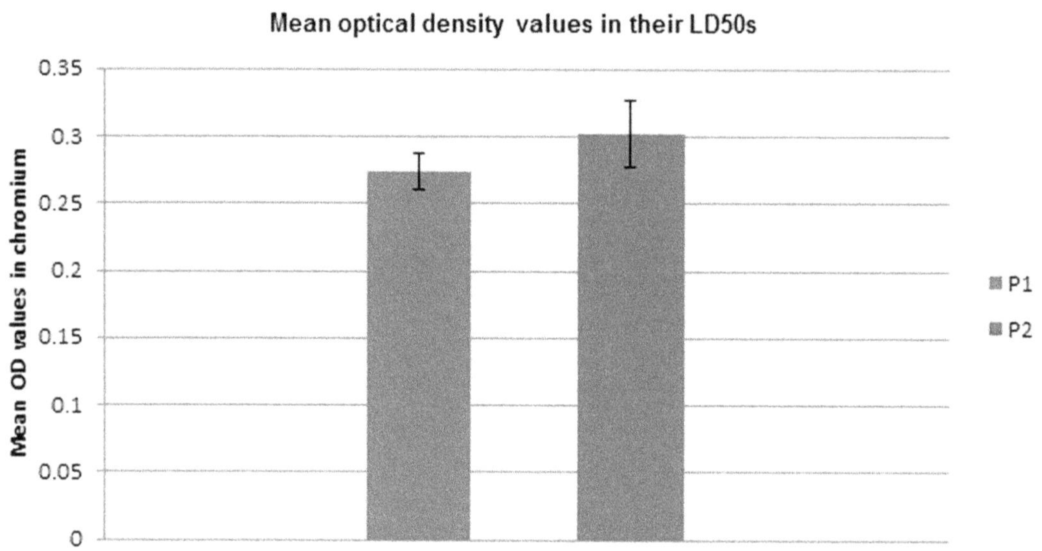

Agarose gel

Figure 3. Quality check on 1.2% agarose gel showing single 1500 bp of 16S rDNA amplicon. Lane 1: 100bp DNA ladder; Lane 2: 16S rDNA amplicon

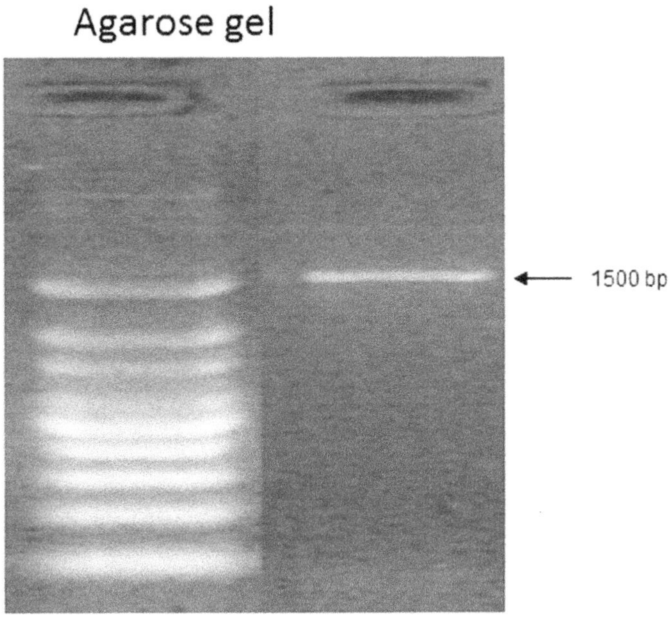

Phylogenetic reading of P1

The bootstrap consensus tree inferred from 1000 replicates is taken to represent the evolutionary history of the taxa analyzed Branches corresponding to partitions reproduced in less than 50% bootstrap replicates are collapsed. The evolutionary distances were computed using the Jukes-Cantor method and are in the units of the number of base substitutions per site. The rate variation among sites was modeled with a gamma distribution (shape parameter = 1). The analysis involved 16 nucleotide sequences. Codon positions included were 1st+2nd+3rd+Noncoding. All positions containing gaps and missing data were eliminated. There were a total of 1409 positions in the final dataset. Evolutionary analyses were conducted in MEGA5 .

So, according to 16s rRNA and blast data analysis P1 is identified as *Bacillus cereus.*

- **For P2:** The culture, which was labeled as **P2** showed similarity with *Enterobacter* sp. based on nucleotide homology a Phylogenetic analysis.

Figure 4. Evolutionary Relationship: The evolutionary history was inferred using the Neighbor-Joining method

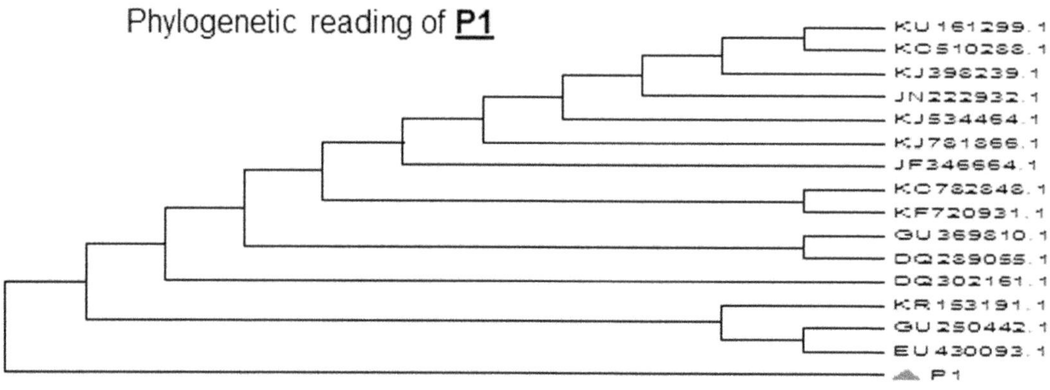

Agarose gel

Figure 5. Quality check on 1.2% agarose gel showing single 500 bp of 16S rDNA amplicon. Lane 1: 100bp DNA ladder; Lane 2: 16S rDNA amplicon

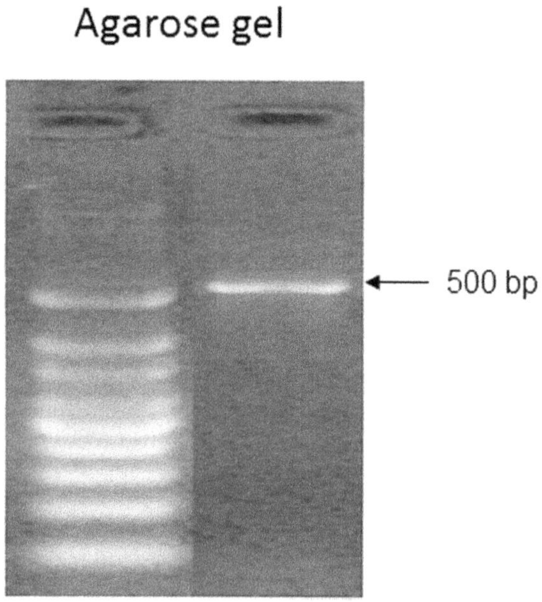

Phylogenetic reading of P2

Figure 6. Evolutionary Relationship: The evolutionary history was inferred using the Neighbor-Joining method

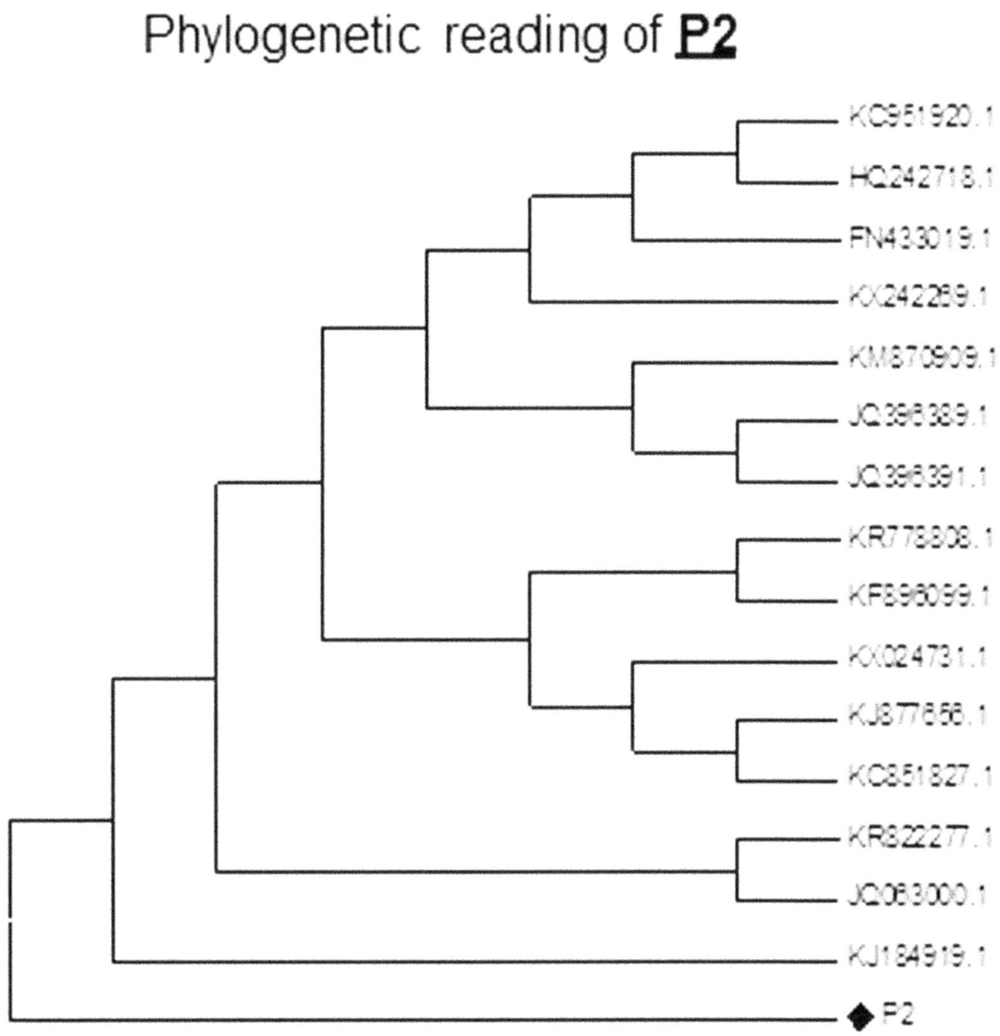

The bootstrap consensus tree inferred from 1000 replicates is taken to represent the evolutionary history of the taxa analyzed Branches corresponding to partitions reproduced in less than 50% bootstrap replicates are collapsed. The evolutionary distances were computed using the Jukes-Cantor method and are in the units of the number of base substitutions per site. The rate variation among sites was modeled with a gamma distribution (shape parameter = 1). The analysis involved 16 nucleotide sequences. Codon positions included were 1st+2nd+3rd+Noncoding. All positions containing gaps and missing data were eliminated. There were a total of 1409 positions in the final dataset. Evolutionary analyses were conducted in MEGA5 .

So, according to 16s rRNA and blast data analysis P2 is identified as *Enterobacter* sp.

Table 6. List of types of fatty acids present in P1

Branched-odd	Branched-even	Anti-iso	Normal
13:0 iso, 15:0 iso 15:1 w5c, 17:0 iso 17:1 iso w5c 17:1 iso w10c	12:0 iso 14:0 iso 16:0 iso	17:0 anteiso 15:0 anteiso 17:1 anteiso A, 13:0 anteiso	12:0, 14:0, 15:0, 16:1 w7c alcohol, 16:0 N alcohol 16:1 w11c, 16:0, 15:0 2OH, 17:0, 18:0, 18:1 w9c 18:3 w6c (6,9,12)

❖ FAME Analysis:

- **P1**:

ECL Deviation: 0.002 Reference ECL Shift: 0.004 Number Reference Peaks: 14
Total Response: 228031 Total Named: 226743
Percent Named: 99.44% Total Amount: 219783
Matches:
Library Sim Index Entry Name
RTSBA6 6.21 0.759 Bacillus-cereus-GC subgroup A
0.529 Bacillus-thuringiensis-israelensis
So, according to FAME result P1 is *Bacillus-cereus*-GC subgroup A

- **P2:**

ECL Deviation: 0.005 Reference ECL Shift: 0.004 Number Reference Peaks: 13
Total Response: 768591 Total Named: 700709
Percent Named: 91.17% Total Amount: 674981
Matches:
Library Sim Index Entry Name
RTSBA6 6.21 0.513 Enterobacter-hormaechei
0.331 Enterobacter-cloacae
0.310 Escherichia-coli-GC subgroup A (DNA homology with Shigella)

❖ Cell morphology and Motility study:

The cell morphology, gram character and motility of P1 and P2 are as given below.

Figure 7. Sherlock Sampler report of different peaks of fatty acids for P1

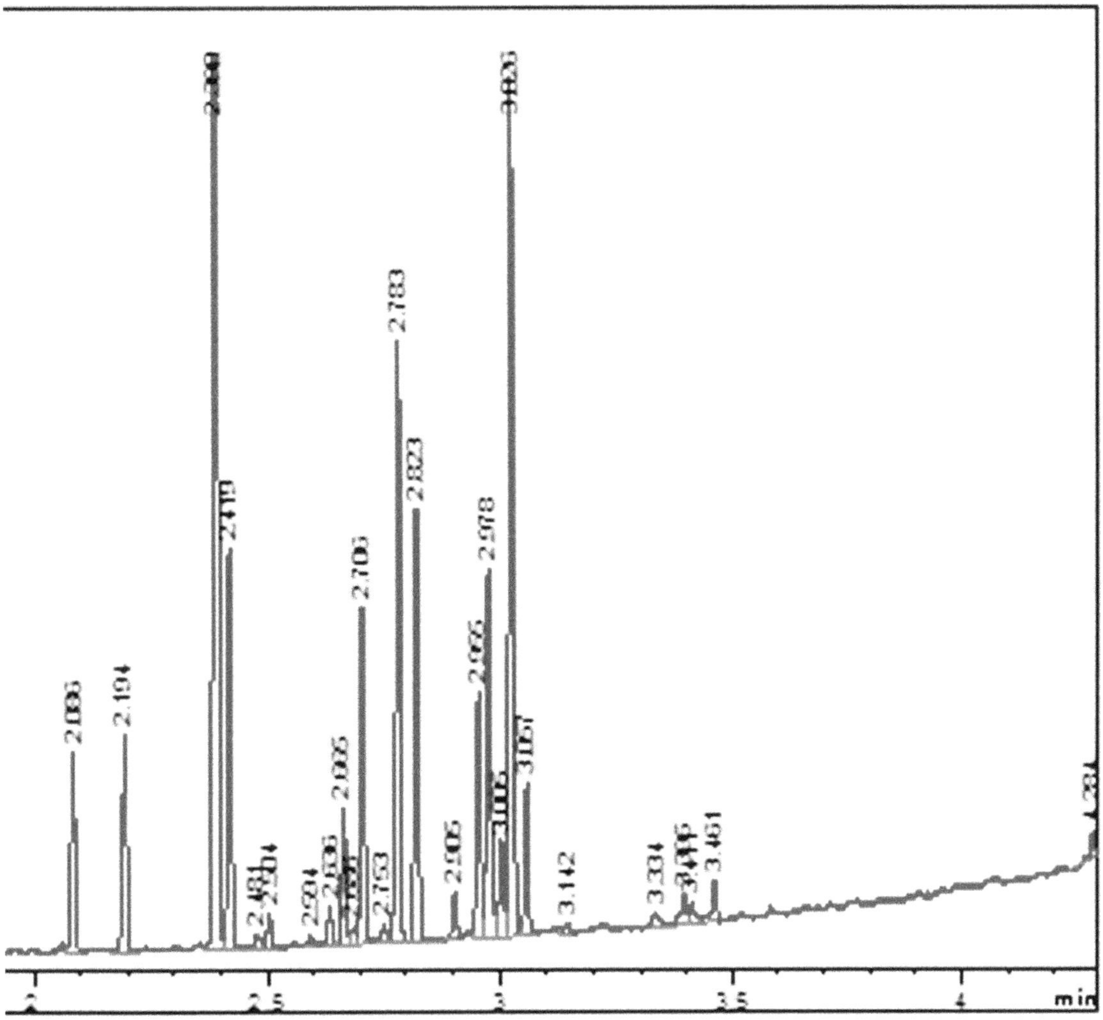

Table 7. List of types of fatty acids present in P2

Branched-odd	Branched-even	Anti-iso	Normal
13:0 iso 3OH, 15:1 iso F 15:0 iso 17:1 iso w5c 17:1 iso ω5c, 17:0 iso 3OH	14:0 iso	13:0 anteiso 15:0 anteiso 17:0 anteiso	10:0 3OH, 11:0, 12:0, 12:0 2OH, 12:0 3OH 13:0, 14:0, 14:02OH, 15:0, 15:1ω8c 15:1ω6c, 15;1ω5c, 16:1ω7c alcohol, 16;1ω5c 16;0, 16:0 3OH, 17:0, 17:1 ω8c, 17:0 cyclo 18:0, 18:1 2OH, 19:0, 19:0 cyclo ω8c, 20:0 20:4 ω6,9,12,15c, 20:1ω7c

Figure 8. Sherlock Sampler report of different peaks of fatty acids for P2

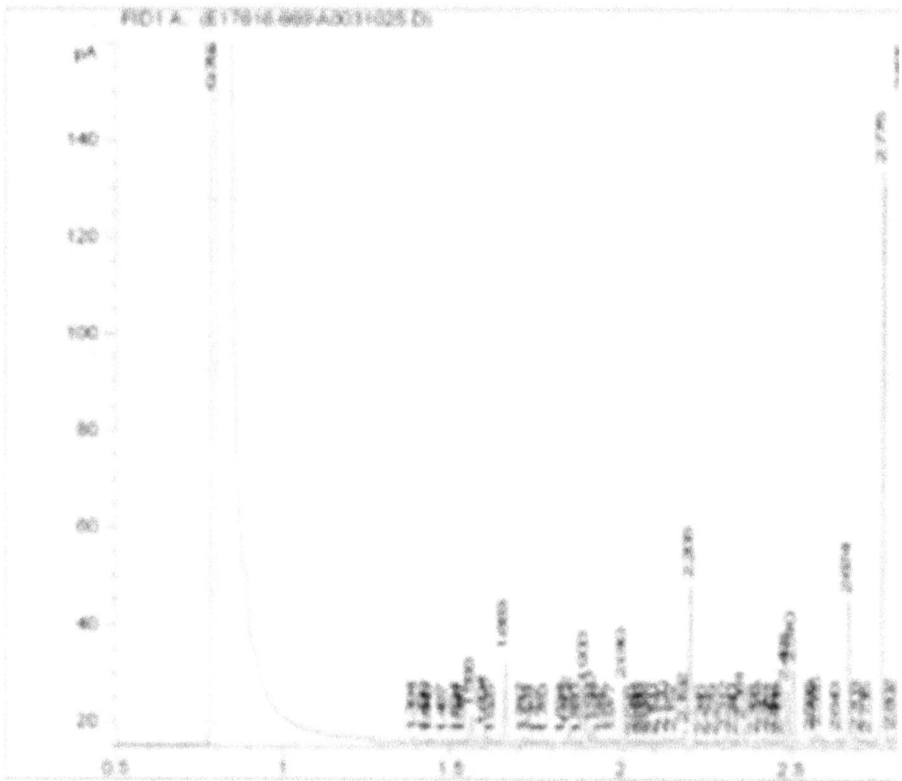

Table 8. Motility, Gram character and cell shape/arrangement of P1 and P2

Sample		Cell Feature	
	Motility	Gram Nature	Shape/Arrangement
P1	Motile	Gram Positive	Bacilli
P2	Motile	Gram Negative	Rod

Table 9. Biochemical test result of P1 and P2

Colony No.	I	MR	VP	Cit	TSI				Carbohydrate Fermentation Test					Oxi	Cat	Urease
					Butt	Slant	H_2S	Gas	Glu	Xyl	Lact	Mannitol	Suc			
P1	-	+	-	+	Acidic	Acidic	-	+	-	-	-	-	-	+	+	+
P2	-	+	-	+	Acidic	Alkaline	+	-	A+G	A+G	A+G	A+G	A+G	-	+	+

+ indicates Positive result, **-** indicates negative result, **A** indicates acid and **G** indicates Gas production

❖ Various biochemical character and antibiotic sensitivity of isolated strains:

Table 10. Biochemical test results of P1 and P2

	Gen	PI	LZ	CIP	TEI	VA	CD	AMC	E	P	OX	CEP	CTX	LE	AT	AK	IPM	CAZ
P1	+	-	-	+	-	-	-	-	-	-	-	-	+	+	-	+	+	-
P2	+	+	+	+	+	+	-	-	-	-	-	-	-	+	+	+	+	-

Gentamycin, 10mcg, Piperacillin=100 mcg, Linzolid=30 mcg, Ciprofloxacin=5 mcg, Teicoplanin=30 mcg, Vancomycin= 30 mcg, Clindamycin=2 mcg, Amoxyclav=30 mcg, Erythromycin=15 mcg, Oxacillin= 1 mcg, Penicillin G= 10 units, Cefotaxime=30 mcg, Levofloxacin=5mcg, Aztreonam=30 mcg, Imipenam=10 mcg, Amikacin=30 mcg, Ceftazidime=30 mcg, Cephalothin=30mcg.

- Study of growth patterns and bioaccumulation of samples in different stressed condition

i) pH:

a) For P1

Figure 9. i) Growth Pattern of P1 in pH5, ii) Growth Pattern of P1 in pH7, iii) Growth Pattern of P1 in pH9, iv) % of chromium uptake of P1 in above said pHs is shown with standard error values. In this case 47x10^6 cells were inoculated.

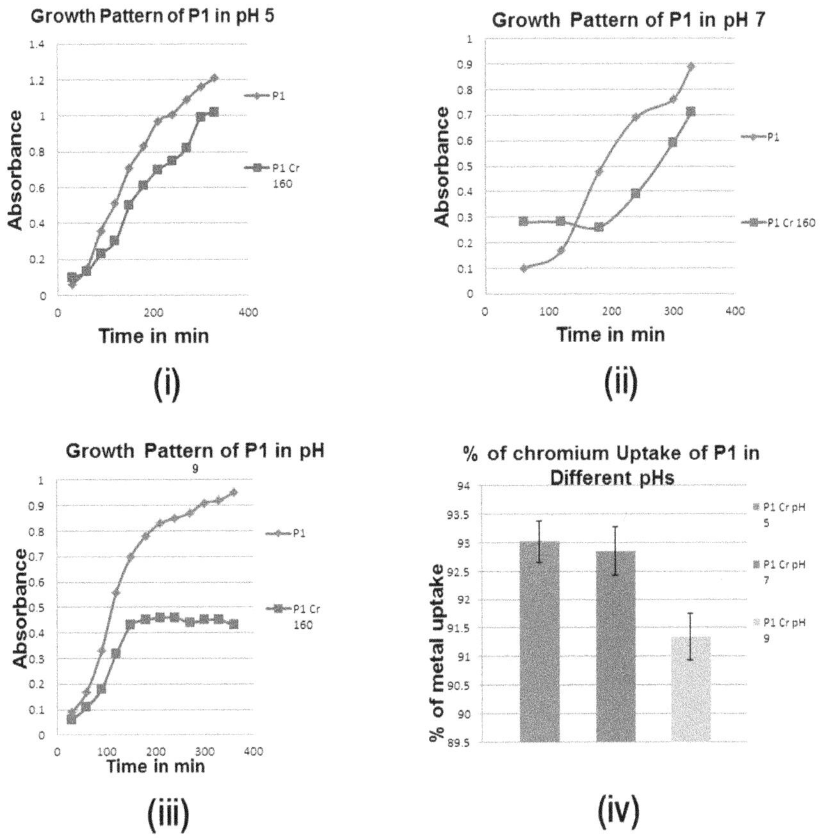

b) For P2

Figure 10. i) Growth Pattern of P2 in pH5, ii) Growth Pattern of P2 in pH7, iii) Growth Pattern of P2 in pH9, iv) % of chromium uptake of P2 in above said pHs is shown with standard error values. In this case 47x106 cells were inoculated.

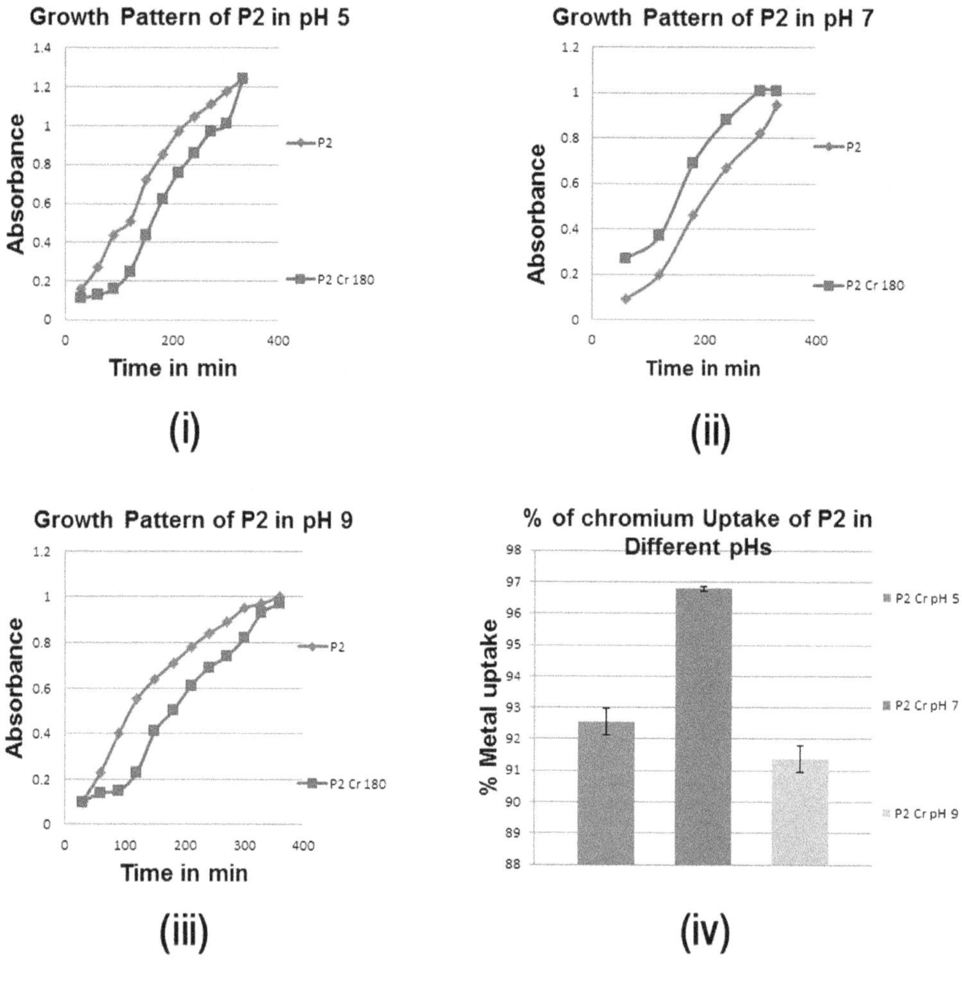

ii) Temperature

a) For P1

Figure 11. i) Growth Pattern of P1 in 30⁰C, ii) Growth Pattern of P1 in 37⁰C, iii) Growth Pattern of P1 in 22⁰C, iv) % of chromium uptake in above said temperatures is shown with standard error values. In this case 47x10⁶ cells were inoculated. In all cases same neutral pH have been maintained.

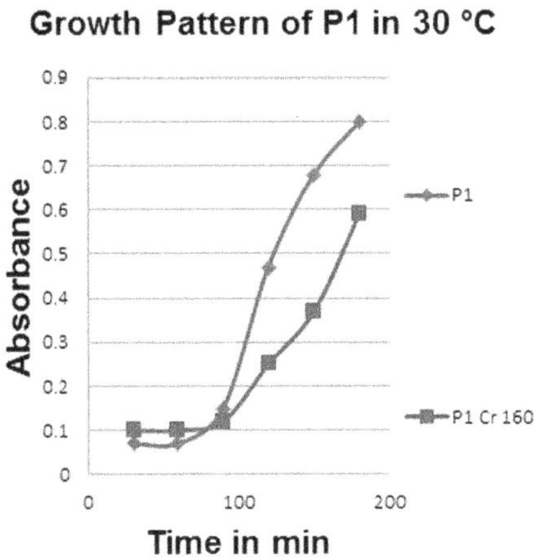

(i)

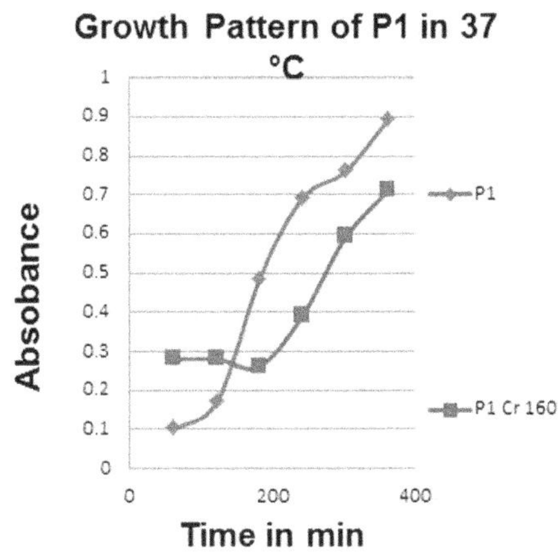

(ii)

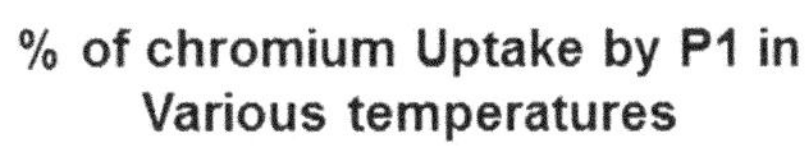

b) For P2

Figure 12. i) Growth Pattern of P2 in 300C, ii) Growth Pattern of P2 in 370C, iii) Growth Pattern of P2 in 220C, iv) % of chromium uptake by P2 in above said temperatures is shown with standard error values. In this case 49x10^6 cells were inoculated. In all cases same neutral pH have been maintained.

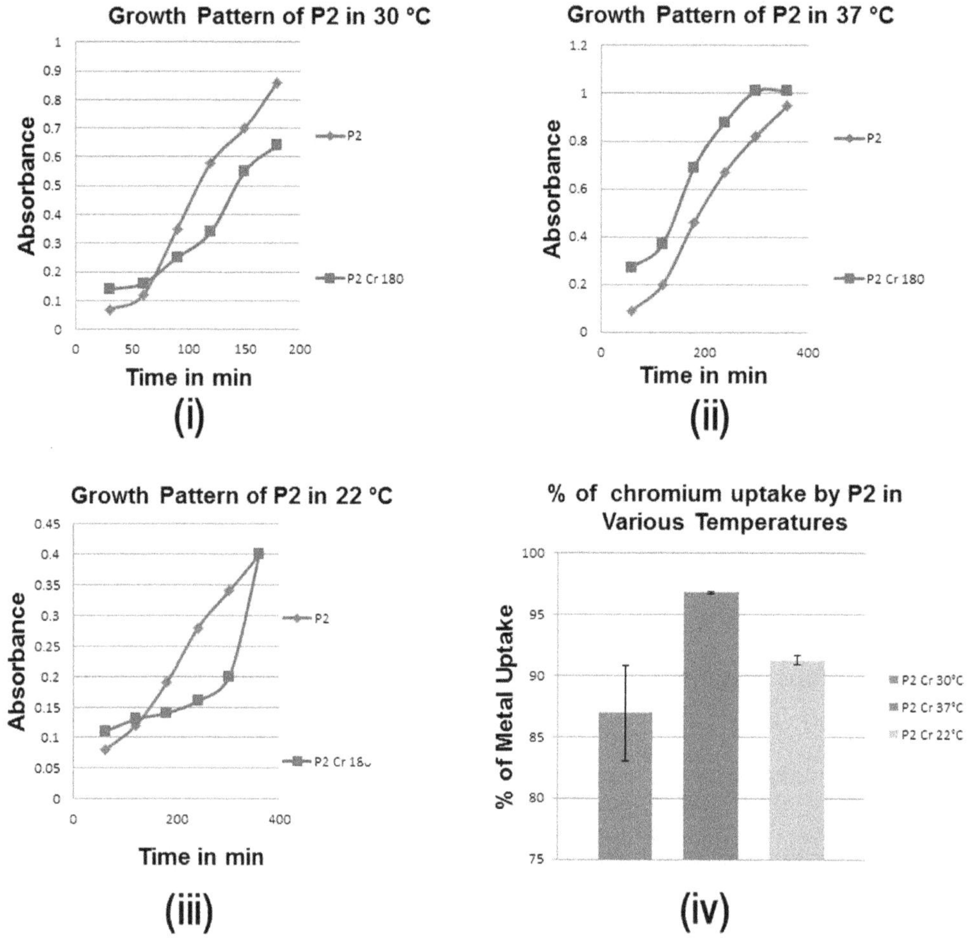

iii) Biomass

a) For P1

Figure 13. i) Growth Pattern of P1 in 400µl biomass, ii) Growth Pattern of P1 in 700µl biomass, Growth Pattern of P1 in 1000µl biomass, iv) % of chromium uptake by P1 in above said biomass categories are shown with standard error values. In these cases same neutral pH and same 37°C temperature have been maintained.

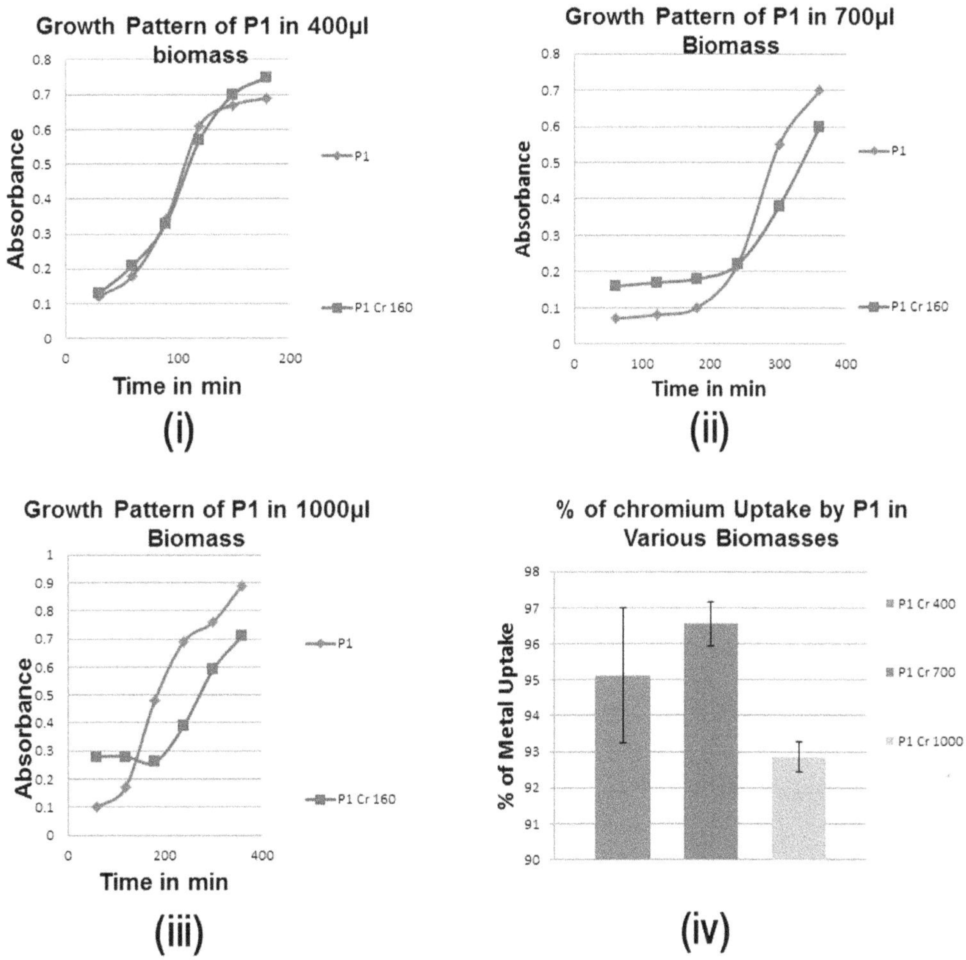

b) For P2

Figure 14. i) Growth Pattern of P2 in 400µl biomass, ii) Growth Pattern of P2 in 700µl biomass, Growth Pattern of P2 in 1000µl biomass, iv) % of chromium uptake by P2 in above said biomass categories are shown with standard error values. In these cases same neutral pH and same 37°C temperature have been maintained.

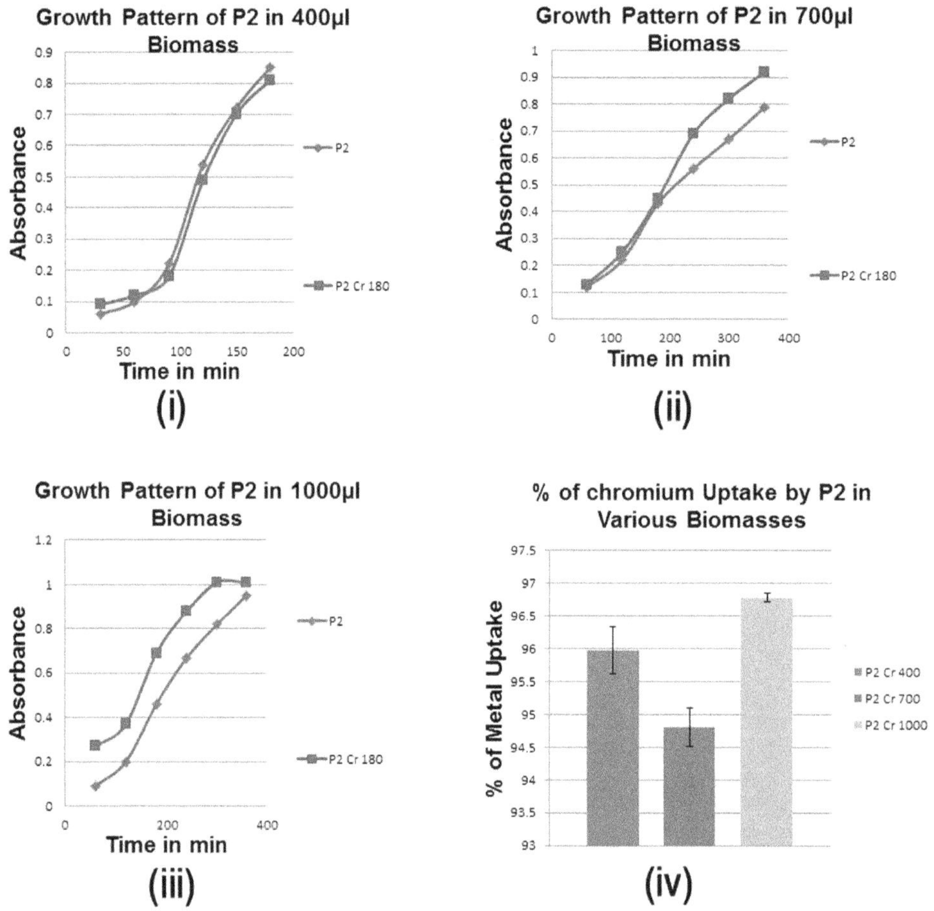

❖ Antagonism Assay:

Table 11. Antagonistic study of P and P2 is revealed that P1 and P2 namely Bacillus cereus and Enterobacter hormaechei are not antagonistic to each other so they will be used as consortium to minimize the chromium toxicity by reducing the toxic, soluble Cr(VI) to nontoxic, insoluble Cr(III)

Antagonism Assay		
Strain No.	**P1**	**P2**
P1	X	-
P2	-	X

FUTURE RESEARCH DIRECTIONS

In future this work will be progressed. Immediate next the TEM-EDX and SEM-EDX will be done for identification and location of metals deposited in bacteria. It will also reveal that whether there will be any change in bacterial morphology for deposition of metal. Then antioxidant tests will be done like Total antioxidant assay, Superoxide Dismutase Assay, Glutathione Reductase assey, Catase and Oxidase assay etc. By this assay it will reveal that whether there is any type of antioxidant present or not.By this study we can assay that SOD is present and by which it can catalyze the dismutation of superoxides. By this Assay we can measure the concentration of total GSH (reduced and Oxidized) in the samples. Catalase detoxifies the H_2O_2 into H_2O and O_2. So this assay is also important. To analyse O_2 as electron accepter this analysis is important. In the meantime SDS gel electrophoresis of isolated protein in stressed condition is analysed with references of control. In this examination it will reveal that whether there is any new type of protein is expressed or not and whether they are up or down regulated. If there will be any novel type of protein expressed then MALDI-MS will be done for quantification of these proteins. Though these bacteria are metal tolerant then they possess metal reductase. The next examination will be metal reductase assay for chromium. Whether there will be any chelation or not for deposition is also be examined by FTIR method. At last development of Consortia of the Selected Bacteria grown in the best possible conditions will be done for bioremediate the toxic metal chromium.

CONCLUSION

Initially eight bacterial strains were identified from the industrial sludge samples. Then according to highest chromium tolerance two were selected from eight designated as P1 and P2 which are Gram positive, motile *Bacillus cereus* and Gram negative, motile *Enterobacter* sp. respectively. According to FAME analysis P1 contains 15:0 iso in highest percentage (31.71%) followed by 17:0 iso (11.28%) as branched-odd fatty acids which were characteristic of Bacillus. After matching with Sherlock sample library RTSBA6 6.21 the highest similarity with Bacillus cereus so P1 is Bacillus cereus GC subgroup A. P2 contains 16:0 in highest percentage (22.55%) followed by 14:0 (5.32%) which is characteristic for Enterobacter. Finally After matching with Sherlock sample library RTSBA6 6.21 P2 is Enterobacter hormaechei. Chromium LD50 of Bacillus cereus and Enterobacter hoemaechei are 160 ppm and 180ppm respectively therefore in 160 ppm half of the Bacillus cereus have survived and in 180ppm half of the Enterobacter population have survived. Now all the experiments were done in this parameter. The strains both are indole negative, therefore they have no tryptophanase enzyme. They can produce various acids like formic, acetic, lactic and succinic acid etc but cannot produce acetoin. Again they can produce Citrase enzyme by which they can utilize citrate as sole source of carbon by breaking down it to oxaloacetate and acetic acid. As they are aerobic organisms naturally they produce catalase enzyme. Among the two Bacillus cereus contains cytochrome c oxidase or indophenol oxidase which transport electrons from donor compounds to acceptors. They also use urease to break the amide bonds in carbon and nitrogen compounds with liberation of ammonia. Enterobacter can ferment every sugar but Bacillus can't. Both strains bioabsorb chromium in stressed conditions significantly. It is observed that they absorb more chromium in low pH than high pH. Near about 93% chromium absorption is occurred in pH7 but 91% absorption in pH9 by both. In neutral pH Enterobacter is more efficient absorber. In both pHs both the strains have growth curve in lower margins compared with control. Both the strains are efficient chromium absorber in 37°C, near about 95% chromium is absorbed in this condition but in other temperatures lower than 37°C the absorption is lower than the above. In biomass category both strains are efficient in every parameter. Over 90% absorption is occurring in every biomass parameter. Both strains are also not antagonistic to each other. So it is revealed that both can be used as bacterial consortium for chromium bioremediation from industrial sludge.

In this study it is observed that two strains *Bacillus cereus* GC subgroup A MCC 3868 and *Enterobacter hormaechei* MCC 3788 can uptake chromium metal more or less 90% and above in every case described above. They take up chromium in same rate even in stressed condition, so they are fit for every condition and for that they can be very efficient tool for environmental cleanup. Both the strains are also not antagonistic and for that bacterial consortium can be made by them. Both are equally useful for bioremediation so, in consortia they can work together. In these perspectives these findings are unique and beneficial for environment.

ACKNOWLEDGMENT

I am very grateful to Rev. Dr. Dominic Savio, S.J. Principal of St Xavier's college, Kolkata for allowing me to join as a scholar in his premiere institute.

My sincere thanks also go to Dr. Sudeshna Shyam Chowdhury, head of the Department of Microbiology, St. Xavier's College, kolkata for her support to every scholars and allow me to access this very rich laboratory.

I would like to express my sincere gratitude to the editors Dr. Satarupa Dey and Dr. Biswaranjan Acharya and full team of IGI-Global for giving me this great opportunity to submit a paper to their publication.

Finally, I am very grateful to the all non-teaching staffs of Microbiology Department, St. Xavier's College, Kolkata for their untiring help by which we can do our PhD work peacefully. I have to thank all my research colleagues with whom I used to share my place in scholar's lab. Without their help I cannot complete my work.

REFERENCES

Abbas, Rafatullah, Ismail, & Lalung. (2014). Isolation, Identification, and Characterization of Cadmium Resistant *Pseudomonas* sp. M3 from Industrial Wastewater. *Journal of Waste Management*. . doi:10.1155/2014/160398

Arévalo-Rangel, Cárdenas-González, Martínez-Juárez, & Acosta-Rodríguez. (2013). *Hexavalent Chromate Reductase Activity in Cell Free Extracts of Penicillium sp*. Hindawi Publishing Corporation. doi:10.1155/2013/909412

Arora. (2012). Microbial Biotechnology: Energy and Environment. *CABI, 2012 - Science*.

Babak, L., Šupinova, P., Zichova, M., Burdychova, R., & Vitova, E. (2012). Biosorption of Cu, Zn and Pb by thermophilic bacteria –effect of biomass concentration on biosorption capacity. *Acta Universitatis Agriculturae et Silviculturae Mendelianae Brunensis*, *40*(5), 2012. doi:10.11118/actaun201260050009

Barałkiewicz, D., & Siepak, J. (1999). Chromium, Nickel and Cobalt in Environmental Samples and Existing Legal Norms. *Polish Journal of Environmental Studies*, *8*(4), 201–208.

Barkay, T., & Schaefer, J. (2001). Metal and radionuclide bioremediation: Issues, considerations and potentials. *Current Opinion in Microbiology*, *4*(3), 318–323. doi:10.1016/S1369-5274(00)00210-1 PMID:11378486

Breil, C., Vian, M. A., Zemb, T., Kunz, W., & Chemat, F. (2017). "Bligh and Dyer" and Folch Methods for Solid–Liquid–Liquid Extraction of Lipids from Microorganisms. Comprehension of Solvatation Mechanisms and towards Substitution with Alternative Solvents International. *Journal of Molecular Sciences*, *2017*(18), 708. doi:10.3390/ijms18040708 PMID:28346372

Bruins, M.R., Kapil, S., & Oehme, F.W. (2000). Microbial resistance to metals in the environment. *Ecotoxicology and Environmental Safety, 45*(3), 198-207.

De, J., Sarkar, A., & Ramaiah, N. (2006). Bioremediation of toxic substances by mercury resistant marine bacteria. *Ecotoxicology (London, England)*, *15*(4), 385–389. Advance online publication. doi:10.100710646-006-0066-4 PMID:16673165

De Jaysankar, N. R., & Vardanyan, L. (n.d.). Detoxification of toxic heavy metals by marine bacteria highly resistant to mercury. *Marine Biotechnology*, *5*, 185–193.

Gadd, G. M. (2009). Heavy Metal Pollutants: Environmental and Biotechnological Aspects. Encyclopedia of Microbiology, 321-334. doi:10.1016/B978-012373944-5.00149-8

Gupta, Tuohy, O'Donovan, & Lohani. (n.d.). *Biotechnology of Bioactive Compounds: Sources and Applications*. Wiley & Blackwell.

Gupta, Schmoll, Maki, Tuohy, & Mazutti. (n.d.). *Applications of Microbial Engineering*. CRC Press Taylor & Francis Group.

Gupta, R., Sundarrajan, S., & Priya, M. (2015, January-March). Isolation and Characterization of Heavy Metal Tolerant Bacterial Isolates VITNJ12 and VITNJ13 from Paper Mill Effluent, Erode District, Tamilnadu, India. *Int. J. Drug Dev. & Res.*, *7*(1), 145–149.

Gurave, Korde, Dhas, & Disale. (2015). Isolation and identification of heavy metal resistant bacteria from petroleum soil of Loni, Ahmednagar, Pelagia Research Library. *European Journal of Experimental Biology, 5*(12), 6-11.

Hall, C., Brachat, S., & Dietrich, F. S. (2005). Contribution of horizontal gene transfer to the evolution of Saccharomyces cerevisiae. *Eukaryotic Cell, 4*(6), 1102–1115. doi:10.1128/EC.4.6.1102-1115.2005 PMID:15947202

Heyrman, J., Mergaert, J., Denys, R., & Swings, J. (1999). *The use of fatty acid methyl ester analysis (FAME) for the identification of heterotrophic bacteria present on three mural paintings showing severe damage by microorganisms Federation of European Microbiological Societies.* Elsevier Science.

Jarosławiecka & Piotrowska-Seget. (2014). Lead resistance in micro-organisms. *Microbiology, 160*, 12–25.

Joshi. (2018). *Biosorption a green approach for heavy metal remedial from water and waste waters.* doi:10.26479/2018.0401.05

Joshi, B. H., & Modi, K. G. J. (2013, April-June). Soil screening and characterization of heavy metal resistant bacteria for its prospects in bioremediation of contaminated. *Environ. Res. Develop. Journal of Environmental Research and Development, 7*(4A).

Kanmani, Aravind, & Preston. (2012). Remediation of chromium contaminants using bacteria. *International Journal of Environmental Science and Technology, 9*(1), 183-193.

Nriagu' & Pacyna. (1988). Quantitative assessment of worldwide contamination of air, water and soils by trace metals. *Nature, 333*.

Pathak, R. K., & Dikshit, A. K. (2012, June). Effect of Various Environmental Parameters on Biosorptive Removal of Atrazine from Water Environment. *International Journal of Environmental Sciences and Development, 3*(3).

Rial, Vázquez, & Murado. (n.d.). *Effects of three heavy metals on the bacteria growth kinetics. A bivariate model for toxicological assessment.* Academic Press.

Sasser. (1990, May). *Identification of Bacteria by Gas Chromatography of Cellular Fatty Acids Revised February 2001.* Academic Press.

Satyanarayana, T. (2005, July). Extremophilic microbes. *Diversity and Perspectives Current Science, 89*(1), 10.

Sekora, Lawrence, Agudelo, Van Santen, & McInroy (2009). Using FAME Analysis to Compare, Differentiate, and Identify Multiple Nematode Species. *Journal of Nematology, 41*(3).

Sengupta, D., Mitra, A. K., Choudhury, S. S., & Chandra, A. (2014). Isotherm Study in Arsenic Tolerant Bacteria Isolated from Arsenic Affected Area in West-Bengal, India. *IOSR Journal of Environmental Science, Toxicology and Food Technology, 8*(1), 8-19.

Sharmili, A. S., & Ramasamy, P. E. (2016). Fatty Acid Methyl Ester (FAME) Analysis of Moderately Thermophilic Bacteria Isolated from the Coramandal Coast, Chennai, Tamilnadu. *European Journal of Experimental Biology, 6*(5).

Tarangini, K., & Gyana, R. (2009, January-March). Optimization of heavy metal biosorption using attenuated cultures of *Bacillus subtilis* and *Pseudomonas aeruginosa. Journal of Environmental Research and Development*, *3*(3).

Umrania, V. V. (2006). Bioremediation of toxic heavy metals using acidothermophilic autotrophes. *Bioresource Technology*, *97*(10), 1237–1242. doi:10.1016/j.biortech.2005.04.048 PMID:16324838

Volesky, B., & Holan, Z. R. (1995). Biosorption of heavy metals. *Biotechnology Progress*, *11*(3), 235–250. doi:10.1021/bp00033a001 PMID:7619394

Wang, Tay, Tay, & Hung. (n.d.). *Environmental bio engineering*. HUMANA Press.

Chapter 7
Phytoremediation:
A Modern Approach

Suparna Pal
https://orcid.org/0000-0002-6975-4156
Lady Brabourne College, Kolkata, India

ABSTRACT

This chapter includes the sources of cadmium and chromium contamination of soil and various detrimental effects on plants and animals. Ecofriendly approach of soil clean up by phytoremediation is the main focus of the author. Heavy metal-induced oxidative stress of plants and their detoxification potentiality has been discussed here to create a wholesome idea about the basic and acute need of phytoremediation. Both enzymatic and non-enzymatic antioxidative defense mechanisms and various other biochemical parameters of metal hyperaccumulator plants are mentioned.

INTRODUCTION

Heavy metal contamination of soil is a global problem due to its negative impact on every component of the ecosystem, threatening the health of vegetation, wildlife and human beings. Bioavailability and bioaccumulation of different heavy metals in aquatic and terrestrial ecosystem are of tremendous global significance as they mainly accumulate in the soil, ground and bottom sediments of seas and oceans and have a long term effect on the biotic factor of this world. There is an absolute need of affordable, environment friendly and sustainable technological solution. Now a day's chemical decontamination of agricultural lands are getting replaced by ecofriendly bioremediation process. Phytoremediation is the foremost attribute of bioremediation. Phytoremediation is an eco-friendly and cost-effective technology for the remediation of heavy metal contaminated soil and water through implication of plants ability to accumulate heavy metals in their harvestable shoot parts. The prerequisite of phytoremediation is identification of native heavy metal tolerant plants with metal tolerance and detoxification capacity and considerable amount of metal uptake as well as accumulation potentiality. By bioremediation any metal contaminated land of India can be converted into agricultural land.

DOI: 10.4018/978-1-7998-4888-2.ch007

Heavy metals interfere and affect biochemical and physiological processes such as photosynthesis, respiration, cell elongation, plant water relationship, nitrogen metabolism and mineral nutrition. Heavy metal induced soil pollution is anthropogenic in origin, such as by the residues of metalliferous mining technology, heavy automobile traffic, smelters, household and industrial wastes (Clement et al. 2007). Depending on the type of industries in the vicinity agricultural lands get contaminated with different metals of no biological roles such as As, Cd, Cr, Pb, Ni etc are deposited in the soil. Among these metals, Cd and Cr are the two most important toxic pollutants affecting both animal and plant physiology extremely. Having maximum industrial usage, Cd and Cr pollution of soil and water is an alarming problem in urban and semi urban areas of India and is of international concern also (Bah et al. 2011). Heavy metal exposure occurs significantly by occupational exposure. As these two metals are higly used in industries (tanning, electroplating, mineral fertilizers, Ni- Cd battery production, paints used for glass and ceramics, soft drink plants) almost 65% of industrial workers and local people living in the close vicinity of these industrial areas are regularly exposed to the hazards of these two toxic metals (Sethi et al. 2006). India, being a third world country, here acute emphasis has been given to the development of industrial hub for the employment of huge population. But the protection and restoration of the environment has not been given same importance or weight age and is almost overlooked.

BACKGROUND

Cadmium

Divalent heavy metal cation Cd with long biological half life is one of the most hazardous elements and a major contaminants due to its greater water solubility and higher phytotoxicity (Clemens 2006). Cd is ubiquotus and biologically non essential. According to the Agency for Toxic Substances and Disease Registry (2007), the position of Cd is 7th in the list of "Top twenty hazardous substances, 2007". Unlike Cu and Fe; Cd is a non redox metal, unable to participate in Fenton reactions but it leads to the formation of reactive oxygen species indirectly by interfering antioxidative defense system of plants (Cargnelutti et al. 2006). Being highly mobile Cd is can be readily taken up by plant roots and transferred to shoot. It can enter the food chain and become detrimental to human and animal health (Chen et al. 2007). Various soil parameters such as pH, redox potential, and rhizosphere chemistry determine the Cadmium bioavailability. Soluble Cd could enter roots and prefers apoplastic or symplastic pathway (Redjala et al. 2009). Being non biodegradable Cd tends to accumulate in atmosphere causing ecological risks. Daily consumption of Cd contaminated foods is a great threat to human health. In Japan, Cd contaminated rice caused Itai Itai disease near Jinzu River basin in the middle of the 20 th century. The risk to health from certain elements in food can be assessed by comparing estimates of dietary exposures with the Provisional Tolerable Weekly Intakes (PTWI) and Provisional Maximal Tolerable Daily Intakes (PMTDI) recommended by the Joint Expert Committee on Food Additives (JECFA) of the Food and Agriculture Organisation (FAO) of the United Kingdom and the World Health Organisation (WHO). For Cd the PTWI set up by JECFA was 7 µg/kg body weights. Recently according to JECFA (2011) the previous PTWI of 7 µg/kg of body weight has been withdrawn, and a PTMI (Provisional Tolerable Monthly Intakes) of 25 µg/kg of body weight has been established because of cadmium's exceptionally long half-life .

Source and Effect of Cd

Atmospheric Cd is mainly anthopogenic in origin. Various human activities such as use of phosphate fertilizers, disposal of house hold and municipal wastes, effluents of cadmium nickel battery manufacturing unit and metal smelting industries contribute excess accumulation of Cd in environment. Cadmium chloride is used in electroplating, dyeing, mirrors, analytical chemistry, vacuum tubes, and lubricants in production of cadmium-containing stabilizers and pigments. Food is the main source of cadmium poisoning for non-occupationally exposed people. Agricultural crops grown in contaminated soil or irrigated with polluted water may have higher risk of Cd accumulation (Rezvani and Zaefarian 2011), resulting its entry into the food chain (Uraguchi et al. 2006; Wei et al. 2008).

Cd toxicity leads to leaf chlorosis, abnormalities in nitrogen metabolism and respiration, compromised growth and even death. Cd toxicity decreases stomatal density and conductance to CO_2 (Baryla 2001), diminishes water and nutrient uptake and distribution. Cd inhibits seed germination, root growth, cell wall modification, mitotic index of cell, also induces damage to different cellular components such as membrane, proteins, DNA. In Cd accumulator or hyperaccumulator plants, Cd tends to be sequestered in apoplast or in vacuoles, attributed to Cd tolerance (Ma et al. 2005). Most prominent and established reason behind Cd toxicity is the production and/or excessive accumulation of ROS (Gratao et al. 2005). Reactive oxygen species, such as singlet O_2, superoxide radical (0_2^-), H_2O_2 are produced in significant quantities in subcellular compartments or organelles due to the partial reduction of O_2 during photosynthesis and respiration under heavy metal stressed condition. These ROS are detrimental to biomolecules such as proteins, membrane lipids and nucleic acids (Cho UH 2005). Among heavy metals Cd and Pb are known to interrupt the photosynthetic electron transport chain which leads to the generation of superoxide radicals and singlet O_2, thus enhance the peroxidant status of the cell by reducing the antioxidant glutathione pool. Cd contamination also increases lipid peroxidation i.e oxidative degradation of lipids of biological membranes. Here free radicals (ROS) steals electron from lipids in cell membrane causing cell damage. PUFA (poly unsaturated fatty acid) is mainly affected. End product of lipid peroxidation malondialdehyde (MDA) reacts with deoxyadenosine / deoxyguanosine forming DNA adducts. According to the International Agency for Research on Cancer, Cd is suspected as mutagenic, teratogenic and human carcinogen (1993). Cd directly affects bone and kidney function) of general population of industrial areas (Bernard 2008).

Chromium

Cr is the 21 st most abundant element in Earth crust. Cr is ubiquitous, found in all phases of environment, air, water and soil. Naturally occurring Cr is composed of three stable isotopes; ^{52}Cr, ^{53}Cr, ^{54}Cr. In environment normal soil contains Cr from 10-50 mg/kg; fresh water (rivers and lakes) Cr concentration range from 20-117 µg/lt. Cr has no verified biological role and has been classified as nonessential for mammals (Kristin 2010). According to Agency of Toxic Substances and Disease Registry (2007) Cr occupies 77 th position in the list of most hazardous substances of the world. Cr exists in two different stable oxidation states in soil, trivalent Cr (III) and hexavalent Cr(VI) but they differ in terms of mobility, bioavailability and toxicity. In most cases, the Cr(III) is the dominating species. Being more mobile, Cr VI is more phytotoxic in comparision to Cr III. Cr(VI) forms chromate and dichromate, which are highly water soluble. On other hand, trivalent Cr is less soluble in water and is required in trace amounts as an inorganic nutrient for sugar and lipid metabolism in animals. Cr is phytotoxic either at all concentrations or above certain threshold levels. Water insoluble Cr III is not considered as health hazards, but Cr VI is extremely toxic and carcinogenic. Agricultural lands in the vicinity of Cr releasing sources are unproductive and unfertile.

Source and Effect of Cr

As Cr and its compounds have multifarious industrial use, huge amount of Cr compound are discharged as liquid, solid and gaseous waste into the environment and ultimately have significant adverse biological and ecological effects. Cr is extensively employed in leather processing and finishing, electroplating cleaning agent, catalytic manufacture, in the production of chromic acid and as a colouring agent of soft drinks (Shanker 2005). Untreated effluents of these industries are the prominent sources of Cr in the atmosphere. Hexavalent Cr is used in industry for metal plating, cooling tower water treatment, wood preservation, paint and plastic industry. Primer paint containing Cr (VI) is used in automobile refinishing applications. Leather tanning industry is the main source for high influx of Cr to the biosphere, accounting for 40% of total industrial use. Annual contribution of tanning industry in India is almost 2000-32000 tons Cr. Cr (VI) has higher solubility and thus bioavailability is more toxic at lower concentrations than Cr(III), which tends to form stable complexes in the soil. Being toxic and non essential, plants do not possess any specific mechanism for Cr uptake. So uptake of this heavy metal occurs through carrier of other essential metals for plant metabolism. Metabolism driven process is needed for Cr (VI) uptake, but Cr (III) is passively taken up, thus Cr (VI) has an easy entry into the plant system (Shanker 2005). Cr get highly accumulated in roots than shoots as Cr is immobilized in the vacuoles of the root cells (Shanker et al. 2005). When both Cr (VI) and Cr (III) penetrate the endodermis through symplast, the hexavalent Cr is very often reduced to trivalent Cr which is retained in the root cortex cells under low concentration of Cr (VI) which is the cause of the lower toxicity of Cr III. Though India doesn't has any permissible scale for Cr, but according to the European Commission Director General of Environment (2010) the maximum allowable limits of Cr in soil for most of the European countries vary within the range of 30-400 mg/kg. Cr is toxic to higher plants at 100 μM/ kg dry weight (Davis et al 2002) and detrimental after crossing this level. Cr is strong oxidant with higher redox potential accounting for rapid induction via generation of ROS and its resultant toxicity (Shanker et al. 2005). Cr phytotoxicity results inhibition of seed germination. Cr affects photosynthesis in terms of CO_2 fixation, electron transport, photophosphorylation and enzymatic activities (Horcsik 2007). Cr inhibits both PS I and PS II. Cr poisoning causes disorganization of chloroplast ultrastructure, inhibition of electron transport process and a diversion of electron from electron donating side of PS I resulting huge decrease in photosynthesis rate. Cr affects both light and dark reaction by inhibiting Hill reaction. Cr reduces the size of peripheral part of the antenna complex resulting the decrease in total chloroplhyll, chlorophyll a/b ratio and carotenoid (Panda and Chowdhury 2004). Cr degrade $^{\delta}$ aminolevulinic acid dehydratase, an important enzyme for chlorophyll biosynthesis (Vajpayee 2001). Cr participates in fenton reactions, providing its redox characters. Production of H_2O_2, OH^-, O_2^- under Cr stress has been demonstrated in many plants generating oxidative stress leading to the damage of DNA, proteins, pigments as well as initiating lipid peroxidation (Chowdhury and Panda 2004; Sinha et al. 2005). Cr hampers respiration by binding cytochrome oxidase. Cr stress imposes a negative effect on Fe absorption (Shanker et al. 2005). Cr inhibits plasma membrane H^+ ATPase activity (Shanker 2005), thus affecting the uptake of different mineral. Reduction in N P K also reduces root growth and impaired penetration of root into the soil due to Cr toxicity (Khan et al. 2007). Higher Cr concentration hampers nitrogen metabolism by inhibiting reductase activity (Chowdhury and Panda 2004). The genotoxicity of Cr (VI) is well documented (Pal and Kundu 2015) in Aligator weed and various other plants. The acute toxicity of Cr is due to its strong oxidation properties. It damages kidney, liver and blood cell, resulting haemolysis, renal and liver failure. The carcinogenicity of chromate dust is known for long time. Continuous exposure to chromate dust cause dermatitis and skin ulcer known as chromate ulcer.

Perspective of Phytoremediation

The remediation of metal contaminated soil, especially agricultural land is of urgent need as metals will persist indefinitely in the environment due to its non biodegradability. The term phytoremediation ("phyto" means plant and Latin suffix "remedium" means clean up or restore) actually refers to different types of plant based technologies which deals with the use of native weeds or genetically engineered plants for cleaning up environment. Restoration of soils, polluted with toxic metals is a major global necessity. Remediation and treatment of Cd, Cr contaminated soil has become an important problem that need to be solved urgently in urban and semi urban industrial areas of India. In past metal pollution of soil have been treated using laborious physical and expensive chemical processes which are not too sustainable and environment safe. The chemical methods for treating contaminated soils include chlorination, precipitation, flocculation, sedimentation, neutralization, equalization and chemical oxidation. Recently more attention has been paid to phytoremediation technologies i.e cleaning up of terrestrial contaminated areas as well as aquatic bodies from heavy metals and organic contaminants by using green plants (Rezvani and Faezeh 2011; Sun et al. 2005). Still now soil dressing and soil washing have practiced to remove Cd contamination but all civil engineering technique is costly and impractical for remediation of large areas. So phytoremediation being a new method for rehabilitating contaminated soil is of global interest. This method has been proven successful because the in situ use of plants for environmental restoration is cost effective, less hazardous in comparison to chemical cleaning up processes and ecofriendly (Gratao et al. 2005).

The key demand and the proper implementation of phytoremediation process depends on the screening and selection of a plant species that have an natural capacity to absorb and accumulate metals at higher concentration, almost 100 times greater than ordinary plants (Baker 1989) so that it can maximize the contaminant removal. This type of plant is known as hyperaccumulator. By definition hyperaccumulator is that plant which can accumulate exceptionally high quantities of one or more metals without manifesting any toxic symptoms, at least 100 mg/kg (0.01% dry weight) Cd, As and some other trace metals, 1000 mg/kg (0.1% dry weight) Co, Cu, Cr, Ni and Pb and 10,000 mg/kg (1% dry weight) Mn and Ni (Reeves and Baker 2000). An appropriate plant for phytoremediation must also have high biomass, faster propagation rate and ability to translocate the target metal into plant shoot. The slow growth rate and low biomass of the hyperaccumulators are documented almost in all literature. There are four standards to judge hyperaccumulators including the threshold value standard (metal concentration in shoot > 100 mg/kg for Cd), BCF standard >1, upto 50-100 (Brooks 1998), TF standard (ratio of metal concentration in shoot/root)> 1 (Wci ct al. 2004) and high capacity to tolerate metal toxicity. Almost 400 hyperaccumulators species have been identified and documented worldwide (Reeves 2003). Few examples are, *Leersia hexandra* (Zhang et al. 2007) Cr hyperaccumulator; *Chengiopanax sciadophylloides*, Mn hyperaccumulator; *Potentilla griffithii* (Qiu et al. 2006) Zn hyperaccumulator; *Thalspi caerulescens* (Baker et al. 2000); *Thalspi praeox* (Torla et al. 2006); *Rorippa globosa*, *Arabidopsis halleri* (Kupper et al. 2000); *Sedum alfredii* (Yang et al. 2004) Cd hyperaccumulator. The indigenous plants have been preferred more as a candidate of phytoremediation purpose as it demands less genetic alteration and adapts easily in native climate conditions and seasonal cycle.

In comparison with crops, weed plants usually display few inherent endurance feature to hypertolerate and higher capacity to absorb heavy metals. Moreover weeds have faster propagation rate and their biomass can increase sharply within a short span of time (Wei et al. 2005). In that point of view weed is an important as well as ideal natural resources of hyperaccumulator to phytoremediate the polluted soil. Inspite of the presence of such hyperaccumulator, large scale application of phytoremediation is not up to the mark due to lower above ground biomass, slow growth and long maturity phase of some hyperaccumulators. Slow progress in cultivating perfect hyperaccumulator by using genetic engineering is also affecting the validity of this method.

In phytoremediation the special plant used may clean up soil by extraction, volatilization, translocation or stabilization (Trotta et al. 2006). So the method of phytoremediation has been categorized in four different plant based technologies each with different modes of phytoremedial action. These include:

Phytoextraction: It is the most common phytoremedial technique. It denotes the use of appropriate plants to absorb metals from soil and translocate them to the harvestable shoots where they accumulate. Thereafter the biomass of the plants seasonally harvested and removed from contaminated sites until the metal load in the soil reduced to an accepted level (Baker et al. 1989; Kramer 2010). It is advantageous as it is cost effective, in situ and non destructive. Application of phytoextraction can reduce phytoavailable metals in the soil. The time required for remediation in this process depends on the severity of soil metal contamination and the tenure of the growing season and the efficiency of the plant candidate itself. The duration may vary from 1 to 20 years. Phytoextraction is feasible only to sites with low or moderate levels of metal pollution.

Rhizofiltration: This process involves growing plants hydrophonically and transfering them into metal contaminated soil where plants absorb and accumulate the metals in their roots and shoots. Then the entire plants are harvested for disposal. The success of rhizofiltration depends on the considerable amount of target metal accumulation and tolerant capacity of the plant candidate. The plant must produce significant amounts of root biomass and root surface area. Sunflower (*Helianthus annus* L.) and Indian mustard are the most promising terrestrial candidates for metal removal in water. Roots of Indian mustard are effective to remove Cd, Cr, Pb, Zn, Ni and sunflower removes Pb, Cs, Sr from hydroponic solutions.

Phytostabilization: Unlike other phytoremediative process, the target of phytostabilization is to stabilize the metal contaminants of a site instead of to remove it. Plants suitable for this technique should be poor translocator of metal to above ground tissue. It has advantages over other soil clean up techniques as it is less expensive, less environmentally evasive, easy to implement and offers aesthetic values. The bioconcentration factor (BCF) and translocation factor (TF) are usually used to estimate plant capability to tolerate and accumulate heavy metals. The BCF is the ratio that represents the metal concentration of the plant tissue to the soil and TF is the ratio of metal concentration in plant shoots to the roots. Plants exhibiting a shoot BCF > 1 are suitable for phytoextraction, and plants with a root BCF > 1 and TF< 1 have the potential for phytostabilization (Rafati et al. 2011).

Phytovolatilization: Some metals (As, Hg and Se) may exist as gas in the air. Some native plants or genetically modified plants are capable of absorbing elemental forms of these metals from soil, biologically convert them to gaseous species within the plant and release them in the atmosphere. Se and Hg contaminated soils are cleaned up by this method.

Tolerance Mechanism of Plants to Heavy Metals

Heavy metals tolerance in plants implies the ability to survive in a soil that is toxic to other plants. The tolerance capacity of plants to heavy metals depends on an interdisciplinary communication of physiological and molecular mechanisms. Plants have adapted different detoxification methods that include various antioxidant pathways (Pandey et al. 2005) which are proven to be sufficient enough for protection from oxidative damage. The most important are low molecular weight non enzymic antioxidants such as ascorbic acid, glutathione, thiole, α tocopherol, protective pigments such as carotenoids (Tausz 2003) and accumulation of free proline. Free proline accumulation is one of the common reported metabolic alteration of plants exposed to heavy metals. Proline accumulation causes reduced damage to membrane and proteins. Proline acts as osmoprotectants, inhibitor of lipid peroxidation and protein stabilizer also. It reduces metal stress specifically Cd stress not by scavenging Cd but by eliminating Cd induced free radical damage. As mentioned before glutathione or non protein thiol group is directly involved in the synthesis of Cd binding peptide phytochelatin and thus develops Cd tolerance to plants. It is reported that species with higher level of SH compound are more tolerant to heavy metal than those non-SH species. Under severe metal stressed condition production of ROS far exceeds the optimum potential of the antioxidant system to detoxify them resulting oxidative damage.

Heavy metals induced oxidative damage is mitigated by the upregulation of antioxidative enzymes such as catalase (CAT), peroxidase (POX), superoxide dismutase (SOD), ascorbate peroxidase (APX) and glutathione reductase (GR) that serve as the first line of defense system and provide tolerance capacity to the plant. SOD is the main free radicals scavenging enzyme and serves as key antioxidants that catalyzes dismutation of superoxides (O_2^-) into O_2 and H_2O_2. Peroxidase converts H_2O_2 into O_2 and H_2O, thus reducing the free radical burden of cell. GR is a flavoenzyme that catalyzes the NADPH dependent conversion of glutathione disulphide (GSSG) to glutathione (GSH) which in turn act as the precursor of metal chelator protein. This reaction is mandatory to maintain a proper GSH/GSSG ratio in cell. APX is another antioxidative enzyme that detoxify peroxides such as H_2O_2 into water using ascorbate as substrate (reducer). They facilitate the transfer of electron from ascorbate to peroxide resulting the formation of harmless dehydroascorbate and water. APX is the integral member of glutathione –ascorbate cycle. The enhanced production of SOD, GR, APX serves as an enormous antioxidative protection against metal stress.

Chelation of Cd, Cr in the cytosol by two classes of metal ligand peptides, Phytochelatin (PC) and Metallothionins (MT) is a very important mechanism of metal detoxification and tolerance. PC are a family of metal complexing peptides that have a general structure (γ-Glu-Cys)n – Gly, where n=2-11, and are rapidly induced in plants under heavy metal stress. PCs are synthesized non transitionally using glutathione as a substrate by PC synthase, an enzyme which activation is the primary line of biochemical defense againt metaltoxicity in plants. Cd accumulation is accompanied by a rapid induction of PC biosynthesis that is capable of Cd sequestration in the vacuole. MTs are gene encoded polypeptide. Plant is also capable of vacuolar compartmentalization to restrict the toxicity of heavy metals.

Though Cd is biologically non essential, it is still observed that plants can accumulate this heavy metal, sometimes in excess. *Chromolaena odorata,* a siam weed was reported as a hyperaccumulator, (Tanhan et al. 2007) accumulating 102.3 mg/kg Cd in shoot and 1440.9 mg/kg Cd in roots. Eight crop species were studied for accumulation and tolerance potential to Cd (Uraguchi et al.2006). Seedlings were treated with different concentrations of Cd for 4 weeks and it was observed that among these eight crops *Avena strigosa* and *Crotalaria juncea* possess the greater potential for Cd accumulation and tolerance. It was reported that under high concentration of Cd and Cr there is an increase in phytochelatin synthase in *Cannabis sativa* L, suggesting an active molecular mechanism to avoid cell damage. Cd uptake and accumulation in *Rorippa globosa* were facilitated by the presence of As in soil (Yuebing et al. 2007). *Pistia stratoites* is a well known aquatic macrophyte capable of accumulating wide range of heavy metals (Sinha 2005). Its potential to tolerate higher metal concentration by the synthesis of metal binding peptides has been used for the treatment of urban sewage (Zemmels 2005). Narain and his group (2011) explored the phytoaccumulation capacity of water hyacinth to clean and improve water quality of natural water bodies polluted with industrial and municipal effluents. Their report revealed the average removal efficiency of this plant, which were 80.26% for Cr and 71.28% for Cd. Singh and his colleagues (2010) surveyed the accumulation, translocation and subsequent uptake of Cd, Zn, Pb, Cu, Ni, Mn and Fe in eleven native plants grown in field polluted with fly ash of a thermal power station. Their report showed highest accumulation of Cd in *Typha sp.* Pandey (2006) studied the accumulation capacity of *Spinacia oleracea* and *Raphanus sativus* grown in the field contaminated with of Cd, Cr, Cu, Ni and Zn from nearby electroplating industry's waste discharge. *S. oleracea* showed higher accumulation of Cd (155 μg/gm) and Cr (302 μg/gm) than *R. sativus*.

The interaction of metal ions with soil matrix is one of the important criteria for metal accumulation and uptake by the plants. Soil pH regulates metal availability and controls the metal uptake by the roots. Low pH facilitates metal solubility in normal agricultural soil. The retention of metal to soil organic matter is weaker at low pH, that makes the metal more bioavailable to the plant for root absorption during ascent of sap (Prasad and Freitas 2003). The high pH value could have accounted for a low transfer of metal from soil to plants.

Some common and often unspecific symptoms, of metals phytotoxicity are: growth inhibition, nutrient imbalance, disturbances in the ion and water regime (Gyuricza et al. 2010), photosynthetic impairment and genotoxicity (Monteiro et al. 2010). The investigation of Zengin and Munzuroglu (2005) on bean seedlings (*Phaseolus vulgaris* L.) grown in Hoagland solution treated with different concentrations of Pb, Cu, Cd and Hg revealed progressive decline in total chlorophyll content with increasing heavy metal concentration, but sharp increase in proline, tocopherol and ascorbic acid. Sandalio and his colleagues (2001) reported that the Cd induced oxidative stress caused growth inhibition as well as reduction in transpiration, photosynthetic rate and chlorophyll content in *Pisum sativum*. Enhanced lipid peroxidation and antioxidative activity of GR, CAT, SOD was also reported. Enhancement of lipid peroxidation is the measurement of oxidative damage of biomembrane with increasing concentration of Cd and Cr is generally due to the formation of ROS. The generation of ROS may be simply due to the shift in the balance between their productivity and functioning of the antioxidant system comprising both enzymatic antioxidants (SOD, CAT, GR, APX) and non enzymatic scavengers such as glutathione, carotenoids, ascorbate and proline. Martin and his colleagues (2011) studied the effect of Cd (10-100μM) on growth parameters, chlorophyll and proline content, enzymatic antioxidative response and lipid peroxidation of Tobacco (*Nicotiana tabacum*). Results showed gradual degradation of chlorophyll content with the increase in Cd concentration and enhanced lipid peroxidation which are indicative of oxidative stress but hyper activity of GPX, SOD scavenging H_2O_2 is attributed the Cd tolerance ability. Srivastava and his coworkers (2011) exposed Soybean (*Glycine max*) to different level of Cd to evaluate growth inhibition in terms of shoot, root lengths and dry weight. Their result showed huge deterioration in chlorophyll content, nitrate reductase activity and total protein contents, while elicitation in both non enzymatic (ascorbate, GSH) and enzymatic (CAT, SOD, APX) antioxidants was also reported. The investigation of Vassilev and Lidon (2011) revealed enhanced membrane lipid peroxidation and K^+ leakage, chlorophyll degeneration, diminished content of soluble protein, glutamic acid in Barley (*Hordeum vulgare*) under 54 μM Cd treatment. Increase in stress related free amino acid was also reported. A comparison of different heavy metal (Co, Ni, Cd, Cr and Pb) toxicity on visible foliar symptoms and some antioxidant properties in sunflower plant was studied by Gopal and Khurana (2011) which revealed that the degree of oxidative damage evaluated by the appearance of visible toxic symptoms and modification in biochemical parameters were found to be in the order Cd> Cr> Ni> Co> Pb. Glutathione alleviates tolerance capacity of plants to ROS by participating in the detoxification of ROS generated by Cd and Cr. Increased GR activity in roots exposed to Cd was also reported in plants including radish, soyabean, sugarcane and *Arabidopsis* (Skorzynska Pilot et al. 2004). Enhanced GR activity in the root and shoot of alfa alfa under Cd exposure was reported by Sorbrino-Plata and coworkers (2009). Shekar and his coworkers (2011) reported that lower concentration of Cd treatment enhanced percentage of total chlorophyll content, but higher concentration showed inhibitory effect. Species with higher level of SH compounds were reported to be more tolerant to heavy metal stress than those non SH groups (Clemens 2006). Bharadwaj and his coworkers (2009) reported the reduction in *Phaseolus vulgaris* seed germination percentage under Pb and Cd combined stress. Pigments, total soluble sugar, starch content, soluble protein decreased with increase in metal concentration. ROS are produced in the young senescing leaf excessively under metal stress and are detoxified by enzymic (CAT, APX, GPX, SOD, GR) and nonenzymic (ascorbate, GSH, alpha tocopherol) antioxidant systems. Metal accumulators were naturally furnished with a higher level of antioxidative enzyme activity for detoxification of ROS. The investigation (Wang et al. 2008) revealed that the activity of SOD and CAT were enhanced rapidly in metal accumulator species in response to Cd and Cr than nonaccumulator species. When the production of ROS is more than the antioxidative potential of the plant, damages occur. Aerobic organisms have developed enzymatic as well as nonenzymatic antioxidants to combat this oxidative stress, among these, most important are low molecular antioxidants such as ascorbic acid, glutathione, thiol, tocopherols and carotenoids (Tausz et al. 2003). Among the enzymatic scavengers SOD, CAT, APX and GR are most effective and participate in a highly equipped detoxification system named the ascorbate-glutathione cycle/ Halliwell-Asada cycle (Noctor et al. 2002). Dey and his co workers (2007) reported the excess production of MDA due to lipid peroxidation under Pb and Cd stress reflecting damaging effect of heavy metals on wheat seedlings. Bah and his group (2011) studied the modification of antioxidant system in *Typha angustifolia* after 30 days exposure to Cd, Cr and Pb and reported the enhanced activity of SOD, CAT, APX under heavy metal stress. Pandey and Singh (2012) reported that pea plants exposed to different Cd concentration showed growth inhibition and induction of chlorosis and necrosis of young leaves but significant enhancement in SOD activity and low activity of POD and CAT explains the accumulation of H_2O_2 in cell causing oxidative damage.

Some plants have the ability to bioaccumulate a considerable concentration of Cd and Zn in leaves without any visible toxic symptoms due to compartmentalization of sequestered metal and strong internal detoxification. Vacuolar compartmentalization and complexation with organic acids decreased the toxicity of the heavy metal. The distribution and subcellular localization of Cd and Zn in Cd hyperaccumulating ecotype of *Thalspi caerulescens* was reported by Ma and her colleagues (2005) that explained its detoxification capacity. Investigation of Chakroun et al. (2010) regarding the accumulation of Pb, Cd and Cu in *Vicia faba* and *Hordeum vulgare* inside a mining district proved that the cultivated areas inside a mining district is the acute source of contamination through food chain. Cd tremendously affects chromosome morphology. A study was carried out by Muneer and his co workers (2011) on root tips and leaves of *Vigna radiata* treated with different levels of Cd. Plant tissue was analyzed for chromosome abnormalities. Various laggered chromosomes, anaphase bridge, undistributed chromosome were observed. To avoid deadlier effect of heavy metal stress plant possess homeostatic appliances that permit them to keep optimum concentration of essential metal ions in cellular compartments and to minimize the detrimental effect. Not only resistance to toxic heavy metals, but metal cation homoestasis is essential for plant nutrition also. Many plant transporters have been identified at molecular levels.

Redox behavior of Cr is attributed to the direct involvement of Cr inducing oxidative stress that initiates degradation of photosynthetic pigments and causing severe damage to cell membrane due to lipid peroxidation (Panda and Chowdhury 2004). Subrahmanyum (2008) reported increased lipid peroxidation rate in *Triticum aestivum* under Cr (VI) stress and showed the reduction of photosynthesis rate under higher concentration of hexavalent Cr which is due to the suppression of light capture efficiency of PS II and interruption of electron transport rate. Accumulation of Cr by plants can reduce growth, pigment content, induce chlorosis, and change enzymatic function. According to few workers, growth of aerial part growth is immensely affected by Cr in rice (Singh et al. 2006) wheat, oat and sorghum (Lopez-Luna et al. 2009), where the length of the shoots are compromised. Photosynthetic pigment is the main parameter affected by Cr when plants or algae were treated with high dosage (Vernay et al. 2007, Subrahmanyam 2008, Rodriguez et al. 2011). Acute inhibition of RuBisCO activity was reported by Dhir et al. (2009) when aquatic macrophytes was exposed to Cr contaminated waste water of electroplating unit. He suggested that this reduced activity probably due to a exchange of Mg^{2+} in the active site of RuBisCO subunits by metal ions; deterioates RuBisCO function resulting a shift in the enzyme's activity from carboxylation to oxygenation. The study carried out by Yildiz and Terzi (2012) on the effect of different concentration of Cr(VI) on dehydrogenase activity, total soluble protein, MDA and antioxidative enzymes of two barley cultivars revealed that decrease in dehydrogenase activity and protein content as well as increase in proline content, lipid peroxidation and SOD activity may be indicative of oxidative stress induced by Cr (VI). Verney and his coworkers (2008) studied the impact of Cr on PS II activity, proline content and alkaloid production in *Datura innoxia* and reported the down regulation of PS II function under Cr stress. Cr accumulation and its effects on other mineral elements in *Amaranthus viridis* were investigated by Zou and his group (2006) using inductively coupled plasma atomic emission spectrometry (ICP-AES). They declared *A. viridis* is not a hyperaccumulator, inspite of having ability to accumulate pretty high amount of soil Cr. Cr has acute genotoxic effect on DNA. Cr also induces mutation in genetic material and form significant amounts of hydroxyl radicals that trigger DNA alterations and other effects (Salnikow and Zhitkovich, 2007). In addition with degradation of phosphodiester bond of nucleic acid, ROS can also affect Mitogen-Activated Protein Kinsases (MAPK), which cause the deregulation of cell proliferation (tumor inducing effect), causing mutagenic effect (Beyersmann and Hartwig 2008). From strong correlative and mechanistic experimental evidence, including work with transgenic plants and algae the involvement of metal induced proline in metal stress defence is pretty well established. Theriappan and his coworkers (2011) investigated the germination of Cauliflower (*Brassica oleracea*) seeds in the presence of both heavy metals (Cd, Zn, Hg) and salinity stress. Handique and Handique (2009) investigated Pb, Hg and Cd induced differential accumulation of proline in lemon grass (*Cymbopogon flexnosus*) and reported that proline accumulation was metal specific, organ specific and dose dependent (linear). Young leaves were better proline accumulator than older leaves. Investigation of Dinakar and his colleagues (2009) on ground nut (*Arachis hypogea*) seedling under Cd stress showed a remarkable increase in proline content and other enzymatic antioxidant in Cd treated plant in comparison to control, despite of a sharp decline in growth parameters and nitrate reductase activity. Combined stress of Pb, As, Cd on Sal (*Shorea robusta*) seedling (Pant et al. 2011) caused significant increase in free proline content with the increasing concentration of heavy metals. The strongest effect of Cd was followed by Pb and

As. Investigation of Ergun and Oncel (2012) revealed enhanced accumulation of free proline and soluble phenolics in response to the toxic effects of Pb, Cd and Zn. Majumder and Kundu (2018) established positive correlation of Cd toxicity of rice and proline production and also reported the genotoxic effect of Cd on rice by RAPD and GST analysis.

Identification of Hyperaccumulator Plants to Serve Phytoremediation

The identification and selection of a promising hyperaccumulator plant is the key step and most important approach for successful phytoremediation. Hyperaccumulators accumulate significant amount of concerned metals in their above ground tissue irrespective of soil metal concentration. Almost 500 plant species of 101 families has been identified as heavy metal accumulators till date (Kramer 2010). These are Asteraceae, Brassicaceae, Cryophyllaceae, Cyperaceae, Fabaceae, Lamiaceae, Poaceae, Violaceae etc. Most of the metal hyperaccumulators reported yet are from Brassicaceae family (Kramar 2010). Alarming Cd and Cr pollution forced to screen for hyperaccumulator plants suitable for phytoremediation. Three Cd and Zn hyperaccumulator have been reported, *Thalspi caerulescence* (Baker et al. 2000) and *Sedum alfredii* (Yang et al. 2004) for Cd and *Arabidopsis halleri* for Zn (Kupper et al. 2000). Wei et al. (2008) reported the Cd hyperaccumulator properties of *Taraxacum monogoliccum* and *Rorippa globosa. Leersia hexandra,* was reported (Zhang et al. 2007) as a hyperaccumulator of Cr. The hyperaccumulation of arsenic has been reported in only two species of Brassicaceae (Karimi 2009) besides *Pteris vitata* (Ma et al. 2001). A Pb hyperaccumulator, perennial shrub *Sesbania drummondii* with high biomass yield was recorded from United State (Sahi et al. 2002). *Crotalaria juncea* and *C. dactylon* have been reported as suitable candidate for Cr and Ni remediation (Saraswat and Rai 2009). Metal accumulator plants though useful to phytoextract metal contaminants from soil but have many shortcomings such as low biomass, edible nature and difficult to harvest. Ghosh and Singh (2005) evaluated the phytoextraction potential of commonly found high biomass, non edible and harmless weed species and compared with two accumulator plants (*Brassica juncea* and *Brassica campestris*). They identified *Ipomea carnea* as a promising candidate for Cr phytoextraction. Due to considerably high above ground biomass production and Cd accumulation in shoot, *Brassica napus* has been declared as a potential candidate for Cd phytoextraction by Selvam and Wong (2008). A study was carried out by Rezvani and Zaefarian (2011) to investigate the growth, bioaccumulation and translocation factor of Cd and Pb in *Aeluropes littoralis*. The enhanced translocation of Cd to the shoot of this plant indicates its great performances for phytoextraction and was introduced as Cd hyperaccumulator. Mazid and his coworkers (2011) reported *Acacia mangium* with higher bioconcentration and translocation factor as an efficient phytoremediator for Cd, Cu and Zn contaminated soil to ameliorate soil pollution. Wang and his colleagues (2010) reported the successful use of *Paulownia fortunei* for the phytoremediation of many Pb and Zn mine tailing.

Hyperaccumulators denotes a group of plant species with genetically inherited traits of metal uptake, hyperaccumulation and tolerance. From last few decades molecular tools have facilitated the understanding of metal hyperaccumulation physiology. Research is going on to identify and clone the genes responsible for metal accumulation, detoxification and tolerance in plant tissues. Transgenic approaches was successfully applied to create genetically engineered hyperaccumulators to promote phytoextraction of metals (Cd, Pb, Cr, As). Implementation of metal transporter genes, improved production of metal detoxifying chelators metallothioneins and phytochelatins are very common types of genetic alteration (Kotrba and Najmanova 2009). Non protein thiols (NPTs) consist of an excess amount of Cys sulfydryl residues and have paramount importance in heavy metal tolerance and detoxification. Reduced form of glutathione

(GSH) is one of the most important components of NPT metabolism. GSH, a sulfur containing tripeptide (γ-Glu- Cys-Gly), is very important for heavy metal sequestration, protecting cells from oxidative damage. GSH is the primary precursor of phytochelatin (PC), the metal chelating peptide involved in heavy metal scavenging and sequestration. Various workers had pretty well established the pivotal roles of GSH and PC synthase in heavy metal tolerance in Cd sensitive *Arabidopsis* mutants. Lee and his coworkers (2003) over expressed an *Arabidopsis* PC synthase (*AtPCS1*) gene in transgenic *Arabidopsis* with the aim of over synthesis of PC, resulting increased metal uptake and enduarance. Sun and his group (2010) reported that the variation in phytochelatin production in two Cd treated species *Rorippa globosa* and *R. islandica* may be used as biomarker of Cd hyperaccumulation and the synthesis of PCS for the enhanced uptake of Cd. Reisinger and his colleagues (2008) compared wild type plant with γ-ECS and GS transgenic regarding the Cd, Cr and As tolerance and accumulation capacity and has reported that over expression of γ-ECS and GS is attributed the enhanced production of PC and GSH which is a master strategy for the development of genetically modified plants with an enhanced phytoremediation capacity. Gasic and Korban (2007) developed transgenic Indian mustard (*Brassica juncea*) expressing *AtPCS* 1 gene, showed increased tolerance and accumulation capacity to Cd and Zn. A hydroponic experiment was conducted by Zeng and his coworkers (2012) to determine the possible effect of exogenous glutathione in rice seedling under Cr stress and revealed that addition of GSH in *in vitro* culture solution obviously alleviated the activity of antioxidative enzymes and reduction in MDA accumulation. Kupper and Kochain (2010) reported the differential cellular expression of ZNT1 and ZNT5, members of ZIP gene family, a novel transporter of plant under Cd, Fe, Mn and Zn toxicity. Investigation of Shahzad and his group (2010) revealed that Metal Tolerance Protein 1 (MTP1) is one of the most important genes present in *A. hallari* that encodes a Zn^{2+}/H^{+} antiporter involved in cytoplasmic Zn detoxification. The genetic analysis of Cd tolerance and hyperaccumulation was decoded in *Arabidopsis halleri* (Bert et al. 2003). Their results suggested that Cd tolerance is administered by more than a cluster of gene.

Other than hyperaccumulators, Cd and Cr tolerant plants have been identified that are important for their detoxification mechanism and alleviating antioxidative defenses. Aquatic macrophytes were reported to be the potential candidate for heavy metals scavenging from aquatic bodies. Odjegba et al. (2007) investigated and explained metal tolerance capacity of *Eichhornia crassipes* and *Pistia stratoites* by its enhanced activity of antioxidative isozymes (CAT, POX, SOD). The multiple metal uptake, accumulation and metal translocation factor from soil (TF) of three wild macrophytes (*Ipomea* sp., *Eclipta* sp. and *Marselia* sp.) have been investigated by Gupta and his group (2008) and reported that *Ipomea* sp. showed TFS> 1 for Cd, Cu, Mn and Zn, while *Eclipta* sp. and *Marselia* sp. showed TFS > 1 for Fe, Cu and Cd. Pal and Kundu (2011, 2014) reported Cd, Cr accumulation capacity of *Alternanthera philoxeroides* both *in situ* and *ex situ* study. Phytoremedial potential of *Amatanthus spinosus* was established (Huang 2019) by pot culture experiment where bioconcentration factor of Cd was found >1 in root than shoot. *Portulaca olaracea* also proven to be a good candidate for soil phytoremediation of industrial areas co-contaminated with Zn, Cd, Cr, As (Tiwari et al. 2008). Pal (2018) investigated the phytoremedial potential of few weeds collected from Cd, Cr contaminated tannery areas of Kolkata and reported that amaranthaceae family is better accumulator of Cr and Cd among which *Alternanthera philoxeroides* has best potentiality of metal uptake and tolerate than other weeds of Kolkata.

CONCLUSION

Phytoremediation is a vast domain to explore yet. Lots of genetically modified plants are getting regenerated with phytochelatin synthetase and MT 3 gene to become a better candidate of phytoextraction to act as hyperaccumulator. This plant based technique is a boon to the environment due to its remarkably non toxic approach rather than the bane of chemical soil cleans up.

REFERENCES

Bah, A. M., Dai, H., Zhao, J., Sun, H., Cao, F., Zhang, G., & Wu, F. (2011). Effect of cadmium, chromium and lead on growth, metal uptake and antioxidative capacity in *Typha angustifolia. Biological Trace Element Research*, *142*(1), 77–92. doi:10.100712011-010-8746-6 PMID:20552296

Baker, A. J. M. (1981). Accumulators and excluders - strategies in the response of plants to heavy metals. *Journal of Plant Nutrition*, *3*(1-4), 643–654. doi:10.1080/01904168109362867

Baker, A. J. M., & Brooks, R. R. (1989). Terrestrial higher plants which hyperaccumulate metallic element-A review of their distribution, ecology and phytochemistry. *Biorecovery*, *1*(4), 811–826.

Baker, A. J. M., McGrath, S. P., Reeves, R. P., & Smith, J. A. C. (2000). Metal hyperaccumulator plants: a review of the ecology and physiology of a biochemical resource for phytoremediation of metal polluted soils. In N. Terry & G. Bauelos (Eds.), *Phytoremediation of contaminated soil and water* (pp. 85–107). Lewis Publishers.

Baryla, A., Carrier, P., Franck, F., Coulomb, C., Sahut, C., & Havaux, M. (2001). Leaf chlorosis in oilseed rape plants (*Brassica napus*) grown on cadmium polluted soil: Causes and consequences for photosynthesis and growth. *Planta*, *212*(5-6), 696–709. doi:10.1007004250000439 PMID:11346943

Bernard, A. (2008). Biomarkers of metal toxicity in population studies:Research potential and interpretation issues. *Journal of Toxicology and Environmental Health*, *71*(18), 1259–1265. doi:10.1080/15287390802211885 PMID:18654897

Bert, V., Meerts, P., & Laprade, P. S. (2003). Genetic basis of cadmium tolerance and hyperaccummulation in *Arabidopsis halleri. Plant and Soil*, *249*(1), 9–18. doi:10.1023/A:1022580325301

Berti, W. R., & Cunningham, S. D. (2000). Phytostabilization of metals. In I. Raskin & B. D. Ensley (Eds.), *Phytoremediation of toxic metals: Using Plants to Clean up the environment* (pp. 71–88). Wiley and Sons Inc.

Beyersmann, D., & Hartwig, A. (2008). Carcinogenic metal compounds: Recent insight into molecular and cellular mechanisms. *Archives of Toxicology*, *82*(8), 493–512. doi:10.100700204-008-0313-y PMID:18496671

Bhardwaj, P., Chaturvedi, A., & Prasad, P. (2009). Effect of enhanced lead and cadmium in soil on physiological and biochemical attributes of *Phaseolus vulgaris* L. *Nature and Science*, *7*(8), 63–75.

Cargnelutti, D., Almeri Tabaldi, L., Spanevello, R. M., de Oliviera Jucoski, G., Battisti, V., Redin, M., Blanco Linares, C. E., Dressler, V. L., de Moraes Flores, E. M., Texeira Nicoloso, F., Morsch, V. M., & Chitolina, M. R. (2006). Mercury toxicity induces oxidative stress in growing cucumber seedlings. *Chemosphere*, *65*(6), 999–1006. doi:10.1016/j.chemosphere.2006.03.037 PMID:16674986

Chakroun, H. K., Souissi, F., Bouchardon, J., Moutte, J., Faure, O., Remon, E., & Abdeljaoued, S. (2010). Transfer and accumulation of lead, zinc, cadmium and copper in plants growing in abandoned mining district area. *African Journal of Environmental Science and Technology*, *4*(10), 651–659.

Chen, F., Wu, F. B., Dong, J., Vincze, E., Zhang, G., Wang, F., Huang, Y., & Wei, K. (2007). Cadmium translocation and accumulation in developing barley grains. *Planta*, *227*(1), 223–232. doi:10.100700425-007-0610-3 PMID:17713783

Cho, U. H., & Seo, N. H. (2005). Oxidative stress in *Arabidopsis thaliana* exposed to cadmium is due to hydrogen peroxide accumulation. *Plant Science*, *168*(1), 113–120. doi:10.1016/j.plantsci.2004.07.021

Choudhury, S., & Panda, S. K. (2004). Toxic effects, oxidative stress and ultrastructural changes in moss *Taxithelium nepalense* (Schwaegr) Broth. under chromium and lead phytotoxicity. *Water, Air, and Soil Pollution*, *167*(1), 73–90.

Clemens, S. (2006). Toxic metal accumulation, responses to exposure and mechanisms of tolerance in plants. *Biochimie*, *88*(11), 1707–1719. doi:10.1016/j.biochi.2006.07.003 PMID:16914250

18. Clement, R., Paredes, C., & Bernal, M. P. (2007). A field experiment investigating the effects of olive husk and cow manure on heavy metal availability in a contaminated calcareous soil from Murica (Spain). *Agriculture, Ecosystems & Environment*, *188*(1-4), 319–326. doi:10.1016/j.agee.2006.06.002

Danh, L. T., Truong, P., Mammucari, R., & Foster, N. (2009). Vetiver grass, Vetiveria zizanioides: A choice plant for phytoremediation of heavy metals and organic wastes. *International Journal of Phytoremediation*, *11*(8), 664–691. doi:10.1080/15226510902787302 PMID:19810597

Davies, F. T. Jr, Puryear, J. D., Newton, R. J., Egilla, J. N., & Saraiva Grossi, J. A. (2002). Mycorrhizal fungi increase Cr uptake by sunflower plants: Influence of tissue mineral concentration, growth and gas exchange. *Journal of Plant Nutrition*, *25*(11), 2389–2407. doi:10.1081/PLN-120014702

Dey, S., Dey, J., Patra, S., & Pothal, D. (2007). Changes in the antioxidative enzyme activities and lipid peroxidation in wheat seedlings exposed to cadmium and lead stress. *Brazilian Journal of Plant Physiology*, *19*(1), 53–60. doi:10.1590/S1677-04202007000100006

Dhir, B., Sharmila, P., Saradhi, P. P., & Nasim, S. A. (2009). *Physiological and antioxidant responses of Salvinia natans exposed to Chromium-rich wastewater*. Ecotoxicol. doi:10.1016/j.ecoenv.2009.03.015

Dinakar, N., Nagajyothi, P. C., Suresh, S., Damodharam, T., & Suresh, C. (2009). Cd induced changes on proline, antioxidant enzymes, nitrate and nitrite reductases in *Arachis hypogea* L. *J. Environ. Biol.* 30(2): 289-294. *Environ. Safe.*, *72*, 1790–1797.

Gasik, K., & Korban, S. S. (2007). Expression of *Arabidopsis* phytochelatin synthase in Indian mustard (*Brassica juncea*) plants enhance tolerance for cadmium and zinc. *Planta*, *225*(5), 1277–1285. doi:10.100700425-006-0421-y PMID:17086401

Ghosh, M., & Singh, S. P. (2005). Comparative uptake and phytoextraction study of soil induced chromium by accumulator and high biomass weed species. *Applied Ecology and Environmental Research, 3*(2), 67–79. doi:10.15666/aeer/0302_067079

Gopal, R., & Khurana, N. (2011). Effect of heavy metal pollutants on sunflower. *African Journal of Plant Science, 5*(9), 531–536.

Gratao, P. L., Prasad, M. N. V., Cardoso, P. F., Lea, P. J., & Azevedo, R. A. (2005). Phytoremediation: Green technology for the clean up of toxic metals in the environment. *Brazilian Journal of Plant Physiology, 17*(1), 53–64. doi:10.1590/S1677-04202005000100005

Gupta, S., Nayek, S., Saha, R. N., & Satpati, S. (2008). Assessment of heavy metal accumulation in macrophyte, agricultural soil and crop plants adjacent to discharge zone of sponge iron factory. *Environ. Geol, 55*(4), 731–739. doi:10.100700254-007-1025-y

Gyuricza, V., Fodor, F., & Szigeti, Z. (2010). Phytotoxic Effects of Heavy Metal contaminated Soil Reveal Limitations of Extract-Based Ecotoxicological Tests. *Water, Air, and Soil Pollution, 210*(1-4), 113–122. doi:10.100711270-009-0228-0

Gyuricza, V., Fodor, F., & Szigeti, Z. (2010). Phytotoxic Effects of Heavy Metal contaminated Soil Reveal Limitations of Extract-Based Ecotoxicological Tests. *Water, Air, and Soil Pollution, 210*(1-4), 113–122. doi:10.100711270-009-0228-0

Handique, G. K., & Handique, A. K. (2009). Proline accumulation in lemongrass (*Cymbopogon flexuosus* Stapf.) due to heavy metal stress. *Journal of Environmental Biology, 30*(2), 299–302. PMID:20121035

Horcsik, Z., Kovacs, L., Laposi, R., Meszaros, I., Lakatos, G., & Garab, G. (2007). Effect of Cr on photosystem 2 in the unicellular alga, *Chlorella pyrenoidosa*. *Photosynthetica, 45*(1), 65–69. doi:10.100711099-007-0010-8

Huang, Y., Xi, Y., Gan, L., Johnson, D., Wu, Y., Ren, D., & Liu, H. (2019). Effect of lead and cadmium on photosynthesis in Amaranthus spinosus and assessment of phytoremediation potential. *International Journal of Phytoremediation, 21*(10), 1041–1049. doi:10.1080/15226514.2019.1594686 PMID:31020865

Hussain, D., Haydon, M. J., Wang, Y., Wong, W., Sherson, S. M., Young, J., Camakaris, J., Harper, J. F., & Cobbett, C. S. (2004). P-Type ATPase heavy metal transporters with roles in essential zinc homeostasis in *Arabidopsis*. *The Plant Cell, 16*(5), 1327–1339. doi:10.1105/tpc.020487 PMID:15100400

Karimi, N., Ghadederian, S. M., Raab, A., Fledmann, J., & Meharg, A. A. (2009). An arsenic accumulating, hypertolerant *Brassica isatis*, Cappadocica. *The New Phytologist, 184*(1), 41–47. doi:10.1111/j.1469-8137.2009.02982.x PMID:19656300

Khan, M. H. (2007). Induction of oxidative stress and antioxidant metabolism in *Calamus tenius* leaves under chromium and zinc toxicity. *Indian Journal of Plant Physiology / Official Publication of the Indian Society for Plant Physiology, 12*, 353–359.

Kotrba, P., Najmanova, J., Macek, T., Ruml, T., & Mackova, M. (2009). Genetically modified plants in phytoremediation of heavy metal and metalloid soil and sediment pollution. *Biotechnology Advances, 27*(6), 799–810. doi:10.1016/j.biotechadv.2009.06.003 PMID:19567265

Kramar, U. (2010). Metal hyperaccumulation in plants. *Annual Review of Plant Biology*, *61*(1), 517–534. doi:10.1146/annurev-arplant-042809-112156 PMID:20192749

Kramer, U. (2005). Phytoremediation: Novel approaches to cleaning up polluted soils. *Current Opinion in Biotechnology*, *16*(2), 133–141. doi:10.1016/j.copbio.2005.02.006 PMID:15831377

Kristin, B. R., Love, S., Nicholas, R. R., McAdory, D., Halder, S. S., Kern, N., Kent, J., & Strickland, J. (2010). Chromium is not an essential trace element for mammals. Effect of a low chromium diet. *Journal of Biological Inorganic Chemistry*, *16*(3), 3–81.

Kupper, H., & Kochain, L. V. (2010). Transcriptional regulation of metal transport genes and mineral nutrition during acclimatization to cadmium and zinc in the Cd/Zn hyperaccumulator, *Thalspi caerulescens* (Ganges population). *The New Phytologist*, *185*(1), 114–129. doi:10.1111/j.1469-8137.2009.03051.x PMID:19843304

Kupper, H., Lombi, E., Zhao, F. J., & McGrath, S. P. (2000). Cellular compartmentation of cadmium and zinc in relation to other elements in the hyperaccumulator *Arabidopsis halleri*. *Planta*, *212*(1), 75–84. doi:10.1007004250000366 PMID:11219586

L'opez-Luna, J., Gonz'alez-Ch'avez, M. C., Esparza-Garc'ıa, F. J., & Rodr'ıguez-V'azquez, R. (2009). Toxicity assessment of soil amended with tannery sludge, trivalent chromium and hexavalent chromium, using wheat, oat and sorghum plants. *Journal of Hazardous Materials*, *163*(2-3), 829–834. doi:10.1016/j.jhazmat.2008.07.034 PMID:18814962

Lee, S., Moon, J. S., Ko, T. S., Petros, D., Goldbrough, P. B., & Schuyler, S. K. (2003). Overexpression of *Arabidopsis* phytochelatin synthase paradoxically leads to hypersensitivity to cadmium stress. *Plant Physiology*, *131*(2), 656–663. doi:10.1104/pp.014118 PMID:12586889

Lopez-Luna, J., Gonzalez-Chavez, M. C., Esparza-Garcia, F. J., & Rodriguez-Vazquez, R. (2009). Toxicity assessment of soil amended with tannery sludge, trivalent chromium and hexavalent Chromium, using wheat, oat and sorghum plants. *Journal of Hazardous Materials*, *163*(2-3), 829–834. doi:10.1016/j.jhazmat.2008.07.034 PMID:18814962

Ma, J. F., Ueno, D., Zhao, F. J., & McGrath, S. P. (2005). Subcellular localization of cadmium and zinc in the leaves of a Cd- hyperaccumulating ecotype of *Thlaspi caerulescens*. *Planta*, *220*(5), 736. doi:10.100700425-004-1392-5 PMID:15517354

Majid, N. M., Islam, M. M., Justin, V., Abdu, A., & Ahmadpour, P. (2011). Evaluation of heavy metal uptake and translocation by *Acacia mangium* as a phytoremediator of copper contaminated soil. *African Journal of Biotechnology*, *10*(42), 8373–8379. doi:10.5897/AJB11.392

Martins, L. L., Mourato, M. P., Cardoso, A. I., Pinto, A. P., & Mota, A. M. (2011). Oxidative stress induced by Cd in *Nicotiana tabaccum* L.: Effects on growth parameters, oxidative damage and antioxidant responses in different plant parts. *Acta Physiologiae Plantarum*, *33*(4), 1375–1383. doi:10.100711738-010-0671-y

Monteiro, M. S., Rodriguez, E., Loureiro, J., Mann, R. M., Soares, A. M. V. M., & Santos, C. (2010). Flow cytometric assessment of Cd genotoxicity in three plants with different metal accumulation and detoxification capacities. *Ecotoxicology and Environmental Safety*, *73*(6), 1231–1237. doi:10.1016/j.ecoenv.2010.06.020 PMID:20663557

Muneer, S., & Qadri, T. N., & Mahmooduzaffar, S. T. O. (2011). Cytogenetic and biochemical investigations to study the response of *Vigna radiata* to cadmium stress. *African Journal of Plant Science*, *5*(3), 183–192.

Narain, S., Ojha, C. S. P., Mishra, S. K., Chaube, U. C., & Sharma, P. K. (2011). Cadmium and Chromium removal by aquatic plants. *International Journal of Environmental Sciences*, *1*(6), 1297–1304.

Noctor, G., Gomez, L., Vanacker, H., & Foyer, C. H. (2002). Interactions between biosynthesis, compartmentation and transport in the control of glutathione homeostasis and signaling. *Journal of Experimental Botany*, *53*(372), 1283–1304. doi:10.1093/jexbot/53.372.1283 PMID:11997376

Pal & Kundu. (2014). Study of metal resistance potential of the Cd, Cr tolerant Alligator Weed. *Journal of Stress Physiology & Biochemistry*, *10*(1), 244–261.

Pal. (2018). Comparative Estimation of Cadmium, Chromium Resistance Capacity of *Amaranthus viridis, Tianthema portulacastrum, Alternanthera philoxeroides* collected from contaminated site and Assessment of phytoremedial potential. *Environment and Ecology*, *36*(2), 409-417.

Pal and Kundu. (2011). A Field Study of Chromium and Cadmium uptake, accumulation and tolerance in Alligator Weed [*Alternanthera philoxeroides* (Mart.) Griseb]. *Proceeding of National Academy of Sciences, Section B: Biological Science, India*, *81*(4), 404–410.

Pandey, N., & Singh, G. K. (2012). Studies on antioxidative enzymes induced by cadmium in pea plants (*Pisum sativum*). *Journal of Environmental Biology*, *33*, 201–206. PMID:23033681

Pandey, S. N. (2006). Accumulation of heavy metal (Cd, Cr, Cu, Ni, Zn) in *Raphanus sativus* L. and *Spinacea oleracea* L. plants irrigated with industrial effluents. *Journal of Environmental Biology*, *27*(2), 381–384. PMID:17436528

Pandey, V., Dixit, V., & Shyam, R. (2005). Antioxidative responses in relation to growth of mustard (*Brassica juncea* cv. Pusa Jai Kishan) plants exposed to hexavalent chromium. *Chemosphere*, *61*(1), 40–47. doi:10.1016/j.chemosphere.2005.03.026 PMID:16157168

Pandey, V., Dixit, V., & Shyam, R. (2009). Chromium effect on ROS generation and detoxification in pea (*Pisum sativum*) leaf chloroplasts. *Protoplasma*, *236*(1-4), 85–95. doi:10.100700709-009-0061-8 PMID:19582547

Pant, P. P., Tripathi, A. K., & Dwivedi, V. (2011). Effects of some heavy metals on some biochemical parameters of Sal (*Shorea robusta*) seedlings at nursery level, Doon valley. *Indian Journal of Agricultural Sciences*, *66*(2), 45–51. doi:10.1080/09766898.2011.11884667

Prado, F. E., Prado, C., Rodriguez-Montelongo, L., Gonzalez, J. A., Pagano, E. A., & Hilal, M. (2010). Uptake of Chromium by *Salvinia minima*: Effect on plant growth, leaf respiration and carbohydrate metabolism. *Journal of Hazardous Materials*, *177*(1-3), 546–553. doi:10.1016/j.jhazmat.2009.12.067 PMID:20080338

Prasad, M. N. V., & Freitas, H. M. O. (2003). Metal hyperaccumulation in plants- biodiversity prospecting for phytoremediation technology. *Electronic Journal of Biotechnology*, *6*(3), 275–321. doi:10.2225/vol6-issue3-fulltext-6

Qiao, J. T., Liu, T. X., Wang, X. Q., Li, F. B., Lv, Y. H., & Liu, C. P. (2017). Simultaneous alleviation of cadmium and arsenic accumulation in rice by applying zero-valent iron and biochar to contaminated paddy soils. *Chemosphere*, *195*, 260–271. doi:10.1016/j.chemosphere.2017.12.081 PMID:29272795

Qiu, R. L., Fang, X. H., Tang, Y. T., Du, S. J., Zeng, X. W., & Brewer, E. (2006). Zinc hyperaccumulation and uptake by *Potentilla griffithi* hook. *International Journal of Phytoremediation*, *8*(4), 299–310. doi:10.1080/15226510600992865 PMID:17305304

Rafati, M., Khorasani, N., Moattar, F., Shirvany, A., Moraghebi, F., & Hosseinzadeh, S. (2011). Phytoremediation potential of *Populus alba* and *Morus alba* for cadmium, chromuim and nickel absorption from polluted soil. *International Journal of Environmental of Research*, *5*(4), 961–970.

Redjala, T., Sterckeman, T., & Morel, J. L. (2009). Cadmium uptake by roots: Contribution of apoplast and of high and low affinity membrane transport systems. *Environmental and Experimental Botany*, *67*(1), 235–242. doi:10.1016/j.envexpbot.2009.05.012

Reeves, R. D. (2003). Tropical hyperaccumulators of metals and their potential for phytoextraction. *Plant and Soil*, *249*(1), 57–65. doi:10.1023/A:1022572517197

Reeves, R. D., & Baker, J. M. (2000). Metal accumulating plants. In H. Raskin & B. D. Ensley (Eds.), *Phytoremediation of toxic metals: Using plants to clean up the environment* (pp. 193–230). John Wiley & Sons Inc.

Reisinger, S., Schiavon, M., Terry, N., Elizabeth, A. H., & Pilon, S. (2008). Heavy metal tolerance and accumulation in Indian mustard (*Brassica juncea*) expressing bacterial γ- glutamyl cysteine synthetase or glutathione synthetase. *International Journal of Phytoremediation*, *10*(5), 1–5. doi:10.1080/15226510802100630 PMID:19260225

Rezvani, M., & Zaefarian, F. (2011). Bioaccumulation and translocation factor of cadmium and lead in *Aeluropus littorali. A.J.A.E*, *2*(4), 114–119.

Rizwan, M., Ali, S., Ali, B., Adrees, M., Arshad, M., Hussain, A., Zia ur Rehman, M., & Waris, A. A. (2019). Zinc and iron oxide nanoparticles improved the plant growth and reduced the oxidative stress and cadmium concentration in wheat. *Chemosphere*, *214*, 269–277. doi:10.1016/j.chemosphere.2018.09.120 PMID:30265934

Rodriguez, E., Santos, C., Lucas, E., & Pereira, M. J. (2011). Evaluation of Chromium (VI) toxicity to *Chlorella vulgaris* Beijerinck cultures. *Fresenius Environmental Bulletin*, *20*, 334–339.

Sahi, S.V., Bryant, N.L, Sharma, N.C., & Singh, S.R. (2002). Characterization of a Lead hyperaccumulator shrub, *Sesbania drummondii. Environ. Sci. Technol., 36* 4676-4680.

Salnikow, K., & Zhitkovich, A. (2007). Genetic and epigenetic mechanisms in metal carcinogenesis and cocarcinogenesis: Nickel, Arsenic, and Chromium. *Chemical Research in Toxicology, 21*(1), 28–44. doi:10.1021/tx700198a PMID:17970581

Sandalio, L. M., Dalurzo, H. C., Gomez, M., Romero-Puertas, M. C., & del Rio, L. A. (2001). Cadmium induced changes in the induced change in the growth and oxidative metabolism of pea plants. *Journal of Experimental Botany, 52*(364), 2116–2126. doi:10.1093/jexbot/52.364.2115

Saraswat, S., & Rai, J. P. N. (2009). Phytoextraction potential of six plant species grown in multimetal contaminated soil. *Chemistry and Ecology, 25*(1), 1–11. doi:10.1080/02757540802657185

Sethi, P. K., & Khandeiwal, D. J. (2006). Cadmium exposure: Health hazards in developing countries. *Medical Toxicology*, (2), 14–15. doi:10.1007/BF03161007 PMID:18072106

Shahzad, Z., Gosti, F., Frerot, H., Lacombe, E., & Rossens, N. (2010). The five AhMTP1 zinc transporters undergo different evolutionary fates towards adaptive evolution to Zn tolerance in *Arabidopsis halleri. PLOS Genetics, 6*(4), e1000911. doi:10.1371/journal.pgen.1000911 PMID:20419142

Shanker, A. K., Cervanter, C., & Loza-Tavera, H. (2005). Chromium toxicity in plants. *Environment International, 31*(5), 739–753. doi:10.1016/j.envint.2005.02.003 PMID:15878200

Shekhar, C., Sammaiah, D., Rambabu, M., & Reddy, K. J. (2011). Effect of cadmium on tomato growth and yield attributes. *J. Microbiol. Biotech. Res, 1*(3), 109–112.

Singh, R., Singh, D. P., Kumar, N., Bhargav, S. K., & Barman, S. C. (2010). Accumulation and translocation of heavy metal in soil and plant from fly ash contaminated areas. *Journal of Environmental Biology, 31*, 421–430. PMID:21186714

Sobrino-Plata, J., Ortega-Villasante, C., Laura Flores-Cáceres, M., Escobar, C., Del Campo, F. F., & Hernández, L. E. (2009). Differential alteration of antioxidant defences as bioindicators of mercury and Cd toxicity in alfa alfa. *Chemosphere, 77*(7), 946–954. doi:10.1016/j.chemosphere.2009.08.007 PMID:19732935

Srivastava,, R., Khan,, R., & Manzoor,, N., & Mahmooduzzafar. (2011). Responses of cadmium exposures on growth, physio-biochemical characteristics and the antioxidative defence system of soybean (*Glycine max* L.). *Journal of Phytology, 3*(10), 20–25.

Subrahmanyam, D. (2008). Effect of Cr toxicity on leaf photosynthetic characteristics and oxidative changes in wheat. *Photosynthetica, 46*(3), 339–340. doi:10.100711099-008-0062-4

Sun, R., Jin, C., & Zhou, Q. (2010). Characteristics of cadmium accumulation and tolerance in *Rorippa globosa* (Turcz.) Thell., a species with some characteristics of cadmium hyperaccumulation. *Plant Growth Regulation, 61*(1), 67–74. doi:10.100710725-010-9451-3

Sun, R., Jin, C., & Zhou, Q. (2010). Characteristics of cadmium accumulation and tolerance in *Rorippa globosa* (Turcz.) Thell., a species with some characteristics of cadmium hyperaccumulation. *Plant Growth Regulation, 61*(1), 67–74. doi:10.100710725-010-9451-3

Sun, Y. B., Zhou, Q. X., Sun, F. H., & Jin, C. X. (2007). Antioxidative defense and proline/ phytochelatin accumulation in a newly discovered Cd- hyperaccumulator, *Solanum nigrum* L. *Environmental and Experimental Botany*, *60*(3), 468–476. doi:10.1016/j.envexpbot.2007.01.004

Tanhan, P., Kruatrachu, M., & Pokethitiyook, P. (2007). Uptake and accumulation of Cadmium, Lead and Zinc by Siam weed [Chromolaena odoruta King and Robinson]. *Chemosphere*, *68*(2), 323–329. doi:10.1016/j.chemosphere.2006.12.064 PMID:17280700

Tantrey, M. S., & Agnihotri, R. K. (2010). Chlorophyll and proline content of gram (*Cicer arietinum* L.) under cadmium and mercury treatments. *Research Journal of Agricultural Science*, *1*(2), 119–122.

Tausz, M., Wonisch, A., Grill, D., Morales, D., & Jimenez, M. S. (2003). Measuring antioxidants in tree species in the natural environment: From sampling to data evaluation. *Journal of Experimental Botany*, *54*(387), 1505–1510. doi:10.1093/jxb/erg175 PMID:12759368

Theriappan, P., Gupta, K. A., & Dhasarathan, P. (2011). Accumulation of proline under salinity and heavy metal stress in Cauliflower seedlings. J. *Appl. Sci. Environ. Manage*, *15*(2), 251–255. doi:10.4314/jasem.v15i2.68497

Tiwari, K. K., Dwivedi, S., Singh, N. K., Rai, U. N., & Tripathi, R. D. (2009). Chromium (VI) induced phytotoxicity and oxidative stress in pea (*Pisum sativum* L.): Biochemical changes and translocation of essential nutrients. *Journal of Environmental Biology*, *30*, 389–394. PMID:20120464

Trotta, A., Falaschi, P., Cornara, L., Minganti, V., Fusconi, A., Drava, G., & Berta, G. (2006). Arbuscular mycorrhizae increase the arsenic translocation factor in the As hyperaccumulating fern *Pteris vitata* L. *Chemosphere*, *65*(1), 74–81. doi:10.1016/j.chemosphere.2006.02.048 PMID:16603227

Uraguchi, S., Watanabu, I., Kiyono, M., & Kuno, K. (2006). Characteristics of Cadmium accumulation and tolerance in novel Cd- accumulating crops, *Avena strigosa* and *Crotalaria juncea*. *Exp. Botany*, *57*(12), 2955–2965. doi:10.1093/jxb/erl056 PMID:16873452

Vajpayee, P., Rai, U. N., Ali, M. B., Tripathi, R. D., Yadav, V., & Sinha, S. N. (2001). Chromium induced physiological changes in *Vallisneria spiralis* L. and its role in phytoremediation of tannery effluents. *Bulletin of Environmental Contamination and Toxicology*, *67*, 246–265. PMID:11429683

Vassilev, A., & Lidon, F. (2011). Cadmium induced membrane damage and changes in soluble protein and free amino acid contents in young barley plants. *Emer. Journal of Food & Agriculture*, *23*(2), 130–136.

Verbruggen, N., Hermans, C., & Schat, H. (2009). Molecular mechanisms of metal hyperaccumulation in plants. *The New Phytologist*, *181*(4), 759–776. doi:10.1111/j.1469-8137.2008.02748.x PMID:19192189

Vernay, P., Gauthier-Moussard, C., Jean, L., Bordas, F., Faure, O., Ledoigt, G., & Hitmi, A. (2008). Effect of chromium species on phytochemical and physiological parameters in *Datura innoxia*. *Chemosphere*, *72*(5), 763–771. doi:10.1016/j.chemosphere.2008.03.018 PMID:18472140

Wang, J., Weihua, L., Zhang, C., & Shisheng, K. (2010). Physiological responses and detoxific mechanisms to Pb, Zn, Cu and Cd in young seedlings of *Paulownia fortunei*. *Journal of Environmental Sciences (China)*, *22*(2), 1916–1922. doi:10.1016/S1001-0742(09)60339-9 PMID:21462710

Wei, S., Zhou, Q., & Saha, U. K. (2008). Hyperaccumulative characteristic of weed species to heavy metals. *Water, Air, and Soil Pollution*, *192*(1-4), 173–181. doi:10.100711270-008-9644-9

Wei, S. H., Zhou, Q. X., & Wang, X. (2004). A newly discovered Cd- hyperaccumulator *Solanum nigrum* L. *Chinese Science Bulletin*, *49*(24), 2568–2573.

Yang, X. E., Long, X. X., Ye, H. B., He, Z. L., Calvert, D. V., & Stoffella, P. J. (2004). Cadmium tolerance and hyperaccumulation in a new zinc hyperaccumulating plant species (*Sedum alfredii Hance*). *Plant and Soil*, *259*(1-2), 181–189. doi:10.1023/B:PLSO.0000020956.24027.f2

Yannarelli, G. G., Fernandez-Alvarez, A. J., Santa-Cruz, D. M., & Tomaro, M. L. (2006). Glutathione reductase activity and isomorph in leaves and roots of wheat plants subjected to cadmium stress. *Phytochemistry*, *68*(4), 505–512. doi:10.1016/j.phytochem.2006.11.016 PMID:17174990

Yildiz, M., & Terzi, H. (2012). Effect of hexavalent chromium stress in roots of Cr- tolerant and Cr-sensitive barley cultivars. *Acta Agronomica Hungarica*, *60*(1), 29–36. doi:10.1556/AAgr.60.2012.1.4

Yuebing, S., Qising, Z., Shuhe, W., & Liping, R. (2007). Growth responses of newly discovered Cd-hyperaccumulator *Rorippa globosa* and its accumulation characteristics of cadmium and arsenic under joint stress of cadmium and arsenic. *Environmental Sciences*, *1*(1), 107–113.

Zayed, A. M., & Terry, N.Agency for Toxic Substances and Disease Registry. (2003). Chromium in the environment: Factors affecting biological remediation. *Plant and Soil*, *249*(1), 139–156. doi:10.1023/A:1022504826342

Zemel, M. B., Richards, J., Milstead, A., & Campbell, P. (2005). Effects of calcium and dairy on body composition and weight loss in African- American adults. *Obesity Research*, *13*(7), 1213–1225. doi:10.1038/oby.2005.144 PMID:16076991

Zeng, F., Qiu, B., Wu, X., Niu, S., & Zhang, G. (2012). *Glutathione mediated alleviation of Cr toxicity in rice plants*. Researchgate.

Zengin, F. K., & Munzuroglue, O. (2005). Effect of some heavy metals on content of chlorophyll, proline and some antioxidant chemicals in bean (*P. vulgaris* L.) seedlings. *Acta Biologica Cracoviensia*, *47*(2), 157–164.

Zhang, X. H., Liu, J., Huang, H. T., Chen, Z., & Wang, D. Q. (2007). Chromium accumulation by the hyperaccumulator plant *Leersia hexandra* Swartz. *Chemosphere*, *67*(6), 1138–1143. doi:10.1016/j.chemosphere.2006.11.014 PMID:17207838

Zoronda, P., Vazquez, S., Esteban, E., Fernandez, M., & Carpena, R. (2002). Cdamium stress in nodulated white lupin: Strategies to avoid toxicity. *Plant Physiology and Biochemistry*, *40*(12), 1003–1009. doi:10.1016/S0981-9428(02)01464-X

Chapter 8
Ricinus communis: A Potent Lead (Pb) Accumulator

Raikamal Pal
Shri Shikshayatan College, India

ABSTRACT

Contamination of soil and ground water with heavy metals is a great threat to human health, vegetation, and wildlife. Pb is the second most hazardous substance according to ATSDR. The main sources of Pb entering an ecosystem are atmospheric Pb (mainly from automobile emission), paint chips, fertilizers, and pesticides and Pb acid batteries or other industrial Pb products. Phytoremediation could provide sustainable techniques for metal remediation. Roots of Ricinus communis were found to accumulate maximum amount of Pb (275.12mg/kg dry wt,). Depending on soil Pb content, the concentration of Pb in shoots of Ricinus communis *also varied. In most cases only a small part of Pb was translocated in the aerial parts. In 95% of the plant samples collected, the root Pb concentration are much greater than those of the shoot lead content, indicating low mobility of Pb from roots to the shoots. Their ability to accumulate higher amounts of Pb in their roots and considering their rapid growth rate and biomass, this plant has the potential for removal of Pb from contaminated soil.*

INTRODUCTION

Indiscriminate use of different heavy metals has been increased due to rapid urbanization.Heavy metals cannot be destroyed or degraded as they occur as natural constituent of earths crust. These heavy metals enter the body system through food, air, and water and bio-accumulate over a period of time. (UNEP/GPA,2004).

In todays industrialized society heavy metals are ubiquitous environmental contaminants. Heavy metal pollution in soil differ from air or water pollution as heavy metals retain much longer than any other component of the biosphere. (Lasat., 2002)

Heavy metal contaminants in soils emitted through metalliferous mining and smelting, metallurgical industries, sewage sludge treatment, warfare and military training, waste disposal sites, agricultural fertilizers and electronic industries (Alloway1995). For example, mine tailings rich in sulphide minerals

DOI: 10.4018/978-1-7998-4888-2.ch008

may form Acid Mine Drainage (AMD) through reaction with atmospheric oxygen and water, and AMD contains elevated levels of metals that could be harmful.

Wild life and human exposed to high level of these heavy metals has adverse effects on both.Natural and anthropogenic activities both are responsible for the heavy metal emission into the environment. Mining operations are the main anthropogenic sources which causes heavy metal emission.(Battarbee et al.,1988; Nriagu, 1989). Even long after mining activities have ceased, the emitted metals continue to persist in the environment. Heavy metals are emitted both in elemental and compound (organic and inorganic) forms in the environment. Various former and present mining sites,foundries and smelters, combustion by products are the anthropogenic sources of emission. These metals dissolve with rain water leached out in sloppy areas, and are carried by acid water downstream or run-off to the species in water or as an integral part of suspended sediments (dissolved species in water have the greatest potential of causing the most deleterious effects). These heavy metal rich sediments may then be accumulated in river bed sediments or seep into the underground water and thus contaminate water from underground sources, particularly wells; and the extent of contamination will depend on the nearness of the well to the mining site. Wells which are located near mining sites have been reported to contain heavy metals at levels that exceed drinking water criteria (Garbarino et al., 1995; Peplow, 1999).

Table 1. United State Environmental Protection Agency (USEPA) maximum contamination levels for heavy metal concentration in air, soil and water

Heavy Metal	Max.Conc.in air (mg/m^3)	Max.Conc. in Sludge (Soil)(mg/kg or ppm)	Max.Conc in Drinking water (mg/L)	Max Conc, in water supporting aquatic life(mg/L or ppm)
Cd	0.1-0.2	85	0.005	0.008^{δ}
Pb	--	420	0.01^{π}(0.0)	0.0058^{δ}
Zn^2	1.5*	7500	5.00	0.0766^{δ}
Hg	--	<1	0.002	0.05
Ca	5	Tolerable	50	Tolerable > 50
Ag	0.01	--	0.0	0.1
As	--	--	0.01	--

(Value in bracket is the desirable limit; WHO ; 1adapted from U.S. – OSHA; 2 EPA, July 1992; _USEPA, 1987; Georgia Code,1993; Florida Code, 1993; Washington Code, 1992; Texas Code, 1991; North Carolina, 1991; *1 for chloride fume, 5 for oxide fume; - - no guideline available).

In the 3rd world countries importance has been given mainly for the establishment of the industries but the issues of protection of environment remain neglected. Thus a number of factories were developed in a unplanned manner, it increases generation affluents from the factories. These affluents are often mixed with heavy metals.The problems of urbanization, population explosion and the increased use of automobiles have become very common. It is well known that environmental pollution is a product of urbanization and technology, and other associated factors of population density.

Depending on the type of industries in the vicinity different metals such as As, Pb, Cd, Cu, Cr, Ni etc are deposited in the soil. Among these some metals are needed for biological function such as Cu, Zn whereas Pb, Cr, As, Hg have no known biological role. All these metals, when present in very low concentration have no or little effect on living organism, but when the level crosses the threshold value these metals affect the biological system. Among these heavy metals Pb, Cd, Cr are more harmful to both plants and animals and these heavy metals are mostly widespread. According to ATSDR (2005) Pb, Cd, Cr stands 2nd, 7th and 77th in position respectively.

BACKGROUND

Lead (Pb)

Pb imposes serious threat to the health of children and wildlife as it is an extremely toxic heavy metals (EPA 2005). The main sources of Pb poisoning include lead paint and old gasoline spills (PbBrCl, 2Pb-BrCl.NH4Cl) resulting in dust and soil contamination of food and water (Xintaras 1992)

Physical Properties

Lead is a main-group element with the symbol **Pb** (from Latin: *Plumbum*) and atomic number 82. Lead has the highest atomic number of all the stable elements, 82 and its atomic weight is 207.2.

Pb is dense, ductile, very soft, highly malleable, bluish white metal with poor electrical conductivity. Metallic lead has a bluish white to silvery shine after been freshly cut, but it soon turns to a dull greyish colour when exposed to air . Liquid lead has a shiny silvery lustre.

Metallic lead is attacked (oxidized) only superficially by air, forming a thin layer of lead oxide that protects it from further oxidation. The metal is not attacked by sulfuric or hydrochloric acids. It dissolves in nitric acid with the evolution of nitric oxide gas to form dissolved Pb(NO3)2.

Exposure Routes

Lead is a common environmental pollutant. Household dust, soil, water, and commercial products, lead in air are the routes of exposure of lead. Environmental contamination includes industrial use of lead, such as is found in facilities that process lead-acid batteries, lead wire or pipes, and metal recycling and foundries. Battery recycling workers are at high risk for lead exposure. Occupational exposure is the main cause of lead poisoning. Facilities that produce a variety of lead-containing products; these include radiation shields, ammunition, certain surgical equipment's, plumbing, circuit boards, jet engines, and ceramic glazes are the causes of lead exposure for the people working in those industries. Lead miners and smelters, plumbers and fitters, auto mechanics, glass manufacturers, construction workers, battery manufacturers and recyclers, firing range instructors, and plastic manufacturers are at great risk for lead exposure in addition. Other occupations that present lead exposure risks include welding,

Lead paint is a major route of lead exposure in children as some lead compounds are colourful and are widely used in paint. Deteriorating lead paint can produce dangerous lead levels in household dust and soil. Deteriorating lead paint and lead-containing household dust are the main causes of chronic lead poisoning. Colourful toys used by children are extremely harmful for them as deteriorating lead can be easily ingested. However, removing lead paint from dwellings, e.g. by sanding or torching, can create lead-containing dust and fumes. Special precautions must be taken when removing lead paint.

Contaminated food, water or alcohol are also the source of lead exposure. Even ingestion of certain home remedy medicines may also expose people to lead or lead compounds. Fruits and vegetables grown in lead contaminated sometimes may be the source of lead ingestion in human. Soil is contaminated through particulate accumulation from lead in pipes, lead paint and residual emissions from leaded gasoline that was used before the Environment Protection Agency issue the regulation around 1980.

Inhalation is the second major pathway of exposure, especially for workers in lead-related occupations. Almost all inhaled lead is absorbed into the body, the rate is 20–70% for ingested lead; children absorb more than adults.

Effects of Lead Toxicity on Plants

Among the heavy metals, lead is a potential pollutant that readily accumulates in soils and sediments. Lead gets easily absorbed and accumulated in different plant parts although it is not an essential element for plants. Depending on different parameters such as pH, particle size and cation exchange capacity of the soils as well as by root exudation and other physico-chemical parameters determines the uptake of lead by plants. A number of toxicity symptoms in plants e.g. stunted growth, chlorosis and blackening of root system is caused by excess Pb accumulation Pb inhibits photosynthesis, upsets mineral nutrition and water balance, changes hormonal status and affects membrane structure and permeability. Mechanisms of Pb-detoxification include sequestration of Pb in the vacuole, phytochelatin synthesis and binding to glutathione and amino acids etc. Pb tolerance is associated with the capacity of plants to restrict Pb to the cell walls, synthesis of osmolytes and activation of antioxidant defence system. Remediation of soils contaminated with Pb using phytoremediation and rhizofiltration technologies appear to have great potential for cleaning of Pb contaminated soils.

Effects of Lead Toxicity on Animals

Lead can also be found listed as a criteria pollutant in the United States Clean Air Act section 108. Lead that is emitted into the atmosphere can be inhaled, or it can be ingested after it settles out of the air. It is rapidly absorbed into the bloodstream and is believed to have adverse effects on the central nervous system, the cardiovascular system, kidneys, and the immune system.

Lead is a poisonous metal that can damage nervous connections (especially in young children) and cause blood and brain disorders. Long-term exposure to lead or its salts (especially soluble salts or the strong oxidant PbO2) can cause nephropathy, and colic-like abdominal pains. The effects of lead are the same whether it enters the body through breathing or swallowing. Lead can affect almost every organ and system in the body. The main target for lead toxicity is the nervous system, both in adults and children.

The Importance and Hazard of Lead as a Rhizospheric Contaminant

Lead imposes a serious threat to the health of children and wildlife as it is an extremely toxic heavy metal. (EPA 2005). The main sources of Pb poisoning include lead paint and old gasoline spills (PbBrCl, 2PbBrCl.NH4Cl) resulting in dust and soil contamination of food and water (Xintaras 1992).

Elemental Pb is insoluble and the most water soluble forms of Pb compounds are lead acetate (2 mg/ml), lead chloride (0.009 mg/ml), and lead nitrate (5 mg/ml). Atmospheric Pb mostly exists as PbSO4 and PbCO3. Many plants have a strategy of Pb exclusion such as *Thlaspi praecox,* which hyperaccumulates Cd and Zn but exclude Pb (Vogel-Mikus et al. 2005).

Pb isa hyperaccumulator as it is cited in different references.. It has been reported that *Sesbania drummondii,* a leguminous shrub, and several *Brassica* species can accumulate significant amounts of Pb in their roots (Blaylock et al. 1997; Sahi et al. 2003; Wang et al. 2001), and *Piptathertan miliacetall,* a grass, accumulates Pb directly correlating to soil concentrations without symptoms of toxicity for 3 weeks (Garcia et al. 1998). Sahi et al. (2002) have noted that S. *drummondii* can tolerate Pb levels up to 1500mg/L and accumulate 40g/kg shoot dry weight. *Brassica juncea* showed reduced growth at 645 mg/gm Pb in the soil substrate, but can accumulate 34.5 gm/kg shoot dry weight, although significant shoot accumulation is not observed until Pb reaches saturation levels in the roots. Most of the shoot accumulation was found in stems and not leaves suggesting that Pb is relatively insoluble (Kumar et al. 1995).

Metal-contaminated soil can be remediated by chemical, physical or biological techniques (McEldowney et al. 1993). Chemical and physical treatments irreversibly affect soil properties, destroy biodiversity and may render the soil useless as a medium for plant growth. These remediation processes can be costly. Phytoextraction is one of the lowest cost techniques for contaminated soil remediation among the listed remediation techniques. There is a need to develop suitable cost-effective biological soil remediation techniques to remove contaminants without affecting soil fertility.

For metal remediation phytoremediation could provide a sustainable technique. The idea that plants can be used for environmental remediation is very old and cannot be traced to any particular source.

OBJECTIVE OF THE STUDY

Phytoremedial potential of wild plants growing in Pb contaminated sites were studied in this chapter. Pb is chosen as the metal of interest because it is listed as the 2nd most hazardous substance in the list of CERCLA(ATSDR2007). The aim of the study was to find out a plant which is tolerant and can accumulate considerable amount of Pb in its parts, so that it can assist in remediation of contaminated soil. For this the desirable qualities of the candidate plant should have the following properties:

1. The plant should have high biomass.
2. Plant should have rapid growth rate.
3. Can tolerate high amount of Pb in soil.
4. Can accumulate significant amount of Pb in its parts.

MATERIAL AND METHOD

Field Work

Eleven different contaminated sites in and around Kolkata were surveyed which were variously contaminated either by some industrial pollutants or by some anthropogenic activities. This eleven sites are spread in three districts Kolkata, Howrah and North 24-Pargans, These three districts have various large and small scale industries and population pressure is maximum in these areas. Sites were selected by taking into account the probable presence of Pb contamination in these areas. Pb was selected to study accumulation pattern in plant and soil from contaminated sites because Pb stands second most hazardous substance according to ATSDR (2007) as well as it was not possible to study all the heavy metals present in those contaminated site due to time constrains. So the study was focused on Pb accumulation pattern and tolerance mechanism of the plants grown luxuriantly in these six contaminated sites. So, the abundance of different plant growing in those areas were studied.

Determination of Relative Abundance

***Quadrate Study*:** All the species of higher plants of any community can be classified in one or other life form. The ratio of life forms of different species in term of no. of percentages in any floristic community is called biological spectrum or phytoclimatic spectrum.

Method: A given plot of area in eleven locations was chosen and was studied for biological spectrum. The area was about 4mt X 4mt. (Raunkiar, 1934)

Table 2. Location of Study Sites

Location No.	Area of Collection	Sources of contaminants
I	Paint Industries, Howrah	Effluents and chemical waste from industries.
II	Habra I, North 24 Parganas	All kind of Anthropogenic and environmental waste.
III	Habra II, North 24 Parganas	-Do-
IV	Habra III, North 24 Parganas	-Do-
V	Habra IV, North 24 Parganas	-Do-
VI	Habra V, North 24 Parganas	-Do-
VII	Battery production plant, Shyamnagar	Mainly effluents from a battery manufacturing unit.
VIII	Tannery Industries, East Kolkata	Chemical waste from leather tanning unit.
IX	Keshtopur Canal,Kolkata	All possible human and animal waste, along with the vehicular discharge.(Municipal sewage)
X	Bantola Leather Industries	Chemical used in leather industries.
XI	EM Bypass, Kolkata	Road side eateries, automobile discharges and waste from few human settlings.

Kolkata North 24 Parganas Howrah

Biomass Measurement

Biomass of most abundant species was measured by weighing method. It has been shown that the biomass of certain plants such as *Ricinus communis* is much greater than those of other species which were seem to be most abundant by quadrate method. As higher biomass is needed for more accumulation of toxic metal i.e. lead and thus it helps in phytoremediation process, therefore in most cases *Ricinus communis* has been selected for metal assay.

Figure 1. **Area map showing study sites of three different districts**

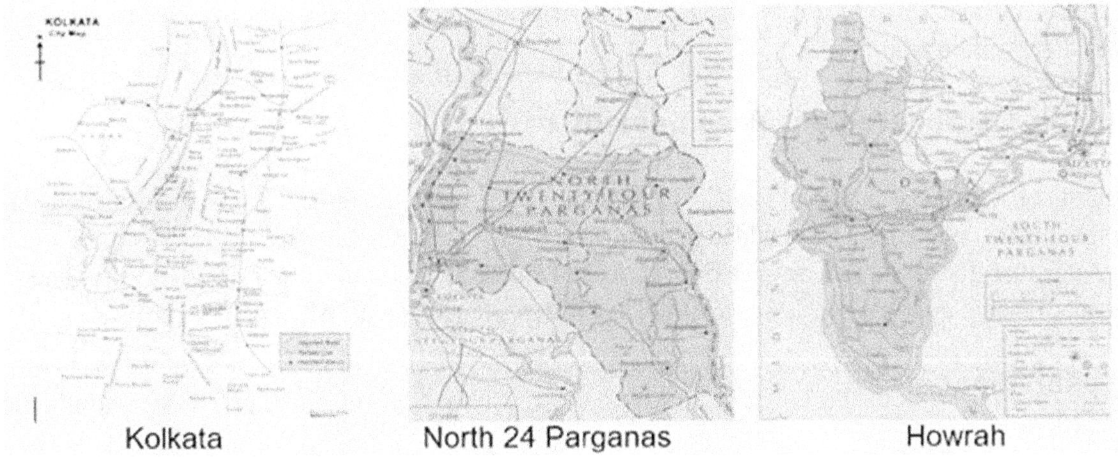

Collection of Sample

Samples were collected in zip lock packets to eliminate possibilities of incorrect sampling. Plants were collected in the shrub stage . Leaves of the shrub stage were collected as follows: Old /senescent leaves, young leaves and emergent leaves .Root were also collected 20cm below ground level.

Digestion of Soil and Plant Sample

Protocol for digestion of soil and plant sample for heavy metals

- In 0.5gm of oven dried soil & tissue, 12ml concentrated HNO3 were added and allowed to stand overnight.
- Heated in hot plate until production of red nitrogen dioxide (NO2) fumes stopped and a watery mass is left.
- 4ml of perchloric acid was added after the beaker was cooled in room temperature.
- Heated again at 70°C to 80°C and evaporated until a white mass is left.
- It was filtered through Whatman 42 ash less filter paper and the volume was made up to 50ml with double distilled water / deionized water.

Lead was estimated in the samples at the Department of Geology, Jadavpur University by AAS analysis following digestion using the method described in the Perkin Elmer handbook for Atomic Absorption Spectroscopy.

The results of metal assay have been given in table no.14. From the result *bioaccumulation factor* and *translocation factor* was determined. A plants ability to accumulate metals from soils can be estimated using the *bioaccumulation factor*, which is defined as the ratio of metal concentration in the shoots to that in soil.

A plants ability to translocate metals from the roots to the shoot is measured using the translocation factors, which is defined as the ratio of metal concentration in the shoots to the roots.

Soil Analysis: Soil parameters such as organic carbon, cation exchange capacity, electric conductivity and pH were measured.

Organic Carbon Content Measurement:

1gm of soil was taken in 500ml conical flask. To it 10 ml 1N Potassium dichromate was added. Then 20 ml concentrated H2SO4 was added in each flask.

After that, the flask was incubated for at least 1 hour. After 1hour 200ml distilled water and 5ml phosphoric acid and 1ml DPA were added. Then it was titrated against Mohr salt.

Measurement of CEC

25ml of 0.05 (N) HCl was added to 1gm of soil sample. After vigorous shaking with glass rod it was left for 20 minutes. Then it was centrifuged for 10minutes at 3000-4000 rpm. The supernatant was discarded. Then distilled water was added to the tube, and filtered with Whatman No 1 filter paper. Now few drops of Ag NO3 was added to it to check the Cl removal. If it becomes turbid (Cl present) the filtrate was discarded, boiling water was added to the filter paper and the test was repeated again. After complete removal of Cl the filter paper was taken, to it 25 ml Barium acetate was added and the filter paper was washed into the beaker so that the entire filtrate comes into the beaker. Then it was incubated for 1 hour, to it 10 drops of Thymol Blue was added, and titrated against 0.05 (N) NaOH.

Measurement of pH

1gm of soil was dissolved in 30 ml distilled water. Then it was kept for 15 minutes. After that, pH of that solution was measured.

RESULT AND DISCUSSION

Study of Relative Abundance

Observation: The following tables and graphical representation shows the frequency of different groups of plants.

From quadrate study it was found that the castor oil plant, *Ricinus communis*, is a plant species of the family Euphorbiaceae, the sole member of the genus *Ricinus*, castor oil plant can reach a height of 2-3 meter in a year. Castor establishes itself easily as apparently native plant and can often be found on waste land. It is a fast growing perennial shrub. This fast growing plant has been found to grow luxuriantly in severely heavy metal contaminated soils of Kolkata and suburbs.

Table 3. Showing frequency and % of biomass of plants from Location 1 (Shalimar)

Plant Species	Family	No. of Plants	Weight of plants	Frequency of Plants(A)	Percentage of Biomass(B)
Tilanthera Sp	Amaranthaceae	45	0.34	71.42	1.6
Ricinus communis	Euphorbiaceae	5	15.5	7.9	72.85
Ipomoea sp	Convovulaceae	7	0.74	11.1	3.5
Solanum sp	Solanaceae	6	4.64	9.5	21.85

A=Frequency of plants =x/åX where, x =no. of plants for a particular sp.
åX=total no of plants in a(4mt x 4mt) plot
B= % of biomass =w/åWx100 where, w=cumulative weight of the particular sp.
åW=total weight of plants

Table 4. Showing frequency and % of biomass of plants from Location 2 (Habra I)

Plant Species	Family	No. of Plants	Wt of plants	Frequency of Plants	% of Biomass
Solanum sp	Solanaceae	10	9.5	41.66	43.7
Eupatorium sp	Asteraceae	5	0.157	35.71	0.72
Ricinus communis	Euphorbiaceae	3	12	04.16	54.64
Sida sp	Malvaceae	6	0.09	25.00	0.43

Table 5. Showing frequency and % of biomass of plants from Location 3 (Habra II)

Plant Species	Family	No. of Plants	Wt. of plants	Frequency of Plants	% of Biomass
Oxalis sp	Oxalidaceae	40	0.143	66.66	0.5
Abutilon indica	Malvaceae	6	10.9	10.00	38.07
Parthenium hysterophorus	Asteraceae	10	0.719	16.66	2.5
Ricinus communis	Euphorbiaceae	4	16.8	6.66	58.37

Table 6. Showing frequency and % of biomass of plants from Location 4 (Habra III)

Plant Species	Family	No. of Plants	Wt. of plants	Frequency of Plants	% of Biomass
Tilanthera sp	Amaranthaceae	25	0.31	60.97	1.7
Croton sp	Euphorbiaceae	8	0.76	19.51	4.1
Solanum nigrum	Solanaceae	4	5.2	9.7	28.0
Cephalandra sp	Cucurbitaceae	2	0.186	4.87	1.0
Ricinus communis	Euphorbiaceae	2	12	4.87	64.5

Table 7. Showing frequency and % of biomass of plants from Location 5(Habra IV)

Plant Species	Family	No. of Plants	Wt. of plants	Frequency of Plants	% of Biomass
Ricinus communis	*Euphorbiaceae*	3	12.3	5.17	91.7
Tilanthera sp	*Amaranthaceae*	40	0.34	68.96	2.54
Croton sp	*Euphorbiaceae*	10	0.62	17.24	4.7
Sida sp	*Malvaceae*	5	0.12	8.62	0.9

Table 8. Showing frequency and % of biomass of plants from Location 6 (Habra V)

Plant Species	Family	No. of Plants	Wt of plants	Frequency of Plants	% of Biomass
Oxalis sp	*Oxalidaceae*	25	0.102	47.16	0.4
Abutilon indicum	*Malvaceae*	5	8.44	9.43	33
Ricinus communis	*Euphorbiaceae*	3	15.6	5.66	60.79
Cyperus sp	*Cyperaceae*	20	0.205	37.73	0.8

Table 9. Showing frequency and % of biomass of plants from Location 7 (Exide Shyamnagar)

Plant Species	Family	No . of Plants	Wt of plants	Frequency of Plants	% of Biomass
Sida sp	*Malvaceae*	7	0.067	15.21	1.68
Abutilon indica	*Malvaceae*	5	1.428	10.86	35.7
Tilanthera sp	*Amaranthaceae*	30	0.12	65.21	3.0
Solanum nigrum	*Solancaeae*	4	2.4	8.69	59.52

Table 10. Showing frequency and % of biomass of plants from Location 8 (Tannery)

Plant Species	Family	No. of Plants	Wt of plants	Frequency of Plants	% of Biomass
Abutilon sp	Malvaceae	6	5.66	35.29	27.27
Ricinus communis	Euphorbiaceae	3	12.6	17.64	60.60
Sida sp	Malvaceae	8	2.50	47.05	12.12

Table 11. Showing frequency and % of biomass of plants from Location 9 (Kestopur)

Plant Species	Family	No. of Plants	Wt. of plants	Frequency of Plants	% of Biomass
Clerodendron inerme	Verbenaceae	5	9.07	14.28	40.89
Tilanthera sp	Amaranthaceae	25	0.226	71.42	1.02
Solanum nigrum	Solanaceae	2	1.45	5.71	6.54
Ricinus communis	Euphorbiaceae	3	11.4	8.57	51.53

Table 12. Showing frequency and % of biomass of plants from Location 10(Bantola)

Plant Species	Family	No. of Plants	Wt.of plants	Frequency of Plants	% of Biomass
Cephalandra indica	Cucurbitaceae	2	0.54	22.22	1.50
Sida sp	Malvaceae	4	2.16	44.44	6.01
Ricinus communis	Euphorbiaceae	3	12	33.33	92.48

Table 13. Showing frequency and % of biomass of plants from Location 11 (Science City)

Plant Species	Family	No. of Plants	Wt of plants	Frequency of Plants	% of Biomass
Tilanthera sesillis	Amaranthaceae	45	0.46	46.39	4.10
Tridax procumbens	Asteraceae	10	2.5	10.3	22.83
Ricinus communis	Euphorbiaceae	2	8.2	2.06	73.05

DETERMINATION OF BIOMASS

The biomasses of different plants are given below.

Next, relative biomass of the plants were measured (Table 14) and a comparative account of percentage of biomass from different contaminated sites (Fig 2) indicates that *R.communis* has the highest amount of biomass among all these plants species. (51.53% in Lx1 to 92.48% in Lx2).L

Figure 2. Comparative account of % of biomass from different contaminated locations

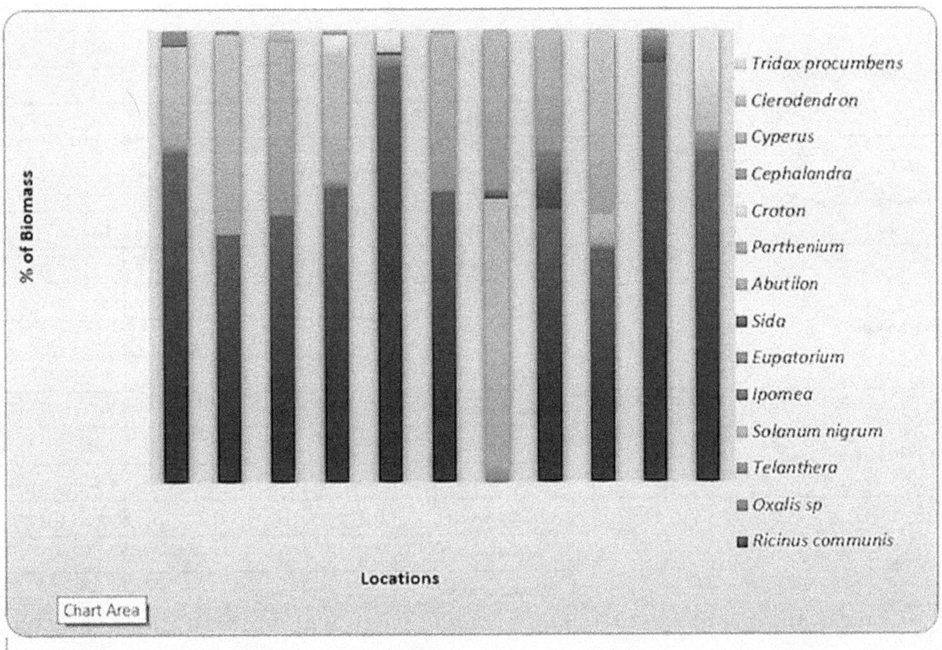

Table 14. Relative height and biomass of plants collected from contaminated sites

Location	Name of the Plant *Species*	Plant Height (inches)	Biomass (Kg)
I (Paint Industries, Howrah)	*Ricinus communis*	47	3.16
	Solanum nigrum	12	0.6
	Tilanthera sp	6	0.02
	Ipomoea sp	16	2.0
II (Habra I, North 24 Parganas)	*Ricinus communis*	59	3.0
	Eupatorium sp	18.3	1.3
	Solanum sp	16	0.8
	Sida sp	11	0.5
III (Habra II, North 24 Parganas)	*Ricinus communis*	63	4.2
	Oxalis sp	3	0.2
	Abutilon indica	30	0.7
	Parthenium hysterophorus	35	0.6
IV (Habra III, North 24 Parganas)	*Ricinus communis*	44	3.3
	Tilanthera sp	8	0.04
	Croton sp	41	2.0
	Solanum nigrum	21	0.8
V (Habra I V, North 24 Parganas)	*Ricinus communis*	61	4.2
	Tilanthera sp	11	0.2
	Croton sp	34	0.6
	Sida sp	12	0.3
VI (Habra V, North 24 Parganas)	*Ricinus communis*	67	5.1
	Cyperus sp	17	0.3
	Oxalis sp	5	0.2
	Abutilon indicum	25	0.8
VII (Battery production plant, Shyamnagar)	*Solanum nigrum*	11	0.5
	Tilanthera sp	6	0.02
	Sida sp	24	0.5
	Abutilon indica	31	1.2
VIII (Tannery Industries, East Kolkata)	*Abutilon sp*	34	1.0
	Sida sp	20	0.5
	Ricinus communis	60	4.3
IX (Keshtopur Canal, Kolkata)	*Ricinus communis*	56	3.7
	Clerodendron inerme	41	2.7
	Tilanthera sp	6	0.2
	Solanum nigrum	34	0.8
X Bantola Leather Industries	*Ricinus communis*	58	5.0
	Cephalandra indica	41	0.7
	Sida sp	23	0.86
XI (EM Bypass, Kolkata)	*Ricinus communis*	49	4.2
	Tilanthera sesillis	9	0.3
	Tridax procumbens	20±1,b	0.2±0.1,a

Table 15. Lead concentration (mg /Kg) in soil, root and shoot samples from different lead contaminated sites

Location	Name of the Plants species	Amount of Lead (Pb) content (mg/Kg)			Translocation Factor (TF) Shoot/root	Bio Accumulation Factor (BF) Shoot/Soil
		Soil	Root	Shoot		
I	*Ricinus communis*	2484*	275.12**	71.87**	0.26	0.028
	Solanum nigrum	Do	Nil	8.73	Nil	0.003
	Clerodendron inerme	Do	5.714	7.97	1.39	0.003
II	*Ricinus communis*	120.42*	48.94**	6.7	0.13	0.055
III	*Ricinus communis*	69.88	16.46**	14.77	0.89	0.211
IV	*Ricinus communis*	46.9	31.18**	5.16	0.16	0.110
V	*Ricinus communis*	55.52	4.62	4.2	0.90	0.075
VI	*Ricinus communis*	39.23	Nil	2.2	Nil	0.056
VII	*Solanum nigrum*	1400*	6.43	7.5	1.16	0.005
	Tilanthera sp	Do	0.122	5.10	41.83	0.003
VIII	*Tilanthera sp*	128*	334.16**	4.3	0.012	0.033
	Ricinus communis	Do	53**	2.4	0.04	0.0003
IX	*Ricinus communis*	14.5	2.69	7.22	2.77	0.497
X	*Ricinus communis*	3.67	Nil	8.5	Nil	2.31
XI	*Ricinus communis*	13.78	11.46**	2.31	0.20	0.167

** Lead concentration is than more than normal in these samples
[Normal concentration of lead is 10mg/kg] (Kabata,Pendias,2001)

Focus of the study was to find out a suitable plant which can be utilized for remediation of Pb contaminated soil. For a potential candidate to be a phytoremediating *sp*ecies the plant s should have a) high biomass, b) rapid growth, c) abundance of growth in metal contaminated sites, d) tolerance and accumulation capacity of metals.

CONCENTRATION OF LEAD IN SOIL

Concentration of Pb in soil varied from location to location (Table -5) depending on the nature of the surrounding environment. Maximum Pb concentration in soil was identified from location I (2484mg/Kg soil), located in the vicinity of paint factory . The site was used as a dumping ground for the factory waste, as Pb is main component of paint, waste from this factory released huge load of Pb in soil no other sites studied recorded such a high amount of Pb in soil. The next highest concentration in soil was recorded from location VII (1400mg/kg soil),near a car-battery manufacturing unit. Lowest concentration of soil Pb was recorded from location X (3.67 mg/kg soil) housing a leather processing unit .Similar leather located near East Kolkata (location VIII) recoded much higher amount of Pb concentration in soil (128 mg/kg soil) .This clearly showed that effluents treatment can reduce the pollution level much significantly. According to Kabata-Pendias (2001) the allowable limit of Pb in uncontaminated soil is 10mg/kg. Taking this value as reference, all the locations (except LX) can be termed as contaminated. From the study it is observed that LI and LVII are highly contaminated, LVIII and II are moderately contaminated and LIII,V, IV, VI, IX and XI are poorly contaminated by Pb. The results clearly indicate that Pb is mainly added to soil by industrial activities.

CONCENTRATION OF Pb IN PLANTS:

Plants (most frequent with significant biomass) were collected from all the eleven experimental sites and assayed for the presence of Pb in them.In most of the sites *Ricinus communis* plants showed highest biomasss and found to be the most significant species of that quadrat (except LX). Therefore Pb accumulation potential of *R. communis* and the next dominant members *Tilanthera philoxiroides* and *Solanum nigrum* were measured. It is observed from the study that Pb accumulation in plants is not dependant on soil Pb content linearly, rather it depends on some other edaphic factors (such as soil pH, organic carbon content etc.) regulating the bio-availability of Pb to the plants. Roots of *Ricinus* was found to accumulate more Pb in comparison with the shoot. Roots of *Ricinus* were found to accumulate maximum amount of Pb (275.12mg/kg dry weight) (Table 14). In LI no detectable amount of Pb was recorded in roots of *Ricinus*. Depending on soil Pb content the concentration of Pb in shoots also varied. In most cases only a small part was translocated in the aerial parts . In 95% of plant samples, the root Pb concentration are much greater than those of the shoot Pb contents, indicating low mobility of Pb from the roots to the shoots. Maximum shoot Pb accumulation was also observed (71.87mg/kg dry weight) in plants collected from LI . According to Baker this plant cannot be considered as a hyper accumulator, as these plants cannot accumulate more than 1000mg/kg dry weight of Pb in their aerial parts, b)do not have a TF value more than 1, c) the concentration of Pb in shoots is not 50- 100 times more than that in plants from non-polluted areas (5mg/kg).

Other than *Ricinus, Solanum nigrum* and *Tilanthera* sp., were also collected from LI, LVII and LVIII, but it was observed that, Pb accumulation in this two plants were not very significant. In LI where soil Pb content were very high, *Solanum* roots showed no accumulated Pb and *Tilanthera* showed very low amount of Pb accumulation (5.7mg/kg dry weight) which are significantly less than the *Ricinus* roots. Reevis and Brooks (1983) reported only four hyperaccumulator species of Pb. Among them *Thalspi rotundifolium* spp. *Cepaeifolium* were found to accumulate 8200mg/kg Pb in shoots. Bary and Clark (1978) found Pb concentration up to 20,000 mg/kg in shoots of *Minuarita verna* grown on a metal mining complex in Yorkshire . But Sieghardt (1987) reported that only 814mg/kg Pb was found to accumulate in the shoots of same sp. While Pb concentration in roots were more, This findings clearly shows that Pb accumulation by the plants is an innate quality of the plants and different population of the same sp may show variable accumulation potential.

SOIL PROPERTIES

As metal bioavailabilty dependant on other edaphic factors, soil pH, organic carbon (OC) content, EC and cation exchange capacities (CEC) of all the experimental sites as well as soil used in pots for growing *Ricinus* plants in Departmental garden were determined. The different parameters clearly indicate how these factors regulate the Pb bio availability to the plants.

In the experimental sites soil pH varies from slightly acidic (5.4in L VI) to neutral (7.9 in LI). Salinity of the soil is reflected by the EC value and value greater than 4 is considered as saline. Here lowest EC value (0.26)was recorded from L V and highest EC value (3.16) was recorded from L III . Organic carbon content values varies from a lowest 0.60 (LI) to a highest 6 (L VI & LIV) (Table-6) . CEC values also varied from sites to sites (6m.e/100gm -49.8m.e/100gm).The combination of elevated soil pH and high organic content may have played a role in limiting the metal availability, resulting in low uptake of Pb by the plants (Jung and Thornton, 1996, Rosselli et.al 2003). From the data it is observed that soil pH values are positively correlated for translocation of Pb in both in root (r=+ 0.42) and in shoot (r=+0.38) .Whereas CEC values are also positively correlated for translocation of Pb in both in root (r=+ 0.21) and in shoot (+0.31) but OC value are negatively correlated for translocation in root (r= – 0.36) and in shoot r= - 0.30 .Soil used in pots to grow experimental plants showed near neutral pH (7.3), organic carbon content was less (0.25%) and CEC value was 6. Low pH and high organic content increase bioavailability of Pb to the plant.

Table 16. Showing soil properties of contaminated sites

Location	pH	EC(ms)	Organic Carbon (%)	CEC (m.e./100gm)
Control	7.4±0.1,cd	1.20±0.1,d	0.42±0.28,a	7±1,a
I	7.7±0.2,cd	2.16±0.1,e	0.7±0.1,b	37.54±0.02,f
II	6.74±0.1,bc	0.423±0.001,b	0.834±0.001,a	16±1,c
III	5.86±0.01,ab	3.16±0.1,g	5±1,b	28.85±0.035,d
IV	5.63±0.15,a	0.54±0.01,c	6±1,b	9.7±0.2,b
V	8.00±1,d	0.26±0.1,a	0.845±0.001,a	49.6±0.15,g
VI	5.4±0.01,a	3.03±0.01,f	6±1,b	35.5±0.173,e

Phytoextraction Efficiency of *Ricinus communis*

Assuming that Pb phytoextraction follows a linear pattern, the quantity of Pb extracted per hectare per year (QPb: kg $Pbha^{-1}y^{-1}$) can be expressed as

$QPb = (10^{-3} X\ bDW\ X\ D)\ X\ (10^{-6}\ X\ [PB]DW)\ X\ C$

[bDw – Dry weight of plant biomass per plant (g $plant^{-1}$DW) ; D – Density of plant per hectare ;

[Pb]DW – Total Pb concentration measured in shoots (mg Pb kg^{-1} DW) ; C – Number of plantsper year](Arshad et al,2008)

By applying this equation, from the present study it was calculated that the quantity of Pb extracted per hectare per year (QPb: kg $Pbha^{-1}y^{-1}$) is **3.426625 kg $Pbha^{-1}y^{-1}$** considering the parameters from Location I because it has highest Pb concentration in soil (2483 mg/kg); (bDw = 3.16 kg/plant, D = 3125 plants/hectare, [Pb]DW= 347 mg Pb kg^{-1} DW, C = 1plant /year).Using an average soil density of $1.2x10^3/m^3$, the weight corresponding the top soil sheet of 1 m^2 surface would be 120kg, therefore the quantity of Pb that correspond to 1 hectare of the contaminated top soil termed S (total Pb quantity in

Figure 3. comparative account of soil properties of different Pb contaminated sites
**The data represents mean plus minus sign SD of three independent replicas. Values carrying different letters are significantly different at $p<0.05$. Means sharing letters within a column are not significantly different by Tukey multiple range test($P<0.05$).*

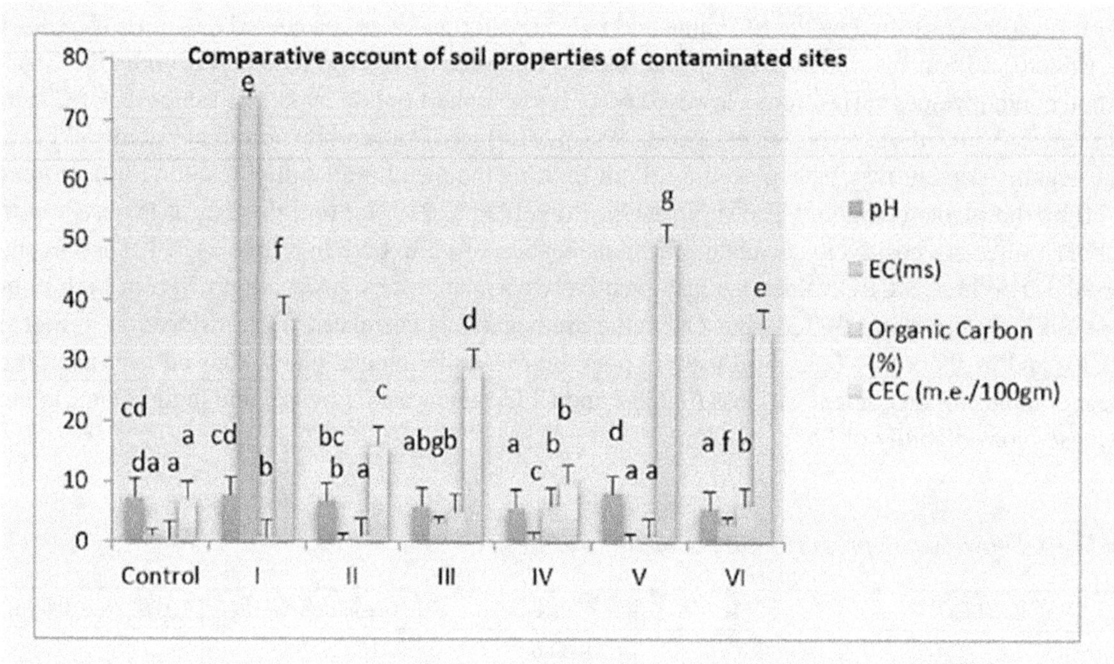

the top soil) is 2970Kg Pb, therefore the no. of years necessary for the *Ricinius* plants to perform total Pb extraction was calculated by the following formula $t= S \times 10^3/QPb$. Therefore the time estimated for total soil remediation is **865.89 years** for *R.communis* grow in L I.

CONCLUSION

In this study accumulation, tolerance and phytoremedial potential of wild plants growing in Pb contaminated sites were studied. Pb is chosen as the metal of interest because it is listed as the 2nd most hazardous substance in the list of ATSDR 2007. For this reason eleven Pb contaminated sites were screened and after quadrat study it was found that *Ricinus communis* is the most commonly growing plant in all these sites, it also has rapid growth rate and have high biomass. That is why *R. communis* was selected as the most suitable plant for phytoremediation. It was found from the study that the plant *R. communis* can tolerate very high amount of Pb in soil (2484 mg/kg). As the plants do not accumulate more than 100mg/kg dry weight Pb in their shoots they cannot be considered as hyperaccumulator but their ability to accumulate high amount of Pb in their roots and considering their rapid growth rate and biomass these plants have the potential for removal of Pb from Pb contaminated soil.

REFERENCES

Agency for Toxic Substances and Disease Registry (ATSDR). (2007). *Toxicological Profile for Arsenic*. U.S. Department of Health and Human Services, Public Health Services.

Bary, S. A. S., & Clark, C. (1978). Problems in interpreting the relationship between the amounts of Pb in plants and soil on metalliferous wastes. *The New Phytologist*, *81*(3), 773–783. doi:10.1111/j.1469-8137.1978.tb01652.x

Battarbee, R., Anderson, N., Appleby, P., Flower, R.G., Fritz, S., Haworth, E., Higgit, S., Jones, Reiser, A., Munro, M.A., Natkanski, J., Oldfield, F., Patrick, S.T., Richardson, N., Rippey, B., & Stevenson, A.C. (1988). *Lake Acidification in The United Kingdom*. ENSIS.

Blaylock, M. J., & Huang, J. W. (2000). Phytoextraction of metals. In I. Raskin & B. D. Ensley (Eds.), *Phytoremediation of toxic metals: using plants to clean-up the environment* (pp. 53–70). John Wiley & Sons, Inc.

BlaylockM.J.SaltD.E.DushenkovS.ZakharovaO.GussmanC.;

Garbarino, J. (1995). *Raising Children in a Socially Toxic Environment*. Jossey-Bass Publishers.

Jung, M. C., & Thornton, I. (1996). Heavy metal contamination of soils and plants in the vicinity of a lead–zinc mine Korea. *Applied Geochemistry*, *11*(1-2), 53–59. doi:10.1016/0883-2927(95)00075-5

Kapulnik, Y., Ensley, B. D., & Raskin, I. (1997). Enhanced accumulation of Pb in Indian mustard by soil-applied chelating agents. *Environmental Science & Technology*, *31*(3), 860–865. doi:10.1021/es960552a

Lasat, M. M. (2002). Reviews and Analyses: Phytoextraction of Toxic Metals- A Review of Biological Mechanisms. *Journal of Environmental Quality*, *31*, 109–120. PMID:11837415

Lasat, M. M., Fuhrmann, M., Ebbs, S. D., Cornish, J. E., & Kochian, L. V. (1998). Phytoremediation of a radiocesium contaminated soil: Evaluation of cesium-137bioaccumulation in the shoots of three plant species. *Journal of Environmental Quality*, *27*(1), 165–168. doi:10.2134/jeq1998.00472425002700010023x

McEldowney, S., Hardman, D. J., & Waite, S. (1993). Treatment Technologies. In S. McEldowney, D. J. Hardman, & S. Waite (Eds.), *Pollution Ecology and Biotreatment*. Longman Scientific & Technical.

Nriagu, J. O. (1989). A global Assessment of Natural Sources of Atmospheric Trace Metals. *Nature*, *338*(6210), 47–49. doi:10.1038/338047a0

Peplow, D. (1999). *Environmental Impacts of Mining in Eastern Washington, Centre for Water and Watershed Studies Fact Sheet*. University of Washington.

Raskin & Ensley (Eds.). (n.d.). *Phytoremediation of toxic metals: using plants to clean-up the environment*. New York: John Wiley & Sons, Inc.

Raunkiaer, C. (1934). *The life forms of plants and statistical plant geography*. Clarendon Press.

Reeves, R. D., & Brooks, R. R. (1984). Studies on metal uptake by plants from serpentine and non-serpentine populations of Thlaspi goesingense H'al'acsy (Cruciferae). *The New Phytologist*, *98*(1), 191–204. doi:10.1111/j.1469-8137.1984.tb06108.x PMID:29681122

Rosselli, W., Keller, C., & Boschi, K. (2003). Phytoextraction capacity of tree growing on a metal contaminated soil. *Plant and Soil, 256*(2), 265–272. doi:10.1023/A:1026100707797

Sahi, C., Agarwal, M., Reddy, M. K., Sopory, S. K., & Grover, A. (2003). Isolation and expression analysis of salt stress-associated ESTs from contrasting rice cultivars using a PCR-based subtraction method. *Theoretical and Applied Genetics, 106*(4), 620–628. doi:10.100700122-002-1089-8 PMID:12595990

Sieghardt, H. (1987). Schwermetall- und N¨ahrelementgehalte von Pflanzen und Bodenproben schwermetallhaltiger Halden im Raum Bleiberg in Ka¨rnten (O¨ sterreich): I. krautige Pflanzen. Z. Pflanzener¨ahr. *Bodenk, 150*(3), 129–134. doi:10.1002/jpln.19871500302

United Nations Environmental Protection/Global Program of Action. (2004). *Why the marine Environment needs Protection from Heavy metals, Heavy metal 2004.* UNEP/GPA Coordination office. http://www.oceansatlas.org/unatlas/uses/uneptextsph/wasteph/2602gpa

USEPA. (2001). Microwave assisted digestion of sediments.Sludge,soils and oils. Official Methods/ USEPA Methods, 3051, 846.

Vogel-Mikus, K., Drobne, D., & Regvar, M. (2005). Zn, Cd and Pb accumulation and arbuscular mycorrhizal colonisation of pennycress Thlaspi praecox Wulf. (Brassicaceae) from the vicinity of a lead mine and smelter in Slovenia. *Environmental Pollution, 133*(2), 233–242. doi:10.1016/j.envpol.2004.06.021 PMID:15519454

Xintaras, C. (1992). *Analysis Paper: Impact of Lead-Contaminated Soil on Public Health.* U.S. Department of Health and Human Services. Public Health Service Agency for Toxic Substances and Disease Registry.

Chapter 9
Environmental Fluoride:
Impact on Humans, Animals and Its Remediation Strategies

Junaid Ahmad Malik
https://orcid.org/0000-0003-4411-2015
Government Degree College, Bijbehara, India

ABSTRACT

Noteworthy multi-disciplinary undertakings have been taken to investigate the effects of natural fluoride ion (F) contamination since the preceding century. Fluoride is a hazard to the earth and human prosperity. Developed and developing countries are standing up to such enormous extents of issues in light of fluoride in the drinking water. Human use to fluoride has bourgeoned since World War II, on account of fluoridated water and toothpaste just as to the normal defilement by huge ventures, from aluminium to pesticides, where fluoride is an essential mechanical concoction similarly as a waste product. The chapter deals with the proportion of fluoride in nature and its impact on human prosperity, generally on the brain, endocrine system, thyroid, pineal gland, immune system, reproductive system, and organ systems. High assemblies of F in soil may really bargain the life of plants, obliterate soil microbial development, upset the soil environment, and cause soil and water defilement. This chapter further emphasizes various biological approaches for the remediation.

INTRODUCTION

The high calibre of ways of life and the wellness of nature are immediately related to one another. The expanding frequency of animal and human medical problems because of modern contamination and anthropogenic changes has pulled in global intrigue and endeavours to discover new solutions for higher control and support the ecological measure (Ahmed, 2007; Tsiros et al., 1998). Poisonous contaminations may moreover be discharged to the earth by utilizing the air, soil, and water. Stack emanations to the air may furthermore add contamination to the soil which may moreover collect in plants by means of their ground rules. These contaminants amass in the food chain and afterward affect individuals and

DOI: 10.4018/978-1-7998-4888-2.ch009

untamed life. Because of variations in home grown and anthropogenic exercises, universal contamination is developing and prompting sickness of the environment with metals, non-metals, natural chemical assemblies, and inorganic mixtures. The first supporters of this defilement are pesticides, sewage removal, bug sprays, herbicides, and the uncontrolled release of squanders. In the industrialized world, a gigantic segment of the populace is revealed every day to an assortment of concoction substances and poisonous metals which are unsafe for human wellness (Figure 1,3). Combinations with the fluorine component are altogether used in almost each organic industry, and air contamination through the fluoride ion (F) is colossal on the earth. In spite of the fact that F has an anticaries impact when applied topically to the teeth, fluorine isn't a key trace element and isn't significant for the improvement of sound teeth and bones (Chouhan and Flora, 2010; SCHER, 2011). Inordinate F admission may moreover antagonistically influence the wellness of animals and plants (Koblar et al, 2011). Fluoride is dispensed generally all through soils, plants, and animals, and is thought to be a quintessential component in animals. Fluoride has a significant capacity in bone mineralization and development of dental lacquers. Fluoride, when ate up in deficient segments (under 0.5 ppm), motives medical problems, for example, dental caries, absence of arrangement of dental veneer, and diminished bone mineralization, explicitly among youths (WHO, 1996). Interestingly, when fluoride is fed on in extra (more than 1 ppm), medical problems may additionally result, which likewise influence the young and old (WHO, 1996). At more noteworthy fluoride fixations, metabolic procedures are influenced in humans, and overexposed people can likewise experience the ill effects of skeletal or dental fluorosis, non-skeletal appearances, or combos of these ailments (Susheela et al., 1993). The occurrence and seriousness of fluorosis depends upon the fluoride fixation in air, soil or water, and the level of introduction to these levels (Table 1).

Among the three sorts of typical media (air, soil, and water), groundwater is the central inception of fluoride collaborator in humans and animals. To sustain life, freshwater must be continually available to individuals. Since the beginning, individuals have relied upon groundwater as a wellspring of drinking water, and even today, most of the people depend on tons of groundwater for persistence. Groundwater includes 97% of all-out freshwater, and in various areas, groundwater sources build up the single greatest available supply of fresh drinking water (WHO, 2004). Exactly when the paces of groundwater extraction outperform stimulate rates, weariness of this significant resource occurs, with a resultant confinement of the drinking water supply.

Fluoride is one of the most important among all the groundwater contaminants and is of concern basically considering the way that it has both short and long stretch effects on human prosperity. Additionally, in various areas of the world, it is very difficult to keep up the vital good ways from introduction to fluoride. Regardless of the way that groundwater quality may be incapacitated by various ordinary constituents, fluoride is among the most prominent toxins, since its geogenic foundation stage renders it so across the board. Fluorosis has been chronicled to have conveyed immense prosperity incapacitation to the inhabitants of more than 25 nations over the world (UNICEF, 1999). The number of people who experience the evil impacts of fluoride ache is growing at sensible amounts. It has been evaluated that over 200 million people overall are at risk for fluorosis (UNICEF, 1999). In India, about 80% of domiciliary freshwater needs in country zones and half in urban areas are met by groundwater. The individuals who rely upon tube wells for freshwater are under hazard from proceeding with introduction to affluence of fluoride, arsenic, iron, nitrate, and saltiness. The degrees of normal fluoride that occur in groundwater stretch out from 0.5 to 48 ppm, or more (Susheela et al., 2003). The closeness of even low degrees of fluoride in groundwater, when joined with a general transcendence of wretchedness, may turn disastrous, especially for kids in urban and semi-urban zones of India or elsewhere. Over the span of recent decades, the inescapability and earnestness of fluorosis has extended radically in India, coming to approach toward scourge levels. At present, 20 of 35 states and Union Territories are going up against preposterous fluoride affirmations, while new areas are also being affected by this issue. Taking into account excess fluoride in groundwater, WHO diminished beyond what many would consider possible in drinking water of fluoride in India from 1.5 to 1.0 ppm in the year 1998 (UNICEF, 1999). Fluoride can be recognized in drinking water just by lab assessment, since water sullied with fluoride is grim, dull, and unscented. Hence, sufferers find that they have been using fluoride-polluted water just after indications of fluoride disease surfaces (Figure 3).

The essential objective of this chapter is to endow pursuers with a layout of the key factors that are at risk for fluoride presentation, including the issues achieved by such introduction, and a portrayal of how fluoride is ingested and utilized in individuals. The level of toxicity rendered by the F in different living forms demands effective ways for the ecological removal and remediation of such toxicant. Some of the known and adopted strategies of its removal include physiochemical and biological approaches. We have thus underscored, in this review, portions of the study of fluorine, the source and occasion of fluoride in the earth, and how it is appropriated in the environment and the strategies for its amelioration and removal.

THE CHEMISTRY OF FLUORINE

Fluorine is an element of the halogen group and is set at VIIA in the periodic table. Fluorine exists as a light yellow green, impactful gas, whose atomic number is 9 and sub-atomic weight is 18.998 g/mol. It is univalent, and in various exacerbates, the stereochemistry of the fluoride particle appears like that of OH, considering the way that their ionic radii are comparative (F 133 pm, OH 140 pm) (O'Donnell, 1975). It is the most electronegative element in the periodic table (4.0 in the Pauling scale) (Neumuller, 1981). Inorganic fluoride is either free or cross-section bound in minerals, or covalently bound in various inorganic or natural blends. Since the electronic setup of fluorine is $1s^2,2s^2,2p^5$, fluorine needs to increment only a solitary electron in its outside shell to shape a stable inorganic fluoride particle. Doing so empowers the atom to achieve a dormant gas arrangement. The high reactivity (i.e., oxidizing limit) of fluorine on a very basic level outcomes from its high electro-cynicism, its shockingly low separation vitality, and the incredible bond quality of the mixes it structures. Considering its high reactivity, fluorine never exists free in nature anyway is continually present in a got structure together with various parts. It reacts vivaciously with one another besides the inactive gases. The ecological wealth of inorganic fluoride is expanding, on the grounds that it is progressively discharged into the earth as waste from use in mechanical and dental wellbeing items.

SPATIAL DISTRIBUTION OF FLUORINE

The total convergence of F in the soil is by and large gotten from the parent material and thusly its dispersal configuration in soil is related to the system of soil organization. The ordinary content of F in the worldwide soil has been evaluated as 329 μg/mg. The most negligible F content are typically present in sandy soil in tolerably tight spots, while the higher centralizations of F are found in soil from suffered mafic rocks and in substantial mud soil (earth soil). The pH of the soil, clay and natural carbon content are the prime determinants of soil F content (Kumar et al., 2016). F enters the soil through different ways, for instance, dry installation, rainfall, and with adulterated litter where it is held expeditiously. The held F fabricates the full scale dissolvable F fixation in the earth, impacts the pH of soil, and can get together with deadly parts, for instance, aluminium and substantial metals. F can exist as the free fluoride particle (F^-) or structure buildings and edifices with components, for instance, iron (Fe), boron (B), calcium (Ca), and aluminium (Al), with the structures of Al and F being commonly inescapable (Domingos et al, 2003).

EFFECTS OF FLUORINE POLLUTION

The free fluorine generated by enterprises pollutes the air as well as, with ground water tainted through fluorine-containing mineral stores, contaminates plants, crops, soil, vegetables, and freshwater bodies. The utilization of these F^- tainted items and sources may bring about harmful belongings on the fitness of both the humans and other fauna (Choubisa and Choubisa, 2016). In spite of the fact that the topical utilization of F on teeth is viewed as having positive impacts, the proof for any beneficial impacts from foundational retention is currently all around considered to be feeble (SCHER, 2011).

ENVIRONMENTAL SOURCES AND MOBILITY OF FLUORIDE

Natural Sources

In nature, fluorine is in a great extent found in sedimentary phosphate rock stores and minerals (Kirk-Othmer, 1996). The major fluorine-containing minerals are fluorspar or fluorite, fluorapatite, and cryolite. For mechanical purposes, the most noteworthy wellspring of fluorine is mineral calcium fluoride (fluorite or fluorspar, CaF_2), which has a fluorine substance of 49% (Fuge, 1988). Fluorite, modernly known as fluorspar, is regularly connected with quartz, calcite, dolomite, or barite (Prud'homme, 1990). Regardless of the way that the greatest proportion of fluoride exists as fluorapatite $[Ca_5(OH,F)(PO_4)]_3$, this mineral is burrowed primarily for its low inorganic fluoride (4% fluorine) (Kirk-Othmer, 1996). By assessment, cryolite (Na_3AlF_6) is extraordinary (54% of F). Despite CaF_2, fluorapatite, and cryolite, different various silicates, for instance, topaz ($Al_2SiO_4(OH,F)_2$), oxides, carbonates, sulfates, phosphates, sellaite (MgF_2), and sodium fluoride or villiaumite, contain minor proportions of inorganic fluoride (Fuge, 1988). Host minerals, for instance, mica (layer silicates), amphiboles (chain silicates), apatite, and tourmaline, and soils, for instance, montmorillonite, kaolinite, and bentonite, moreover contain inorganic fluoride (Reimann and Decaritat, 1998). Fluorine may occur in limestone that is connected with tremolite, actinolite, and pyroxene, where fluoride obsessions may show up at a level of 0.4–1.2%. A bit of the fluorine may be accessible in earth material admixed with suffered limestone.

Anthropogenic Sources

In developing countries, the native soil fluoride content is vehemently affected by the use of composts and by declaration of present day airborne poisons. A couple of anthropogenic sources upgrade the fluoride substance of soil. Noteworthy spread wellsprings of fluoride fuse coal expending, oil refining, steel creation, substance age, soil creation, Al cleaning, glass and clean gathering, square and aesthetic collecting, scattering of fluoride-containing fertilizers and pesticides, wastes from sewage and slops, production of uranium hexafluoride (UF_6), and uranium trifluoride (UF_3) from the nuclear industry (Neumuller, 1981; Fuge, 1988; Fuge and Andrews, 1988). Phosphate containing fertilizers, especially the super-phosphates, are possibly without a doubt the most noteworthy wellsprings of fluoride debasement to provincial territories. A reiterated use of rock phosphate that contains a couple of percent of fluoride was appeared to basically lift the fluoride substance of soils (Omueti and Jones, 1977). Normal increments of phosphatic fertilizer (50–100 kg P_2O_5/ha/yr) may lift soil fluoride content by 5–10 ppm/yr (Gilpin and Johnson, 2014). Rock phosphates for the most part contain generally 3.5% of fluoride; phosphatic manures contain some place in the scope of 1.5 and 3.0% fluoride (McLaughlin et al., 2006). Labile and water-dissolvable fluoride obsessions appear, apparently, to be for the most part impacted by mechanical tainting (Polomski et al., 2012; Haidouti, 2001). The surface soil in the locale of block field endeavours found to contain water-dissolvable fluoride (1:1) ran from 0.59 to 2.74 ppm; $CaCl_2$ extractable fluoride went from 0.69 to 3.18 ppm, however the mean supreme fluoride fixation moved from 322 to 456 µg/g (Jha et al., 2014). The deferred usage of super-phosphate can similarly effectsly affect soil fluoride content (Kudzin and Pashova, 2009).

DISTRIBUTION AND MOBILITY OF FLUORIDE IN SOILS

The fluoride inalienable to soil is commonly delimited inside a few minerals, most usually apatite [especially fluorapatite ($Ca_5(PO_4)_3F$, fluorite (CaF_2), cryolite (Na_3AlF_6)], types of topaz ($Al_2(SiO_4)F_2$), and inside micaceous soil minerals. Fluoride is additionally existing in soil as explicitly and vaguely adsorbed particles and mixes (Bowen, 2006; Pickering, 2015). The intuitive soil fluoride level is boundlessly dependent on the parental material from which the dirt is shaped, while it's flexibly in the soil profile is a job of soil-framing systems, of which the degree of enduring and earth content are the most extreme articulated. Extremely basic soil minerals, for example, biotite, muscovite, and hornblende, may contain as much as a few percent of fluoride and in this way are the principal establishment of fluoride in soil. Fluoride has stayed evaporated from the shallow skylines of most soils; subsequently, it isn't dumbfounding that natural issue has a low family relationship for fluoride. Omueti and Jones (2012) gave the scope of fluoride fixation in natural matter of the shallow skyline to be as low as 0.03–0.12 ppm. In practically all investigations of the fluoride substance of soils that are unadulterated, high inconstancy has been depicted. The normal fluoride substance of soil, around the world, had stayed determined to be 320 ppm (Kabata-Pendias and Pendias, 2004). The all out fluoride focus in typical arable soils ranges from 150 to 360 ppm in any case can reach up to 620 ppm. Various writers relate the fluctuation of absolute soil fluoride substance to the molecule size of the soil, specifically; expanding measures of complete fluoride are related with developing earth content (Robinson and Edington, 1946; Omueti and Jones, 2012). Under ordinary conditions, it is likewise basic for absolute soil fluoride substance to increment with soil profile profundity. This may result from the low partiality that fluoride has for natural issue (Omueti and Jones, 2012). Expanding soil fluoride content, with expanding profundity, may likewise be owed to long haul descending development of fluoride through the soil profile. The labile types of soil fluoride are differently assessed to be as water-dissolvable, corrosive extractable, and sap extractable fluoride mixes. Portable or water-dissolvable fluoride is easily adsorbed by earth and phosphates. At lower pH levels, sorption reductions because of the improvement of solvent Al–F species, for example, $(AlF)^{2+}$ and $(AlF_2)^+$

edifices (Barrow and Ellis, 2006; Wenzel and Blum, 2002). Conversely, in soils having high pH and low degrees of shapeless Al species, dirt tops and OM for the most part sorb slight F (Omueti and Jones, 2012). At antacid pHs, the expanded negative surface charge results in repugnance of anionic F. Under such conditions, the significant maintenance system is the trading of F with OH gatherings of nebulous materials, for example, Al hydroxide. For this situation, the precious stone cross section OH of earth minerals is traded by F, resultant in a synchronized arrival of Al and Fe. Another F retaining apparatus includes F precipitation as CaF_2 in calcareous soils (Slavek et al., 1996). At nonpartisan to soluble pH levels, F exists overwhelmingly as the F^- ion, and at a pH of < 5.5, the fluoride is complexed with Al (Wenzel and Blum, 2002). The adsorption of F over a scope of 2–16 mg/L was very much marked by Langmuir isotherms and the pH of the dirt arrangement stage appropriately expanded, despite the fact that the upsurge in OH–was simply a minor portion of the adsorbed F. The request in the capacity of different materials to assimilate F was as per the following: $Al(OH)_3$ encourage on bentonite> $Al(OH)_3$ >> hydrated hallosite and dried out halloysite > a feebly acidic soil >> kaolinite > gibbsite > antacid soil > goethite > bentonite and vermiculite.

Al(OH)3 has an incredibly high F adsorption estimations. The F adsorption happens primarily by trade with OH groups from $Al(OH)_3$, and fundamental Al polymers adsorbed on mineral surfaces, as opposed to through trade with precious stone cross section OH gathering of soil minerals. In calcareous soils, the development of somewhat doable CaF_2 and F combinations with Al, Fe, and Si is answerable for the low movement of this component. In sodic soils, conversely, elevated levels of transferable Na impacts intensified dissolvability of F. Chhabra et al. (1999) had likewise portrayed an immediate increment of water-soluble fluoride with the upsurge of interchangeable sodium rates.

DISTRIBUTION OF FLUORIDE IN PLANTS AND SOIL

Fluoride isn't a significant plant constituent yet is imperative to the life and living organisms including humans. Nevertheless, constant ingestion of exorbitant quantities of fluoride may prompt the "fluorosis," while problematic levels in the food can have a correspondingly dangerous impact in different manners. Thusly, the fluoride content in plants is of consideration not solely to humans yet additionally for other livestock animals (Keerthisinghe et al., 2001).

Fluoride Absorption and Mobility in Plants

The accessibility of fluoride to plants is regularly not related to the total fluoride or dissolvable fluoride substance of the soil in which they develop. Regardless, under certain soil and plant conditions, the F element of plants seems to reflect its occasion in soils (Shupe and Sharma, 2006). Bieliyakova (2007) gave the extent of fluoride in plant trash to fluoride in soils as 0.2 and 0.6 for settled and emblematic vegetation, respectively. These characteristics show a tolerably low F bioavailability to plants. The dissolvable F part in soil is taken up idly by roots and clearly is successfully moved in plants. In spite of the way that it has been shown that plants can take up F successfully from defiled soils, the bioavailability of dissolvable fluoride is of extensively less significance than that from airborne mixes. Toward the day's end, when fluoride is accessible as both air pollution and a soil poison, the F take-up by plants from air is significantly more basic than is the take-up from soil. A couple of factors impact plant accumulating of airborne fluoride, yet the most explained are air fluoride obsession and the term of presentation.

Vaporous fluoride enters the leaf through stomata and a short time later separates in the water that infiltrates the cell walls. The regular movement of water in a leaf is toward the goals of most prominent dissipation, which are the leaf edges and tips. Passed on by water, fluoride moves in the edges and tips, so these domains are usually the first to show visual harm. Generally, leaves are the most delicate when they are fledgling and up 'til now emerging. Once thoroughly made, leaves may be ordinarily impenetrable to F. Introduction to a high F obsession causes putrefaction of leaf parts or even the whole leaf.

Acute Level of fluoride in soil

Protracted fluoride take-up is risky to plants and animals, along these lines it is amiable to restrict the paces of fluoride storing up in soils. It is an inconvenient task to develop a single upper limit fixation for F poisonousness to plants, in light of the fact that different soils change in their ability to hold fluoride. Accordingly a high union of fluoride is immovably held in a fine-completed indistinct Al hydroxide-rich soil. Regardless, a much lower obsession is held in a coarse-completed calcareous soil that is low in ill-defined Al blends and as such may hurt plants. Plants are known to take up F; nonetheless how much this happens differs among plant species (Brewer, 1966). In soil arrangement, differing concoction kinds of fluoride exists. In acidic soils, a colossal part of fluoride occurs as AlF^{2+} (Manoharan et al., 2006). The Al–F species are non-lethal to plants at lower applications; this species is smoothly toxic at higher obsessions, dependent upon species (Manoharan, 2006; McLaughlin et al., 2006). Stevens et al. (1997) showed that Al–F species are less noxious than are Al^{3+}, $Al(OH)^{2+}$, and $Al(OH)^{2+}$ species to tomato and oats.

Around the globe, at present, there are no standards or even recommendations regarding what the commendable farthest compasses of F in soils should be. The clarification may be the nonattendance of the data open for F substance of soils worldwide and besides the wide course of action that exists in uncovered the levels of fluoride in soils.

FLUORIDES IN GROUNDWATER

Water is a key trademark asset for continuing with life and is among nature's most enormous blessings. At the point when saw as a wearisome and unlimited asset, today, water routinely depicts the constraints of human, social, and cash related improvement for a district. The basic wellspring of freshwater for supporting life on earth is groundwater. Tragically, groundwater is either being progressively exhausted for water game plan of harvests, ebb and flow, or different uses, or is getting spoiled by different harmful substances. The nearness of fluoride as a contaminant of groundwater has become a general issue, since it is ordinarily found in groundwater sources. The issue of high fluoride content in groundwater assets is significant, in light of both toxicological and geo-basic concerns. The fluoride that defiles groundwater gets for the most part from essential contemplates yet is influenced by the influence of adjacent and local geography and the nearness of certain hydro-geochemical conditions. The major wellspring of fluoride in groundwater is fluoride-bearing minerals that exist in rocks and soils. The misery and fluid isolating structures that happen in soils expect a critical movement in picking the extents of fluoride that scopes groundwater. The different parts that direct the presence of fluoride into water from fluoride-bearing minerals are (I) the compound bit of water, (ii) the nearness and responsiveness of fluoride minerals to water, and (iii) the contact time between the source mineral and water (Keller, 1979).

Everything considered, water quality (e.g., pH, hardness, and ionic quality) in addition acknowledge a basic crusade by affecting mineral dissolvability, complexation, and sorption/exchange responses (Apambire et al., 2007). The fundamental state of groundwater favors expanded dissolvability of fluoride-bearing minerals. Alkalinity prepares fluoride from fluorite with precipitation of calcium carbonate, considering the way that the dissolvability of CaF_2 increments with an augmentation in $NaHCO_3$ (Handa, 1975; Saxena and Ahmed, 2001).

$$CaF_2 + 2NaHCO_3 = CaCO_3 + 2Na^+ + 2F^- + H_2O + CO_2$$

The above condition plainly shows the procedures that could control negative (among fluoride and calcium) and positive connections (among fluoride and bicarbonate) when both are in contact with one another. Water tests in which fluoride levels surpass 5 mg/L are oversaturated with respect to fluorite. When fluorite arrives at balance, calcite is evacuated by precipitation, which enables the fluoride fixation to expand (Kim and Jeong, 2005). In groundwater, the regular convergence of fluoride relies upon the geologic, concoction, and physical attributes of the springs, porosity and the corrosiveness of the soils and rocks, the temperature, the activity of other compound components, and the profundity of the wells. In normal water, the fluoride structures solid edifices with Al, and along these lines, fluorine science is to a great extent directed by Al fixation and pH level (Skjelkvale, 1994).

Underneath pH 5, fluoride is fundamentally complexed with Al, dominatingly with the AlF^{2+} complex, and henceforth the grouping of free fluoride is diminished to low levels. As the pH develops, the Al–OH structures request over the Al–F structures, and the free fluoride level increases. Fluoride happens at some level in essentially all groundwater, in any case the exposure found in most consumable waters is under 1 mg/L (Hem, 2015). It has been speculated that fluoride-bearing minerals are typically just sparingly water dissolvable, beside villiaumite, and these minerals discharge fluoride to water bit by bit (Cronin et al., 2000; Saxena and Ahmed, 2003). The pace of fluorite separating might be quicker in sodium bicarbonate-containing waters, and the presence of fluoride from soil minerals depends unequivocally upon the pH level (Apambire et al., 2007; Saxena and Ahmed, 2003). The most unprecedented blend of fluoride in groundwater is ordinarily obliged by the dissolvability of fluorite (Handa, 1975; Apambire et al., 2007; Cronin et al., 2000; Saxena and Ahmed, 2003; Chae et al., 2017). When as far as possible for fluorite (CaF_2) is emanated to, a converse relationship will exist among fluoride and calcium presentations. Prior assessments have uncovered that there is a close association between high fluoride content and delicate, antacid (i.e., sodium bicarbonate) groundwater that is worn out of calcium (Handa, 1975; Whittemore et al., 2013; Chae et al., 2017). Molten rocks that have been formed from unbelievably developed magmas are a rich wellspring of fluorine bearing minerals. The plagioclase bit of fluid rocks is ordinarily high in albite, the sodium-rich compounds (Hyndman, 1985). Subsequently, the groundwater in contact with these stones is typically delicate and calcium lacking, which thinks about higher fluoride measures when congruity with fluorite is cultivated (Ozsvath, 2006).

High fluoride obsessions (up to 30 mg/L) can in like manner result from anion interchange (OH^- for F^-) with certain soil minerals, suffered mica, and oxyhydroxides that are ordinarily found in superfluous soils and alluvial rocks (Whittemore et al., 2013; Apambire et al., 2007; Warren et al., 2005). Research focus examinations have shown that aluminum hydroxides have an especially high fluoride exchange limit (Cronin et al., 2000). It has been found in past research that a close relationship exists between pH levels and fluoride fixation. Sometimes, the effect of living course of action time conveys a prompt association between fluoride obsessions and the significance at which a water test was assembled (Hudak and Sanmanee, 2003; Edmunds and Smedley, 2005; Kim and Jeong, 2005; Chae et al., 2017). The effect of air on fluoride obsessions in groundwater is, all things considered, attributed to precipitation, and to resuscitate rates and groundwater stream (Edmunds and Smedley, 2005). Domains of high precipitation, for instance, sodden tropical areas, are increasingly opposed to have high fluoride obsessions in groundwater, considering the way that dissolvable particles, for instance, fluoride are depleted out and debilitated. Of course, some dry conditions are noted for having high fluoride content, considering the way that the low paces of groundwater resuscitate lead to deferred water–mineral association and higher salinities; such invigorate improves mineral crumbling (Handa, 1975). Air can in like manner sway separated fluoride levels. Temperature straightforwardly influences the dissolvability of fluorine-bearing minerals. For example, the parity predictable for fluorite increases from 10–10.80 at 10oC to 10–10.57 at 25oC (Edmunds and Smedley, 2005), which considers commonly 30% more fluoride to separate in debilitate plans (Table 1). It is assessed that about 200 million people, addressing 25 nations around the globe, are sufficiently introduced to F to place them at risk for having fluorosis. High fluoride levels have been represented to exist in groundwater tests taken from gigantic bits of Africa, China, the Middle East, and southern Asia (India and Sri Lanka).

The raised fluoride levels of water in China recognised more than 26 million people facing dental fluorosis, and an additional one million people encountering skeletal fluorosis, in 2004. At Yellowstone National Park in the United States, the proportion of fluoride found in underground springs ran from 25 to 50 ppm (Neuhold and Sigler, 1960). In Mexico, 5,000,000 people (about 6% of the masses) are impacted by using fluoride-contaminated groundwater (UNICEF, 1999). Poland, Finland, and the Czech Republic have levels of fluoride in drinking water as high as 3 ppm (Czarnowski et al., 2006; Lahermo et al., 1990). In the Ethiopian Rift Valley, fluoride obsessions in the extent of 1.5–177 ppm have been represented (Kloos et al., 1993). Other truly affected domains join the dry bits of northern China (Inner Mongolia), African countries like Ivory Coast, Senegal, North Algeria, Uganda, Ethiopia, northern Mexico, and central Argentina (WHO, 2005). Besides, Mambali (1982) declared the recurrence of fluorosis related with excess fluoride confirmation from groundwater in a couple of African countries. These countries address domains of the Rift Valley known to be among the most genuinely fluoride-affected countries on the planet. In various fluorotic zones, people, particularly kids, face transportability inconveniences from destroying skeletal fluorosis (Mjengera and Mkongo, 2003). India furthermore faces this proportionate issue.

Fluorosis, in India, was first recognized by Short et al. (1937) at Prakasam, Andhra Pradesh. Around then the disease was overwhelming in only 4 states, to be explicit Andhra Pradesh, Tamil Nadu, Punjab, and Uttar Pradesh. Fluorosis during 1999 existed as endemic in 17 states of India (UNICEF, 1999). This is examined that 62 million Indian people, with 6 million youths, experience the evil impacts of fluorosis that results from eating up fluoride-corrupted water (Susheela et al., 2003). In India, the Bureau of Indian Standards has suggested the most remote purpose of fluoride in consumable water to be some place in the scope of 1.0 and 1.5 mg/L (BIS 1991).

TOXICITY OF FLUORINE IN BIRDS AND ANIMALS

The fluoride ion (F^-) isn't viewed as fundamental for human growth, development and improvement (SCHER, 2011) including for the advancement of solid teeth and bones, and the interminable ingestion of fluoride at significant levels (over 6 mg/day) can be poisonous to animals and humans and cause dental, skeletal, and non-skeletal fluorosis (WHO, 2002). Fluorosis for the most part happens in two structures: (I) endemic fluorosis brought about by drinking water or devouring nourishment with a high F substance and (ii) modern fluorosis coming about because of introduction to air containing an extensive F content. Fluorosis, in humans, animals, and flying animals, influences the skeletal parts of the body as well as influences delicate tissues, e.g., cerebrum, liver, kidney, thyroid, and spinal line. Anjum et al. (2014) explored the impact of a high convergence of F on hepatic and renal catalysts in four groups of domestic chickens, A, B, C, and D, getting 0, 10, 20, and 30 μg/g of NaF by body weight, separately, on a week by week reason for about a month. Alkaline phosphatase (ALP), aspartate aminotransferase (AST), alanine aminotransferase (ALT), and bilirubin were resolved as markers of liver capacity, while uric corrosive was utilized as a parameter for renal capacity. The outcomes demonstrated high qualities for every one of the parameters ($p<0.05$) in the F-treated groups showing that F affected hepatic and renal capacity in the uncovered birds (Anjum et al., 2014). Albeit all animals are helpless to high portions of F, the resilience level changes starting with one animal varieties then onto the next. Important wellsprings of F for terrestrial animals are drinking water, soil, and vegetation polluted with F discharged by various exercises, for example, volcanic emissions and industrial activities (Ranjan and Ranjan, 2015). The digestion of fluorine in animals is like that of humans (Figure 1,3).

Figure 1. Fluorine absorption and distribution in animals/mammals (Source: Mustafa et al., 2017)

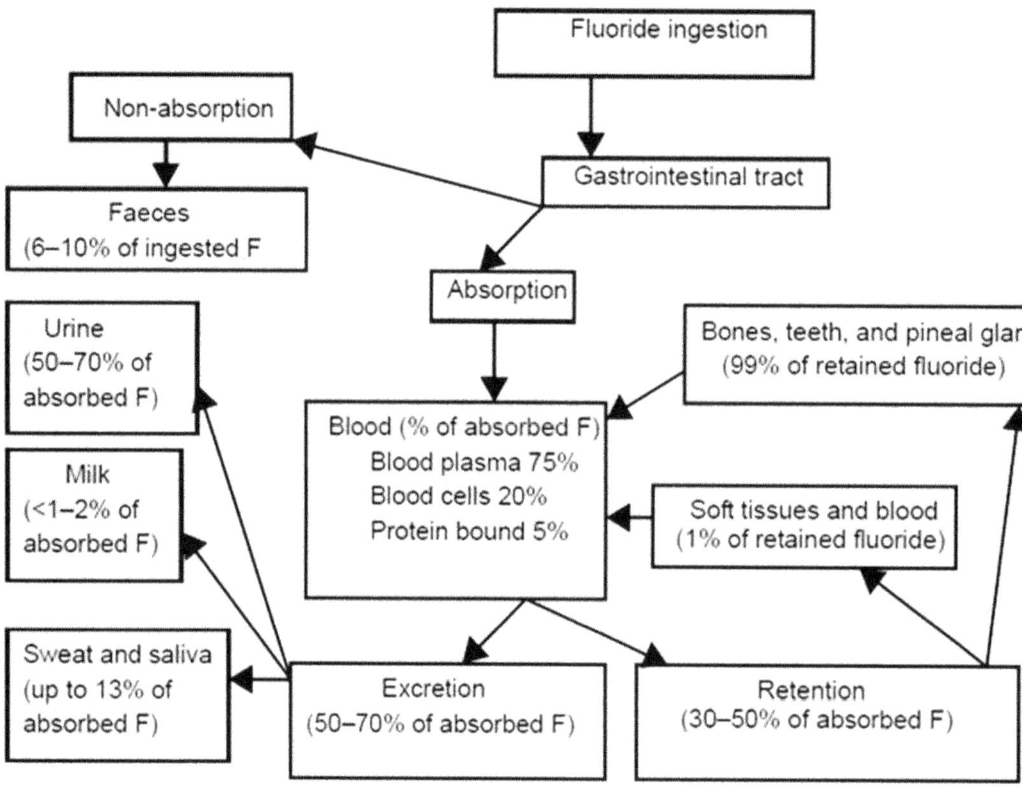

Among the terrestrial vertebrates, herbivores are more defenceless than carnivores and other animals. Domestic and wild herbivores are increasingly presented to natural F pollution since they are non-selective eaters and can devour sullied feed, water, and search. Cattle and sheep have pulled in more consideration from specialists around the world, maybe because of their huge populaces and their more noteworthy economic importance. Be that as it may, different animals, including water bison, ponies, goats, pigs, and wild cervids, can likewise experience the ill effects of F-lethality normally (Ranjan and Ranjan, 2015). Skeletal, non-skeletal, and dental fluorosis have been contemplated in wild oxen (*Bubalus bubalis*), camels (*Camelus dromedarious*), jackasses (*Equus asinus*), ponies (*Equus caballus*), and cows (*Bos taurus*) (Choubisa, 2010).

METABOLISM OF FLUORIDES IN HUMANS

The measure of fluoride that an individual takes in every day is substantially less than the sum that is taken in when devoured through food and water. The purpose behind this is air fluoride levels are normally under 1.0 µg/m3. The most elevated characteristic water level of fluoride announced was 2800 ppm (WHO 2004). In India, centralizations of 0.1–0.3 ppm of fluoride are accounted for in water (Das et al., 1981; Singh et al., 2011). Surface water fixations for the most part go from 0.01 to 0.3 ppm, though saltwater holds a greater fluoride level i.e., 1.2–1.5 ppm (WHO 2002) (Table 1).

Table 1. Fluoride exposure and human health effects

F conc. (mg/l)	Health effects
0.5	Vulnerable to dental caries
0.5-1.5	Strengthens bones/teeth
1.5-4.0	Children more vulnerable to dental fluorosis
4.0	Vulnerable to dental/skeletal fluorosis
10	Severe skeletal fluorosis, perchance cancer

[a]**Source**: Dissanayake (1991)

Effect on the Brain and Central Nervous System

On the premise on data to a great extent got from histological, chemical and molecular examinations, it is obvious that fluorides can meddle with the elements of the brain and the body by immediate and aberrant methods. A couple of epidemiologic investigations of Chinese populaces have announced IQ shortfalls in kids presented to fluoride at 2.5 to 4 mg/L in drinking water. Despite the fact that the examinations needed adequate detail for the board of trustees to completely evaluate their quality and pertinence to U.S. populaces, the consistency of the outcomes seems critical enough to warrant extra research on the impacts of fluoride on insight. Fluorides additionally increment the generation of free radicals in the brain through a few distinctive organic pathways. These progressions have a bearing on the likelihood that fluorides demonstration to build the danger of building up Alzheimer's disorder. Investigations of populaces presented to various concentrations of fluoride ought to be attempted to assess neurochemical changes that might be related with dementia. Thought ought to be given to surveying impacts from interminable exposure, impacts that may be deferred or happen late throughout everyday life, and individual helplessness (US Food and Drug Administration, 1999).

Effect on Endocrine System

In outline, proof of a few sorts shows that fluoride influences ordinary endocrine capacity or reaction; the impacts of the fluoride-actuated changes shift in degree and kind in various people. Fluoride is along these lines an endocrine disruptor in the wide feeling of changing typical endocrine capacity or reaction, albeit most likely not in the feeling of mirroring an ordinary hormone. The components of activity stay to be worked out and seem to incorporate both immediate and backhanded systems, for instance, direct incitement or restraint of hormone discharge by obstruction with second envoy work, circuitous incitement or hindrance of hormone emission by impacts on things, for example, calcium parity, and restraint of fringe chemicals that are essential for initiation of the typical hormone (Figure 3). A portion of these [endocrine] impacts are related with fluoride consumption that is feasible at fluoride fixations in drinking water of 4 mg/L or less, particularly for little youngsters or for people with high water admission. A large number of the impacts could be viewed as subclinical impacts, implying that they are not unfavourable health impacts. In any case, late work on fringe hormonal lopsided characteristics and endocrine-disturbing synthetic substances showed that unfavourable health impacts, or expanded dangers for creating antagonistic impacts, may be related with apparently minor disparities or perturbations or annoyances in hormone concentrations. Further research is expected to investigate these conceivable outcomes (American Academy of Pediatric Dentistry, 1994).

Effect on the Thyroid

A few lines of data show an impact of fluoride presentation on thyroid capability. It is hard to anticipate precisely what impacts on thyroid capacity are likely at what grouping of fluoride presentation and under what conditions. In people, impacts on thyroid capacity were related with fluoride exposures of 0.05-0.13 mg/kg/day when iodine admission was sufficient and 0.01-0.03 mg/kg/day when iodine admission was deficient. Admission of supplements, for example, calcium and iodine regularly isn't accounted for in investigations of fluoride impacts. The impacts of fluoride on thyroid capacity, for example, may rely upon whether iodine admission is low, satisfactory, or high, or whether dietary selenium is sufficient (Figure 3).

Effect on the Pineal Gland

The single animal investigation of pineal capacity demonstrates that fluoride introduction brings about changed melatonin generation and adjusted planning of sexual development. Regardless of whether fluoride influences pineal capacity in people stays to be illustrated. The two investigations of menarcheal age in people show the plausibility of prior menarche in certain people presented to fluoride, yet no conclusive explanation can be made. Late data on the job of the pineal organ in people recommends that any operator that influences pineal capacity could influence human wellbeing in an assortment of ways, remembering impacts for sexual development, calcium digestion, parathyroid work, postmenopausal osteoporosis, malignancy, and mental sickness (Figure 3).

Fluoride's Effect on Insulin Secretion/Diabetes

The accomplishment from the exposed investigations is that sufficient fluoride introduction appears to acknowledge increases in blood glucose or crippled glucose flexibility in specific individuals and to manufacture the earnestness of specific sorts of diabetes. All things considered, obstructed glucose absorption appears, apparently, to be connected with serum or plasma fluoride unions of about 0.1 mg/L or progressively critical in the two creatures and people. In like manner, diabetic individuals will much of the time have higher than conventional water admission, and in this way, will have higher than run of the mill fluoride utilization for a given centralization of fluoride in drinking water. A normal 16-20 million people in the U.S. have diabetes mellitus; subsequently, any activity of fluoride presentation in the improvement of upset glucose assimilation or diabetes is possibly basic (American Academy of Pediatrics Committee on Nutrition, 1995).

Effect on Immune System

Regardless, patients who live in either a falsely fluoridated set-up or a set-up where the drinking water typically contains fluoride at 4 mg/L have all congregated fluoride in their skeletal structures and possibly have outstandingly high fluoride fixations in their bones. The bone marrow is where safe cells create and that could impact humoral resistance and the age of antibodies to outer synthetic compounds. There is no uncertainty that fluoride can impact the cells related with giving resistant reactions. The request is what degree, expecting any, of the people eating up drinking water containing fluoride at 4.0 mg/L all the time will have their insusceptible systems exchanged off? Not a lone epidemiologic assessment has investigated whether fluoride in the drinking water at 4 mg/L is connected with changes in safe limit (O'Reilly and Featherstone, 1987). Nor has any examination dissected whether a person with an immunodeficiency disease can suffer fluoride ingestion from drinking water. It is enthusiastic that careful biochemical examinations be coordinated to make sense of what fluoride focuses occur in the bone and including interstitial fluids from prologue to fluoride in drinking water at up to 4 mg/L, since bone marrow is the wellspring of the forebears that produce the resistant framework cells

Effect on the Gastrointestinal System

The various fluoridation studies about in the past neglected to thoroughly test for changes in GI manifestations and there are no examinations on drinking water containing fluoride at 4 mg/L in which GI side effects were painstakingly archived. GI impacts seem to have been once in a while assessed in the fluoride supplement studies that pursued the previous ones during the 1950s and 1960s. There are a couple of case reports of GI upset in subjects presented to drinking water fluoridated at 1 mg/L (Figure 3). Those impacts were seen in just few cases, which propose extreme sensitivity. In any case, the accessible information are not strong enough to decide if that is the situation.

Effect on the Kidney

The kidneys shoulder an essential job in counteracting the development of over the top fluoride in the body. Among healthy people, the kidneys discharge around half of the day by day fluoride consumption. In any case, among people with kidney ailment, the kidneys' capacity to discharge turns out to be

Figure 2. Dental Fluorosis (Source: Fluoridation Forum Report, 2002, Page-126)

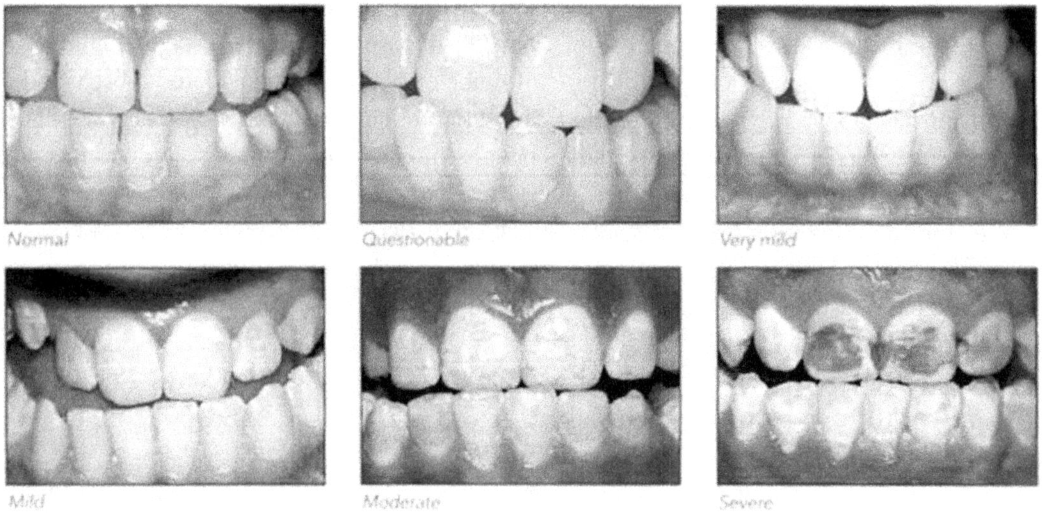

uniquely hindered, bringing about a development of fluoride inside the body. It is very much perceived that people with kidney ailment have an increased defencelessness to the aggregate harmful impacts of fluoride. Of specific concern is the potential for fluoride, when collected in the skeletal framework, to cause, or compound, renal osteodystrophy - a bone infection usually found among individuals with cutting edge kidney ailment. What's more, fluoride has been absolutely appeared to harm kidney work at high portions over transient exposures in both animals and humans. The effect of low dosages of fluoride, given over extensive stretches of time, has been insufficiently examined. An ongoing animal study, directed by researchers at the US Environmental Protection Agency revealed that presentation to only 1 ppm fluoride caused kidney harm in rodents on the off chance that they drank the water for an all-encompassing timeframe, while another investigation from China found an expanded pace of kidney illness among people devouring more than 2 ppm (Seppäet al., 1994). Consequently, the unfavourable impacts to kidney work that fluoride causes at high portions over brief timeframes may likewise be recreated with little dosages whenever devoured over significant stretches of time (Figure 3).

Effects on the Teeth

As indicated by the recent consensus opinion of the dental research community, fluoride's primary - if not sole-advantage to teeth originates from TOPICAL application to the outside surface of teeth, not from ingestion. Maybe as anyone might expect, thusly, tooth rot rates have declined at comparable rates in every single western nation in the last 50% of the twentieth century - independent of whether the nation fluoridates its water or not (Figure 2). Today, tooth rot rates all through mainland Western Europe are as low as the tooth rot rates in the United States - regardless of a significant dissimilarity in water fluoridation predominance in the two districts. Inside nations that fluoridate their water, late enormous scale studies of dental wellbeing - using current logical techniques not utilized in the early studies from the 1930s-1950s - have discovered little distinction in tooth decay, including " baby bottle tooth decay

Figure 3. Fluoride metabolism and its effects (Source: Waugh et al., 2019)

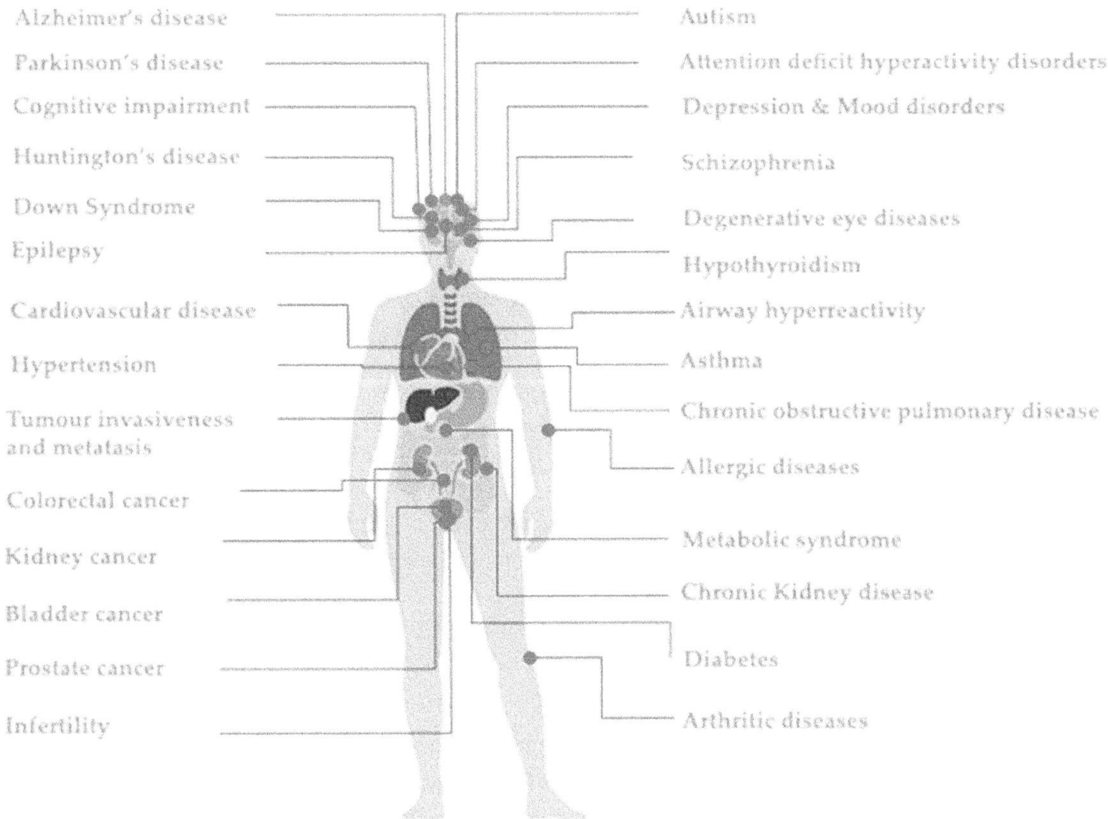

", among fluoridated and un-fluoridated communities (American Dental Association, 2003; Centers for Disease Control and Prevention, 2001).

Fluoride and Cancer

Fluoride seems to have a potential to start or advance malignant growths, especially of the bone, however the proof to date is speculative and blended. As noted above, osteosarcoma is of specific worry as a potential impact of fluoride in view of (1) fluoride testimony in bone, (2) the mutagenic impact of fluoride on bone cells, (3) animal results depicted above, and (4) pre-1993 distribution of some positive, just as negative, epidemiologic reports on relationship of fluoride introduction with osteosarcoma chance (Figure 3). Osteosarcoma displays the best from the earlier believability as a potential malignancy target site as a result of fluoride's deposition in bone, the NTP animal study discoveries of borderline expanded osteosarcomas in male rodents, and the known mitogenic impact of fluoride on bone cells in culture (National Cancer Institute, 2001). Standards of cell science demonstrate that upgrades for quick cell division increment the dangers for a portion of the isolating cells to get harmful, either by initiating irregular changing occasions or by exposing threatening cells that already were in non-dividing states. Fluoride presentation has likewise been connected to bladder malignant growth - especially among labourers presented to abundance fluoride in the working environment. As per the US National Research

Council, ——further look into on a potential impact of fluoride on bladder malignant growth hazard ought to be directed.

BIOREMEDIATION STRATEGIES FOR ENVIRONMENTAL F

The procedure lies in splitting down this ecological contaminant by embracing cost-effective, eco-accommodating, savvy and present day advances. Organic procedures, viz. bioremediation including the utilization of bacteria, microbes, fungi, and higher plants that grasps favourable choice to oversee F contamination, recuperate sullied soil and develop vegetation. The effectiveness of indigenous characteristic operators might be upgraded, enhanced and preferred over the perilous concoctions in sustainable agriculture and future.

Physical Methods

A few physiochemical methods, for example, reverse osmosis, electrodialysis adsorption ion exchange, Nalgonda strategy, biosorption, nanofiltration, and so on., have recently been accounted for F expulsion (Mukherjee and Halder, 2016; Sehn, 2008).

Biological Methods

Biological strategies deal with the association of living life forms like plants and microorganisms or their items for the management/expulsion of contaminants from the environs. The idea of bioremediation is to a great extent executed in various manners which helps in the change of natural contaminants into less difficult structures, for example, carbon dioxide and water through the escalated activity of plants and microbes as a feature of their metabolic procedures (Sarkar et al., 2006). These organisms have the power to tolerate various poisons by obtaining strategies like bioaccumulation, biotransformation, biosorption, and so on (Mukherjee et al., 2017). Scientists have revealed both aerobic and anaerobic dilapidation corridors for the mineralization and adjustment of natural contaminants in the air.

Biotic managements are extra profitable due to their cost-viability, operative candour and smaller amount of mud formation. The utilization of natural and biological sources makes the technique ecologically courteous and extremely good for comprehensive managements (Doble and Kumar, 2005).

Phytoremediation and Phytostabilization

It is intervened by plants that have capacity to occupy the debased soil. The instant activity of root exudates immobilizes F by catching in the soil frameworks, in this way constraining its solvency. Additionally, the roots forestall the movement of F because of flattening and disintegration by adsorption and precipitation, respectively, inside the root zone (Pollard et al., 2014). Fluoride tolerant plants may be able to deactivate it to a much extent than the sensitive ones (Table 2). Possible arrangements incorporates moving towards F obtuse metabolic pathways; complexation of F with natural elements and their expulsion from the site of chemical hindrance; response with cationic situates; sequestration in vacuoles and translocation to the leaf surfaces, and so on (Mukherjee et al., 2017). Quickly developing high biomass crop with a broad root framework and indicating least noxiousness could be raised to remediate F from the environment (Baunthiyal and Ranghar, 2013). Correspondingly, the marine plant species can be handily used to remediate F from the contaminated water bodies. Baunthiyal and Sharma (2012) exhibited the capability of eight tree types of semi-dry area for hyperaccumulation of F, out of which Prosopis juliflora (Sw.) DC., was appeared to expel huge measure of F from groundwater and soil. Santos-Díaz and Zamora-Pedrazaa (2010) revealed the three species, *Camellia japonica* L., *Pittosporum tobira* Thunb. and *Saccharum officinarum* L. had the option to evacuate F productively. *S. officinarum* demonstrated greatest removal capacity, proposing the initiation of some detoxification procedure in the cell to withstand F measures (Table 2).

*Table 2. List of some fluoride ihyperaccumulator plant species (**Source**: Katiyar et al., 2020)*

Species	Accumulated fluoride
Spinacia oleracea L.	1.7 g/kg
Amaranthus L.	20.9 g/kg
Abelmoschus esculentus L.	0.43 g/kg
Vachellia tortilis (Forssk.)	592.24 μg/g
Prosopis juliflora (Sw.) DC.	852.01 μg/g
Brassica oleracea L.	12.91 mg/mL
Portulaca grandiflora Hook.	22.96 mg/mL
Camellia sinensis L.	1442 mg/kg
Saccharam officinarum L.	1012 mg/kg
Nerium oleander L.	3.7 mg/g
Portulaca oleracea L.	2.4 mg/g
Pogonatherum crinitum (Thunb.) Kunth	2.8 mg/g

Microbial Remediation

Soil restoration subsequent to evacuating its poisons and toxins by utilizing microscopic organisms or fungi is named as microbial remediation. These organisms use the contaminants and afterward debase them for energy and multiplication. Microbial remediation is accomplished thru three procedures which are Natural attenuation (naturally with indigenous soil microorganisms), Biostimulation (in presence of outer supplements) and Bioaugmentation (utilization of externally presented microorganisms).

CONCLUSION

This chapter contemplates about the significance of fluorine and talks about a high admission of F, by means of ingestion or breath from an assortment of sources, may cause lethality in people and animals, including dental, skeletal, and non-skeletal fluorosis. The F danger can be intense or ceaseless relying upon the level and the span of the F-consumption. An adjustment in the ideological way to deal with fluoride use for dental caries anticipation is elementary in the worldwide public health community. The investigations in the literature show that the nearness of a high concentration of F in soil influences plants and aquatic life and prompts soil and water contamination. Plant species with vulnerability to F contamination in soil might be radically harmed. Also, F contamination may devastatingly affect the microbial movement in soil and upset the soil biology. While numerous physiochemical procedures are known, bioremediation is a secure and inventive strategy for remediating this toxicant. It envisages on the process of changing the toxins into soluble and bioavailable ones, which subsequently expedites phytoremediation. As epidemiological investigations keep on refining our comprehension of the dose-response relationship and field studies better portray the regions of conceivably high fluoride, endeavours to decrease fluoride related medical issues should turn out to be progressively powerful.

REFERENCES

Ahmed, D. E. (2007). Environmental pollution and its impact on human and animal health: A review. *The Sudan J Vet Res*, *22*, 37–46.

American Academy of Pediatric Dentistry. (1994). Special issue: Reference manual 1995. *Pediatric Dentistry*, *16*(special issue), 1–96.

American Academy of Pediatrics Committee on Nutrition. (1995). Fluoride supplementation for children: Interim policy recommendations. *Pediatrics*, *95*, 777. PMID:7724324

American Dental Association. (2003). *Fluoridation Facts. National and International Organizations that Recognize the Public Health Benefits of Community Water Fluoridation for Preventing Dental Decay*. Author.

Anjum, M. K., Mughal, S. M., Sayyed, U., Yaqub, A., Khaliqu, A., & Rashid, A. M. (2014). Influence of increasing fluoride dose rates on selected liver and kidney enzymes profile in domestic chicken (Gallus domesticus). *Journal of Animal & Plant Sciences*, *24*(1), 77–80.

Apambire, W. B., Boyle, D. R., & Michel, F. A. (2007). Geochemistry, genesis, and health implications of fluoriferous groundwaters in the upper regions of Ghana. *Environ Geol*, *33*(1), 13–24. doi:10.1007002540050221

Barrow, N. J., & Ellis, A. S. (2006). Testing a mechanistic model. III The effect of pH on fluoride retention by a soil. *Journal of Soil Science*, *37*(2), 287–293. doi:10.1111/j.1365-2389.1986.tb00030.x

Baunthiyal, M., & Ranghar, S. (2013). Accumulation of fluoride by plants: Potential for phytoremediation. *Clean Soil Air Water*, *43*(1), 127–132. doi:10.1002/clen.201300353

Baunthiyal, M., & Sharma, V. (2012). Phytoremediation of fluoride contaminated water and soil: A search for fluoride hyperaccumulators. *J Agri Tech*, *8*, 1965–1978.

Bieliyakova, T. M. (2007). Fluorine in soils and plants as related to endemic fluorosis. *Почвоведение*, *8*, 55.

BIS. (1991). *Indian standard specification for drinking water. IS: 10500*. Bureau of Indian Standards.

Bowen, H. J. M. (2006). *Trace elements in biochemistry*. Academic.

Brewer, R. F. (1966). Fluorine. In H. D. Chapman (Ed.), *Diagnostic criteria for plants and soils* (pp. 180–196). University of California.

Cao, J., Zhao, Y., & Liu, J. (1997). Brick tea consumption as the cause of dental fluorosis among children from Mongol, Kazak and Yugu populations in China. *Food and Chemical Toxicology*, *35*(8), 827–833. doi:10.1016/S0278-6915(97)00049-5 PMID:9350228

Centers for Disease Control and Prevention (CDC). (2001). Recommendations for Using Fluoride to Prevent and Control Dental Caries in the United States. *MMWR*, *50*(RR-14), 1–42. PMID:11521913

Chae, G. T., Yun, S. T., Mayer, B., Kim, K. H., Kim, S. Y., Kwon, J. S., Kim, K., & Koh, Y. K. (2017). Fluorine geochemistry in bedrock groundwater of South Korea. *The Science of the Total Environment*, *385*(1–3), 272–283. doi:10.1016/j.scitotenv.2007.06.038 PMID:17655916

Chhabra, R., Singh, A., & Abrol, I. P. (1999). Fluorine in sodic soils. *Soil Science Society of America Journal*, *44*(1), 33–36. doi:10.2136ssaj1980.03615995004400010008x

Choubisa, S. L. (2010). Natural amelioration of fluoride toxicity (fluorosis) in goats and sheep [scientific correspondence]. *Current Science*, *99*(10), 1331–1332.

Choubisa, S. L., & Choubisa, D. (2016). Status of industrial fluoride pollution and its diverse adverse health effects in man and domestic animals in India. *Environmental Science and Pollution Research International*, *23*(8), 7244–7254. doi:10.100711356-016-6319-8 PMID:26903127

Chouhan, S., & Flora, S. J. (2010). Arsenic and F: Two major ground water pollutants. *Indian Journal of Experimental Biology*, *48*, 666–678. PMID:20929051

Cronin, S. J., Manoharan, V., Hedley, M. J., & Loganathan, P. (2000). Fluoride: A review of its fate, bioavailability, and risks of fluorosis in grazed-pasture systems in New Zealand. *New Zealand Journal of Agricultural Research*, *43*(3), 295–321. doi:10.1080/00288233.2000.9513430

Czarnowski, W., Wrzesniowska, K., & Krechniak, J. (2006). Fluoride in drinking water and human urine in Northern and Central Poland. *The Science of the Total Environment*, *191*(1-2), 177–184. doi:10.1016/0048-9697(96)05259-X PMID:8885430

Das, D. K., Burman, G. K., & Kidwai, A. L. (1981). Chemical composition of monsoon rainwater over Bhopal, Madhya Pradesh during 1977 and 1978. *Mausam (New Delhi)*, *32*, 221–228.

Doble, M., & Kumar, A. (2005). *Biotreatment of industrial effluents*. Elsevier, Butterworth-Heinemann.

Domingos, M., Klumpp, A., Rinaldi, M. C. S., Modesto, I. F., Klumpp, G., & Delitti, W. B. C. (2003). Combined effects of air and soil pollution by fluoride emissions on *Tibouchina pulchra* Cogn., at Cubatão, SE Brazil, and their relations with aluminium. *Plant and Soil*, *249*(2), 297–308. doi:10.1023/A:1022800225753

Edmunds, W. M., & Smedley, P. L. (2005). Fluoride in natural waters. In O. Selinus (Ed.), *Essentials of medical geology* (pp. 301–329). Elsevier.

Fuge, R., & Andrews, M. J. (1988). Fluorine in U.K. environment. *Environmental Geochemistry and Health*, *10*(3–4), 96–104. doi:10.1007/BF01758677 PMID:24213649

Fung, K. F., Zhang, Z. Q., Wong, J. W. C., & Wong, M. H. (1999). Fluoride contents in tea and soil from tea plantations and the release of fluoride into tea liquor during infusion. *Environmental Pollution*, *104*(2), 197–205. doi:10.1016/S0269-7491(98)00187-0

Gilpin, L., & Johnson, A. H. (2014). Fluorine in agricultural soils of Southern Pennsylvania. *Soil Science Society of America Journal*, *44*(2), 255–258. doi:10.2136ssaj1980.03615995004400020010x

Haidouti, C. (2001). Fluoride distribution in soils in the vicinity of a point emission source in Greece. *Geoderma*, *49*(1-2), 129–138. doi:10.1016/0016-7061(91)90096-C

Handa, B. K. (1975). Geochemistry and genesis of fluoride containing ground waters in India. *Ground Water*, *13*(3), 275–281. doi:10.1111/j.1745-6584.1975.tb03086.x

Hem, J. D. (2015). U.S. Geological Survey Water-Supply, Paper: Vol. 2254. The study and interpretation of the chemical characteristics of natural water (3rd ed.). Academic Press.

Hudak, P. F., & Sanmanee, S. (2003). Spatial patterns of nitrate, chloride, sulfate, and fluoride concentrations in the woodbine aquifer of north-central Texas. *Environmental Monitoring and Assessment*, *82*(3), 311–320. doi:10.1023/A:1021946402095 PMID:12602633

Hyndman, D. W. (1985). *Petrology of igneous rocks* (2nd ed.). McGraw-Hill.

Jha, S. K., Nayak, A. K., Sharma, Y. K., Mishra, V. K., & Sharma, D. K. (2014). Fluoride accumulation in soil and vegetation in the vicinity of brick fields. *Bulletin of Environmental Contamination and Toxicology*, *80*(4), 369–373. doi:10.100700128-008-9391-z PMID:18345473

Kabata-Pendias, A., & Pendias, H. (2004). Elements of group VII. In Trace elements in soils and plants. CRC.

Katiyar, P., Pandey, N., & Sahu, K. K. (2020). Biological approaches of fluoride remediation: Potential for environmental clean-up. *Environmental Science and Pollution Research International*, *27*(12), 1–12. doi:10.100711356-020-08224-2 PMID:32146673

Keerthisinghe, G., McLaughlin, M. J., & Freney, J. R. (2001). Use of gypsum, phosphogypsum and fluoride to ameliorate subsurface acidity in a pasture soil. In R. J. Wright, V. C. Aligar, & R. P. Murrmann (Eds.), Plant soil interactions at low pH (pp. 509–517). Academic Press.

Keller, E. A. (1979). *Environmental geology*. Charles & Merril.

Kim, K., & Jeong, G. Y. (2005). Factors influencing natural occurrence of fluoride-rich ground waters: A case study in the southeastern part of the Korean Peninsula. *Chemos*, *58*(10), 1399–1408. doi:10.1016/j.chemosphere.2004.10.002 PMID:15686758

Kirk-Othmer. (1996). *Encyclopedia of chemical technology* (3rd ed., Vol. 10). Wiley.

Kloos, H., Tekle-Haimanot, R., Kloos, H., & Zein, A. H. (Eds.). (1993). *Fluorosis. The ecology of health and disease in Ethiopia*. West View Press.

Koblar, A., Tavcar, G., & Ponikvar-Svet, M. (2011). Effects of airborne F on soil and vegetation. *Journal of Fluorine Chemistry*, *132*(10), 755–775. doi:10.1016/j.jfluchem.2011.05.022

Kudzin, Y. K., & Pashova, V. T. (2009). Fluorine content of soils and plants after prolonged application of fertilisers. *Soils and Fertilizers*, *33*, 451.

Kumar, B., Naaz, A., Shukla, K., Narayan, C., Singh, G., Kumar, A., Ramanathan, A. L., & Anshumali. (2016). Spatial variability of fluorine in agricultural soils around Sidhi District, Central India. *Journal of the Geological Society of India*, *87*(2), 227–235. doi:10.100712594-016-0391-z

Lahermo, P., Ilmasti, M., Juntunen, R., & Taka, M. (1990). *The geochemical atlas of Finland. Part 1: the hydro-geochemical mapping of Finnish groundwater*. Geological Survey of Finland.

Mambali, S. S. (1982). Necessity and efforts made to map the occurrence of fluoride within the country, Tanzania Mainland. A report the workshop on domestic water health standards with emphasis on fluoride Arusha, Tanzania Ministry of Water and Energy.

Manoharan, V., Loganathan, P., Parfltt, R. L., & Tillman, R. W. (2006). Changes in soil solution composition and aluminium speciation under legume-based pastures in response to long-term phosphate fertiliser applications. *Australian Journal of Soil Research*, *34*(6), 985–998. doi:10.1071/SR9960985

McLaughlin, M. J., Simpson, P. G., Fleming, N., Stevens, D. P., Cozens, G., & Smart, M. K. (2006). Effect of fertiliser type on cadmium and fluorine concentrations in clover herbage. *Australian Journal of Experimental Agriculture*, *37*(8), 1019–1026. doi:10.1071/EA96117

Mjengera, H., & Mkongo, G. (2003). appropriate defluoridation technology for use in fluorotic areas in Tanzania. *Physics and Chemistry of the Earth*, *28*(20-27), 1097–1104. doi:10.1016/j.pce.2003.08.030

Mukherjee, S., & Halder, G. (2016). Assessment of fluoride uptake performance of raw biomass and activated biochar of Colocasia esculenta stem: Optimization through response surface methodology. *Environmental Progress & Sustainable Energy*, *35*(5), 1305–1316. doi:10.1002/ep.12346

Mukherjee, S., Sahu, P., & Halder, G. (2017). Microbial remediation of fluoridecontaminated water via a novel bacterium *Providencia vermicola* (KX926492). *Journal of Environmental Management*, *204*, 413–423. doi:10.1016/j.jenvman.2017.08.051 PMID:28915476

National Cancer Institute. (2001). *Cancer Facts: Fluoridated Water*. http://fluoride.oralhealth.org/ papers/2001/ncicancerfacts101000.htm

Neuhold, J. M., & Sigler, W. F. (1960). Effects of sodium fluoride on carp and rainbow trout. *Transactions of the American Fisheries Society*, *89*(4), 358–370. doi:10.1577/1548-8659(1960)89[358:EOSFOC]2.0.CO;2

Neumuller, O. A. (1981). Rommps Chemie Lexicon (8th ed., vol. 2). Franck'sche Verlagshandlung.

O'Donnell, T. A. (1975). Fluorine. In *Comprehensive inorganic chemistry* (Vol. 2). Pergamon Press.

O'Reilly, M. M., Featherstone, J. D. (1987). Demineralization and remineralization around orthodontic appliances: an in vivo study. *Am J Orthod Dentofacial Orthop.*, *92*(1), 33-40.

Omueti, J. A. I., & Jones, R. L. (1977). Fluoride adsorption by Illinois soil. *Journal of Soil Science*, *28*(4), 564–572. doi:10.1111/j.1365-2389.1977.tb02264.x

Omueti, J. A. I., & Jones, R. L. (2012). Fluorine distribution with depth in relation to profile development in Illinois. *Soil Science Society of America Journal*, *44*(2), 247–249. doi:10.2136ssaj1980.03615995004400020007x

Ozsvath, D. L. (2006). Fluoride concentrations in a crystalline bedrock aquifer, marathon county, Wisconsin. *Environ Geol*, *50*(10), 132–138. doi:10.100700254-006-0192-6

Pickering, W. F. (2015). The mobility of soluble fluoride in soils. *Environ Pollut Ser B*, *9*(44), 281–308.

Pollard, A. J., Reeves, R. D., & Baker, A. J. M. (2014). Facultative hyperaccumulation of heavy metals and metalloids. *Plant Science*, *218*, 8–17. doi:10.1016/j.plantsci.2013.11.011 PMID:24467891

Polomski, J., Fluhler, H., & Blaser, P. (2012). Fluoride induced mobilization and leaching of organic matter, iron and aluminium. *Journal of Environmental Quality*, *11*(3), 452–456. doi:10.2134/jeq1982.00472425001100030027x

Prud'homme, M. (1990). Fluorspar. In: Canadian mineral year book, 1989 energy, mines, and resources. Academic Press.

Ranjan, R., & Ranjan, A. (2015). *Fluoride toxicity in animals. Springer Briefs in Animal Sciences (part of an E-book collection)*. Springer International Publishing.

Reimann, C., & Decaritat, P. (Eds.). (1998). *In: Chemical elements in environment. Fact sheets for the geochemist and environmental scientist*. Springer.

Robinson, W. D., & Edington, G. (1946). Fluorine in soils. *Soil Science*, *61*(5), 341–353. doi:10.1097/00010694-194605000-00001

Santos-Díaz, M. S., & Zamora-Pedrazaa, C. (2010). Fluoride removal from water by plant species that are tolerant and highly tolerant to hydrogen fluoride. *Fluoride*, *43*, 150–156.

Sarkar, M., Banerjee, A., Pramanick, P. P., & Sarkar, A. R. (2006). Use of laterite for the removal of fluoride from contaminated drinking water. *Journal of Colloid and Interface Science*, *302*(2), 432–441. doi:10.1016/j.jcis.2006.07.001 PMID:16899254

Saxena, V. K., & Ahmed, S. (2001). Dissolution of fluoride in groundwater: A water reaction study. *Environ Geol*, *40*(9), 1084–1087. doi:10.1007002540100290

Saxena, V. K., & Ahmed, S. (2003). Inferring the chemical parameters for the dissolution of fluoride in ground water. *Environ Geol*, *43*(6), 731–736. doi:10.100700254-002-0672-2

Scientific Committee on Health and Environmental Risks (SCHER). (2011 May 16). *Opinion of critical review of any new evidence on the hazard profile, health effects, and human exposure to fluoride and the fluoridating agents of drinking water*. Brussels, Belgium: Directorate General for Health and Consumers, European Commission.

Sehn, P. (2008). Fluoride removal with extra low energy reverse osmosis membranes: Three years of large scale field experience in Finland. *Desalination*, *223*(1-3), 73–84. doi:10.1016/j.desal.2007.02.077

Seppä, L., Pöllänen, L., & Hausen, H. (1994). Caries-preventive effect of fluoride varnish with different fluoride concentrations. *Caries Research*, *28*(1), 64–67. doi:10.1159/000261623 PMID:8124700

Short, H. E., McRobert, T. W., Bernard, A. S., & Mannadinayer, A. S. (1937). Endemic fluorosis in the Madras Presidency. *The Indian Journal of Medical Research*, *25*, 553–561.

Shupe, J. L., & Sharma, R. P. (2006). Fluoride distribution in natural ecosystem and related effects on wild animals. In D. D. Hemphill (Ed.), *Trace substances in environmental health* (Vol. 10, p. 137). University of Missouri.

Singh, S. P., Khare, P., Satsangi, G. S., Lakhani, A., Kumari, S. S., & Srivastava, S. S. (2011). Rainwater composition at a regional representative site of a semi-arid region of India. *Water, Air, and Soil Pollution*, *127*(1/4), 93–108. doi:10.1023/A:1005295215338

Skjelkvale, B. L. (1994). Factors influencing fluoride concentrations in Norwegian Lakes. *Water, Air, and Soil Pollution*, *77*, 151–167.

Slavek, J., Farrah, H., & Pickering, W. F. (1996). Interaction of clays with dilute fluoride solutions. *Water, Air, and Soil Pollution*, *23*(2), 209–220. doi:10.1007/BF00206977

Stevens, D. P., McLaughlin, M. J., & Alston, A. M. (1997). Phyto-toxicity of aluminium–fluoride complexes and their uptake from solution culture by *Avena sativa* and *Lycopersicon esculentum. Plant and Soil*, *192*(1), 81–93. doi:10.1023/A:1004224526067

Susheela, A. K. (2003). *Treatise on fluorosis* (2nd ed.). Fluorosis Research & Rural Development Foundation.

Susheela, A. K., Kumar, A., Bhatnagar, M., & Bahadur, R. (1993). Prevalence of endemic fluorosis with gastrointestinal manifestations in people living in some North-Indian villages. *Fluoride*, *26*(2), 97–104.

Tsiros, T. X., Haidouti, C., & Chronopoulou, A. (1998). Airborne F contamination of soil and olive trees near an aluminium plant: Measurement and simulation. *J Environ Sci Health*, *33*(7), 1309–1324. doi:10.1080/10934529809376790

UNICEF. (1999). States of the art report on the extent of fluoride in drinking water and the resulting endemicity in India. Report by Fluorosis and Rural Development Foundation for UNICEF.

US Food and Drug Administration. (1999). Anticaries drug products for over-the-counter human use. *Code of Federal Regulations, 21*(CFR Part 355), 280-5.

Wang, L. F., & Huang, J. Z. (1995). Outline of control practice of endemic fluorosis in China. *Social Science & Medicine*, *41*(8), 1191–1195. doi:10.1016/0277-9536(94)00429-W PMID:8578341

Warren, C., Burgess, W. G., & Garcia, M. G. (2005). Hydro-chemical associations and depth profiles of arsenic and fluoride in quaternary loess aquifers of Northern Argentina. *Mining Magazine (London)*, *69*(5), 877–886. doi:10.1180/0026461056950295

Wenzel, W. W., & Blum, W. E. H. (2002). Fluorine speciation and mobility in contaminated soils. *Soil Science*, *153*(5), 357–364. doi:10.1097/00010694-199205000-00003

Whittemore, D. O., MacFarlane, P. A., Doveton, J. H., Butler, J. J., Chu, T., Bassler, R., Smith, M., Mitchell, J., & Wade, A. (2013). The Dakota aquifer program annual report, FY92. Kansas. *Geological Survey Open-File Report (United States)*, 93–1.

WHO. (1996). *Guidelines for drinking water quality. Recommendations* (Vol. 1). World Health Organisation.

WHO. (2002). *Fluorides: Environmental health criteria 227*. Geneva: World Health Organization. Available from: http://www.inchem.org/documents/ehc/ehc/ehc227.htm#1.5

WHO. (2004). *Guidelines for drinking water quality* (3rd ed.). World Health Organisation.

WHO. (2005). *Water-related diseases – fluorosis, the disease and how it affects people*. World Health Organization.

Chapter 10
Cadmium- and Lead-Tolerant PGPRs as Proficient Toxicity Alleviators for Agricultural Crops

Amit Kumar Pal
Department of Botany, University of Kalyani, India

Anjan Hazra
Agricultural and Ecological Research Unit, Indian Statistical Institute, Kolkata, India

Chandan Sengupta
Department of Botany, University of Kalyani, India

ABSTRACT

Agricultural lands are being polluted with different contaminants due to various anthropogenic activities like toxic discharge from Ni-Cd battery industry, tannery industry, alloying of metals like steel, application of agrochemicals, etc. Cadmium and lead contamination in agricultural land are directed towards global food insecurity. Bioremediation, stress alleviation, and phytostimulation by Cd and Pb tolerant PGPR is a promising eco-friendly method to develop sustainable agricultural system. At present, cadmium and lead-tolerant plant growth promoting rhizobacteria (PGPR) can be a sustainable option for heavy metal-contaminated agricultural lands. PGPRs such as Bacillus, Bradyrhizobium, Enterobacter, Klebsiella, Micrococcus, Pseudomonas, Ralstonia, *etc. can survive the metal stress and stimulate the plant growth under Cd and Pb contaminated condition by direct or indirect plant growth promoting ability. So, these PGPRs could be exploited as biofertilizers and bioremediators under Cd or Pb stressed conditions for futuristic agricultural development.*

DOI: 10.4018/978-1-7998-4888-2.ch010

INTRODUCTION

Various metals such as lead, zinc, cadmium, nickel, copper and mercury are continuously being added to our soil ecosystem through different agrochemical usage, industrial waste disposal, application of urban sewage sludge, vehicle exhausts, and waste incineration and from many anthropogenic sources (Noumavo et al. 2016). These contaminations in our agricultural soils cause threat to food safety. Plants uptake different metals into their body and retains for a long days. As a result thesc mctals can transferred to the higher tropic level like human and other animals.

Cadmium inhibits plant growth, hampers plant water relationship and ion metabolism, inhibits chlorophyll biosynthesis, inhibits many enzymes like Fructose bis-phosphatase, Fructose 6 phosphate kinase, Phosphoenolpyruvate carboxylase, NADP+ glyceraldehyde-3-phosphate dehydrogenase Ribulose-1,5-bisphosphate carboxylase oxygenase, and Carbonic anhydrase (Krantev, Yordanova, and Popova 2006; Popova et al. 2009). Cd act as a potent nephrotoxin and class-I carcinogen in animals. Lead (Pb) persists in the environment for a long time and causes anemia, reproductive impairment, renal failure, neurodegenerative damage etc. (Eslami et al. 2011). Pb badly affects plants in the seed germination, biomass production, root-shoot growth, chlorophyll content and ion distribution (Trvedi and Erdei 1992; He 1990).

These heavy metals affect a considerable harmful effect on environment, soil ecosystems and human health due to their mobility and solubility (Kabata-Pendias 1992). Frequently, the soil may be contaminated as much as a harmful waste (Berti and Jacob, 1996). In developing countries, heavy metals contamination in soil is a much talkative concern (Yanez *et al.*, 2002; Pramanik et al. 2016; Pal and Sengupta, 2019). It is now important to remediate the contaminated soil to develop suitable agricultural land.

Cause of Soil Contamination by Heavy Metals

The heavy metal can be contaminated in the environment by means of various man made or by natural occurrences. Sewage sludge are beneficially used as a common and useful disposal mechanism in agricultural land, but these sewage sludge are normally contaminated with various heavy metals like Cd, Pb, Ni, Cu, Cr etc (Singh *et al.*, 2004). These metals commonly leached or channeled with soil particles. Such metals produced some long term deleterious effects on the plants or animals. Though sewage sludge have some advantages as it contained some available nutrients for crop yield but these short term beneficial effects of sewage sludge can be ignored for their long term toxic effects on plant yield and production. Cadmium and lead present in the sewage sludge accumulated in the plant body, transferred into higher tropic level and ultimately into human (Logan and Chaney, 1983). According to Benitez at al. (2001) higher concentration of heavy metals like Cd, Mn and Zn are deposited in agricultural soil after using sewage sludge as land application (Benítez *et al.*, 2001).

Heavy metal contamination of soil from atmosphere of many industries like sulfuric acid plants, energy plants, paint industry, metallurgy and production of construction materials etc. In many developing countries disposal of garbage and dumping of waste along the road side is one of the major subjects of concern in aspect of soil pollution. According to Agritas and Kilicel, (1999), many toxic metals coming from human activities are finally gathered and ultimately pollute soil (Agirtas and Kilicel, 1999). These toxic metals can transfer to the plant body that grows in the polluted soil (Ogunsola *et al.*, 1993) and finally they transfer into the animal body (Kilice, 1999). Gasoline and lubricating oil contains different heavy metals like Pb, Zn, Cu etc. These heavy metals are deposited in soil and vegetation beyond the considerable limit alongside the highway (Nyangababo and Hamya, 1986). Many studies revealed that in north America and Europe, heavy metal contamination occur from many industry and from traffic in urban region (Tijhuis *et al.*, 2002; Imperato *et al.*, 2003; Madrid *et al.*, 2004; Chirenje *et al.*, 2004; Crnković *et al.*, 2006).

In recent days chemical fertilizers, pesticides and mulch has tremendous important in case of agricultural production and plant yield as these fertilizers enhance plant yield very quickly (Zhang and Zhang, 2007; Zhang *et al.*, 2011). In agricultural field phosphate fertilizer are used in high amount. Phosphate fertilizer is produced from phosphate rocks which are good source of cadmium. So, phosphate fertilizer is contaminated with cadmium. The higher amount of heavy metals present in fertilizers are as follows- phosphoric fertilizer > compound fertilizer > potash fertilizer > nitrogen fertilizer (Boyd, 2010). Normally, lower concentration of heavy metal is not harmful for plants and humans but different anthropogenic sources like urban effluent waste disposal and mining gradually increase the heavy metal concentration in agricultural field. Normally, chemical pesticide are composed of organic-inorganic compound or only minerals but some pesticides composed of some heavy metals like Cu, Hg, Zn as well as Pb and Cd (Arao *et al.*, 2010).

There are no advantageous physiological activity of many metals such as cadmium and lead for plants. Application of some phosphate fertilizers in the soil ultimately adds different toxic elements such as Cd, F, Hg, and Pb (Raven *et al.*, 1998) which were taken by many plants such as radish (*Raphanus sativus* L.), lettuce (*Latuca sativa* L.), and garden peas (*Pisum sativum* L.) (Reuss *et al.*, 1978) etc.

Cd and Pb Toxicity on Plants

There are some heavy metals which are known to be essential to man and other animals for medicine and other dietary requirement at the regulatory limits. These bio- essential heavy metals are Fe, Zn, Mg and Ca. When the concentration of these heavy metals exceed from their regulatory limits, they shows toxicity or poisoning (Marschner, 1995). Zinc is known to be masculine element, which is essential for balancing the copper level in the body and for reproductive mechanism (Nolan, 1983). Zinc is essential for the co-factor of dehydrogenating enzymes (Holum, 1983). Moreover Zn deficiency is the reason for producing anemia and retardness (McCluggage, 1991). Calcium (Ca) is essential for developing stronger bones and teeth (Holum, 1983). In blood plasma and other body fluids, magnesium (Mg) presents as a very important electrolyte. It is required for various concentrations with age and sex. For pregnant women, its requirement is highest (Vormann, 2003; Rude and Gruber, 2004).

There are some other heavy metals which have no known use in plant or animal body. Even these heavy metals like Cd, Pb and their methylated form are very toxic from a very low concentration (Holum, 1983; Nolan, 1983; Ferner, 2001; Young, 2005; Smiri et al. 2010). Cadmium produced toxic symptoms even in very low concentration. Diseases like osteomalacia, osteoporosis and continuous fractures, myocardic dysfunctions, increased blood pressure, various pulmonary diseases like bronchiolitis, emphysema, alveolitis and renal damage etc are the cause of cadmium toxicity (McCluggage, 1991; Hellström et al., 2001; European Union, 2002; Young, 2005). Pb is known to be very toxic and deleterious among the other heavy metals because it harm kidney, Central Nervous System (CNS), cardiovascular system (Ogwuebgu and Muhanga, 2005), gastrointestinal tract (GIT), urinary tract (INECAR, 2000; Ferner, 2001) etc (Abbas et al. 2014; Aoshima 2016).

In case of plants, some heavy metals such as Cu, Co, Mo, Mn, Fe, Ni and Zn are not toxic in low condition. Moreover, they are necessary for the plant development and health at a low concentration, though the very high concentration can lead to poisoning (Garrido *et al.*, 2005; Rascio *et al.*, 2011). Nickel is essential for urease activity in the leaves. Under nickel deficiency condition, urease activity hampers and urea deposited in leaves resulting leaf tip necrosis (Taiz and Zeiger, 2006). High concentration of nickel cause much abnormality such as chlorosis, hamper CO_2 absorption as well as exchange of other gases, damage water absorption, production of free radical and reactive oxygen species (Ali *et al.*, 2009).

Some heavy metals are unnecessary for plants such as Cd, Hg, Pb or Se as they don't have any function in plant physiology (Garrido *et al.*, 2002; Rascio *et al.*, 2011). Cadmium is the noxious toxin for plants at a very low concentration. Because of high solubility, very low concentration of cadmium in soil ranging from 0.32 to 1mM considered to be highly toxic (Wagner, 1993). Cadmium alters the nutrient availability for plants such as calcium (Ca), potassium (K), copper (Cu), magnesium (Mg), manganese (Mn), zinc (Zn), and nickel (Ni) (Moreno *et al.*, 1999). Cadmium inhibit nutrient absorption, chlorophyll metabolism and structure, carbon metabolism enzymes, photo system II and many other physiological processes resulting lower growth rate and biomass as well as DNA damage (Liang, 2005; Karantev, 2006; Popova, 2009).

Higher lead concentration in soil inhibit many plant physiological mechanisms such as photosynthesis, water absorption, produced wilting of older leaves, reduced growth rate with foliar stunting and short root browning (Bhattacharyya *et al.*, 2008; Eslami *et al.* 2011). These metals accumulation in plants is the major subject of concern due to their risk to supplying metal toxicity through food chain (Jordao *et al.*, 2006).

Bioremediation and PGPR

Phyto-remediation for heavy metals is one of the modern low cost techniques where growing metal accumulating higher plants are used to eliminate the contaminants from the heavy metal polluted soil (Brooks, 1998). Different kind of plants are used for the bioremediation of heavy metals from different type of heavy metal polluted soil such as Indian mustard (Blaylock *et al.*, 1997), maize (Huang and Cunningham, 1996) and many others can accumulate a good amount of heavy metals in their body. These plants are also known as hyper accumulator plants (Dushenkov *et al.*, 1995). This phytoremediation technique is most flourishing and comparatively cost effective which can be ecologically advanced and immensely accepted for heavy metal abolition. But there are also some constrains to use those hyper accumulator plants for remediate the contaminated soil because the growth of most of these plants are slow and time consuming though there are some high biomass producing hyperaccumulator plant are used now-a-days such as tomato (*Lycopersicon esculentum*), sunflower (*Helianthus annuus*), Indian mustard (*Brassica juncea*), and yellow poplar (*Liriodendron tulipifera*) (de Mello-Farias *et al.*, 2011).

Recently scientist give attention for the microorganism for their immense role in the removal of metals from the heavy metal polluted soil, and their potentiality in development of plant-microbe interaction in normal soil ecological condition as well as in very ruthless condition like metal contamination etc (Khan and Lee, 2013; Pramanik et al. 2016). Many microorganisms can not only survive under heavy metal stress condition but also they can extract these heavy metals from the polluted soil environment. In that scenario, especially heavy metal tolerant plant growth promoting rhizobacteria (PGPR) can be a sustainable option for heavy metal-contaminated agricultural lands (Khan and Lee 2013). These metal tolerant bacteria may have some outstanding role for positive growth promotion of plants. On the basis of this knowledge plant growth promoting rhizobacteria is a most promising option for not only elimination of heavy metal but also their negative impact on plants which ultimately boost the plant's growth. Many researchers stated that plant growth promoting rhizobacteria (PGPR) can stimulate phytoextraction of cadmium and many other heavy metals from contaminated soil and re-vitalize the soil (Sessitsch*et al.*, 2013). Wu *et al.* (2006) reported that *Azotobacter chroococcum, Bacillus megaterium* HKP-1, *Bacillus mucillaginosus* HKK prevented lead toxicity and stimulated the growth of *Brassica juncea.* Dary*et al.* (2010) showed increments in plant biomass, nitrogen content and stabilized the plants against cadmium and lead stress by inoculating *Bradyrhizobium* sp., *Pseudomonas* sp., *Ochrobactrum cytisi.* Tripathi *et al., (2005)* observed that *Pseudomonas putida* KNP9 abridged the Cd and Pb uptake in *Vigna radiate* plant. Similar type of growth enhancement of *Glycine max, Vigna radiata, Triticum vulgare* under many metal stresses were observed by inoculation of bacteria like *Pseudomonas* sp. (Gupta *et al.*, 2002). Many researcher exploited different bacteria for the plant growth enhancement under heavy metal stress condition (Belimov *et al.*, 2005; Ma *et al.*, 2011; Pramanik et al. 2016).

Furthermore, phytoextraction can increase its efficiency by application of PGPR (Sessitsch *et al.*, 2013). A wide range of soil bacteria can tolerate heavy metals and they can mobilize or immobilize heavy metals within their environment (Gadd, 1990). Delorme *et al* and also other researchers observed that a large no of metal resistant bacteria were present in the rhizosphere of *Thalpsic aerulescens* (Delorme *et al.*, 2001), *Alyssum bertolonii* (Mengoni *et al.*, 2001) or *Alyssum murale* (Abou-Shanab *et al.*, 2003) which are hyperaccumulator plants. Burd *et al* inoculated Indian mustard and canola (*Brassica campestris*) seeds with the PGPR strain *Kluyvera ascorbata* SUD165, which produces siderophore and some enzymes like 1-aminocyclopropane- 1-carboxylate (ACC) deaminase. They observed that PGPR strain protected the plant for their growth under different heavy metal stress such as Ni, Pb and Zn (Burd *et al.*, 1998). Accrording to Belimov *et al* metal resistant PGPR producing ACC deaminase enhanced the growth of the rape plant (canola; *Brassica napus*) under cadmium contaminated soil (Belimov *et al.*, 2001). Under cadmium stress condition some PGPR capable of producing auxin and fixing of nitrogen stop the cadmium uptake and ultimately produced higher growth of barley plants (Belimov *et al.*, 2001; Pishchik *et al.*, 2002; Pramanik et al. 2016). Many metal resistant rhizobacteria suppress the metal mobility to the plants grown in heavy metal contaminated soil by releasing metal chelating agents and phosphate solubilization of soil by acidification (Abou-Shanab *et al.*, 2003). According to many researcher phytohormone produced by PGPR like indol acetic acid (IAA) insist the plants for producing defense mechanism against heavy metal stress (Ahmad *et al.*, 2008). Many scientists observed that PGPR produced much other type of plant beneficial products such as vitamins, enzymes, siderophores and different type of antibiotics (Burd *et al.*, 2000; Glick, 2001). PGPR can indirectly inhibit the ethylene synthesis under heavy metal stress condition by producing ACC deaminase which leads the plants as unaffected. Many PGPR can produce siderophore as well as some stable compound with metals like Cu, Cr, Cd, Zn, Pb, etc and immobilized those metals for plant availability. These results were revealed by many workers. Dary *et al.* (2010) showed that cadmium, lead and copper uptake reduced by application of PGPR on *Lupinus cuteus* plant (Dary *et al.*, 2010). *Bacillus mycoides*, and *Micrococoous roseus* like PGPRs can survive under high cadmium stress condition as well as lead, nickel, zinc etc and they can survive on the medium HEPES-MES (Motesharezadeh, 2008). PGPR can eliminate the metal phytotoxicity by different mechanisms such as biosorption, by which an elevated concentration of heavy metals can be accumulated by some autonomous non-metabolic pathway methods (Motesharezadeh, 2008). *Pseudomonas putida* KNP9 detoxify the effect of Cd and Pb on the *Vigna radiata* which reduced the uptake of Cd and Pb and ultimately improved plant development under greenhouse condition (Tripathi *et al.*,

2005). PGPRs such as *Bacillus, Bradyrhizobium, Enterobacter, Klebsiella, Micrococcus, Pseudomonas, Ralstonia, Serratia, Stenotrophomonas, Variovorax* etc. can survive the metal stress and stimulate the plant growth under heavy metal contaminated condition by direct or indirect plant growth promoting ability (Treesubsuntorn et al., 2018).

For these reasons many heavy metal tolerant PGPRs can be an alternative and futuristic preference to develop the sustainable agricultural system under heavy metal-contaminated condition (Khan and Lee 2013). Cadmium and lead tolerant bacteria survive the metal stress by direct or indirect processes such as biosorption, metal transformation through oxidative/reduction/enzymatic reduction, metal volatilization and bio precipitation (Pramanik et al. 2016) and intracellular bioaccumulation (Chen et al. 2016; Treesubsuntorn et al. 2018). Many heavy metal resistant PGPR can stimulate phytoextraction (Ahmad et al. 2016; Płociniczak et al. 2013; Sarathambal et al. 2017) whereas some repress the metal mobility to the plant parts by releasing metal chelating agents in soil (Abou-Shanab et al. 2008; Idris et al. 2004; Tripathi et al. 2005).

Mechanism of PGPR Action

Plant associated bacteria can be grouped into three broad categories i.e. deleterious, beneficial and neutral groups (Dobbelaere *et al.*, 2003). Beneficial bacteria for the growth of plants are normally called as plant growth promoting rhizobacteria (PGPR). They can grow freely in the soil or in symbiotic relationship with some plant roots (Kloepper *et al.*, 1989). Some report depicted that PGPRs may settle within the plant root tissue or may present on the root surface. These free living PGPRs can act as bio fertilizer for plant growth promotion by different way. There are different mechanisms of plant growth promoting ability of PGPRs, these are production of different phytohormone such as indol acetic acid (IAA), Phos-

Figure 1. Plant growth enhancement with influence of PGPR under heavy metal stress (Zaidi et al., 2006)

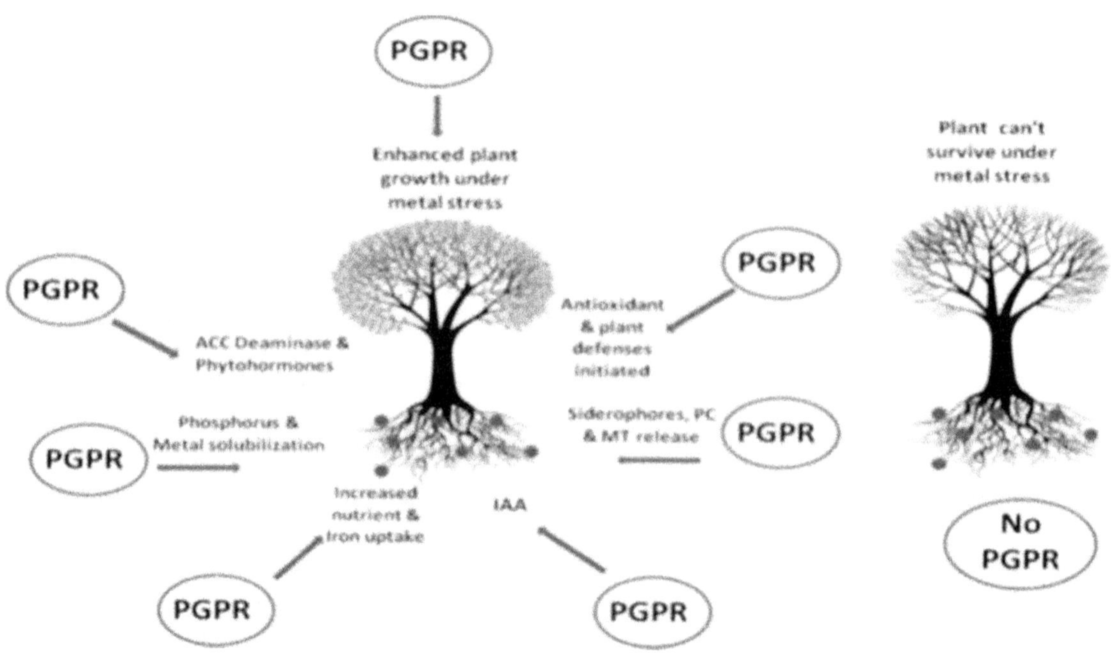

Table 1. Alleviation of heavy metals stress in plants by different PGPRs

PGPR	Heavy metals	Plants	PGPR function	Investigated by
Achromobacter xylosoxidans strain Ax10	Cu	*Brassica juncea*	Notably enhanced Cu uptake by plants and raised the root or shoot length and fresh or dry weights	Ma et al. (2009)
Azotobacter chroococcum HKN-5	Pb	*Brassica juncea*	Eliminated Pb toxicity and enhanced Plant growth	Wu et al. (2006)
Bacillus licheniformis NCCP-59	Ni	*Oryza sativa*	Enhanced seed germination under Ni stress and alleviated toxicity	Jamil et al. (2014)
Bacillus mucillaginosus HKK, *Bacillus megaterium* HKP-1,	Zn, Pb	*Brassica juncea*	Plant growth enhancement and prevent metal stress	Wu et al. (2006)
Bacillus sp. PSB10	Cr	*Cicer arietinum*	Prominently enhanced nodulation, growth, seed production and grain protein amount. Abridged the uptake of Cr in shoots, roots and grains	Wani and Khan (2010)
Bacillus subtilis SJ-101	Ni	*Brassica juncea*	Facilitated Ni accumulation	Zaidi et al. (2006)
Bacillus weihenstephanens is Strain SM3	Ni, Cu, Zn	*Helianthus annuus*	Enhanced plant biomass and accumulation of Zn and Cu in the shoot and root systems, further enhanced soluble Ni, Zn and Cu concentration in soil.	Rajkumar et al. (2008)
Pseudomonas sp., Ochrobactrum Cytisi, Bradyrhizobium sp.750	Cu, Cd, Pb	*Lupinus luteus*	Improved plant biomass, enhanced nitrogen (N_2) content, increased phytostabilization potential	Dary et al. (2010)
Brevibacillus sp.	Zn	*Trifolium repens*	Improved plant yield and nutrition. Reduced plant Zn content.	Vivas et al. (2006)
Kluyvera ascorbata SUD165	Ni, Pb Zn	*Brassica napus, Solanum lycopersicum*	No augmentation of metal uptake in comparison to non-inoculated plants. Reduction in decrease of metal stress	Burd et al. (2000)
Mesorhizobium sp. RC3	Cr (VI)	*Cicer arietinum*	Improved nodule numbers, dry weight, seed yield and grain protein compared to un-inoculated treatments. N_2 content in roots and shoots enhanced by 46% and 40%	Wani et al. (2008)
Microbacterium oxydans AY509223 (RS)	Ni	*Alyssum murale*	Enhanced phytoextraction of Ni	Abou shanab et al. (2006)
Ochrobactrum sp. , Bacillus cereus	Cr (VI)	*Vigna radiata*	Cr toxicity was reduced by reduction of Cr (VI) to Cr (III) in seedlings.	Faisal and Hasnain (2006)
Pseudomonas aeruginosa Strain MKRh3	Cd	*Vigna munga*	Reduced accumulation, increased yield	Ganesan (2008)
Pseudomonas fluorescens, Pseudomonas aeruginosa, Ralstonia metallidurans	Pb, Cr	*Zea mays*	Supported plant growth, aided soil metal mobilization and increased Pb and Cr uptake	Braud et al. (2009)
Pseudomonas putida KNP9	Cd, Pb	*Vigna radiata*	Enhanced plant growth reduced Cd and Pb uptake.	Tripathi et al. (2005)
Pseudomonas sp.	Cr, Cd, Ni	*Glycine max, Vigna radiata, Triticum vulgare*	Development of higher plant growth under different metal stress.	Gupta et al. (2002)
Sinorhizobium sp. Pb002	Pb	*Brassica juncea*	Effectiveness of Pb phytoextraction improved	Di Gregorio et al. (2006)
Azomonas sp. RJ4, Xanthomonas sp. RJ3, RJ10, Bacillus sp. RJ31, Pseudomonas sp.	Cd	*Brassica napus*	Enhanced plant growth and Cd accretion elevated.	Sheng and Xia (2006)

phate solubilization and competition with other deleterious rhizobacteria via production of iron chelating compound siderophore which are available for plants only (Kohler *et al.*, 2009). PGPR promote plant growth by two methods either in i) direct way or ii) indirect way. (Kloepper *et al.*, 1989; Glick, 1995).

The direct plant growth promoting ability promotes the growth by either by producing some substances or by promoting the growth by uptake of nutrients from the plant's environment (Glick *et al.*, 1995). The of the PGPR's different direct growth promoting abilities are i) nitrogen fixation ii) phosphate solubilization iii) siderophores (iron chelating compound) production iv) decreasing ethylene concentration v) phytohormones production (Kloepper*et al.*, 1989; Glick, 1995).

Indirect mechanism for the growth enhancement of plants by PGPRs are i) they can produce different antibiotic ii) PGPR develops induced systemic resistance by competing other pathogenic bacteria for the available root habitat iii) by fungal cell wall degrading enzymes chitinase production iv) antifungal compound production v) reduction of rhizospheric iron etc. (Dunne *et al.*, 1993).

Molecular Aspects Behind Metal Tolerance of Bacteria

Certain bacterial taxa has evolved with exposure to various heavy metals and acquired many metal-resistant genes for their adaptability in that habitat. Many of such operon-clustered metal-resistant genes has been reported in bacterial systems, such as *chrA*, *cadB*, *merA*, *pbrA* and *ars* have been reported for chromium, lead, cadmium, mercury and arsenic tolerance, respectively (Das *et al.*, 2016). For example, a *Pseudomonas putida* strain chromosome was identified, which consists of two contrary transcribed genes *cadA* and *cadR* those involves a Cd-transporting ATPase sand a Cd-responding transcriptional regulatory protein respectively (Lee *et al.*, 2001). Another metal tolerance responsible gene *PbrA*, has been reported to encode the PbrA protein, which on supplementation of other ATPases (*CadA* and *ZntA*), exports lead, cadmium and zinc ions to the periplasm (Hynninen *et al.*, 2009). Moreover, many of the novel metal-resistant genes in bacterial system are yet to be explored for subsequent application in the bioremediation practices. Information of these genes, collectively are presently being recommended for further exploitation in environmental bioremediation program. The reciprocal association between potential, genetically engineered microbes and their host plants can also improve the effectiveness of bioremediation process of metal contaminated sites (Azad *et al.*, 2014). High end biotechnological approaches such as genetic engineering, gene editing etc. should be adopted for manipulation of bacterial genome toward enhancement of their toxic metal detoxification ability. Thus an integrated approach of microbiology and biotechnological advancement might lead to development of promising bioremediations for futuristic sustainable agriculture in toxic metal prone land situation.

CONCLUSION

PGPRs are the perfect alternative for the chemical fertilizers that hamper the soil texture and fertility over time. PGPRs enrich available soil nutrients which is cost effective. PGPR mediated phytoextraction of heavy metals show the ways to reduce soil pollution. Heavy metal tolerant PGPRs having different plant growth promoting abilities can induce higher plant vigor under heavy metal stress conditions. Overall, these PGPRs might be used to develop a futuristic sustainable agricultural practice.

REFERENCES

Abou-Shanab, R. A., Angle, J. S., Delorme, T. A., Chaney, R. L., Van Berkum, P., Moawad, H., Ghanem, K., & Ghozlan, H. A. (2003). Rhizobacterial effects on nickel extraction from soil and uptake by Alyssum murale. *The New Phytologist*, *158*(1), 219–224. doi:10.1046/j.1469-8137.2003.00721.x

Abou-Shanab, R. A. I., Angle, J. S., & Chaney, R. L. (2006). Bacterial inoculants affecting nickel uptake by Alyssum murale from low, moderate and high Ni soils. *Soil Biology & Biochemistry*, *38*(9), 2882–2889. doi:10.1016/j.soilbio.2006.04.045

Abou-Shanab, R. I., Delorme, T. A., Angle, J. S., Chaney, R. L., Ghanem, K., Moawad, H., & Ghozlan, H. A. (2003). Phenotypic characterization of microbes in the rhizosphere of Alyssum murale. *International Journal of Phytoremediation*, *5*(4), 367–379. doi:10.1080/15226510309359043 PMID:14750563

Abyar, H., Safahieh, A., Zolgharnein, H., & Zamani, I. (2012). Isolation and identification of Achromobacter denitrificans and evaluation of its capacity in cadmium removal. *Polish Journal of Environmental Studies*, *21*(6), 1523–1527.

Agirtas, M. S., & Kilicel, F. (1999). Determination of Cu, Ni, Mn and Zn pollution in soil at the shore of Van Lake with Flame Atomic Spectrophotmetry. *Bulletin of Pure and Applied Science*, *18*, 45–47.

Ahmad, F., Ahmad, I., & Khan, M. S. (2008). Screening of free-living rhizospheric bacteria for their multiple plant growth promoting activities. *Microbiological Research*, *163*(2), 173–181. doi:10.1016/j.micres.2006.04.001 PMID:16735107

Ahmad, I., Akhtar, M. J., Asghar, H. N., Ghafoor, U., & Shahid, M. (2016). Differential effects of plant growth-promoting rhizobacteria on maize growth and cadmium uptake. *Journal of Plant Growth Regulation*, *35*(2), 303–315. doi:10.100700344-015-9534-5

Ali, S. Z., Sandhya, V., Grover, M., Kishore, N., Rao, L. V., & Venkateswarlu, B. (2009). Pseudomonas sp. strain AKM-P6 enhances tolerance of sorghum seedlings to elevated temperatures. *Biology and Fertility of Soils*, *46*(1), 45–55. doi:10.100700374-009-0404-9

Aoshima, K. (2016). Itai-itai disease: Renal tubular osteomalacia induced by environmental exposure to cadmium—historical review and perspectives. *Soil Science and Plant Nutrition*, *62*(4), 319–326. doi:10.1080/00380768.2016.1159116

Arao, T., Ishikawa, S., Murakami, M., Abe, K., Maejima, Y., & Makino, T. (2010). Heavy metal contamination of agricultural soil and countermeasures in Japan. *Paddy and Water Environment*, *8*(3), 247–257. doi:10.100710333-010-0205-7

Azad, M. A. K., Amin, L., & Sidik, N. M. (2014). Genetically engineered organisms for bioremediation of pollutants in contaminated sites. *Chinese Science Bulletin*, *59*(8), 703–714. doi:10.100711434-013-0058-8

Benitez, E., Romero, E., Gomez, M., Gallardo-Lara, F., & Nogales, R. (2001). Biosolids and biosolids-ash as sources of heavy metals in a plant-soil system. *Water, Air, and Soil Pollution*, *132*(1-2), 75–87. doi:10.1023/A:1012012924151

Berti, W. R., & Jacob, L. W. (1996). Chemistry and phytotoxicity of soil trace elements from repeated sewage sludge application. *Journal of Environmental Quality*, *25*(5), 1025–1032. doi:10.2134/jeq1996.00472425002500050014x

Bhattacharyya, P., Chakrabarti, K., Chakraborty, A., Tripathy, S., & Powell, M. A. (2008). Fractionation and bioavailability of Pb in municipal solid waste compost and Pb uptake by rice straw and grain under submerged condition in amended soil. *Geosciences Journal*, *12*(1), 41–45. doi:10.100712303-008-0006-9

Blaylock, M. J., Salt, D. E., Dushenkov, S., Zakharova, O., Gussman, C., Kapulnik, Y., Ensley, B. D., & Raskin, I. (1997). Enhanced accumulation of Pb in Indian mustard by soil-applied chelating agents. *Environmental Science & Technology*, *31*(3), 860–865. doi:10.1021/es960552a

Boyd, R. S. (2010). Heavy metal pollutants and chemical ecology: Exploring new frontiers. *Journal of Chemical Ecology*, *36*(1), 46–58. doi:10.100710886-009-9730-5 PMID:20108028

Braud, A., Jézéquel, K., Vieille, E., Tritter, A., & Lebeau, T. (2006). Changes in extractability of Cr and Pb in a polycontaminated soil after bioaugmentation with microbial producers of biosurfactants, organic acids and siderophores. *Water Air and Soil Pollution Focus*, *6*(3-4), 261–279. doi:10.100711267-005-9022-1

Brooks, R. R. (1998). *Plants that hyperaccumulate heavy metals: their role in phytoremediation, microbiology, archaeology, mineral exploration and phytomining*. CAB International.

Burd, G. I., Dixon, D. G., & Glick, B. R. (1998). A plant growth-promoting bacterium that decreases nickel toxicity in seedlings. *Applied and Environmental Microbiology*, *64*(10), 3663–3668. doi:10.1128/AEM.64.10.3663-3668.1998 PMID:9758782

Burd, G. I., Dixon, D. G., & Glick, B. R. (2000). Plant growth-promoting bacteria that decrease heavy metal toxicity in plants. *Canadian Journal of Microbiology*, *46*(3), 237–245. doi:10.1139/w99-143 PMID:10749537

Chen, Y., Chao, Y., Li, Y., Lin, Q., Bai, J., Tang, L., Wang, S., Ying, R., & Qiu, R. (2016). Survival strategies of the plant-associated bacterium Enterobacter sp. strain EG16 under cadmium stress. *Applied and Environmental Microbiology*, *82*(6), 1734–1744. doi:10.1128/AEM.03689-15 PMID:26729719

Chirenje, T., Ma, L. Q., Reeves, M., & Szulczewski, M. (2004). Lead distribution in near-surface soils of two Florida cities: Gainesville and Miami. *Geoderma*, *119*(1-2), 113–120. doi:10.1016/S0016-7061(03)00244-1

Crnković, D., Ristić, M., & Antonović, D. (2006). Distribution of heavy metals and arsenic in soils of Belgrade (Serbia and Montenegro). *Soil & Sediment Contamination*, *15*(6), 581–589. doi:10.1080/15320380600959073

Dary, M., Chamber-Pérez, M. A., Palomares, A. J., & Pajuelo, E. (2010). "In situ" phytostabilisation of heavy metal polluted soils using Lupinus luteus inoculated with metal resistant plant-growth promoting rhizobacteria. *Journal of Hazardous Materials*, *177*(1-3), 323–330. doi:10.1016/j.jhazmat.2009.12.035 PMID:20056325

Das, S., Dash, H. R., & Chakraborty, J. (2016). Genetic basis and importance of metal resistant genes in bacteria for bioremediation of contaminated environments with toxic metal pollutants. *Applied Microbiology and Biotechnology*, *100*(7), 2967–2984. doi:10.100700253-016-7364-4 PMID:26860944

de Mello-Farias, P. C., Chaves, A. L. S., & Lencina, C. L. (2011). Transgenic plants for enhanced phytoremediation–physiological studies. In *Genetic Transformation*. InTech. doi:10.5772/24355

de Souza, J. T., Arnould, C., Deulvot, C., Lemanceau, P., Gianinazzi-Pearson, V., & Raaijmakers, J. M. (2003). Effect of 2, 4-diacetylphloroglucinol on Pythium: Cellular responses and variation in sensitivity among propagules and species. *Phytopathology*, *93*(8), 966–975. doi:10.1094/PHYTO.2003.93.8.966 PMID:18943863

Delorme, T. A., Gagliardi, J. V., Angle, J. S., & Chaney, R. L. (2001). Influence of the zinc hyperaccumulatorThlaspicaerulescens J. & C. Presl. and the nonmetal accumulator Trifoliumpratense L. on soil microbial populations. *Canadian Journal of Microbiology*, *47*(8), 773–776. doi:10.1139/w01-067 PMID:11575505

Di Gregorio, S., Barbafieri, M., Lampis, S., Sanangelantoni, A. M., Tassi, E., & Vallini, G. (2006). Combined application of Triton X-100 and Sinorhizobium sp. Pb002 inoculum for the improvement of lead phytoextraction by Brassica juncea in EDTA amended soil. *Chemosphere*, *63*(2), 293–299. doi:10.1016/j.chemosphere.2005.07.020 PMID:16153689

Dobbelaere, S., Vanderleyden, J., & Okon, Y. (2003). Plant growth-promoting effects of diazotrophs in the rhizosphere. *Critical Reviews in Plant Sciences*, *22*(2), 107–149. doi:10.1080/713610853

Dunne, C., Crowley, J. J., Moënne-Loccoz, Y., Dowling, D. N., & O'Gara, F. (1997). Biological control of Pythiumultimum by Stenotrophomonasmaltophilia W81 is mediated by an extracellular proteolytic activity. *Microbiology*, *143*(12), 3921–3931. doi:10.1099/00221287-143-12-3921

Dushenkov, V., Kumar, P. N., Motto, H., & Raskin, I. (1995). Rhizofiltration: The use of plants to remove heavy metals from aqueous streams. *Environmental Science & Technology*, *29*(5), 1239–1245. doi:10.1021/es00005a015 PMID:22192017

European Union. (2002). *Heavy Metals in Wastes*. European Commission on Environment. https://ec.europa.eu/environment/waste/ studies/pdf/heavy_metalsreport.pdf

Faisal, M., & Hasnain, S. (2006). Growth stimulatory effect of Ochrobactrum intermedium and Bacillus cereus on Vigna radiata plants. *Letters in Applied Microbiology*, *43*(4), 461–466. doi:10.1111/j.1472-765X.2006.01977.x PMID:16965380

Ferner, D. J. (2001). *Toxicity, heavy metals. eMed J, 2(5), 1.* Google Scholar.

Gadd, G. M. (1990). Heavy metal accumulation by bacteria and other microorganisms. *Experientia*, *46*(8), 834–840. doi:10.1007/BF01935534

Ganesan, V. (2008). Rhizoremediation of cadmium soil using a cadmium-resistant plant growth-promoting rhizopseudomonad. *Current Microbiology*, *56*(4), 403–407. doi:10.100700284-008-9099-7 PMID:18239844

Garrido, S., Del Campo, G. M., Esteller, M. V., Vaca, R., & Lugo, J. (2005). Heavy metals in soil treated with sewage sludge composting, their effect on yield and uptake of broad bean seeds (Viciafaba L.). *Water, Air, and Soil Pollution*, *166*(1-4), 303–319. doi:10.100711270-005-5269-4

Glick, B. R. (2001). Phytoremediation: Synergistic use of plants and bacteria to clean up the environment. *Biotechnology Advances*, *21*(5), 383–393. doi:10.1016/S0734-9750(03)00055-7 PMID:14499121

Gupta, A., Meyer, J. M., & Goel, R. (2002). Development of heavy metal-resistant mutants of phosphate solubilizing Pseudomonas sp. NBRI 4014 and their characterization. *Current Microbiology*, *45*(5), 323–327. doi:10.100700284-002-3762-1 PMID:12232661

Hellström, L., Elinder, C. G., Dahlberg, B., Lundberg, M., Järup, L., Persson, B., & Axelson, O. (2001). Cadmium exposure and end-stage renal disease. *American Journal of Kidney Diseases*, *38*(5), 1001–1008. doi:10.1053/ajkd.2001.28589 PMID:11684553

Holum, J. R. (1983). *Elements of General and Biological Chemistry* (6th ed.). John Wiley and Sons.

Huang, J. W., & Cunningham, S. D. (1996). Lead phytoextraction: Species variation in lead uptake and translocation. *The New Phytologist*, *134*(1), 75–84. doi:10.1111/j.1469-8137.1996.tb01147.x

Hynninen, A., Touzé, T., Pitkänen, L., Mengin-Lecreulx, D., & Virta, M. (2009). An efflux transporter PbrA and a phosphatase PbrB cooperate in a lead-resistance mechanism in bacteria. *Molecular Microbiology*, *74*(2), 384–394. doi:10.1111/j.1365-2958.2009.06868.x PMID:19737357

Idris, R., Trifonova, R., Puschenreiter, M., Wenzel, W. W., & Sessitsch, A. (2004). Bacterial communities associated with flowering plants of the Ni hyperaccumulator Thlaspi goesingense. *Applied and Environmental Microbiology*, *70*(5), 2667–2677. doi:10.1128/AEM.70.5.2667-2677.2004 PMID:15128517

Imperato, M., Adamo, P., Naimo, D., Arienzo, M., Stanzione, D., & Violante, P. (2003). Spatial distribution of heavy metals in urban soils of Naples city (Italy). *Environmental Pollution*, *124*(2), 247–256. doi:10.1016/S0269-7491(02)00478-5 PMID:12713924

Institute of Environmental Conservation and Research (INECAR). (2000). *Position Paper Against Mining in Rapu-Rapu*. INECAR, Ateneo de Naga University. www.adnu.edu.ph/Institutes/Inecar/pospaper1.asp

Jamil, M., Zeb, S., Anees, M., Roohi, A., Ahmed, I., ur Rehman, S., & Rha, E. (2014). Role of Bacillus licheniformis in phytoremediation of nickel contaminated soil cultivated with rice. *International Journal of Phytoremediation*, *16*(6), 554–571. doi:10.1080/15226514.2013.798621 PMID:24912242

Jordao, C. P., Nascentes, C. C., Cecon, P. R., Fontes, R. L. F., & Pereira, J. L. (2006). Heavy metal availability in soil amended with composted urban solid wastes. *Environmental Monitoring and Assessment*, *112*(1-3), 309–326. doi:10.100710661-006-1072-y PMID:16404548

Kabata-Pendias, A. (1992). Trace metals in soils in Poland occurrence and behaviour. *Soil Science*, *140*, 53–70.

Khan, A. L., & Lee, I. J. (2013). Endophytic Penicillium funiculosum LHL06 secretes gibberellin that reprograms Glycine max L. growth during copper stress. *BMC Plant Biology*, *13*(1), 86. doi:10.1186/1471-2229-13-86 PMID:23721090

Kilice, F. (1999). Investigation of Toxic Heavy Metals Pollution in the Road Dust a the Centre of Van Turkey. *Bulletin of Pure and Applied Science*, *18c*, 1–4.

Kloepper, J. W., Lifshitz, R., & Zablotowicz, R. M. (1989). Free-living bacterial inocula for enhancing crop productivity. *Trends in Biotechnology*, *7*(2), 39–44. doi:10.1016/0167-7799(89)90057-7

Kohler, J., Hernández, J. A., Caravaca, F., & Roldán, A. (2009). Induction of antioxidant enzymes is involved in the greater effectiveness of a PGPR versus AM fungi with respect to increasing the tolerance of lettuce to severe salt stress. *Environmental and Experimental Botany*, *65*(2-3), 245–252. doi:10.1016/j.envexpbot.2008.09.008

Krantev, A., Yordanova, R., & Popova, L. (2006). Salicylic acid decreases Cd toxicity in maize plants. *General and Applied Plant Physiology*, *1*, 45–52.

Lee, S. W., Glickmann, E., & Cooksey, D. A. (2001). Chromosomal locus for cadmium resistance in Pseudomonas putida consisting of a cadmium-transporting ATPase and a MerR family response regulator. *Applied and Environmental Microbiology*, *67*(4), 1437–1444. doi:10.1128/AEM.67.4.1437-1444.2001 PMID:11282588

Liang, Y., Wong, J. W. C., & Wei, L. (2005). Silicon-mediated enhancement of cadmium tolerance in maize (Zea mays L.) grown in cadmium contaminated soil. *Chemosphere*, *58*(4), 475–483. doi:10.1016/j.chemosphere.2004.09.034 PMID:15620739

Logan, T. J., & Chaney, R. L. (1983). Utilization of municipal wastewater and sludge on land-metals. In *Proceedings of the Workshop on Utilization of Municipal Wastewater and Sludge on Land* (pp. 235-326). University of California Riverside.

Ma, Y., Prasad, M. N. V., Rajkumar, M., & Freitas, H. (2011). Plant growth promoting rhizobacteria and endophytes accelerate phytoremediation of metalliferous soils. *Biotechnology Advances*, *29*(2), 248–258. doi:10.1016/j.biotechadv.2010.12.001 PMID:21147211

Ma, Y., Rajkumar, M., & Freitas, H. (2009). Improvement of plant growth and nickel uptake by nickel resistant-plant-growth promoting bacteria. *Journal of Hazardous Materials*, *166*(2-3), 1154–1161. doi:10.1016/j.jhazmat.2008.12.018 PMID:19147283

Madrid, L., Díaz-Barrientos, E., Reinoso, R., & Madrid, F. (2004). Metals in urban soils of Sevilla: Seasonal changes and relations with other soil components and plant contents. *European Journal of Soil Science*, *55*(2), 209–217. doi:10.1046/j.1365-2389.2004.00589.x

Marschner, H. (1995). Adaptation of plants to adverse chemical soil conditions. *Mineral Nutrition of Higher Plants*.

McCluggage, D. (1991). Heavy metal poisoning. NCS Magazine.

Mengoni, A., Barzanti, R., Gonnelli, C., Gabbrielli, R., & Bazzicalupo, M. (2001). Characterization of nickel-resistant bacteria isolated from serpentine soil. *Environmental Microbiology*, *3*(11), 691–698. doi:10.1046/j.1462-2920.2001.00243.x PMID:11846759

Moreno, J. L., Hernández, T., & Garcia, C. (1999). Effects of a cadmium-contaminated sewage sludge compost on dynamics of organic matter and microbial activity in an arid soil. *Biology and Fertility of Soils*, *28*(3), 230–237. doi:10.1007003740050487

Motesharezadeh, B. (2008). *Study of the possibility of increasing phytoremediation efficiency in heavy metal-contaminated soil by biological factors* (Doctoral dissertation). Univ College Agr Nat Resour, Univ of Tehran, Iran. (in Persian)

Nolan, K. R. (1983). Copper toxicity syndrome. *Journal of Orthomolecular Psychiatry*, *12*(4), 270–282.

Noumavo, P. A., Agbodjato, N. A., Baba-Moussa, F., Adjanohoun, A., & Baba-Moussa, L. (2016). Plant growth promoting rhizobacteria: Beneficial effects for healthy and sustainable agriculture. *African Journal of Biotechnology*, *15*(27), 1452–1463. doi:10.5897/AJB2016.15397

Nyangababo, J. T., & Hamya, J. W. (1986). The deposition of lead, cadmium, zinc and copper from motor traffic on *Brachiariaenimi* and soil along a major Bombo road in Kampala city. *The International Journal of Environmental Studies*, *27*(1-2), 115–119. doi:10.1080/00207238608710280

Ogunsola, O. J., Oluwole, A. F., Obioh, I. B., Asubiojo, O. I., Akeredolu, F. A., Akanle, O. A., & Spyrou, N. M. (1993). Analysis of suspended air particulates along some motorways in Nigeria by PIXE and EDXRF. *Nuclear Instruments & Methods in Physics Research. Section B, Beam Interactions with Materials and Atoms*, *79*(1-4), 404–407. doi:10.1016/0168-583X(93)95373-D

Ogwuegbu, M. O. C., & Muhanga, W. (2005). Investigation of lead concentration in the blood of people in the copper belt province of Zambia. *J Environ*, *1*, 66–75.

Ommati, M. M., Jamshidzadeh, A., Heidari, R., Sun, Z., Zamiri, M. J., Khodaei, F., ... Yeganeh, B. S. (2019). Carnosine and histidine supplementation blunt lead-induced reproductive toxicity through antioxidative and mitochondria-dependent mechanisms. *Biological Trace Element Research*, *187*(1), 151–162. doi:10.100712011-018-1358-2 PMID:29767280

Pal, A. K., & Sengupta, C. (2019). Isolation of Cadmium and Lead Tolerant Plant Growth Promoting Rhizobacteria: Lysinibacillus varians and Pseudomonas putida from Indian Agricultural Soil. *Soil and Sediment Contamination: An International Journal*, *28*(7), 601–629. doi:10.1080/15320383.2019.1637398

Popova, L. P., Maslenkova, L. T., Yordanova, R. Y., Ivanova, A. P., Krantev, A. P., Szalai, G., & Janda, T. (2009). Exogenous treatment with salicylic acid attenuates cadmium toxicity in pea seedlings. *Plant Physiology and Biochemistry*, *47*(3), 224–231. doi:10.1016/j.plaphy.2008.11.007 PMID:19091585

Pramanik, K., Ghosh, P. K., Ghosh, A., Sarkar, A., & Maiti, T. K. (2016). Characterization of PGP traits of a hexavalent chromium resistant Raoultella sp. isolated from the rice field near industrial sewage of Burdwan District, WB, India. *Soil and Sediment Contamination: An International Journal*, *25*(3), 313–331. doi:10.1080/15320383.2016.1137861

Rajkumar, M., Ma, Y., & Freitas, H. (2008). Characterization of metal-resistant plant-growth promoting Bacillus weihenstephanensis isolated from serpentine soil in Portugal. *Journal of Basic Microbiology*, *48*(6), 500–508. doi:10.1002/jobm.200800073 PMID:18785659

Rascio, N., & Navari-Izzo, F. (2011). Heavy metal hyperaccumulating plants: How and why do they do it? And what makes them so interesting? *Plant Science*, *180*(2), 169–181. doi:10.1016/j.plantsci.2010.08.016 PMID:21421358

Raven, P. H., Berg, L. R., & Johnson, G. B. (1998). *Environment* (2nd ed.). Saunders College Publishing.

Reuss, J. O., Dooley, H. L., & Griffis, W. (1978). Uptake of Cadmium from Phosphate Fertilizers by Peas, Radishes, and Lettuce 1. *Journal of Environmental Quality*, *7*(1), 128–133. doi:10.2134/jeq1978.00472425000700010026x

Rude, R. K., & Gruber, H. E. (2004). Magnesium deficiency and osteoporosis: Animal and human observations. *The Journal of Nutritional Biochemistry*, *15*(12), 710–716. doi:10.1016/j.jnutbio.2004.08.001 PMID:15607643

Sarathambal, C., Khankhane, P. J., Gharde, Y., Kumar, B., Varun, M., & Arun, S. (2017). The effect of plant growth-promoting rhizobacteria on the growth, physiology, and Cd uptake of Arundo donax L. *International Journal of Phytoremediation*, *19*(4), 360–370. doi:10.1080/15226514.2016.1225289 PMID:27592507

Sessitsch, A., Kuffner, M., Kidd, P., Vangronsveld, J., Wenzel, W. W., Fallmann, K., & Puschenreiter, M. (2013). The role of plant-associated bacteria in the mobilization and phytoextraction of trace elements in contaminated soils. *Soil Biology & Biochemistry*, *60*, 182–194. doi:10.1016/j.soilbio.2013.01.012 PMID:23645938

Sheng, X. F., & Xia, J. J. (2006). Improvement of rape (Brassica napus) plant growth and cadmium uptake by cadmium-resistant bacteria. *Chemosphere*, *64*(6), 1036–1042. doi:10.1016/j.chemosphere.2006.01.051 PMID:16516946

Singh, K. P., Mohan, D., Sinha, S., & Dalwani, R. (2004). Impact assessment of treated/untreated wastewater toxicants discharged by sewage treatment plants on health, agricultural, and environmental quality in the wastewater disposal area. *Chemosphere*, *55*(2), 227–255. doi:10.1016/j.chemosphere.2003.10.050 PMID:14761695

Smiri, M., Chaoui, A., Rouhier, N., Gelhaye, E., Jacquot, J. P., & El Ferjani, E. (2010). *RETRACTED: Effect of cadmium on resumption of respiration in cotyledons of germinating pea seeds*. Academic Press.

Taiz, L., & Zeiger, E. (2006). *Plant Physiology* (4th ed.). Sinauer Associates.

Tijhuis, L., Brattli, B., & Sæther, O. M. (2002). A geochemical survey of topsoil in the city of Oslo, Norway. *Environmental Geochemistry and Health*, *24*(1), 67–94. doi:10.1023/A:1013979700212

Treesubsuntorn, C., Dhurakit, P., Khaksar, G., & Thiravetyan, P. (2018). Effect of microorganisms on reducing cadmium uptake and toxicity in rice (Oryza sativa L.). *Environmental Science and Pollution Research International*, *25*(26), 25690–25701. doi:10.100711356-017-9058-6 PMID:28480489

Tripathi, M., Munot, H. P., Shouche, Y., Meyer, J. M., & Goel, R. (2005). Isolation and functional characterization of siderophore-producing lead-and cadmium-resistant Pseudomonas putida KNP9. *Current Microbiology*, *50*(5), 233–237. doi:10.100700284-004-4459-4 PMID:15886913

Vivas, A., Biro, B., Ruiz-Lozano, J. M., Barea, J. M., & Azcon, R. (2006). Two bacterial strains isolated from a Zn-polluted soil enhance plant growth and mycorrhizal efficiency under Zn-toxicity. *Chemosphere*, *62*(9), 1523–1533. doi:10.1016/j.chemosphere.2005.06.053 PMID:16098559

Vormann, J. (2003). Magnesium: Nutrition and metabolism. *Molecular Aspects of Medicine*, *24*(1-3), 27–37. doi:10.1016/S0098-2997(02)00089-4 PMID:12537987

Vos, C. M., Tesfahun, A. N., Panis, B., De Waele, D., & Elsen, A. (2012). Arbuscular mycorrhizal fungi induce systemic resistance in tomato against the sedentary nematode Meloidogyne incognita and the migratory nematode Pratylenchus penetrans. *Applied Soil Ecology*, *61*, 1–6. doi:10.1016/j.apsoil.2012.04.007

Wagner, G. J. (1993). Accumulation of cadmium in crop plants and its consequences to human health. *Advances in Agronomy*, *51*, 173–212. doi:10.1016/S0065-2113(08)60593-3

Wani, P. A., & Khan, M. S. (2010). Bacillus species enhance growth parameters of chickpea (Cicer arietinum L.) in chromium stressed soils. *Food and Chemical Toxicology*, *48*(11), 3262–3267. doi:10.1016/j.fct.2010.08.035 PMID:20813149

Wani, P. A., Khan, M. S., & Zaidi, A. (2008). Chromium-reducing and plant growth-promoting Mesorhizobium improves chickpea growth in chromium-amended soil. *Biotechnology Letters*, *30*(1), 159–163. doi:10.100710529-007-9515-2 PMID:17849087

Wu, C. H., Wood, T. K., Mulchandani, A., & Chen, W. (2006). Engineering plant-microbe symbiosis for rhizoremediation of heavy metals. *Applied and Environmental Microbiology*, *72*(2), 1129–1134. doi:10.1128/AEM.72.2.1129-1134.2006 PMID:16461658

Yanez, L., Ortiz, D., Calderon, J., Batres, L., Carrizales, L., Mejia, J., Martínez, L., García-Nieto, E., & Díaz-Barriga, F. (2002). Overview of human health and chemical mixtures: Problems facing developing countries. *Environmental Health Perspectives*, *110*(s6, Suppl. 6), 901–909. doi:10.1289/ehp.02110s6901 PMID:12634117

Young, R. A. (2005). *Toxicity profiles: toxicity summary for cadmium, risk assessment information system. RAIS*. University of Tennessee.

Zaidi, S., Usmani, S., Singh, B. R., & Musarrat, J. (2006). Significance of Bacillus subtilis strain SJ-101 as a bioinoculant for concurrent plant growth promotion and nickel accumulation in Brassica juncea. *Chemosphere*, *64*(6), 991–997. doi:10.1016/j.chemosphere.2005.12.057 PMID:16487570

Zhang, W., & Zhang, X. (2007). A forecast analysis on fertilizers consumption worldwide. *Environmental Monitoring and Assessment*, *133*(1-3), 427–434. doi:10.100710661-006-9597-7 PMID:17286167

Zhang, Y. F., He, L. Y., Chen, Z. J., Zhang, W. H., Wang, Q. Y., Qian, M., & Sheng, X. F. (2011). Characterization of lead-resistant and ACC deaminase-producing endophytic bacteria and their potential in promoting lead accumulation of rape. *Journal of Hazardous Materials*, *186*(2-3), 1720–1725. doi:10.1016/j.jhazmat.2010.12.069 PMID:21227577

Chapter 11
Fungi-Mediated Detoxification of Heavy Metals

Suchhanda Ghosh
Shri Shikshayatan College, India

ABSTRACT

Heavy metal pollution is one of the major environmental problems today. Therefore, the elimination of heavy metal ions from wastewater is important to protect public health. The use of biological material in the removal and recovery of toxic metals from industrial wastes has gained important credibility during recent years. Several microorganisms including bacteria, algae, yeast, and fungi have been reported to effectively accumulate or adsorb heavy metals through biosorption. Fungal biomaterial has been proved to be efficient as a biosorbent. High percentage of the cell wall material and availability of fungal biomass as a by-product of various antibiotic and food industries makes it an obvious choice. Thus, the chapter deals with detoxification of heavy metals from contaminated sources using biomaterials with special reference to fungi.

INTRODUCTION

Rapid improvement in industrialization has made human life comfortable, but it has also brought along with it a disruption of environmental balance. Heavy metals produced as a byproduct of many such industrial processes is toxic, and its accumulation in the environment can lead to severe health hazards in human beings. It can also harm the eco-system by accumulating in the food chain. Environmental heavy metal pollution can be primarily associated with the following causes:

- Seepage and overburdens generated from mines- associated with mining operations
- Effluents produced from electroplating plants
- Effluents produced from Coal-based power plants
- Byproducts produced from nuclear reactors

DOI: 10.4018/978-1-7998-4888-2.ch011

Heavy metals due to their non-biodegradable nature pose severe threat to the environment as it cannot be removed from the system once it enters; these metals may also seep in to the soil contaminating the ground water sources. Hence, removal or de-toxification of the accumulated heavy metal from the environment is a major challenge for the environmental scientists.

Several techniques are available for treatment of the effluents which are physiochemical in nature. However, the major drawbacks of these processes are their high operational cost, high energy consumption and lack of efficiency with respect to complete removal of the metal concerned. The problems mentioned for the physico-chemical methods could be reduced with the use of biological organisms.

Out of the several biological methods known for removal of metals from aqueous solution bioaccumulation and biosorption have been proved to be effective (Volesky and Holan, 1995). However, biosorption using dead biomass have been preferred over bioaccumulation. Biosorption has a few advantages over active cellular accumulation like: absence of toxicity limits, possibility to regenerate and recycle of biomass, easy absorbance and recovery of the sorbed biomaterial.

Several biological substances are being used for the process of biosorption. Among the known biological materials, microbes have gained importance because they are ubiquitous in nature and can be grown and manipulated easily. A number of microbial organisms have been used as a biosorbent.

Fungi pose a suitable material for biosorption among the microbes, as it is a common product of industrial processes like food, brewing and distillation; biomass content of fungal cells are high, it is comparatively easier to handle and can regenerate within a short span . Thus fungi can serve as an effective biomaterial for heavy metal removal from aqueous solution.

HEAVY METAL POLLUTION

Heavy metals occur naturally in the environment like many other metallic elements and have an atomic weight higher than the molecular weight of water. They occur in the earth crust naturally and do not interact with the normal biotic system. However, they may get introduced in to the environment via (1) natural phenomenon like volcanic eruptions, forest fire, deep sea vents etc. and (2) anthropogenic events like mining, industrial effluents, smelters etc. In recent times magnified exposure of the heavy metals are happening in the regular life of the biotic elements through anthropogenic acivities involving their indiscriminate use in industrial, domestic, agricultural, medical and technological sectors.

Heavy metals may have some biological roles and are known as **essential metals** (zinc, nickel, copper). These metals are required in very low concentration and can be detrimental to life forms in a slightly higher concentration. The other heavy metals like lead, cadmium, mercury etc. are not known to have any role related to biological organism and are known as **non-essential metals**.

The heavy metals are systemic toxicants and cause various adverse health issues in human and animals. The adversity however is dependent on the chemical nature, time of exposure and dose of the metal concerned. It has also been reported that co-exposure to metal/metalloid mixture cause more severe effects on human health (Wang et. al., 2008).

CONVENTIONAL METHODS OF HEAVY METAL REMOVAL

There are several known and tested physico-chemical methods of heavy metal removal from the environment like:

Physical Methods: Reverse osmosis, electro-dialysis, ultra-filtration, ion exchange

Chemical Methods: Chemical precipitation, electrochemical treatment, oxidation/reduction

Biological Methods: Phyto-remediation

These well known processes come with some major disadvantages that include high reagent and energy requirement, low cost efficiency, generation of toxic sludge and inability for complete metal removal. There is definitely a need to find out better alternative for these processes. Biosorption of metals using microbial organisms could be a solution to these problems.

BIOSORPTION OF HEAVY METALS

The process of removal of metal or metalloid species, compounds and particulates from solution using a biological material can be defined as biosorption (Gadd, 1993). There are several reports of accumulation and desorption of heavy metals by microorganisms. Different groups of microbes have been used for this purpose, members of bacteria, algae, yeast and fungi either in living state or in dead condition have been used for heavy metal removal (Huang *et. al.*, 1988, Antunes *et. al.*, 2003, Sag *et. al.*, 2003; Prasenjit and Sumathi, 2005, Mala *et. al.*, 2006, Gupta *et. al.*, 2007). Microbial biomass as an adsorbent has gained importance in recent times as a potential alternative technique in contrast to the already existing metal removal processes (Ozturk, 2007). Usage of biological substrate is associated with a number of advantages: (a) these microbes has a diverse kind of biologically active sites for heavy metal binding especially on their cell wall, (b) they are of small and uniform size and (c) there is very less chance of interference in their case as compared to alkali and alkali-earth metals and ion exchange resins (Madrid and Camara, 1997).

Biosorption basically involves the process of adsorption of a dissolved solid (sorbate) from a liquid containing the dissolved solid (solvent) on to a biological material (sorbent). There is a high affinity of the sorbent for the sorbate species facilitating its removal from the aqueous phase by different mechanisms. The absorption of sorbate on to the surface of the sorbent keeps on increasing until it reaches a state of equilibrium that exists between the adsorbed solid present in solution to that present adhered to the sorbent (Das *et. al.*, 2008).

Advantages of Biosorption: Biosorption has some advantages over the conventional metal removal processes (Sahin and Ozturk 2005; Alluri *et. al.* 2007): The major advantages are as follows:

- Cheaper source of biomass
- Multiple heavy metal uptake during metal interaction
- Treatment of a large volume of waste possible with the same set of biomass
- Highly selective for removal of heavy metal
- Is active in different conditions of physical parameters like time, temperature, pH and chemical parameters like interference of co-ions, concentration of sorbate or sorbent etc.
- Easy and cheap recovery of metal from metal loaded biomass
- Reduced production of waste or toxic material

BIOSORPTION MECHANISM: MODES OF METAL UPTAKE

Metal sorption is a complicated mechanism and there are several controlling factors of the process, like, the nature of biomass (living or dead), type of biomaterials, properties of metal solution chemistry, ambient conditions such as pH, temperature, concentration of biomass etc. (Das *et. al.*, 2008).

The process by which the heavy metal adheres to the surface of the biomaterial can be of three types: 1) metabolism dependent biosorption, 2) metabolism independent biosorption and 3) bioaccumulation of metal species (Gadd, 1990; Sag and Kutsal, 2001).

1) Metabolism dependent biosorption

This process is exhibited by living cells. Metal biosorption is active, where the metal ion binds to the cell surface (essentially the cell wall) by a single process or a combination of processes viz. physical adsorption or inorganic micro-precipitation, formation of coordination complex, ion excahange and so on (Volesky, 1990; Wang *et. al.*, 2000). The process involves the association of the heavy metal with the cellular metabolic process of the microorganism, however the metal concerned remains primarily adhered to the cell surface.

2) Metabolism independent biosorption

Metal binding to the cell surface could be passive and can occur in either living or non-living microorganisms. Non-viable biomass exhibits a higher affinity for metallic ions as compared to the living biomass (Ilhan *et. al.*, 2004). The process can be due to ionic interaction or simple physiochemical adsorption. The functional groups present on the cell surface as a mosaic often play the key role in the process the major functional groups reported to be involved include carboxyl (-COOH), phosphate (-PO4), thiol (-SH), amide ($-NH_2$)and hydroxide (-OH) (Volesky, 1990).

3) Metal accumulation

Metals can also be a part of cellular metabolism (Pabst *et. al.*, 2010; Campbell *et. al.*, 2002). Active metal sorption involves metabolic uptake of metal ions into the inner parts of the cell.

FUNGAL BIOMASS AND METAL UPTAKE

Biosorption being a surface phenomenon may depend on polarity and surface area of the biosorbent. A high content of cell-wall material with a large number of sites for binding of metals in fungi thus makes it a suitable choice as a biosorbent. (Gadd, 1990).

The fungal cell wall provides mechanical strength to the cell and it is the interface of the cell with the external environment. It contributes to 30% of the dry mass of the cell. It is an extremely complex structure consisting of an elastic framework. It is like a mosaic of different functional groups.

The fungal cell wall presents a multi laminate, micro fibrillar structure and reveals two phases: i) the outer layer and ii) an inner layer of microfibrillar nature. The cell wall is primarily made up of polysaccharide, which contributes to 80% of the cellular dry weight, in the polysaccharide different proteins are anchored in different ways.

The chief components commonly found in fungal cell wall are as tabulated in Table: 1.

Table 1. Chief Components of Fungal Cell Wall

Component Nature	Component Name
Fibrillar component	Cellulose
	β-glucan
	Chitin
Matricial component	Glucoproteins
	Chitosans
	α-glucan
	Lipids
	Ingorganic salt
	Pigments

The composition of the cell wall material varies between the different fungal species. Studies mediated by electron microscopy reveals that in case of the mycellial members, chitin, is the primary wall component contributing to nearly 30% of the cellular dry weight, whereas, yeast, has a more complex cell wall composition, containing glucan, mannan, proteins and lipid. Different metal binding groups are also known to be present in fungal cell wall (Dhankar and Hooda, 2011).

MODELLING OF BIOSORPTION

Adherence of the metals to the active binding sites on the available biomass continues to an equilibrium state where the amount of metal ions distributed in the sorbent and solution phase maintains a balance, therefore, detailed information on adsorption equilibrium is necessary to understand and optimize the process of metal biosorption.

Isotherm Models

Adsorption equilibrium may be defined as the capacity of the adsorbent for the adsorbate.

To obtain the equilibrium value the amount of metal adhering to the sorbent (qe) is plotted against the final concentration of the metal present in solution (C).

$$Qe=V\,[Ci-C]/S$$

V = volume (L) of solution contacted with the sorbent; Ci = initial concentration of the sorbate (mgL^{-1}) and C = final concentration of the sorbate (mgL^{-1}); S = amount of biosorbent (mgL^{-1}).

Out of different isotherm models available to describe equilibrium sorption distribution, Langmuir and Freudlich models are by far the most commonly used.

Langmuir model: $\mathbf{Q_{eq} = Q_{max}.(b).C_{eq}/1+(b).C_{eq}}$ (Langmuir, 1918)

Freundlich model: $\mathbf{Q_{eq} = K(C_{eq})^{1/n}}$ (Freundlich, 1906)

Where, Qeq is the amount of metal ion biosorbed at equilibrium per unit weight of biomass; Ceq is the metal ion concentration at equilibrium; Qmax and b are Langmuir model constants and K and n are Freundlich model constants.

Sorption Kinetics

The sorption kinetics describes uptake of the solute that finally controls the time of residence of a sorbate at the solid-solution interface. This will ultimately provide information about the pathways, as well as the mechanism of the process (Ho and McKay, 2000). The most commonly used models are:

The pseudo-first-order equation (Lagergren, 1898):

$$\text{Log } (q_e - q_t) = \log q_e - (k_{ad}/2.303)t$$

Where, k_{ad} (min^{-1}) = rate constant of pseudo-first-order adsorption process.

The values of k_{ad} were calculated from the plots of log ($q_e - q_t$) vs. time (t).

The pseudo-second-order equation (Ho and McKay, 1999):

$$t/q_t = 1/kq_e^2 + (1/q_e)t$$

Where, k = rate constant, q_t is the metal uptake capacity at time.

With the help of the above mentioned models the phenomenon of biosorption could be statistically established with respect to its energy consumption and viability.

FACTORS CONTROLLING FUNGAL BIOSORPTION

Metal sorption by microbial biomass is known to be affected by several factors like, nature of the biomass, the type of the metal used and the ambient environmental factors. The major factors influencing the process of biosorption include metal ion concentration, biomass concentration in aqueous phase, temperature and pH (Das *et. al.*, 2008). Growth, nutrition, and age of the biomass, can also influence the process due to changes in cell wall composition, extracellular product formation, cell size, etc.

Temperature

The biosorption reactions are generally exothermic in nature and the adsorption rate has an inverse relation with the incubation temperature. Ahalya *et. al.* (2003) reported that a temperature within a range of 20°C -35°C is most suitable for metal biosorption, higher temperature often affects the cell surface topography and thus reduces its capacity to adhere to metal particles.

pH

pH of the aqueous phase has a very important role to play in the biosorption processes. The process is highly dependent on the changes in pH gradient in almost all kinds of biological systems used for the process including bacteria, algae, and fungi (Das *et. al.*, 2008). Generally, for most of the biomass types, metal uptake declined with the decrease of the pH of the solution from 6.0 to 2.5. Metal removal from solution has been recorded to be negligible at a pH below 2.0. However, a contrasting condition has been reported by RajaRao and Bhargavi (2013), where, metal uptake was augmented with an increase in the pH of the solution from 3.0 to 5.0. pH is also known to be affect the active metal binding sites, solubility of the metal, solution chemistry, activity of the functional groups present on the cell wall and the competition between co-ions. It has been reported that, there is an increase in the density of the negative charge present on the cell surface when there is an increase in the pH of the solution. The reason may be attributed to the process of de-protonation of the active metal binding sites leading to an increased rate of biosorption (Martinez-Juarez *et. al.*, 2012). In addition, optimum value of the pH is a major controlling factor for metal sorption.

Biomass Concentration

Biomass concentration also plays a major role on biosorption of metal from aqueous solution (Modak and Natarajan, 1995). Metal sorption is more with a low density of biomass as compared to higher density at equilibrium; therefore, cellular electrostatic interaction plays a vital role in metal uptake (Gourdon *et. al.*, 1990). When the concentration biomass is low, metal uptake increases, whereas, there is a sharp decline in the rate of biosorption with higher concentration of biomass as crowding leads to interference between the active metal binding sites (Malkov and Nuhoglu, 2005). High biomass concentration often posses restriction over the access of the metal ions to the binding sites (Fourest and Roux, 1992). The initial concentration however, provides an important driving force which helps the metal in solution to overcome all mass transfer resistance between the aqueous and solid phases and hence adsorb to the surface of the biomaterial (Zouboulis *et. al.*, 1997). An optimum concentration of biomass has been reported to facilitate biosorption, however, a concentration higher than that of the optimum may adversely affect biosorption (Gadd and White, 1985).

Metal Ion Concentration

The concentration of the metal ion in solution has an impact on the rate of biosorption. With a high concentration of the solute in solution, the rate of biosorption is augmented. When the initial metal ion concentration is low, due to greater available sites for metal binding the process of sorption becomes independent of the metal concentration in solution. The contrast happens when the metal ion concentration increases in solution, then the attachment of the solute to cell surface is dependent on the initial metal ion concentration. Biosorption of Chromium was reported to increase with an increase in metal concentration from 2 to 6 mM, using *Aspergillus* sp. and *Rhizopus* sp. isolated from waste water (Ahmed *et. al.*, 2005). It is thus absolutely essential to identify the maximum saturation potential of a biosorbent, for which the highest possible intial metal ion concentration has to be determined, for execution of a successful biosorption process.

Pre-treatment of Biomass

Physical pretreatment of the biomass by autoclaving, boiling and drying may interfere with the process of biosorption (Pal *et. al.*, 2006), while treatment with chemicals of alkaline nature has been reported to augment the process of metal sorption (Wang and Chen, 2006). Physical as well as chemical pretreatments are known to affect the cellular permeability and electric potential thereby exposing the metal binding groups making them accessible for the metal ions. Pre treating agents like alkali, acid, detergents and heat have been used for the purpose of cell surface modification (Ahalya *et. al.*, 2003). Presence of both physical and chemical factors can affect biosorption in both positive as well as negative ways. Biosorbents are prepared by pre-treating the biomass by different methods. Effective metal biosorption on to cell surface depends on certain properties of the biomass like, number of active sites present on the cell surface, accessibility of the site for the metal and the chemistry involved in between the metal and the biosorbent (Ahluwalia and Goyal, 2005).

The physical factors affecting the cell-surface modification include heatng/autoclaving, freezing, lyophilization and drying of the cell, whereas, chemicals that affect the surface properties of a cell include detergents, organic solvents, alkali and acids. This kind of pre-treatment modifies the surface of the cell by either masking the functional groups present on the cell surface, removing them completely or by modifying the active metal binding sites (Vieira and Volesky, 2000). Removal of Cd has been studied by nine species of fungi in batch and continuous reactors, where, the pre-treating chemicals were used on the biomass for modification of the active metal binding stuctures viz. carboxyl, amino and phosphate (Huang *et. al.*, 1998). Illhan *et. al.*, (2004) studied the effect of pre-treatment on biosorption capacity of *Penicillium lanosa-coeruleum*, and reported that pre-treatment by heating or by using chemicals like NaOH and detergent augmented biosorption of Pb and Cu whereas, glutaraldehyde pretreatment improved biosorption of Ni. Kogrej and Pavko, (2001), used immobilized *Rhizopus nigricans* for removal of Pb from aqueous solution. Effect of pre-treatment on Pb biosorption capacity was studied by *Aspergillus versicolor*, *Penicillium verrucosum* and *Metarrhizium anisopliae* var. *anisopliae* by Cabuk *et. al.* (2005).

METAL ELUTION POST BIOSORPTION FROM LOADED FUNGAL BIOMASS

Desorption is a process where the metal loaded biomass is eluted and it is made suitable for biosorption again. Desorption is very important when the used biomass is expensive or not readily available. The phenomenon is strongly dependent on the mechanism involved in biosorption and the nature of biosorbent. The eluant should not cause any damage to the biomass; it should also be environmentally compatible and effective. Dilute mineral acids such as sulfuric acid, hydrochloric acid and different organic acids (citric, acetic, gluconic, tartaric) and complexing agents (EDTA, thiosulfate) has been used for the purpose (Akthar *et. al.*, 1996). Interaction of the eluant with the biosorbent material should be restricted as far as possible in order to minimize the damage of the biosorbent and favour its reuse. The technology also ensures the possibility of recovery of valuable metals such as silver, gold, platinum, cadmium etc. which if disposed in to the environment may again get accumulated in the ecosystem and cause the same problem.

Several microbes both in living and non-living forms are capable of sorbing several toxic materials from solution including heavy metals. Use of algae, bacteria and fungi, as biosorbents of heavy metals are ample.

Fungal biomaterial has been proved to be efficient as biosorbent. List of fungal members reported to act as metal biosorbents are presented in Table 2.

Table 2. List of fungal members reported as metal bio-sorbents

Organism	Metal sorbed	Reference
Sacchromyces cerevisae	Mn	Fadel *et. al.*, 2015
S. cerevisae	Pb, Zn, Cr, Co, Cd and Cu	Farhan and Khadom, 2015
Paecilomyces lilacinus *Mucoromycote sp.*	Cd	Xia *et. al.*, 2015
Aspergillus niger	U	Wang *et. al.*, 2017
A. niger, *A. flavus*	Cu and Pb	Iram and Abrar, 2015
A. niger	Cu and Ni	Rao and Bhargavi, 2013
A. niger	Cu and Ni	Javaid *et. al.*, 2011
A. niger	Cr	Sadhana Mala *et.al.*, 2006
A. niger	Cu, Ni, Cr, and Zn	Filipovic-Kovacevic *et. al.*, 2000
A. fumigatus	Pb, Cr, Cd and Zn	Shazia *et. al.*, 2013
A. aculeatus	Cd	Pandey and Banerjee, 2012
A. versicolor	Cr, Cu and Ni	Tastan and Donmez, 2010
A. flavus	Cr	Deepa *et. al.*, 2006
A. foetidus	Cr	Prasenjit and Sumathi, 2005
A. flavus, *A. fumigatus,* *Cladosporium sp.,* *Candida albicans* *Mucor rouxii,* *Helminthosporium sp.,*	Hg	Martinez-Juarez *et. al.*, 2012
Aspergillus sp., *Rhizopus sp.*	Cd, Cr, Co, Cu and Ni	Ahmad *et. al.*, 2005
Aspergillus niger, *Penicillium chrysogenum* *Rhizopus oryzae*	Zn and Co	Tahir *et. al.*, 2017
Aspergillus sp.	Cr and Ni	Congeevaram *et. al.*, 2007
Aspergillus sp.	Cr	Sen and Ghosh Dastidar, 2007
Aspergillus sp.	Cr	Srivastava and Thakur, 2006
P. lanosa-coeruleum	Pb, Cu and Ni	Ilhan *et. al.*, 2004
P. cyclopium	Cu and Co	Tsekova *et. al.*, 2006
Mucor racemosus	Cu, Zn, Pb	El-Morsy *et. al.*, 2013
M. rouxii	Ni, Zn, Pb and Cd,	Yan and Viraraghavan, 2008
Pleurotus eous	Pb, Cr and Ni	Suseem and Mary, 2014
Morganella morganii	Cr	Ergul-Ulger *et. al.*, 2014
Talaromyces helicus	Cu	Romero *et. al.*, 2006
Agaricus microsporous	Cd, Hg and Cu	Garcia *et. al.*, 2005
Phanrochaete chrysosporium	Cr	Marandi, 2011
Phanrochaete chrysosporium	Ni and Pb	Ceribasi and Yetis, 2001
Morteriella sp.	Co	Pal *et. al.*, 2006
Rhizopus oryzae, *R. oligosporus,* *R. arrhizus,* *A. oryzae*	Cd	Yin *et. al.*, 1999
R. arrhizus	Cr and Fe	Sag and Kutsal, 1996
R. stolonifer, *Macrophomina phaseolina*	Pb, Cd, Cu and Zn	Fawzy *et. al.*, 2017
Volvariella volvacea	Pb, Cd, Co, and Cu	Purakayastha and Mitra, 1992

APPLICATIONS OF FUNGAL BIOMATERIAL AS BIOSORBENT

Bioremoval of Cr(VI) has also been executed by development of consortium of the organisms isolated from the Sukinda chromite mines (Samuel *et.al.*, 2012).

COST EFFECTIVENESS OF SORPTION USING FUNGAL BIOMASS

Fungi are naturally available raw material in metal rich soil; they are ubiquitous and play significant role in the ecosystem as a decomposer, nutrient recycler and bio-transformer. The biomass produced in fermentation and other industrial processes can act as a source of very good and cheap biosorbent. A combination of biosorption with metabolically dependent processes like bioreduction and bioprecipitation called intrabiological hybrid technologies can be used in reactor designing. Improved mathematical models justifying the dynamics of the process and computer simulations could also be used in future for better understanding and improvement of the process.

REFERENCES

Ahalya, N., Ramachandra, T. V., & Kanamadi, R. D. (2003). Biosorption of Heavy Metals. *Research Journal of Chemistry and Environment*, *7*, 71–79.

Ahluwalia, S. S., & Goyal, D. (2005). Removal of heavy metals by waste tea leaves from aqueous solution. *Engineering in Life Sciences*, *5*(2), 158–162. doi:10.1002/elsc.200420066

Ahmed, I., Zafar, S., & Ahmed, F. (2005). Heavy metal biosorption potential of *Aspergillus* and *Rhizopus* sp. isolated from waste water treated soil. *Journal of Applied Science & Environmental Management*, *9*, 123–126.

Akhtar, M. N., Sastry, K. S., & Mohan, P. M. (1996). Mechanism of metal ion absorption by fungal biomass. *Biometals*, *9*, 21–28.

Alluri, H. K., Ronda, S. R., Settalluri, V. S., Bondii, J. S., Suryanarayana, V., & Venkateshwar, P. (2007). Biosorption: An eco-friendly alternative for heavy metal removal. *African Journal of Biotechnology*, *6*(25), 2924–2931. doi:10.5897/AJB2007.000-2461

Antunes, W. M., Luna, A. S., Henriques, C. A., & DaCosta, A. C. A. (2003). An evaluation of copper biosorption by a brown seaweed under optimized conditions. *Electronic Journal of Biotechnology*, *6*, 174–184.

Cabuk, A., Ilhan, S., Filik, C., & Caliskan, F. (2005). Pb (2+) biosorption by pretreated fungal biomass. *Turkish Journal of Biology*, *29*, 23–28.

Campbell, P. G. C., Errecalde, O., Fortin, C., Hiriart-Baer, V. P., & Vigneault, B. (2002). Metal bioavailability to phytoplankton applicability of the biotic ligand model. *Comparative Biochemistry and Physiology. Toxicology & Pharmacology: CBP*, *133*(1-2), 189–206. doi:10.1016/S1532-0456(02)00104-7 PMID:12356527

Ceribasi, I. H., & Yetis, U. (2001). Biosorption of Ni(II) and Pb(II) by *Phanerochaete chrysosporium* from a binary metal system- Kinetics. *Water S.A.*, *21*, 15–20.

Congeevaram, S., Dhanarani, S., Park, J., Dexilin, M., & Thamaraiselvi, K. (2007). Biosorption of chromium and nickel by heavy metal resistant fungal and bacterial isolates. *Journal of Hazardous Materials*, *146*(1-2), 270–277. doi:10.1016/j.jhazmat.2006.12.017 PMID:17218056

Das, N., Vimala, R., & Karthika, P. (2008). Biosorption of heavy metals- An overview. *Indian Journal of Biotechnology*, *7*, 159–169.

Deepa, K. K., Sathishkumar, M., Binupriya, A. R., Murugessan, G. S., Swaminathan, K., & Yun, S. E. (2006). Sorption of Cr(VI) from dilute solutions and wastewater by live and pretreated biomass of *Aspergillus flavus*. *Chemosphere*, *62*(5), 833–840. doi:10.1016/j.chemosphere.2005.04.087 PMID:15982702

Dhankar, R., & Hooda, A. (2011). Fungal biosorption- an alternative to meet the challenges of heavy metal pollution in aqueous solutions. *Environmental Technology*, *3*(5), 467–491. doi:10.1080/09593330.2011.572922 PMID:21877528

El-morsy, E. M., Nour El-Dein, M. M., & El-Diamoney, S. M. M. (2013). *Mucor recemosus* as a biosorbant of metal ions from polluted water in northern delta of Egypt. *Mycosphere*, *4*(6), 1118–1131. doi:10.5943/mycosphere/4/6/9

Ergul-Ulger, Z., Ozkan, A. D., Tunca, E., Atasagun, S., & Tekinay, T. (2014). Chromium(VI) biosorption and bioaccumulation by live and acid modified biomass of a novel *Morganella morganii* isolate. *Separation Science and Technology*, *49*(6), 907–914. doi:10.1080/01496395.2013.866681

Fadel, M., Naziha, M. H., Elshafei, M. M., Mostafa, A. H., Ahmed, M. A., & Khater, H. M. (2015). Biosorption of manganese from groundwater by biomass of *Sacchromyces cerevisae*. *Housing and Building National Research Centre Journal*, *13*, 106–113.

Farhan, S. N., & Khadom, A. A. (2015). Biosorption of heavy metals fromaqueous solutions by *Saccharaomyces cerevisae*. *International Journal of Industrial Chemistry*, *6*(2), 119–130. doi:10.100740090-015-0038-8

Filipovic-Kovacevic, Z., Sipos, L., & Briski, F. (2000). Biosorption of chromium, copper, nickel and zinc ions onto fungal pellets of *Aspergillus niger* 405 from aqueous solutions. *Food Technology and Biotechnology*, *38*, 211–216.

Fourest, E., & Roux, J. C. (1992). Heavy metal biosorption by fungal mycelial by-product, mechanisms and influence of pH. *Applied Microbiology and Biotechnology*, *37*(3), 399–403. doi:10.1007/BF00211001

Gadd, G. M. (1990). Heavy metal accumulation by bacteria and other microorganisms. *Experientia*, *46*(8), 834–840. doi:10.1007/BF01935534

Gadd, G. M. (1993). Interactions of fungi with toxic metals. *The New Phytologist*, *124*(1), 25–60. doi:10.1111/j.1469-8137.1993.tb03796.x

Gadd, G. M., & White, C. (1985). Copper uptake by *Penicillium ochro-chloron*: Influence of pH on toxicity and demonstration of energy dependent copper influx using protoplasts. *Journal of General Microbiology*, *131*, 1875–1879.

Garcia, M. A., Alonso, J., & Melgar, M. J. (2005). *Agaricus macrosporus* as a potential bioremediation agent for substrates contaminated with heavy metals. *Journal of Chemical Technology and Biotechnology (Oxford, Oxfordshire), 80*(3), 325–330. doi:10.1002/jctb.1203

Groudev, S. N., & Groudeva, V. I. (1988). Microbiological solubilization of gold. *Symposium of Precious and Rare Metals.*

Gupta, N., Sabat, J., Parida, R., & Kerkatta, D. (2007). Solubilization of tricalcium phosphate and rock phosphate by microbes isolated from chromite, iron and manganese mines. *Acta Botanica Croatica, 66*, 197–204.

Ho, Y. S., & McKay, G. (1999). Pseudo-second order models for adsorption systems. *Journal of Hazardous Materials, 136*(3), 681–689. doi:10.1016/j.jhazmat.2005.12.043 PMID:16460877

Huang, C. P., Westman, D., Quirk, K., & Huang, J. P. (1988). The removal of cadmium (II) from dilute solutions by fungal biomass. *Water Science and Technology, 20*(11-12), 369–376. doi:10.2166/wst.1988.0308

Ilhan, S., Cabuk, A., Filik, C., & Calishan, F. (2004). Effect of pretreatment of biosorption of heavy metals by fungal biomass. *Trakya University Journal of Natural Sciences, 5*, 11–17.

Javaid, A., Bajwa, R., & Manzoor, T. (2011). Biosorption of heavy metals by pretreated biomass of *Aspergillus niger. Pakistan Journal of Botany, 43*, 419–425.

Lagergren, S. (1898). About the theory of so called adsorption of souble substances. *Kungliga Svenska Vetenkapsakademiens Handlingar, 24*, 1–39.

Langmuir, I. (1918). The adsorption of gases on plane surfaces of glass, mica and platinum. *Journal of the American Chemical Society, 40*(9), 1361–1403. doi:10.1021/ja02242a004

Madrid, Y., & Camara, C. (1997). Biological substrates for metal preconcentration and speciation. *Trends in Analytical Chemistry, 16*(1), 36–44. doi:10.1016/S0165-9936(96)00075-1

Mala, J. G. S., Nair, B. U., & Puvanakrishnan, R. (2006). Bioaccumulation and biosorption of chromium by *Aspergillus niger* MTCC 2594. *The Journal of General and Applied Microbiology, 52*(3), 179–186. doi:10.2323/jgam.52.179 PMID:16960334

Malkoc, E., & Nuhoglu, Y. (2005). Investigations of nickel(II) removal from aqueous solutions using tea factory waste. *Journal of Hazardous Materials, 127*(1-3), 120–128. doi:10.1016/j.jhazmat.2005.06.030 PMID:16125314

Marandi, R. (2011). Biosorption of hexavalent chromium from aqueous solution by dead fungal biomass of *Phanerochaete chrysosporium*: Batch and fixed bed studies. *Canadian Journal of Chemical Engineering and Technology, 2*, 8–22.

Martinez-Juarez, V. M., Cardenas-Gonzalez, J. F., Torre-Bouscoulet, M. E., & Acosta-Rodriguez, I. (2012). Biosorption of mercury (II) from aqueous solutions on to fungal biomass. *Bioinorganic Chemistry and Applications, 1*, 1–5. doi:10.1155/2012/156190 PMID:23028382

Modak, J. M., & Natarajan, K. A. (1995). Biosorption of metals using nonliving biomass-a review. *Minerals & Metallurgical Processing*, *12*, 189–196.

Ozturk, A. (2007). Removal of nickel from aqueous solution by the bacterium *Bacillus thuringiensis*. *Journal of Hazardous Materials*, *147*(1-2), 518–523. doi:10.1016/j.jhazmat.2007.01.047 PMID:17320284

Pabst, M. W., Miller, C. D., Dimkpa, C. O., Anderson, A. J., & McLean, J. E. (2010). Defining the surface adsorption and internalization of copper and cadmium in a soil bacterium, *Pseudomonas putida*. *Chemosphere*, *81*(7), 904–910. doi:10.1016/j.chemosphere.2010.07.069 PMID:20797767

Pal, A., Ghosh, S., & Paul, A. K. (2006). Biosorption of cobalt by fungi from serpentine soil of Andaman. *Bioresource Technology*, *97*(10), 1253–1258. doi:10.1016/j.biortech.2005.01.043 PMID:16023340

Pandey, A., & Banerjee, D. (2012). Biosorption of Cd(II) using discarded biomass of *Aspergillus aculeatus* DBF9. *Terestrial and Aquatic Environmental Toxicology*, *6*, 8–13.

Prasenjit, B., & Sumathi, S. (2005). Uptake of chromium by *Aspergillus foetidus*. *Journal of Material Cycles and Waste Management*, *7*(2), 88–92. doi:10.100710163-005-0131-8

Purakayastha, R. P. P., & Mitra, A. K. (1992). Metal uptake by mycelia during submerged Biology growth and by sporocarps of an edible fungus *Volvariella volvacea*. *Indian Journal of Experimental*, *302*, 1184–1187.

RajaRao & Bhargavi. (2013). Studies on biosorption of heavy metals using pretreated biomass of fungal species. *International Journal of Chemistry and Chemical Engineering*, *3*, 171–180.

Romero, M. C., Reinoso, E. H., Urrutia, M. I., & Kiernan, A. M. (2006). Biosorption of heavy metals by *Talaromyces helicus*: A trained fungus for copper and biphenyl detoxification. *Electronic Journal of Biotechnology*, *9*, 221–226. doi:10.2225/vol9-issue3-fulltext-11

Sag, Y., & Kutsal, T. (1996). Fully competitive biosorption of chromium(VI) and iron(III) ions from a binary metal mixture by *Rhizopus arrhizus*: Use of the competitive Langmuir model. *Process Biochemistry*, *31*(6), 573–585. doi:10.1016/S0032-9592(96)00003-9

Sag, Y., & Kutsal, T. (2001). Recent trends in biosorption of heavy metals. *Biotechnology and Bioprocess Engineering; BBE*, *6*(6), 376–385. doi:10.1007/BF02932318

Sag, Y., Tatar, B., & Kutsal, T. (2003). Biosorption of Pb(II) and Cu(II) by activated sludge in batch and continuous-flow stirred reactors. *Bioresource Technology*, *87*(1), 27–33. doi:10.1016/S0960-8524(02)00210-9 PMID:12733571

Sahin, Y., & Ozturk, A. (2005). Biosorption of Cr(VI) ions from aqueous solution by the bacterium *Bacillus thuringiensis*. *Process Biochemistry*, *40*(5), 1895–1901. doi:10.1016/j.procbio.2004.07.002

Sandana Mala, J. G., Unni Nair, B., & Puvanakrishnan, R. (2006). Bioaccumulation and biosorption of chromium by *Aspergillus niger* MTCC 2594. *The Journal of General and Applied Microbiology*, *52*(3), 179–18. doi:10.2323/jgam.52.179 PMID:16960334

Sen, M., & Ghosh Dastidar, M. (2007). Biosorption of Cr(VI) by resting cells of *Aspergillus* sp. *Iranian Journal of Environmental Health Sciences & Engineering*, *4*, 9–12.

Shazia, I. U., Sadia, G. R., & Talat, A. (2013). Bioremediation of heavy metals using isolates of filamentous fungus *Aspergillus fumigatus* collected from polluted soil of Kasur, Pakistan. *International Research Journal of Biological Sciences*, *2*, 66–73.

Srivastava, S., & Thakur, I. S. (2006). Isolation and process parameter optimization of *Aspergillus* sp. for removal of chromium from tannery effluent. *Bioresource Technology*, *97*(10), 1167–1173. doi:10.1016/j.biortech.2005.05.012 PMID:16023341

Suseem, S. R., & Mary Saral, A. (2014). Biosorption of heavy metals using mushroom *Pleurotus eous*. *Journal of Chemical and Pharmaceutical Research*, *6*, 2163–2168.

Tahir, A., Lateef, Z., Abdel-Mageed, A., Sholkamy, E. N., & Mostafa, A. (2017). In vitro compatibility of fungi for the biosorption of zinc(II) and copper(II) from electroplating effluent. *Current Science*, *112*(04), 839–844. doi:10.18520/cs/v112/i04/839-844

Tastan, B. E., & Donmez, G. (2010). Single and binary biosorption of heavy metal ions and a reactive dye by *Aspergillus versicolor. Communications de la Faculty of Sciences University Ankara Series B*, *56*, 13–25. doi:10.1501/Commub_0000000549

Tsekova, K., Christova, D., & Ianis, M. (2006). Heavy metal biosorption sites in *Penicillium cyclopium. Journal of Applied Science & Environmental Management*, *10*(3), 117–121. doi:10.4314/jasem.v10i3.17341

Volesky, B. (1990). *Biosorption of heavy metals*. CRC Press.

Volesky, B., & Holan, Z. R. (1995). Biosorption of heavy metals. *Biotechnology Progress*, *11*(3), 235–250. doi:10.1021/bp00033a001 PMID:7619394

Wang, G., & Fowler, B. A. (2008). Roles of biomarkers in evaluating interactions amoug mixtures of lead, cadmium and arsenic. *Toxicology and Applied Pharmacology*, *233*(1), 92–99. doi:10.1016/j.taap.2008.01.017 PMID:18325558

Wang, J. L., & Chen, C. (2006). Biosorption of heavy metals by *Saccharomyces cerevisiae*: A review. *Biotechnology Advances*, *24*(5), 427–451. doi:10.1016/j.biotechadv.2006.03.001 PMID:16737792

Wang, J. L., Han, Y. J., & Qian, Y. (2000). Progress in metal biosorption by microorganisms. *Journal of Microbiology (Seoul, Korea)*, *27*, 449–452.

Xia, L., Xu, X., Zhu, W., Huang, Q., & Chen, W. (2015). A comparative study on the biosorption of Cd^{2+} onto *Paecilomyces lilacinus* XLA and *Mucoromycote* sp. XLC. *International Journal of Molecular Sciences*, *16*(7), 15670–15687. doi:10.3390/ijms160715670 PMID:26184169

Yan, G., & Viraraghavan, T. (2000). Effect of pretreatment on the biosorption of heavy metals on *Mucor rouxii. Water S.A.*, *26*, 119–123.

Yan, G., & Viraraghavan, T. (2000). Effect of pretreatment on the biosorption of heavy metals on *Mucor rouxii. Water S.A.*, *26*, 119–123.

Yin, P., Yu, Q., Jin, B., & Ling, Z. (1999). Biosorption removal of cadmium from aqueous solution by using pretreated fungal biomass cultured from starch waste water. *Water Research*, *33*(8), 1960–1963. doi:10.1016/S0043-1354(98)00400-X

Zouboulis, A. L., Matis, K. A., & Hancock, I. C. (1997). Biosorption of metals from dilute aqueous solutions. *Seperation and Purification Methodology*, *26*(2), 255–295. doi:10.1080/03602549708014160

Chapter 12
An Eco-Friendly Approach for the Eradication of Heavy Metal Contaminants by Nano-Bioremediation

Chandana Mohanty
https://orcid.org/0000-0002-2107-141X
Kalinga Institute of Industrial Technology (Deemed), India

Sneha Shriparna Satpathy
Kalinga Institute of Industrial Technology (Deemed), India

Sweta Mohanty
Kalinga Institute of Industrial Technology (Deemed), India

ABSTRACT

Nanomaterials manifest distinct physical and chemical properties and have received much attention from researchers in different areas of environmental sciences, specifically in bioremediation. However, bioremediation may not always impart contrivable approaches when subjected to high concentrations of contaminants that are harmful to most microorganisms, which include heavy metals and salts. Nanotechnology on the other hand exhibits a number of potential environmental benefits such as treatment and remediation, pollution prevention, and sensing and detection of pollutants. Nanomaterials used towards bioremediation provide less-toxic effects on indigenous microorganisms and improve microbial biodegradation activity. Credibility of nanotechnology to cut down pollution is in its developing stage and could potentially revolutionize the field of environmental sustainability. Nano-bioremediation is a new emerging technique for remediation of pollutants using biosynthetic nanoparticles.

DOI: 10.4018/978-1-7998-4888-2.ch012

INTRODUCTION

Widespread industrialization, urbanization and modern agricultural practices have unleashed extraordinarily alarming figures of pollutants & contaminants into the environment. An environment can be polluted or contaminated. Elevated concentration of pollutants or contaminants leads to unenviable and abhorrent change in the physical, chemical and biological attributes of the environment. Now the challenging task of the 21st century is to eradicate the contaminants by environmentally sound, viable, non-polluting, and economically compliant technologies (Yadav, Singh, Gupta, & Kumar, 2017). Currently pollution due to heavy metals has caused serious long-term health impact in human beings. The accretion of heavy metals certainly cause various lethal effects in the body by inducing oxidative stress. The main threat to human health is associated with subjection to heavy metals such as Pb, As, Hg, and Cd etc. These metals have been extensively studied and their effect on human health has been regularly reviewed by international bodies such as WHO (Järup, 2003). Heavy metals have a specific density of 5gm/c.m.^{3}. These heavy metals have been used in different areas for thousands of years. Though the adverse effects of heavy metals have been known for quite a long time, exposure to heavy metal still continues and even increasing in some areas. Generation and remissive handling of these toxic materials leads to serious complications, summoning environmental, social and economy of the world. In order to attenuate, such threats efficient strategies need to be implemented to accomplish the environmental sustainability. The field of study focuses on investigating the irradiation of contaminant from the environment is known as environment remediation(Tripathi, Sanjeevi, Anuradha, Chauhan, & Rathoure, 2018).

"Remediate" means to solve the problem and "bioremediation" means the use of different biological agent (like microbes, plants, fungi etc.) to degrade the contaminant, on-site or off-site and convert it into less toxic forms (Rizwan, Singh, Mitra, & Morve, 2014). Although, these bio-based remedial approaches are highly efficient, cost effective and causes less disruption to the environment but its indomitable practices give rise to severe toxic by-product which trigger environmental damage as well as deterioration of biological organism used in this process (Cecchin et al., 2017). To overcome this obstacle, one of an environment-friendly and cost-effective method is the bioremediation of pollutants using bio-nanoparticles. Nano-bioremediation is a highly studied and explode area of remediation of contaminants using nano technologies. Many studies suggested that nano-bioremediation is more effective and significant strategies for the management of heavy metal hazards. In contrast, several conventional strategies like precipitation, ion exchange etc. are also in use for the eradication of these heavy metal but they have certain limitation such as non-specificity, expensive, less efficient and generate toxic by-product (Gaur, Flora, Yadav, Tiwari, & Impacts, 2014). Therefore, nano-bioremediation has the potential not only to reduce the overall costs of cleaning up from large-scale contaminated sites, but it can also reduce clean up time. Further researches in the field of nano-bioremediation should focus on the integrated exploitation of nanoparticles, genetically modified microbes and plants to design eco-friendly, inexpensive, robust and sustainable remediation strategies. In this chapter, we will try to recollect the different aspects of nano-bioremediation with reference to heavy metal contamination and its application to mitigate efficient strategies to achieve environmental sustainability.

Need for New Technique

Traditionally, remediation of heavy-metal-contaminated soils employs either onsite management or excavation and ultimately dumping to a landfill site. This method of subsequent scrapping merely switches the contamination muddle elsewhere in addition to the perils associated with the transportation of contaminated soil and relocation of contaminants and pollutants from the dumped landfill into an adjoining environment. Soil washing for removal of contaminants from the polluted soil is an surrogate approach for the excavation and disposal to landfill. This method is quite expensive and yields a residue rich in heavy metals, which will require further treatment. Accumulation of the heavy metal contaminants leads to several health disorders in human and animals. Therefore, there is a need to mitigate efficient sustainable innovative technologies. The predominant treatment methods such as physical, chemical, physiochemical, and biological methods, though efficient, have several drawbacks which make remediation processes complex (S. R. Kumar & Gopinath, 2017).In addition to this, physical methods used for remediation of metals for example, ion-exchange, membrane filtration, adsorption, precipitation, electro-coagulation, cementation etc. have numerous limitations such as pH sensitivity, use of cost expensive tools, less efficient in higher concentration of contaminants (S. R. Kumar & Gopinath, 2017). Similarly, chemical methods like chelating, reduction, oxidation, chemical washing etc. are followed by using various high cost chemical which deteriorate the environmental balance and can be carcinogenic to mankind (S. R. Kumar & Gopinath, 2017). However, several bio-based approaches are also taking forth in the arena of remediation of heavy metal contaminants. These biological techniques include different living organism to remove the contaminants by utilizing them into their living system (Tripathi et al., 2018; Yadav et al., 2017). The bioremediation solution gained a greater consideration due to its advantages over cost effective and environment friendly approach. Moreover, not all contaminant can be degraded by this method although higher concentration of contaminant led to toxic effect on organism involved in this process (Gaur et al., 2014; Tripathi et al., 2018; Yadav et al., 2017). Among these technologies, phytoextraction is another remediation method involving uptake of heavy metals contaminant through root translocation and also associated with several disadvantages such as metal reclamation, metal phototoxicity, dumping of biomass etc. (S. R. Kumar & Gopinath, 2017; Salt, Smith, & Raskin, 1998). Nevertheless, to enhance the advantageous features of bioremediation strategy, it is combined with nanotechnology and named as nano-bioremediation.

Nano-biotechnology

Nano-biotechnology is a fast-growing discipline of research where small miniature molecules are formulated and applied to study various biological phenomena. It is a multidisciplinary field that brings the science of the almost incomprehensibly small devices closer and closer to reality. The applications of these technology is so numerous that it seems to contribute virtually all fields of science and technology. Nano-biotechnology offers a wide range of uses in medicine and health sciences. Applications towards biological sciences includes various innovations of different drug delivery systems. These drug delivery systems were successfully employed to deliver the therapeutic drug and ultimately helpful to cure various deadly diseases including cancer. Many diseases that do not have cures today may be cured by nanotechnology in the future. These nanoscale drug delivery particles have a wide range of applications as well as enormous potential of being applied for various human healthcare practices (S. R. Kumar & Gopinath, 2017; V. Kumar, Yadav, Biotechnology: International Research in Process, & Technology, 2009; Niemeyer & Medicine, 2006)

Nanoparticle

Nanoparticles are small nanoscale engineered material having at least one dimension sized from 1 to 100 nanometers. Their extreme microscopic dimension, which gives unique advantage that exhibit wide variety of possible applications in biomedical, optical, and electronic fields. They are broadly classified into two categories as organic and inorganic. The organic nanoparticles comprises carbon nanoparticles (fullerenes), while the other class includes magnetic, noble metal (e.g. silver, gold, palladium etc.) and semiconductor (e.g. titanium dioxide and zinc oxide) (Tripathi et al., 2018; Yadav et al., 2017). The formulation of these nanoparticles requires different types of matrix material such as polysaccharides, synthetic polymer and proteins. They are generated by several physical and chemical methods through top down or bottom up system (Garg, Visht, Sharma, & Kumar, 2011).In addition to this, the formulation process of nanoparticle is energy consuming and need various hazardous chemical. In contrast to this, many studies showed the fabrication of nanoparticles by different living organism such as bacteria, algae, fungi, yeast and even from some species of plant (Tripathi et al., 2018; Yadav et al., 2017). This biological synthesizing system of nanoparticles serve as a prospective eco-friendly approach towards attaining a clean and green sustainable remediation procedure.

Salient Characteristics of Nanoparticle

The uniqueness of nanoparticle is based upon its tiny size distribution, shape, specific surface area, topography and aggregation state. Therefore, nanoparticle plays a pivotal role in variety of application including several environmental managerial strategies. The reactivity, mobility, environmental toxicity and persistency are the key features of nanoparticle for assessment of risk related to the ecological disturbance (Ripp & Henry, 2011; Zhuang & Gentry, 2011). Hence the controlled use of these engineered nanoparticle considered as the most promising application for eradication of many hazardous heavy metal contaminant in a clean way. Most of the environmental application of nanotechnology are based on their unique properties of having high surface to mass ratio, increased absorption rate and higher penetration to contamination zone (I. Khan, Saeed, & Khan, 2019).This process involves co-precipitated, aggregation, photo-degradation etc. for the removal of heavy metals such as mercury, lead, thallium, cadmium and arsenic from natural water (I. Khan et al., 2019; Nnaji, Omotugba, Ampitan, Babatunde, & Oluwole, 2019). Due to its exclusively high surface area and efficient binding affinity nanoparticles are found to be an excellent bio-sorbent of various metal contaminants. The adsorption process is considered as an instantaneous remedy for the removal of heavy metal because of its efficient removal mechanism, with no toxic by-product, easy operation and cost effectiveness (S. R. Kumar & Gopinath, 2017; Mahmud, Huq, & Yahya, 2017). In addition to this, several bio-based sorbent materials have been studied with respect to their high capability of waste removal in a wide range of physiological condition. A number of distinctive bio-based adsorbent have been identified which created by various microbes like bacteria, algae and yeast (Yadav et al., 2017). Although they are not yet efficient for the removal of high concertation of heavy metal effluent due to toxic effect of the hazardous contaminants. Therefore, biological adsorbent combined with nanotechnology and form an exclusively promising material with high economic and non-toxic approach towards the remediation of large-scale metal effluent. Moreover, this engineered combinational technique enhances the ability, specificity and rapid removal rate over the traditional adsorption method. In spite of the fact that extensive research had already been done in the arena of bio adsorption but yet this zone needed further more research.

Nano-bioremediation

Nano-bioremediation is an emerging and promising technology for recent era of advancement because of its authentic application in different fields of science (Koul & Taak, 2018). Nano-bioremediation is an integral technique to overcome the disadvantages of conventional methods of remediation and provide a better alternative for the detoxification of environmental problems. This method is concern with the task of removal of organic or inorganic harmful material by lowering the risk of contamination within a short time interval (Singh, Behera, & Kumar, 2020). In addition to this, the method involves three essential features as sensing of pollutant, treatment and remediation and prevention of pollution (Tripathi et al., 2018). Nevertheless, the use of various nanomaterial formulated by using iron, nickel and palladium etc., have characterized as an effective decontaminator for sustainable stabilization of numerous transitional metals such as chromium and arsenic and dehalogenation of persistent organic compounds (Cecchin et al., 2017; Tripathi et al., 2018). The removal of heavy metal is better operated by adsorption process as biosorber are handy, cost effective and have higher efficiency. Some commonly well proved adsorbents are mentioned below.

Classification of Nano-Material Used for Remediation

1. Polymer-based Nanomaterial

Modern era of industrialization led to dispersal of several toxic metal in the environment like Arsenic (As), Lead (Pb), Copper (Cu), Zinc (Zn), Mercury (Hg), Cobalt (Co), Chromium (Cr) etc., (Mahmud et al., 2017). These pollutants are highly toxic and their early exclusion is required which can be effectively done by adsorption on polymer-based nanoparticles. Numerous polymer-based adsorbent materials have been tested to prove the compatibility of these matrix. The significant properties of polymeric nanoparticles depend on the hydrophobicity, electrostatic interactions, bond forming capacity with the ions of the effluents (S. R. Kumar & Gopinath, 2017). Polymer as dendrimers are considered as effective adsorbent for the eradication of heavy metal as well as organic waste. The chemical structure of these dendrimers is made up of hydrophobic internal portion for the attachment of organic compounds and a hydroxyl or amine group on the outer surface for the adsorption of heavy metal. Generally, they are associated with the removal of Cu^{2+} from an aqueous solution by ultrafiltration (S. R. Kumar & Gopinath, 2017). Recently an author evaluated the potential of polyaniline and polypyrrole (PPy), conducting polymeric nanoparticles, in terms of their considerable adsorption efficiency for heavy metals like Chromium, Lead, Zinc, Copper, Nickel etc. from their aqueous solution. The adsorbent material was chemically synthesized by oxidative polymerization in presence of $FeCl_3.6H_2O$, as an oxidant. The surface morphology and other characterization was done by Brunauer-Emmet-Teller (BET) surface analysis, field emission scanning electron microscopy (FESEM) and attenuated total reflectance fourier-transform infrared ATR-(FTIR) spectroscopy. The experiment was done by varying different parameters such as, resident time, pH and quantity of adsorbent used (Mahmud et al., 2017).

2. Metal-based Nanomaterial

As per several literature the metal-based nano-adsorbents are termed as metal organic frameworks (MOFs). They have massive capability to adsorb heavy metal toxicants due to their versatile functional group and tailored sequence. MOFs are designed by combining metal ion precursor (e.g., Fe(III), Cu(II), Ca(II), Al(III), Mg(II), Zn(II), Cd(II), Co(II), Zr(IV), Ln(III), and Ti(III)) and organic ligand (e.g., p-phthalic acid, benzoic acid, pyridine, imidazole, piperazine, amines, carboxylates, sulfonates and phosphates etc.) (Chen et al., 2018; Pi et al., 2018). There successful sorption ability is due to the diverse morphology, enlarged surface area and high porosity (Falcaro et al., 2016; Y. Yang et al., 2017). Metal oxides are much more efficient for the treatment of waste water. Furthermore, scrutinized it for the adsorption of numerous metal ions, comprising Cd(II), Cu(II), Hg(II), La(III), Mn(II) etc. (S. B. Khan, Marwani, Asiri, Bakhsh, & Treatment, 2016). In another research experiment the researcher texturized nano-sized combination of magnetite and TiO_2 nanomaterial for the adsorption of both As(III) and As(V) (Mayo et al., 2007). In addition to this researcher synthesized the Titanium nano-flower with a high surface area and relatively greater adsorption capacity than Titanium nanotube and nanowire. Also, proved its capability for the selective removal of highly toxic metal ion Cd(II) (Huang et al., 2012). Furthermore, researcher synthesized magnesium hydroxide nanotubes to form $Mg(OH)_2/Al_2O_3$ composites array by a standard chemical deposition method inside the pores of porous anodic alumina membranes (AAM) for the elimination of nickel ion from waste water (S. Zhang et al., 2006). Moreover, another experiment was employed by some researcher with the sonochemical synthesis and characterization of ZnO nanomaterial and compared its efficacy from commercially available ZnO(M-ZnO) (Banerjee, Chakrabarti, Maitra, & Dutta, 2012).

3. Iron-based Nanomaterial

Iron based nanomaterial are the best categorized novel nano-sorbent for the exclusion of heavy metal hazard from the polluted environment (**Table 1**). The nanoscale iron material has distinctive notable feature as significant aggregation, co-precipitation of ferrous and ferric salts, easier separation of effluent, re-usability, magnetic property and cost-effectiveness (Cundy, Hopkinson, & Whitby, 2008; Wu et al., 2019). These versatile nanomaterial plays a pivotal role in remediation process by combining with various other adsorbents. Massive examples of combination were demonstrated in different studies. Withal, chitosan-coated Fe_3O_4 nanoparticles were developed by a team of researcher with the reaction of carboxymethylated chitosan and Fe_3O_4 nanoparticles via carbodilmide activation, for the enhanced elimination of arsenic As(III) (Abdollahi, Zeinali, Nasirimoghaddam, Sabbaghi, & Treatment, 2015). In another experiment, researchers insight core shell Fe@ Fe_2O_3 nanowires for the exclusion of anoxic Cr(VI) (Mu, Ai, Zhang, Song, & interfaces, 2015). Researcher synthesized iron-oxide nanoparticles by mediating self-assembly process of ethylene glycol to form a 3D nanostructure and demonstrated the outstanding ability of removal of heavy metals and other contaminants from polluted water (Zhong et al., 2006). Commonly used iron-based nano-adsorbents are nano-sized zero-valent ion (nZVI), iron sulfide nanoparticles, bimetallic Fe nanoparticles, and nano-sized FeO (Ludwig et al., 2007). Amid all, nano-sized zero-valent iron nanoparticles (nZVI) are considered to be thousand times more adaptable due to its high reactivity and larger surface area (Bhandari, 2018). Some more important features of nZVI includes increased transport efficacy, favorable quantum size and deprived standard potential (Tosco, Papini, Viggi, & Sethi, 2014). In context to this, researcher textured nZVI for the arsenic remediation and discussed it as a better agent for sustainable remediation (Kanel, Greneche, Choi, & technology, 2006). Researcher studied the role of nZVI as a good alternative for the abolition of chlorinated compounds, organochloride pesticides, polychlorinated biphenyls, heavy metal ions, and inorganic anions (Karn, Kuiken, & Otto, 2009; Mueller et al., 2012). Mostly iron oxide nanomaterial are constructed for waste water treatment, widely classified in two categories as nano-sorbent and immobilization matrix. Extensive studies have been done to showcase the wide applications of iron-based nanomaterial in waste detoxification while several investigations are yet on experimental stage (S. R. Kumar & Gopinath, 2017). Therefore, furthermore research should be encouraged to elevate the efficiency of this nanomaterial and to make it commercial.

4. Carbon-based Nanomaterial

Carbon-based nanomaterials are considered as a mainstay for the waste treatment regime. A huge number of carbon-based nano-matrix are studied and characterized as nanocrystals and carbon nanotubes to solve a broad range of environmental applications: sorbents, high-flux membranes, depth filters, antimicrobial agents, environmental sensors, renewable energy technologies, and pollution prevention strategies (Mauter, Elimelech, & Technology, 2008). Additionally, iron oxide are combined with activated carbon fibers to form a suitable nano-matrix for the evacuation of arsenic and 17α-ethinyl estradiol (EE2) from water (Hristovski, Nguyen, Westerhoff, & A, 2009). The unique properties of carbon-based nanomaterials are its higher strength, high stiffness, good thermal and electrical conductivities along with tremendous adsorption capacity. Carbon nanotubes have cylindrical pores on their surface from which adsorbent molecules interact. This interaction depends upon pore size and geometry of pores. The cylindrical and spherical pore also facilitate the strong binding of the adsorbent to the matrix. Hence, the carbon nanotubes can adsorb molecules much stronger than activated carbons, which have slit-shaped or wedge-shaped pores (R. T. Yang, 2003). In various previous research these carbon-based nanotubes have been used for the removal of heavy metals like Cr^{3+}, Pb^{2+}, and Zn^{2+} as well as metalloids such as arsenic compounds, organics biological impurities, volatile organic compounds, and dioxins (Li et al., 2003; Rao, Lu, Su, & Technology, 2007). In an experiment the researcher used carbon nanotube sheets for the removal of some divalent heavy metal ions (Cu^{2+}, Zn^{2+}, Pb^{2+}, Cd^{2+}, Co^{2+}) from their aqueous solutions and concluded that these sheets can be employed as adsorbent for water treatment (Tofighy& Mohammadi, 2011). In another study, carbon-encapsulated magnetic nanoparticles are developed via. utilizing a mild-temperature annealing process and justified it as a promising candidate for efficient removal of heavy metal ions from waste water (D. Zhang et al., 2010). A review led the author showcases the multipurpose application of carbon nanotube as sensing and detection of heavy metal. The author also describes the types of carbon nanomaterial as single-walled, multi-walled and nanofiber and high-light the properties, specifically high sorption and pre-concentration of heavy metal ion (Wanekaya, 2011). In another research, researcher textured an advance carbon-based nano-material, termed as carbon nano-onions (CNOs). Like other nano-material CNOs is also found to be synthesized in a cost-effective way using a laser-assisted combustion synthesis process, and characterized for their potential remediation application. CNOs possessed ten times higher sorption capacity than fullerenes for heavy metal pollutants such as Pb^{2+}, Cu^{2+}, Cd^{2+}, Ni^{2+} and Zn^{2+} (Seymour, Su, Gao, Lu, & Li, 2012). Various methods were adopted for synthesis of different biomaterials that have discussed above were summarized in **Table 2**.

5. Single Enzyme-based Nanomaterials

Another effective way of remediation is the use of enzyme-based nanomaterial as they function as biocatalyst in the process of elimination of contaminants. Although, they have some limitation as lack of stability and short catalytic lifetime, their usefulness as cost-effective alternative is inadequate (Bhandari,

2018). Therefore, to increase the stability, longevity, and reusability of the enzymes, they supposed to attach to magnetic iron nanoparticles. If enzymes are attached to the magnetic iron nanoparticles, then we can easily separate the enzymes from reactants or products by applying a magnetic field. The first single enzyme-based nanoparticles (SENs) were assembled by two researchers, using chymotrypsin as a model enzyme (Kim et al., 2005). In addition to this, a researcher used two different catabolic enzymes, trypsin and peroxides, to form core shell magnetic nanoparticles (MNPs). MNP enzyme conjugates were found to be more stable, efficient, and economical (Rizwan et al., 2014). The enzyme-based nanomaterial can also be used to degrade as well as detect higher concentration of pesticide. The combination of biosensor and nanoparticle enhance the respons, sensitivity and selectivity for analytes. In a study a researcher fabricated enzyme-based biosensor using myriad nanomaterial for the detection of pesticide {Rawtani, 2018 #77}.

Production/Fabrication of Biogenic Nanoparticles

Traditionally, nanoparticles were produced only by physical and chemical methods. The call for biosynthesis of nanoparticles was fostered due to the steep exorbitant budget of physical and chemical processes. In the hunt for cost-effective pathways for nanoparticle synthesis, both microorganisms and then plant and plant extracts were explored for their synthesis. Surprisingly, it is possible to acquire nanoparticles from simple microorganisms, like bacteria and up to more evolved plants. Biosynthesis of nanoparticles is a bottom-up approach, wherein, oxidation/reduction is the main reaction taking place. The reason behind reduction of metal compounds into their respective nanoparticles is because of the microbial enzymes or the plant phytochemicals with antioxidant and/or their reducing properties. By the amalgamation of organic and inorganic methods, synthesis of nanoparticles of interest in nanotechnology (design of nano-devices) and medicine (controlled-release of drugs) can be produced. The biogenetic production is now of high interest due to simplicity of the procedures and their versatility (Popescu, Velea, Lőrinczi, & biostructures, 2010).

It is quite surprising that from simple microbes and bacteria up to more evolved plants it is possible to get nanoparticles [1]. The production of nanoparticles is environmentally friendly because this involves natural phenomena that take place in the biological systems. Moreover, the biologically fabricated nanostructures offer substantially different properties: good adhesion, tribologically good properties, optical and electrical properties of high interest in optoelectronics.

With reference to nanoparticle production the term "biogenic" encloses a huge platter of divergent procedures and techniques employed for the production of nanoparticles, stretching from the utilization of plant or cell extracts to diminish dissolved metals into nanoparticles, to the use of microorganisms and their innate abilities for the production of such nanomaterial. The amalgamation of the green biogenic route has various benefits over the traditional nanoparticles fabrication processes. The procedure is lucid, comparatively less complicated, can be scaled up, and is eco-friendly. The most appealing characteristic feature of the biogenic synthesis route is that it can produce non-toxic nanoparticles free from the noxious imperishable commercial chemicals and surfactants that are routinely used in the standardized physical and chemical fabrication units.

By Microbes (Bacteria, Yeast, Fungi)

In general, bioremediation is established on the co-metabolism action of one organism or a consortium of microorganisms. The transformation of contaminants in this process confers a little potential or zero benefit to the cell, and therefore this technique is described as no beneficial biotransformation. Through numerous studies it has been revealed that many organisms like prokaryotes and eukaryotes, have a natural ability to biosorb toxic heavy metal ions.{Coelho, 2015 #78} Single-celled microbes like Diatoms (brown algae) are able to proactively and discreetly design and influence natural nanostructures formation. They have a silica exoskeleton named Frustules, which consists of well-organized SiO2 nanoparticles (size 50-100 nm). Diatoms are unicellular photosynthetic eukaryotic algae which produce intricately structured cell walls made up of nano-patterned silica. Researchers have shown that the freshwater diatom, *stauroneis sp.*, can be used to manufacture Silicon-Germanicum nanoparticles (MubarakAli et al., 2013). These nanoparticles are formed from naturally occurring precursors, under a few hour time limit and within the room temperatures (Popescu et al., 2010).

By Bacteria

Bacteria is predominantly the most favored model organism for the research of biogenic nanoparticle fabrication and its industrial application due to their ease of growth and multiplication. Scientists can skill-fully manipulate them genetically. Bacteria have the potential ability to mobilize and immobilize metals. Certain bacteria can reduce metal ions and are also capable to precipitate metals at nano-meter scale. Bacteria are contemplated as a potential 'bio-factory' for the synthesis of nanoparticles like gold, silver, platinum, palladium, titanium, titanium dioxide, magnetite, cadmium sulphide and so forth. The employment of bacteria as a source of enzymes that can catalyze specific reactions leading to inorganic nanoparticles is a new coherent strategy for the biosynthesis and utilization of enzymes, microbial enzymes, vitamins, polysaccharides, biodegradable polymers, microorganisms, and biological systems for fabrication of nanoparticles.The typical characteristic features of nanoparticles are guarded by providing favourable conditions for the considerable significant parameters which affect the growth and development of the micro-organisms, and their cellular activities and enzymatic processes (optimization of growth and reaction conditions).Thus, more elaborate studies are needed to understand the exact mechanisms of reaction and identify the enzymes and proteins which involve nanoparticle biosynthesis. The large-scale production of nanoparticles using bacteria is fetching huge attention as it does not require any hazardous, toxic and expensive chemical ingredients for synthesis and stabilization processes (Yadav et al., 2017).

By Fungi

Fungi are more efficient for biological synthesis of nanomaterials because of their forbearance to bioaccumulation of metals, high binding ability and intercellular uptake makes. They have been extensively employed for the biosynthesis of nanoparticles. Besides being monodisperse, nanoparticles with well-defined dimensions can be obtained using fungi (Yadav et al., 2017).Compared to bacteria fungi are an excellent source of numerous extracellular enzymes regulating nanoparticle synthesis. Fungi could be employed as a high-yielding source for the mass production of nanoparticles. This is because fungi secrete substantial amount of proteins which directly transcribes to elevated productivity of nanoparticles (Mohanpuria, Rana, & Yadav, 2008).

Interpreting the nature of the biogenic nanoparticle is equally important. Microbiological methods generate nanoparticles at a much slower rate than that observed when plant extracts are used.The production of bioactive nanoparticles from lichen fungi (*Usneea longissima*) in the culture conditions has been documented (Shahi & Patra, 2003). In the biogenesis of metal nanoparticles by a fungus, enzymes produced reduce a salt to its metallic solid nanoparticles through the catalytic effect (Oksanen, Pere, & Paavilainen, 2000). This is one of the crucial stumbling block for the biogenesis of nanoparticles using microbes and must be amended if it aims to compete with other classical methods. For commercial applications, fungi should have certain characteristic features which comprises elevated production of specific enzymes or metabolite, high growth rate, ease in handling large-scale production and cost-effective manufacturing procedures and techniques which dominates over other approaches (Vahabi, Mansoori, & Karimi, 2011). Fungi stands out over other biological systems due to its wide diversity, easy availability, simple culture techniques, reducing the overall time and increasing cost-effectiveness. This, in turn provides an eco-friendly approach for the production of nanoparticles. Genetic engineering techniques can also be applied to refine and enhance the particle properties in near future (Saxena, Sharma, Gupta, & Singh, 2014; Yadav et al., 2017).

By Yeast

Yeast has been maneuvered successfully in the synthesis of Cadmium sulphide (CdS) and Lead sulphide (PbS) nanoparticles. Researcher have shown that *Torilopsis* species is able to synthesize nanoscale PbS (intracellularly) when exposed to aqueous Pb^{2+} ions (Kowshik et al., 2002). Researchers have also established to have found the CdS Quantum dots, formed in *Schizosacharomyces* pombe yeast cells (Kowshik et al., 2002). Recently, elevated proportions of silver (Ag) nanoparticles have been derived using silver tolerant yeast strains MKY3 (Kowshik et al., 2002).

By Plant (Phyto-bioremediation)

With regard to the current concerns involving the environmental contamination, it has led to the development of pertinent technologies to evaluate and analyze the existence and mobility of metals in soil, water, and wastewater. At present, phytoremediation is one of the most potent low-cost technological practice to extract or remove inactive metals and metal pollutants from contaminated soil. Phytoremediation is the employment of plants for the removal contaminants from soil, sediments, and water. It can be also termed as "botanical remediation" or "green remediation". Plants have the natural ability to accumulate, degrade, or concentrate the contaminants from the environment (i.e., soil, water and air). The eradication of toxic metals and other pollutants is the major objective for the implementation of phytoremediation. Plants have extra-ordinary metal-accumulating capacity and hence are also known as hyper-accumulator plants. These plants which are distinguished as metal hyper-accumulators and wild, are able to clean-up contaminating elements which are 10-500 times higher as compared to the ones that are cultivated. Phytoremediation takes the advantage of the extra-ordinarily selective uptake potential of the plant root systems, with other abilities of the entire plant body like translocation, bioaccumulation, and contaminant degradation. There exists numerous species of plants that have been efficient in absorbing contaminants such as lead, cadmium, chromium, arsenic, and various radionuclides from soils. Phytoremediation procedures can be classified or grouped as phyto-extraction, phyto-degradation, phyto-stabilization, phyto-volatilization, rhizofiltration and rhizodegradation.{Salt, 1998 #14} One of such phytoremediation techniques, phyto-extraction, can be used against the eradication of heavy metals from soil using its capability to uptake metals which are essential for plant growth (Fe, Mn, Zn, Cu, Mg, Mo, and Ni). Some metals with unknown biological function (Cd, Cr, Pb, Co, Ag, Se, and Hg) can also be accumulated. Phytoremediation techniques have high potential in the sterilization of the areas that have medium level of contamination and have limited risks.{Srivastav, 2018 #79}

Nano-phytoremediation (integration of nanomaterials along with phytoremediation) is an environment friendly technology which incorporates the utilization of nanoscale materials for the adsorption of pollutants and their degradation and plants used to accumulate the degraded but still pollution-prone matter. It integrates nanotechnology and phyto-technology for the "remediation" of contaminants or pollutants. Use of nanomaterial with phytoremediation can have the potential to increase the decontamination efficiency and turnover than the other phytoremediation process alone. In the present day biogenesis of nanoparticles by plants is garnering much significance due to its less complicated and simple procedure, absence of toxic substances and products, and occurrence of natural capping agents (Gurunathan et al., 2009). The factors which advocate the beneficial utilization of plants for the synthesis of nanoparticles is that they are easily available, safe to handle and possess a broad variability of metabolites that may aid in reduction. Water soluble phytochemicals need brief incubation time for the reduction of metal ions, but on the other hand fungi and bacteria require a much prolonged incubation period. Hence, in comparison to bacteria and fungi, plants are better suitors for the biosynthesis of nanoparticles. Using techniques like plant tissue culture and downstream processing procedures, metallic as well as oxide nanoparticles can be synthesized on an industrial scale once issues such as the metabolic status of the plant are properly addressed. It is evident from compiled information that effect of nanoparticles varies from plant to plant and depends on their mode of application, size, and concentration (Yadav et al., 2017). Though nano-phytoremediation provides numerous benefits, but it also brings in certain drawbacks and limitation, which should be contemplated while seeking to apply this technology for human well-being. If low cost is an edge, then the time required to yield the results can be prolonged and extensive. The concentration of the pollutants or contaminants and the presence of other toxins should be under the tolerance limits of the plant to be employed. At the same time selection of the particular plants with the potential for the remediation of the innumerable heterogonous contaminants is not easy. These constraints and the probable chances of these plants entering in the food chains, should be contemplated, evaluated and transcripted when implementing this technology.

Nano-phytoremediation is a technique which involves nanotechnology and phyto-technology for the remediation of environmental pollutants.

CONCLUSION

Heavy metals (mercury, copper, arsenic, cadmium, chromium, etc.) are predominantly considered as toxic materials and need immediate remediation. Emerging applications of nanomaterial will endeavor to find an effective remediation solution for removing these heavy metals. The nature which comprises of plants, algae, fungi, yeast etc., conducts itself like a huge enormous "bio-laboratory" that contains a large reserve of biomolecules. These naturally occurring biomolecules have been reported to play a proactive part in the synthesis of nanoparticles with distinct shapes and sizes thereby propelling towards the designing of greener, safe and environmentally benign protocols for the biogenesis of nanoparticles. It is expected that applications of nanoparticles will increase inevitably, because of its potent future prospective, and it will play a pivotal role in sustainable development. Recently, the utilization of microorganisms and plants for the production and manufacturing of nanoparticles has evidently established its inherent potential. Simple bacteria to complex eukaryotes have been employed for the synthesis of nanoparticles of desired size and shape. The biogenesis of nanoparticle has proved to be efficiently feasible, eco-friendly, stable, non-toxic and evidently cost effective. The fabrication of green nanoparticles are well suited for the large scale production, with nontoxic plant material that are easily biodegradable. Other biogenically synthesized metals and its oxide nanoparticles have constructively convincing roles in wellbeing of mankind. Thus nano-bioremediation has also been significantly called as a "biological response to environmental abuse". However, several threats correlated prospectively with the diverse nanomaterial is an area to worry about. The various ecological interference and associated health risks may limit the widespread applications of nanomaterial for environmental remediation. Hence, to make this technology more beneficial than damaging, close surveillance and arbitration measures needs to be administered. Though the sphere is still in its infancy, but successful outcomes of late in the genesis and application claim its bright future.

REFERENCES

Abdollahi, M., Zeinali, S., Nasirimoghaddam, S., Sabbaghi, S. J. D., & Treatment, W. (2015). *Effective removal of As (III) from drinking water samples by chitosan-coated magnetic nanoparticles*. Academic Press.

Banerjee, P., Chakrabarti, S., Maitra, S., & Dutta, B. K. J. U. s. (2012). *Zinc oxide nano-particles–sonochemical synthesis, characterization and application for photo-remediation of heavy metal*. Academic Press.

Bhandari, G. (2018). Environmental Nanotechnology: Applications of Nanoparticles for Bioremediation. In Approaches in Bioremediation (pp. 301-315). Springer.

Cecchin, I., Reddy, K. R., Thomé, A., Tessaro, E. F., Schnaid, F. J. I. B., & Biodegradation. (2017). *Nanobioremediation: Integration of nanoparticles and bioremediation for sustainable remediation of chlorinated organic contaminants in soils*. Academic Press.

Chen, K., Sun, Z., Fang, R., Shi, Y., Cheng, H. M., & Li, F. J. A. m. (2018). Metal–Organic Frameworks (MOFs)-. *Derived Nitrogen-Doped Porous Carbon Anchored on Graphene with Multifunctional Effects for Lithium–Sulfur Batteries.*, *28*(38), 1707592.

Cundy, A. B., Hopkinson, L., & Whitby, R. L. J. S. e. (2008). Use of iron-based technologies in contaminated land and groundwater remediation. *RE:view*, *400*(1-3), 42–51. PMID:18692221

Falcaro, P., Ricco, R., Yazdi, A., Imaz, I., Furukawa, S., Maspoch, D., . . . Doonan, C. J. J. C. C. R. (2016). *Application of metal and metal oxide nanoparticles@ MOFs*. Academic Press.

Garg, A., Visht, S., Sharma, P. K., & Kumar, N. J. D. P. S. (2011). *Formulation, characterization and application on nanoparticle: a review*. Academic Press.

Gaur, N., Flora, G., Yadav, M., Tiwari, A. J. E. S. P., & Impacts. (2014). *A review with recent advancements on bioremediation-based abolition of heavy metals*. Academic Press.

Gurunathan, S., Kalishwaralal, K., Vaidyanathan, R., Venkataraman, D., Pandian, S. R. K., Muniyandi, J., . . . Biointerfaces, S. B. (2009). *Biosynthesis, purification and characterization of silver nanoparticles using Escherichia coli*. Academic Press.

Hristovski, K. D., Nguyen, H., Westerhoff, P. K. J. J. o. E. S., & A, H. P. (2009). *Removal of arsenate and 17α-ethinyl estradiol (EE2) by iron (hydr) oxide modified activated carbon fibers*. Academic Press.

Huang, J., Cao, Y., Liu, Z., Deng, Z., Tang, F., & Wang, W. J. C. E. J. (2012). *Efficient removal of heavy metal ions from water system by titanate nanoflowers*. Academic Press.

Järup, L. J. B. m. b. (2003). *Hazards of heavy metal contamination. Academic* Press.

Kanel, S. R., Greneche, J.-M., Choi, H. J. E. s., & Technology. (2006). *Arsenic (V) removal from groundwater using nano scale zero-valent iron as a colloidal reactive barrier material*. Academic Press.

Karn, B., Kuiken, T., & Otto, M. J. E. h. p. (2009). *Nanotechnology and in situ remediation: a review of the benefits and potential risks*. Academic Press.

Khan, I., Saeed, K., & Khan, I. J. A. J. o. C. (2019). *Nanoparticles: Properties, applications and toxicities*. Academic Press.

Khan, S. B., Marwani, H. M., Asiri, A. M., Bakhsh, E. M. J. D., & Treatment, W. (2016). *Exploration of calcium doped zinc oxide nanoparticles as selective adsorbent for extraction of lead ion*. Academic Press.

Kim, B. C., Nair, S., Kim, J., Kwak, J. H., Grate, J. W., Kim, S. H., & Gu, M. B. J. N. (2005). *Preparation of biocatalytic nanofibres with high activity and stability via enzyme aggregate coating on polymer nanofibres*. Academic Press.

Koul, B., & Taak, P. (2018). Chemical Methods of Soil Remediation. In *Biotechnological Strategies for Effective Remediation of Polluted Soils* (pp. 77–84). Springer. doi:10.1007/978-981-13-2420-8_4

Kowshik, M., Ashtaputre, S., Kharrazi, S., Vogel, W., Urban, J., Kulkarni, S. K., & Paknikar, K. J. N. (2002). *Extracellular synthesis of silver nanoparticles by a silver-tolerant yeast strain MKY3*. Academic Press.

Kumar, S. R., & Gopinath, P. (2017). Nano-bioremediation applications of nanotechnology for bioremediation. In *Handbook of Advanced Industrial and Hazardous Wastes Management* (pp. 27–48). CRC Press. doi:10.1201/9781315117423-2

Kumar, V., Yadav, S. K. J. J. o. C. T., Biotechnology: International Research in Process, E., & Technology, C. (2009). *Plant-mediated synthesis of silver and gold nanoparticles and their applications*. Academic Press.

Li, Y.-H., Wang, S., Luan, Z., Ding, J., Xu, C., & Wu, D. J. C. (2003). *Adsorption of cadmium (II) from aqueous solution by surface oxidized carbon nanotubes*. Academic Press.

Ludwig, R. D., Su, C., Lee, T. R., Wilkin, R. T., Acree, S. D., Ross, R. R., . . . technology. (2007). *In situ chemical reduction of Cr (VI) in groundwater using a combination of ferrous sulfate and sodium dithionite: a field investigation*. Academic Press.

Macaskie, L., Creamer, N., Essa, A., Brown, N. J. B., & bioengineering. (2007). *A new approach for the recovery of precious metals from solution and from leachates derived from electronic scrap*. Academic Press.

Mahmud, H., Huq, A., & Yahya, R. (2017). *Polymer-based adsorbent for heavy metals removal from aqueous solution*. Paper presented at the IOP Conference Series: Materials Science and Engineering. 10.1088/1757-899X/206/1/012100

Mauter, M. S., Elimelech, M. J. E. S., & Technology. (2008). *Environmental applications of carbon-based nanomaterials*. Academic Press.

Mayo, J., Yavuz, C., Yean, S., Cong, L., Shipley, H., Yu, W., . . . Materials, T. o. A. (2007). *The effect of nanocrystalline magnetite size on arsenic removal*. Academic Press.

Mohanpuria, P., Rana, N. K., & Yadav, S. K. J. J. o. n. r. (2008). *Biosynthesis of nanoparticles: technological concepts and future applications*. Academic Press.

Mu, Y., Ai, Z., Zhang, L., Song, F. J. A. a. m., & interfaces. (2015). *Insight into core–shell dependent anoxic Cr (VI) removal with Fe@ Fe2O3 nanowires: indispensable role of surface bound Fe (II)*. Academic Press.

MubarakAli, D., Arunkumar, J., Nag, K. H., SheikSyedIshack, K., Baldev, E., Pandiaraj, D., . . . Biointerfaces, S. B. (2013). *Gold nanoparticles from Pro and eukaryotic photosynthetic microorganisms—Comparative studies on synthesis and its application on biolabelling*. Academic Press.

Mueller, N. C., Braun, J., Bruns, J., Černík, M., Rissing, P., Rickerby, D., . . . Research, P. (2012). *Application of nanoscale zero valent iron (NZVI) for groundwater remediation in Europe*. Academic Press.

Nancharaiah, Y., Mohan, S. V., & Lens, P. J. B. t. (2016). *Recent advances in nutrient removal and recovery in biological and bioelectrochemical systems*. Academic Press.

Nnaji, J., Omotugba, S. K., Ampitan, A., Babatunde, K., & Oluwole, E. J. F. J. O. S.-I. (2019). Article. *Review of Nanomaterials for Remediation of Petroleum Impacted Soil and Water*, *3*(4), 276–284.

Pi, Y., Li, X., Xia, Q., Wu, J., Li, Y., Xiao, J., & Li, Z. J. C. E. J. (2018). *Adsorptive and photocatalytic removal of Persistent Organic Pollutants (POPs) in water by metal-organic frameworks (MOFs)*. Academic Press.

Popescu, M., Velea, A., Lőrinczi, A. J. D. j. o. n., & biostructures. (2010). *Biogenic production of nanoparticles*. Academic Press.

Rao, G. P., Lu, C., Su, F. J. S., & Technology, P. (2007). *Sorption of divalent metal ions from aqueous solution by carbon nanotubes: a review*. Academic Press.

Ripp, S., & Henry, T. B. (2011). *Biotechnology and nanotechnology risk assessment: minding and managing the potential threats around us*. ACS Publications.

Rizwan, M., Singh, M., Mitra, C. K., & Morve, R. K. J. J. o. N. (2014). *Ecofriendly application of nanomaterials: nanobioremediation*. Academic Press.

Saxena, J., Sharma, M. M., Gupta, S., & Singh, A. J. W. J. P. S. (2014). *Emerging role of fungi in nanoparticle synthesis and their applications*. Academic Press.

Seymour, M. B., Su, C., Gao, Y., Lu, Y., & Li, Y. J. J. o. N. R. (2012). *Characterization of carbon nano-onions for heavy metal ion remediation.* Academic Press.

Shahi, S. K., & Patra, M. J. R. o. A. M. S. (2003). *Microbially synthesized bioactive nanoparticles and their formulation active against human pathogenic fungi*. Academic Press.

Sharma, S., Tiwari, S., Hasan, A., Saxena, V., & Pandey, L. M. (2018). Recent advances in conventional and contemporary methods for remediation of heavy metal-contaminated soils. *3 Biotech, 8*(4), 216. doi:10.100713205-018-1237-8

Singh, R., Behera, M., & Kumar, S. (2020). Nano-bioremediation: An innovative remediation technology for treatment and management of contaminated sites. In Bioremediation of Industrial Waste for Environmental Safety (pp. 165-182). Springer.

Tofighy, M. A., & Mohammadi, T. J. J. o. h. m. (2011). *Adsorption of divalent heavy metal ions from water using carbon nanotube sheets*. Academic Press.

Tosco, T., Papini, M. P., Viggi, C. C., & Sethi, R. J. J. o. c. p. (2014). *Nanoscale zerovalent iron particles for groundwater remediation: a review*. Academic Press.

Tripathi, S., Sanjeevi, R., Anuradha, J., Chauhan, D. S., & Rathoure, A. K. (2018). Nano-Bioremediation: Nanotechnology and Bioremediation. In Biostimulation Remediation Technologies for Groundwater Contaminants (pp. 202-219). IGI Global.

Vahabi, K., Mansoori, G. A., & Karimi, S. J. I. J. (2011). *Biosynthesis of silver nanoparticles by fungus Trichoderma reesei (a route for large-scale production of AgNPs)*. Academic Press.

Wanekaya, A. K. J. A. (2011). *Applications of nanoscale carbon-based materials in heavy metal sensing and detection*. Academic Press.

Wu, Y., Pang, H., Liu, Y., Wang, X., Yu, S., Fu, D., . . . Wang, X. J. E. p. (2019). *Environmental remediation of heavy metal ions by novel-nanomaterials: a review*. Academic Press.

Yadav, K., Singh, J., Gupta, N., & Kumar, V. J. J. M. E. S. (2017). *A review of nanobioremediation technologies for environmental cleanup: a novel biological approach.* Academic Press.

Yang, R. T. (2003). *Adsorbents: fundamentals and applications*. John Wiley & Sons. doi:10.1002/047144409X

Yang, Y., Dong, H., Wang, Y., Wang, Y., Liu, N., Wang, D., & Zhang, X. J. I. C. C. (2017). *A facile synthesis for porous CuO/Cu2O composites derived from MOFs and their superior catalytic performance for CO oxidation*. Academic Press.

Zhang, D., Wei, S., Kaila, C., Su, X., Wu, J., Karki, A. B., . . . Guo, Z. J. N. (2010). *Carbon-stabilized iron nanoparticles for environmental remediation*. Academic Press.

Zhang, S., Cheng, F., Tao, Z., Gao, F., Chen, J. J. J. o. A., & Compounds. (2006). *Removal of nickel ions from wastewater by Mg (OH) 2/MgO nanostructures embedded in Al2O3 membranes*. Academic Press.

Zhong, L. S., Hu, J. S., Liang, H. P., Cao, A. M., Song, W. G., & Wan, L. J. J. A. M. (2006). *Self-Assembled 3D flowerlike iron oxide nanostructures and their application in water treatment*. Academic Press.

Zhuang, J., & Gentry, R. W. (2011). Environmental application and risks of nanotechnology: a balanced view. In *Biotechnology and Nanotechnology Risk Assessment: Minding and Managing the Potential Threats around Us* (pp. 41–67). ACS Publications. doi:10.1021/bk-2011-1079.ch003

APPENDIX

Table 1. List of Pollutants Remediated by Iron-based Nanoparticles

Carbon tetrachloride	Chrysoidine	Cis-Dichloroethene
DChloroform	Tropaeolin	1,1-Dichloroethane
Dichloromethane	Acid orange	PCBs
Hexachlorobenzene	Mercury	Dioxins
Pentachlorobenzene	Nickle	Pentachlorophenol
Tetrachlorobenzene	Silver	TNT
Dichlorobenzene	Bromoform	Dichromate
Chlorobenzene	Dibromochloromethane	Arsenic
DDT	Tetrachloroethene	Perchlorate
Lindane	Trichloroethene	Nitrate
Orange II	Acid red	Vinyl chloride
Chloromethane	Cadmium	NDMA
Trichlorobenzene		

Table 2. Type of Nanomaterials and Their Synthesis Methods and Examples. Adopted from Rizwan et al., 2014

Nanomaterials	Synthesis methods	Examples
Metal nanoparticles	Photochemical Electrochemical Biochemical Thermochemical	Pt, Rh, Pd, Ir, Ag, Au, Cu, Co, Ni, FeNi, Cu Au, CoNi, CdTe, CdSe, ZnS
Carbon nanomaterial	Arc-discharge Laser ablation Chemical vapor deposition	Cylindrical-nanotube (SWNT, MWNT), fullerenes
Metal-oxide nanoparticles	Hydrothermal Solvothermal Sol-gel Reverse micelles method Electrochemical deposition	ZnO, $Fe2O_3$, Fe_3O_4, MgO, $BaCO_3$, $BaSO_4$, TiO_2
Polymer nanomaterial	Electrochemical polymerization	Nanowire of polypyrrole, polyaniline, poly (3,4 ethylenedioxythiophane), dendrimers (PAMAM)

Chapter 13
Sustainable Treatment of Landfill Leachate Using Constructed Wetlands:
An Eco-Friendly Approach

Vivek Rana
https://orcid.org/0000-0001-9442-4805
Central Pollution Control Board, Ministry of Environment, Forest, and Climate Change, Government of India, Delhi, India

ABSTRACT

Sanitary landfilling is the major method of disposal of municipal solid waste (MSW) in developing countries. The disposal of MSW in landfills generates a large amount of highly toxic leachate, which has high potential hazards for the public, flora, fauna health and ecosystems. Advanced leachate treatment systems using biological and chemical treatment methods are recently implemented in developed countries, but high investment and operating costs restricted their application in most of the developing countries. To overcome this problem, an alternative sustainable treatment technology such as phytoremediation could be beneficial. The constructed wetland treatment system is an economical alternative for leachate treatment using local resources and is an energy-efficient technology. These green systems utilize anaerobic and aerobic reactions to break down, immobilize, or incorporate organic substances and other contaminants from polluted effluent. This chapter highlights the recent advances in the treatment of landfill leachates using constructed wetlands.

INTRODUCTION

Increased urbanization and industrialization lead to a substantial surge in the amount and diversity of solid waste (Khandelwal et al., 2019). The generation of municipal solid waste (MSW) attributes to unplanned development and migration of rural population to urban areas (Aluko et al., 2003; Oloruntade et al., 2013). Solid waste disposal in landfills is a widely adopted method as it is cost-effective but it also

DOI: 10.4018/978-1-7998-4888-2.ch013

creates environmental hazard due to production of leachate (Ziyang et al., 2009). After disposal of MSW in a landfill, the moisture mixes with refuse and produces dark-coloured liquid called leachate which comprise of elevated concentrations of particular contaminants. The high concentration of pollutants in leachate can be attributed to dissolution of waste in a liquid phase, and thereafter its accumulation and percolation (Aziz et al., 2011). Leachate composition is complex which mainly consists of heavy metals, organic matter (biodegradable and refractory to biodegradation), chlorinated organic and inorganic salts, and inorganic compounds (Renou et al., 2008). In a landfill, fresh leachate comprise of high chemical load [high biochemical oxygen demand (BOD_5) and high chemical oxygen demand (COD)] with acidic pH however, with time, it becomes stable. Ammonia is a major pollutant which persists for a long time span in a landfill which does not decrease with the emission of landfill gases (Kjeldsen et al., 2002). The elevated levels of ammonia may diminish biological processes, speed up eutrophication, and reduce level of dissolved oxygen in leachate (Mojiri et al., 2016). In a landfill, the prime objective is to stabilize the MSW using natural metabolic pathways. The various hazards that can be imposed on the environment due to the production of leachate are: (i) contamination of groundwater due to percolation of leachate; (ii) contamination of surface water sources due to unsuitable selection of landfill sites; and (iii) release of greenhouse gases into atmosphere (Bulc, 2006).

Conventionally, treatment of leachate is done through physical, chemical and biological methods (Aziz et al., 2012). Phytoremediation proves advantageous over conventional wastewater treatment technologies due to its cost-effective and environment-friendly nature (Akratos et al., 2018). Phytoremediation is being adapted widely to treat a wide array of inorganic and organic pollutants (McCutcheon and Schnoor, 2004). Constructed wetlands (CWs) utilizes phytoremediation technology along with other physical and chemical processes to treat leachate emanating from MSW landfills, wastewater emanating from pulp and paper industry, textile industry, pharmaceutical industry, sugarcane industry, tannery industry, winery industry, etc. (Davies et al., 2009; Arivoli et al., 2015; Madera-Parra et al., 2015; Vymazal, 2017; Akratos et al., 2018; Sanchez-Galvan and Bolanos-Santiago, 2018; Flores et al., 2019). This chapter highlights the utilization of constructed wetland systems for the treatment of toxic leachate emanating from municipal solid waste landfills.

Municipal Solid Waste (MSW) Generation in India

A swift increase in the quantity of MSW generated has been observed due to increasingly affluent lifestyles, and growth of industrial and commercial sectors. However, the MSW generation rate depends upon the economic development, lifestyle, climate, and urbanization of a nation. In India, handling of MSW is governed by Municipal Solid Waste (Management and Handling) Rules 2000 (Ministry of Environment, Forest & Climate Change) which are further revised in 2016. In India, 90% of the total generated MSW is dumped in open landfills (Thakur et al., 2020). The urban people in India have significantly amplified from 1960 to 2011 as shown in Figure 1.

The Indian MSW contains 40-60% organic content, 30-60% inert material such as glass, 3-6% paper and miscellaneous, and 1% others (Rana et al., 2017)

Figure 1. Expansion of urbanization in India (Adopted and modified from Thakur et al., 2020)

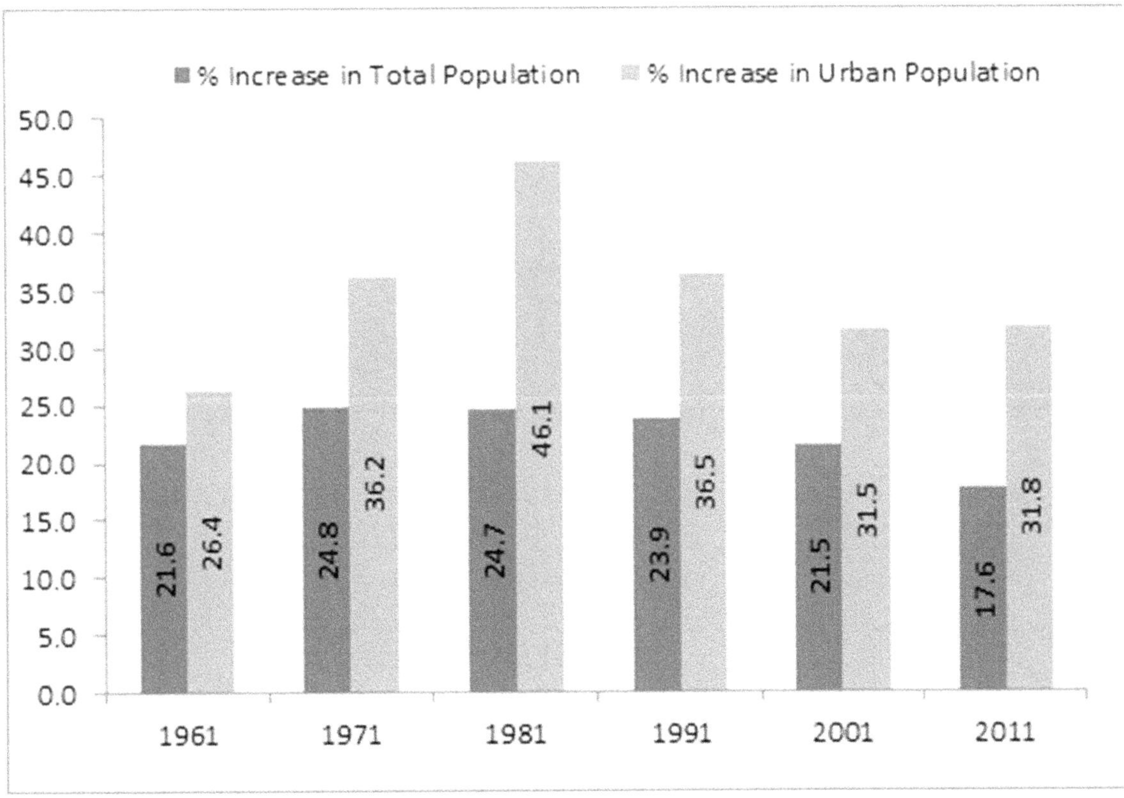

Table 1. Biodegradation phases of MSW in a landfill

Biodegradation phase	Description
Aerobic phase	Aerobic decomposition of readily biodegradable organic matter accompanied with generation of CO_2 and H_2O.
Acidogenic phase	Hydrolysis takes place which encompasses fermentation and activity of acidogenic bacteria. The process produces CO_2, H_2, NH_4^+, and organic acids.
Acetogenic phase	This phase is characterized with slow growth of methanogenic bacteria and is accompanied with production of CO_2, H_2, and acetic acid.
Methanogenic phase	In this phase, the production of methane with landfill gas (composed of 60% CH_4 and 40% CO_2) at a stable rate takes place.
Aerobic phase	Methane production rate decreases and nitrogen appears due to diffusion from the atmosphere (H_2O, CO_2)

Degradation of Municipal Solid Waste In Landfills

After dumping MSW in a landfill, its biodegradation takes place in five phases which are shown in Table 1 (Christensen and Kjeldsen, 1989; Schiopu and Gavrilescu, 2010).

The leachate composition and gas generation in a MSW landfill is shown in Figure 2.

Figure 2. Composition of leachate and gas during different degradation phases in a MSW landfill (Christensen and Kjeldsen, 1989)

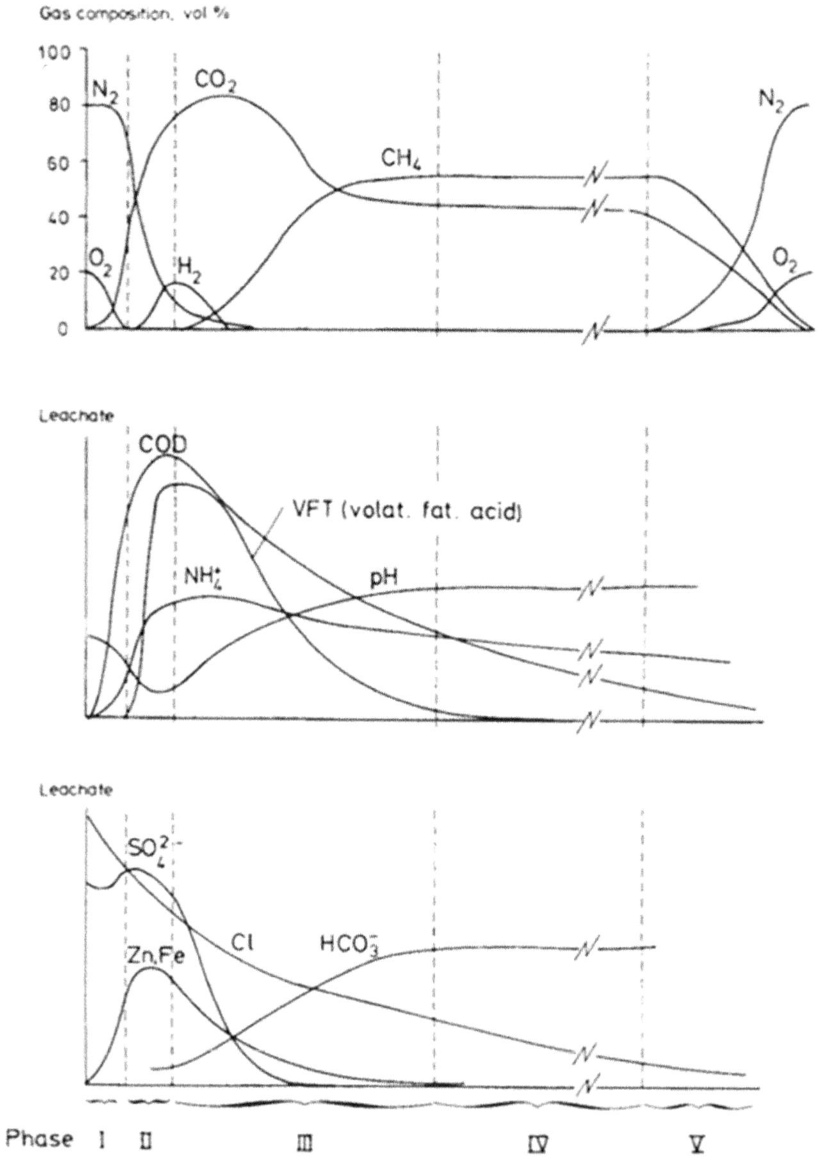

Aerobic Degradation Phase of Municipal Solid Waste in Landfills

The degradation of fresh MSW happens in the presence of oxygen which is supplied by diffusion and rainwater but aerobic degradation has taken place only for a short duration owing to insufficiency of oxygen in the landfill. The decomposition in the presence of oxygen leads to the conversion of proteins into amino acids which are further degraded to nitrates, carbon dioxide, water, and sulfates.

During this phase, (i) transformation of carbohydrates into CO_2 and H_2O; and (ii) hydrolysis of fats into fatty acids and glycerol takes place. They are further converted into simpler catabolites and the cellulose is decomposed into glucose by the extracellular enzymes which consequently lead to the formation of carbon dioxide and water. The significant leachate generation during aerobic degradation does not take place due to its short duration (Christensen and Kjeldsen, 1989).

Anaerobic Degradation Phase of Municipal Solid Waste in Landfills

Anaerobic degradation of MSW in landfills takes place in three phases (Christensen and Kjeldsen, 1989):

i. The first phase is *acid fermentation* which is characterized with a reduction in pH, and elevated concentrations of volatile acids. Also, inorganic ions (such as Cl^-, SO_4^{2-}, Ca^{2+}, Mg^{2+}, Na^+) are also present in substantial amount. In this phase, the drop in redox potential reduces the high concentration of sulfates, generates metal sulfides, and precipitate metals such as iron, manganese, etc. Thereafter, the pH reduces due to two reasons: (i) high production of volatile fatty acids; and (ii) high partial pressure of carbon dioxide. The anaerobic processes are stimulated by a consortium of anaerobes (comprising strictly anaerobic bacteria and facultative anaerobic bacteria). The growth of methane-producing bacteria is facilitated by a reduction in the redox potential by facultative anaerobes. The leachate during acid fermentation is characterized with high BOD_5 values (> 10000 mg/l), high BOD_5/COD ratio (> 0.7) and acidic pH (5-6) and ammonia (500-1000 mg/l). The generation of ammonia is attributed to fermentation and hydrolysis of proteinous compounds.

ii. The second phase is *intermedial anaerobic phase*, which is characterized by a gentle growth of methane-producing bacteria. The development of methane-producing bacteria may be suppressed due to the presence of toxic organic volatile acids in surplus amount (6-16 mg/l). This phase is characterized with a decrease in the concentration of volatile fatty acids, H_2, and CO_2 and an increase in the content of methane gas. Additionally, the biological reduction leads to a reduction in the sulfate concentration. During this phase, the solubility of metals decreases as the conversion of fatty acids increases the alkalinity. Ammonia is released and is not transformed into an anaerobic environment.

iii. The third phase is m*ethanogenic fermentation* phase, which is stimulated by methane-producing bacteria. Methane-producing bacteria tolerate pH between 6 and 8. In this phase, the leachate comprises neutral pH values, low concentrations of volatile acids and total dissolved solids. Leachates produced during this phase are characterized by relatively low BOD and low BOD/COD ratio. Ammonia continues to be released by the first stage acetogenic process. The three characteristic periods according to the BOD_5/COD-ratio are (i) Acid phase (BOD_5/COD: 0.4); (ii) Intermediate phase ($0.4 > BOD_5/COD > 0.2$); and (iii) Methanogenic phase (BOD_5/COD: 0.2).

LANDFILL LEACHATE

The leachate generated in a MSW landfill is a foul-smelling dark brown or black colored liquid. It consists of organic and inorganic material, comprising several refractory organic compounds, inorganic salts, and metal ions. The leachate composition is very complex which contains high levels of contaminants and biological toxicity (Kjeldsen et al., 2002).

Considering a landfill already filled and covered such as a drainage basin, in which both waste layers and the final capping have already been deposited, the amount of leachate is related to the mass balance relative to the water inflows and outflows in the sector:

$$L = P - R + R^{*} - ET + J + IS + IG + \left(\Delta US - \Delta UW\right) + B \tag{1}$$

Where, L = volume of leachate;

P = rainfall;

R = surface runoff;

R* = surface runoff from external areas;

ET = evapotranspiration;

J = irrigation and/or recirculation of leachate;

IS = infiltration water from surface water bodies;

IG = infiltration water from groundwater;

ΔUS = variations of water content in the capping material;

ΔUW = variation of water content in the amount of disposed waste;

B = production or consumption of water associated with the different aerobic and anaerobic biochemical degradation reactions of organic substances.

Characteristics of Landfill Leachate

The toxins in leachate can be categorized as (Kjeldsen et al., 2002):

a. The dissolved organic matter, which is enumerated as COD accompanied with volatile fatty acids and more refractory compounds such as fulvic-like and humic-like compounds.
b. Inorganic macro components such as Ca^{2+}, Mg^{2+}, Na^{+}, K^{+}, NH_4^{+}, Fe^{2+}, Mn^{2+}, Cl^{-}, SO_4^{2-}, and HCO^{3-}.
c. Xenobiotic organic compounds originating from household/industrial chemicals including a variety of aromatic hydrocarbons, phenols, chlorinated aliphatics, pesticides, and plasticizers.
d. Heavy metals such as Cd, Cr, Cu, Pb, Ni, and Zn.

The general characteristics of landfill leachate generated in MSW disposal sites are listed in Table 2.

Table 2. General characteristics of leachate in municipal landfill sites across the globe

Location	Characteristics												Reference
	pH	EC	BOD	COD	TSS	TS	NO_3-N	NO_2-N	NH_4-N	PO_4-P	TOC	Total P	
Bangkok (Thailand)	8.73	6.61	775	2950	-	-	-	-	-	-	-	24.7	Sawaittayothin and Polprasert (2006)
Ankara (Turkey)	-	-	-	4770	-	-	64	3.7	2865	75	-	-	Yalcuk and Ugurlu (2009)
Penang (Malaysia)	8.42	-	686	923.4	685	-	-	-	-	-	-	117	Akinbile et al. (2012)
Nonthaburi (Thailand)	7.7	15.7	-	1370		13545	-	-	-	-	355	-	Ogata et al. (2015)
Isfahan (Iran)	7.95	3.87	461	2301	607	-	-	41.17	-	-	40.4	-	Mojiri et al. (2016)

EC Electrical conductivity; BOD Biochemical oxygen demand; COD Chemical oxygen demand; TSS Total suspended solids; TS Total solids, TOC Total organic carbon; All parameters are expressed in mg/l except pH and EC (mS/cm)

Conventional Treatment Technologies

The treatment and collection of leachate is done by techniques shown in Table 3 depending upon the various parameters.

Table 3. On-site and off-site leachate treatment strategies (Adopted from Schiopu and Gavrilescu (2010) and Townsend et al. (2015))

Treatment technique	Parameters considered
Leachate treatment on-site and discharge	Construction of treatment system on-site
On-site pre-treatment followed by off-site treatment	Some off-site facilities may have limits that require pre-treatment
Transportation and treatment of leachate off-site	Leachate is transported via tanker truck or pumped directly through a pipe to an off-site wastewater treatment facility

The aspects to be considered while applying techniques for the treatment of leachate are (Schiopu and Gavrilescu, 2010; Townsend et al., 2015):

a. Composition, properties, and volume of leachate.
b. Requirement for storage of leachate.
c. Local water standards requirement for discharge.
d. Technical value, ease of implementation, and cost-effectiveness of technologies.

Physical and chemical processes for the treatment of landfill leachate include coagulation/flocculation, adsorption, and chemical oxidation. Coagulation-flocculation includes the reduction of repulsion between particles which results in settling of heavy particles formed by attraction and aggregation (Wiszniowski et al., 2006). In chemical precipitation, the addition of appropriate chemicals for the precipitation of pollutants is carried out which are then separated by sedimentation or filtration. Precipitation is used to remove high strength ammonium nitrogen from leachate (Wiszniowski et al., 2006). Adsorption is used for the removal of hydrophobic substances (Welander and Henrysson, 1998). Chemical oxidation methods such as Fenton's process are used due to advantages such as (i) no mass transfer limitation (homogeneous nature); and (ii) Inexpensive and non-toxic nature of iron and hydrogen peroxide (Lopez et al., 2004; Zhang et al., 2005).

Biological methods applied to treat leachate comprise aerobic (such as trickling filters, membrane bioreactor, sequencing batch reactor, and moving-bed bio-film reactor) as well as anaerobic techniques (such as up-flow anaerobic sludge blanket reactor, and fluidized bed reactor). Among aerobic processes, the sequencing batch reactor proves to be an efficient technique for nitrification-denitrification processes. The process features favored its wide applicability for the treatment of leachate (Kennedy and Lentz, 2000). In a moving-bed bio-film reactor, the suspended porous polymeric carriers are kept in continuous movement in the aeration tank. The active biomass film grows as a biofilm on the surfaces of porous polymeric carriers (Loukidou and Zouboulis, 2001).

The moving-bed bio-film reactor is characterized with (i) short sludge-settling periods; (ii) low sensitivity to toxic compounds; and (iii) removal of both organic and high ammonia in a single process. Among anaerobic processes, an up-flow anaerobic sludge blanket reactor can have (i) high treatment efficiency; and (ii) short hydraulic retention time (Lin et al., 2000).

The problems encountered in the treatment of leachate are (Wang et al., 2018):

i. The high amount of organic matter and toxic pollutants in leachate makes it difficult to achieve discharge standards using a single biochemical or physicochemical process. To overcome this problem, the selection of a sustainable and reasonable technique which encompasses a combination of physicochemical and biochemical processing is the primary challenge.
ii. Absence of an effective leachate treatment technique for 100% removal of ammoniacal nitrogen. Secondly, improving the efficiency of total nitrogen removal is a problem as conventional biological treatment methods are successful in removing ammoniacal nitrogen but does not effectively remove total nitrogen.
iii. Due to substantial changes in leachate quality and quantity during different seasons, the selection of an appropriate combination of techniques to ensure a stable operation is the third challenge in leachate treatment.
iv. The complex nature of pollutants in leachate requires expensive treatment technologies which are a constraint.

PHYTOREMEDIATION: A SUSTAINABLE TECHNOLOGY

Phytoremediation is a bioremediation process that uses vegetation and their associated microbial assemblages for the removal of pollutants from environmental components, such as soil, water, and air. This technique comprises the utilization of a combination of physical, chemical and biological processes to reduce, degrade, or immobilize harmful pollutants existing in the wastewater (Prasad, 2003; Zhang et al., 2007; Mukhopadhyay and Maiti, 2010). Phytoremediation is advantageous over other treatment processes as it is eco-friendly, economically feasible, and efficient. Phytoremediation has emerged as the method of choice for cleaning up a wide array of pollutants from different environmental media. The researchers have demonstrated the efficiency of decentralized wastewater treatment systems i.e. CWs for treatment of domestic and industrial wastewaters (Calheiros et al., 2007; Comino et al., 2013; Vymazal, 2014; Zhang et al., 2014).

Types of Phytoremediation

Phytoremediation functions through different methods (such as phytoextraction, phytostabilization, phytodegradation, rhizodegradation, and phytovolatilization) individually or in combination (Ali et al., 2013; Saxena et al., 2020). Phytoextraction encompasses the uptake of pollutants from soil/water by plant roots and their translocation to and accumulation in above-ground biomass (Rana and Maiti, 2018a). Phytostabilization is defined as the technique which includes the immobilization of pollutants in the soil. The immobilization leads to the reduction in the biological availability of the pollutants

(Clemente, 2019). In phytodegradation, enzymes (such as dehalogenase and oxygenase) generated by plants are involved in the uptake, metabolization, and degradation of organic pollutants. This technique is independent of the rhizospheric microorganisms (Gong et al., 2019). Rhizodegradation includes the degradation of pollutants within the rhizosphere by the microorganisms due to which the biodegradation process is enhanced by the influence of the plant roots (Vishnoi and Srivastava, 2007; Mukhopadhyay and Maiti, 2010). In the phytovolatilization process, the plant roots absorbs pollutants which are transported to leaves by the xylem and is thereafter released in the atmosphere in less toxic forms due to metabolic modification (Mohammed and M-Ridha, 2019).

CONSTRUCTED WETLANDS (CWS)

Landfill leachate treatment using conventional and new physicochemical and biological methods incurs high construction, and operation and maintenance cost. A sustainable low-cost alternative for leachate treatment is constructed wetlands (Yalcuk and Ugurlu, 2009; Ji et al., 2020). CWs are engineered wastewater treatment systems that work on natural processes encompassing sediments, plants, and associated microbial assemblages to remove pollutants from wastewaters (Vymazal, 2005; Rana and Maiti, 2018b). CWs were first developed in 1960 by Dr. K Seidel in Germany from where they gained importance and by 1995; over 200 units had been installed in Europe (mostly in Denmark, Germany and the United Kingdom) and another 200 units in the United States of America. The schematic diagram of implementing a CW technology for the treatment of landfill leachate is shown in Figure 3. CWs are being used for the removal of metals, nutrients, organic matter, hydrocarbons, etc. (Hashmat et al., 2019; Saeed et al., 2019; Tang et al., 2019; Srivastava et al., 2020).

Figure 3. Schematic diagram of implementing a constructed wetland technology for leachate treatment (Adopted from Bulc, 2006)

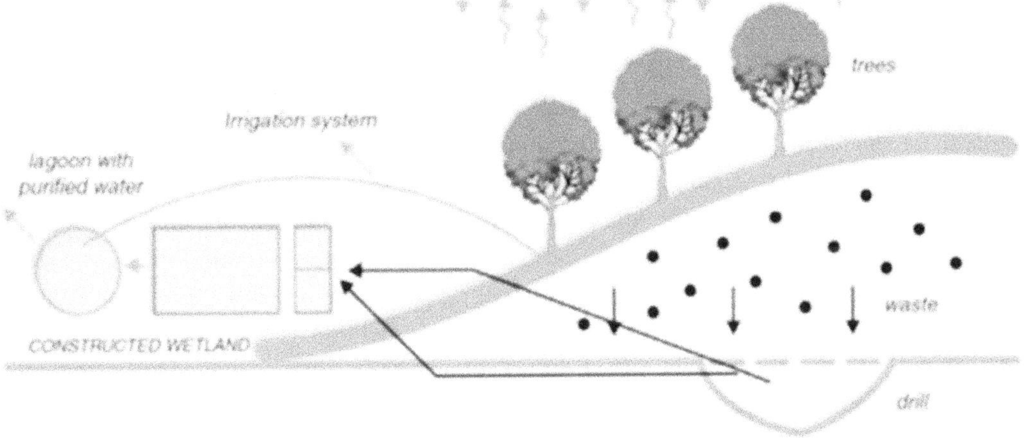

The various merits of using CWs for wastewater treatment are (i) inexpensive installation, operation, and maintenance in comparison to conventional treatment technologies; (ii) utilization of natural processes for wastewater treatment; (iii) reuse and recycling of water is enabled; (iv) simple construction and operation; (v) and reduction in excess sludge production.

However, the various demerits of CW systems are (i) comparatively higher treatment time than other methods; (ii) dependency of treatment efficiency on environmental factors.; (iii) requirement of the large land area is a constraint for CWs; (iv) shock load to the biological components of a constructed wetland due to toxic pollutants; and (v) vague dynamics of the treatment process leading to inaccurate design and operation criteria (Luederitz et al., 2001; Vymazal, 2007).

The fate of pollutants in a CW is driven by (a) settling of suspended particulate matters; (b) microbial breakdown and transformation of pollutants; (c) chemical filtration, precipitation, and adsorption through contact with water, sediment, litter, and plants; (d) predation and natural die-off of pathogens; and (e) plant uptake (Xu and Mills, 2018).

Based on the water surface, the CWs are generally of two types: (i) free water surface type; and (ii) submerged flow type. The submerged flow type CWs can be horizontal or vertical depending upon the wastewater flow regime. Submerged flow wetlands are preferred over free water surface wetlands due to: (i) relatively easy installation; (ii) inexpensive; and (iii) discouragement to the possibility of mosquito breeding that is likely with a free water surface wetland (Vymazal, 2005).

Three types of wetland plants are there: (i) floating-leaved plants, with leaves that grow from the vegetative portions near the bottom of the wetland until floating at the surface; (ii) emergent plants, with all or part of their vegetative and sexually reproductive parts above the water surface; and (iii) submerged plants, that have all portions of the plant underwater, or the weed is dependent upon water for support. The weeds, due to their extensive root system, offer a large surface area for attached microorganisms for an efficient decomposition of the organic matter (Vymazal, 2010).

LEACHATE TREATMENT USING CONSTRUCTED WETLANDS (CWS)

Landfill leachate treatment using CWs is gaining importance in the World. Research is going on in this field in different developing and developed countries. The removal efficiencies of CWs used for leachate treatment are shown in Table 4.

In Nigeria, Aluko and Sridhar (2005) recommended *Ipomoea aquatica* Forssk. for the treatment of landfill leachate using CWs. In Bangkok, Sawaittayothin and Polprasert (2006) reported removal efficiencies of 91% for BOD, 96% for total N and 99.7% for Cd using CWs planted with *Typha angustifolia* L. In Thailand, Chiemchaisri et al. (2009) reported high organic removal efficiencies of more than 90% in terms of BOD and COD was achieved in subsurface horizontal flow CW system. In Turkey, Yalcuk and Ugurlu (2009) reported removal efficiencies of 62.3% for NH_4–N, 35.7% for COD, 52.6% for PO_4–P, and 40% for Fe (III) in a CW planted with *Typha latifolia* L. in Zeolite (clinoptilolite) amended treatment bed. In Bulgaria, Lavrova and Koumanova (2010) reported removal of COD (96%), BOD (92%), NH_3-N (100%), and total P (100%) from landfill leachate by using a vertical flow CW planted with Ph*ragmites australis* L. In France, Grisey et al. (2012) conducted studies where cattail (*Typha latifolia* L.) and reeds (*Phragmites australis* L.) were used for the removal of metals from landfill leachate. In Malaysia, Akinbile et al. (2012) used CWs planted with *Cyperus haspan* L. and the results showed that the CWs with *Cyperus haspan* L. were capable of removing 86.6% of turbidity, 86.6% of color, 98.8%

Table 4. Pollutant removal efficiencies of constructed wetlands for leachate treatment (Adopted and modified from Kumar and Choudhary (2018))

Location	System Configuration	Removal Efficiency (%)					Reference
		Turbidity, Color, Solids, Chloride, Sulfate, Bacteria, Phenol	BOD	COD	Nutrients	Heavy Metals	
Esval (Norway)	HSSF (HRT: 5 days) and SF (HRT: 40 days)	-	91	88	Total nitrogen: 83	-	Maehlum (1995)
Dragonja landfill site on the Adriatic coast	SF, Hydraulic load: 3 cm/day	Bacteria: 85	46	68	Ammoniacal nitrogen: 81	Fe: 80	Bulc et al. (1997)
Ljubljana (Slovenia)	Combined VSSF and HSSF, Plant: Reeds and Cattails, Hydraulic load: 0.5 cm/day	Chloride: 35, Sulfide: 49	59	50	Ammoniacal nitrogen: 51, Phosphate: 53	Fe: 84	Bulc (2006)
Kampung Padang Siding, Ulu Pauh, Perlis (Malaysia)	HSSF, Plant: *Limnocharis flava* (L.) Buchenau and *Scirpus atrovirens* Willd., Flow rate: 0.029 m3/day, HRT: 24 h	-	-	-	Ammoniacal nitrogen: 61.3, Phosphate: 52	-	Kamarudzaman et al. (2011)
Pulau Burung Sanitary Landfill (Malaysia)	HSSF, *Cyperus haspan* L., HRT: 3 week	Turbidity: 86.6, Color: 86.6, TSS: 98.8	78.7	91.8	Ammoniacal nitrogen: 53.8, Total nitrogen: 67, Total phosphorus: 99.7	Fe: 59, Mg: 75, Mn: 70.5, Zn: 89.4	Akinbile et al. (2012)
Isfahan (Iran)	HSSF (Leachate, 20% and domestic wastewater, 80%), Plant: *Typha domingensis* Pers., Optimum contact time: 48.7 h	Phenol: 90.5	-	-	-	Mn: 89.4	Mojiri and Ziyang (2015)
Hisar (Haryana, India)	VSSF (Up flow and down flow), Plant: *Canna* and *Typha* (Mixed culture), HRT: 4, 8, 12 Days	-	-	87.2	Ammoniacal nitrogen: 72.2, Total nitrogen: 74.3, Phosphate: 85.9	-	Singh et al. (2016)
NW Bulgaria	VSSF, Plant: *Phragmites australis* (Cav.) Trin. ex Steud., HLR: 0.38 cm^3 cm^{-2} min^{-1}	-	95.96	94.69	-	-	Lavrova (2016)
San Pedro village (Southwest Colombia)	HSSF, Plant: Polyculture - *Gynerium sagittatum* (Aubl.) P. Beauv., *Colocasia esculenta* (L.) Schott. and *Heliconia psittacorum* L. f., HRT: 7 day, Water inflow 0.5 m^3/day	-	-	67	-	Hg, Pb, and Cd: 10-80	Madera-Parra (2016)
Vellore (Tamil Nadu, India)	HSSF, Plants: Cattail (*Typha latifolia* L.) and *Scirpus californicus* (C.A. Mey.) Steud., HRT: 1-24 h	Turbidity: 84, Total solids: 91	-	82	Ammoniacal nitrogen: 65, Phosphate: 89	-	Mathew et al. (2016)
Malaysia	Adsorption and HSSF, Plant: *Typha domingensis* Pers., Contact time: 50.2 h, Leachate to wastewater mixing ratio - 20%	Color: 90.3	-	86.7	99.2	Ni: 86, Cd: 87.1	Mojiri et al. (2016)
Isfahan (Iran)	HSSF, Plant: *Vetiver*, Flow rate: 27 L/day, HRT: 5day	-	30	34	Ammoniacal nitrogen: 26, Nitrate nitrogen: 40, Total nitrogen: 50	-	Bakhshoodeh et al. (2017)
Japan	VSSF, Plant: *Phragmites australis* (Cav.) Trin. ex Steud.	Phenol: 100	-	-	-	-	Dan et al. (2017a)
Chlewnica (Northern Poland)	Multistage SSF, Plant: *Phragmites australis* (Cav.) Trin. ex Steud., HRT: 1.1-3.5 days	-	95	86.6	Nitrogen: 98.5	-	Wojciechowska (2017)

HSSF Horizontal sub-surface flow; SF Surface flow; SSF Sub-surface flow; VSSF Vertical sub-surface flow; HRT Hydraulic retention time of TSS, 91.8% of COD, 78.7% of BOD, 53.8% of NH_3-N, 99.7% of TP, 67% of TN, 59% of Fe, 75% of Mg, 70.5% of Mn, and 89.4% of Zn. In another study conducted by Madera-Parra et al. (2015) in Columbia, the CWs planted with *Gynerium sagittatum* (Aubl.) Beauv., *Colocasia esculenta* (L.) Schott. and *Heliconia psittacorum* L. f. showed removal efficiencies of 66% for COD, 67% for TKN, 72% for NH^{4+}-N, and 92-98% for heavy metals. Dan et al. (2017b) conducted experiments for the treatment of synthetic leachate using vertical flow CWs planted with *Phragmites australis* (Cav.) Trin. ex Steud. and *Juncus effusus* L. and observed high removal of metals (Zn, Cr, Ni, Cd, Fe, and Pb).

CONCLUSION

Rapid urbanization is increasing the generation of MSW globally. Sanitary landfills are designed for the disposal of MSW wherein leachates are produced due to the mixing of refuse with moisture. Leachates consist of toxic pollutants that may contaminate the environment. Conventional physicochemical and biological methods for the treatment of leachates are expensive and non-efficient. To overcome this problem, phytoremediation is a new process that can be employed (in CWs) for efficient leachate treatment. The CWs are well established for the treatment of municipal wastewater. Researchers have shown that CW technology may prove beneficial for the treatment of toxic leachate. However, long-term dependability on the CW technology for leachate treatment can be revealed by applying this technique on a large-scale.

ACKNOWLEDGMENT

This research received no specific grant from any funding agency in the public, commercial, or not-for-profit sectors.

REFERENCES

Akinbile, C. O., Yusoff, M. S., & Zuki, A. A. (2012). Landfill leachate treatment using sub-surface flow constructed wetland by *Cyperus haspan. Waste Management (New York, N.Y.), 32*(7), 1387–1393. doi:10.1016/j.wasman.2012.03.002 PMID:22456086

Akratos, C. S., Tekerlekopoulou, A. G., & Vayenas, D. V. (2018). Treatment of wastewater from tanneries and the textile industry using Constructed Wetland systems. In Constructed Wetlands for Industrial Wastewater Treatment (pp. 327-341). John Wiley & Sons Ltd. doi:10.1002/9781119268376.ch16

Ali, H., Khan, E., & Sajad, M. A. (2013). Phytoremediation of heavy metals-concepts and applications. *Chemosphere, 91*(7), 869–881. doi:10.1016/j.chemosphere.2013.01.075 PMID:23466085

Aluko, O. O., & Sridhar, C. (2005). Application of constructed wetlands to the treatment of leachates from a municipal solid waste landfill in Ibadan, Nigeria. *Journal of Environmental Health, 67*(10), 58–62. PMID:15991686

Aluko, O. O., Sridhar, M. K., & Oluwande, P. A. (2003). Characterization of leachates from a municipal solid waste landfill site in Ibadan, Nigeria. *Journal of Environmental Health Research, 2*(1), 32–37.

Arivoli, A., Mohanraj, R., & Seenivasan, R. (2015). Application of vertical flow constructed wetland in treatment of heavy metals from pulp and paper industry wastewater. *Environmental Science and Pollution Research International*, *22*(17), 13336–13343. doi:10.100711356-015-4594-4 PMID:25940487

Aziz, S. Q., Aziz, H. A., Yusoff, M. S., & Bashir, M. J. (2011). Landfill leachate treatment using powdered activated carbon augmented sequencing batch reactor (SBR) process: Optimization by response surface methodology. *Journal of Hazardous Materials*, *189*(1-2), 404–413. doi:10.1016/j.jhazmat.2011.02.052 PMID:21420786

Aziz, S. Q., Aziz, H. A., Yusoff, M. S., & Mohajeri, S. (2012). Removal of phenols and other pollutants from different landfill leachates using powdered activated carbon supplemented SBR technology. *Environmental Monitoring and Assessment*, *184*(10), 6147–6158. doi:10.100710661-011-2409-8 PMID:22068314

Bakhshoodch, R., Soltani-Mohammadi, A., Alavi, N., & Ghanavati, H. (2017). Treatment of high polluted leachate by subsurface flow constructed wetland with vetiver. *Civil Engineering (New York, N.Y.)*, *49*, 43–44.

Bulc, T., Vrhovsek, D., & Kukanja, V. (1997). The use of constructed wetland for landfill leachate treatment. *Water Science and Technology*, *35*(5), 301–306. doi:10.2166/wst.1997.0222

Bulc, T. G. (2006). Long term performance of a constructed wetland for landfill leachate treatment. *Ecological Engineering*, *26*(4), 365–374. doi:10.1016/j.ecoleng.2006.01.003

Calheiros, C. S., Rangel, A. O., & Castro, P. M. (2007). Constructed wetland systems vegetated with different plants applied to the treatment of tannery wastewater. *Water Research*, *41*(8), 1790–1798. doi:10.1016/j.watres.2007.01.012 PMID:17320926

Chiemchaisri, C., Chiemchaisri, W., Junsod, J., Threedeach, S., & Wicranarachchi, P. N. (2009). Leachate treatment and greenhouse gas emission in subsurface horizontal flow constructed wetland. *Bioresource Technology*, *100*(16), 3808–3814. doi:10.1016/j.biortech.2008.12.028 PMID:19200716

Christensen, T. H., & Kjeldsen, P. (1989). Basic Biochemical Processes in Landfills. In *Sanitary Landfilling: Process, Technology and Environmental Impact*. Academic Press.

Clemente, R., Arco-Lázaro, E., Pardo, T., Martín, I., Sánchez-Guerrero, A., Sevilla, F., & Bernal, M. P. (2019). Combination of soil organic and inorganic amendments helps plants overcome trace element induced oxidative stress and allows phytostabilization. *Chemosphere*, *223*, 223–231. doi:10.1016/j.chemosphere.2019.02.056 PMID:30784729

Comino, E., Riggio, V., & Rosso, M. (2013). Grey water treated by an hybrid constructed wetland pilot plant under several stress conditions. *Ecological Engineering*, *53*, 120–125. doi:10.1016/j.ecoleng.2012.11.014

Dan, A., Fujii, D., Soda, S., Machimura, T., & Ike, M. (2017a). Removal of phenol, bisphenol A, and 4-tert-butylphenol from synthetic landfill leachate by vertical flow constructed wetlands. *The Science of the Total Environment*, *578*, 566–576. doi:10.1016/j.scitotenv.2016.10.232 PMID:27836343

Dan, A., Oka, M., Fujii, Y., Soda, S., Ishigaki, T., Machimura, T., & Ike, M. (2017b). Removal of heavy metals from synthetic landfill leachate in lab-scale vertical flow constructed wetlands. *The Science of the Total Environment*, *584*, 742–750. PMID:28131455

Davies, L. C., Cabrita, G. J. M., Ferreira, R. A., Carias, C. C., Novais, J. M., & Martins-Dias, S. (2009). Integrated study of the role of *Phragmites australis* in azo-dye treatment in a constructed wetland: From pilot to molecular scale. *Ecological Engineering*, *35*(6), 961–970. doi:10.1016/j.ecoleng.2008.08.001

Flores, L., García, J., Pena, R., & Garfí, M. (2019). Constructed wetlands for winery wastewater treatment: A comparative Life Cycle Assessment. *The Science of the Total Environment*, *659*, 1567–1576. doi:10.1016/j.scitotenv.2018.12.348 PMID:31096366

Gong, Y., Chen, J., & Pu, R. (2019). The enhanced removal and phytodegradation of sodium dodecyl sulfate (SDS) in wastewater using controllable water hyacinth. *International Journal of Phytoremediation*, *21*(11), 1080–1089. doi:10.1080/15226514.2019.1606779 PMID:31044608

Grisey, E., Laffray, X., Contoz, O., Cavalli, E., Mudry, J., & Aleya, L. (2012). The bioaccumulation performance of reeds and cattails in a constructed treatment wetland for removal of heavy metals in landfill leachate treatment (Etueffont, France). *Water, Air, and Soil Pollution*, *223*(4), 1723–1741. doi:10.100711270-011-0978-3

Hashmat, A. J., Afzal, M., Fatima, K., Anwar-ul-Haq, M., Khan, Q. M., Arias, C. A., & Brix, H. (2019). Characterization of hydrocarbon-degrading bacteria in constructed wetland microcosms used to treat crude oil polluted water. *Bulletin of Environmental Contamination and Toxicology*, *102*(3), 358–364. doi:10.100700128-018-2518-y PMID:30542756

Ji, M., Hu, Z., Hou, C., Liu, H., Ngo, H. H., Guo, W., Shaoyong, L., & Zhang, J. (2020). New insights for enhancing the performance of constructed wetlands at low temperatures. *Bioresource Technology*, *301*, 122722. doi:10.1016/j.biortech.2019.122722 PMID:31983576

Kamarudzaman, A. N., Ismail, N. S., Aziz, R. A., & Jalil, M. F. (2011). Removal of nutrients from landfill leachate using subsurface flow constructed wetland planted with Limnocharis flava and Scirpus atrovirens. *2011 International Conference on Environmental and Computer Science*.

Kennedy, K. J., & Lentz, E. M. (2000). Treatment of landfill leachate using sequencing batch and continuous flow upflow anaerobic sludge blanket (UASB) reactors. *Water Research*, *34*(14), 3640–3656. doi:10.1016/S0043-1354(00)00114-7

Khandelwal, H., Dhar, H., Thalla, A. K., & Kumar, S. (2019). Application of life cycle assessment in municipal solid waste management: A worldwide critical review. *Journal of Cleaner Production*, *209*, 630–654. doi:10.1016/j.jclepro.2018.10.233

Kjeldsen, P., Barlaz, M. A., Rooker, A. P., Baun, A., Ledin, A., & Christensen, T. H. (2002). Present and long-term composition of MSW landfill leachate: A review. *Critical Reviews in Environmental Science and Technology*, *32*(4), 297–336. doi:10.1080/10643380290813462

Kumar, I. P., & Choudhary, A. K. (2018). Constructed Wetlands-A Sustainable Solution for Landfill Leachate Treatment. International J*ournal of Latest Technology in Engineering, Management &.Applied Sciences (Basel, Switzerland)*, *7*(6), 101–106.

Lavrova, S. (2016). Treatment of landfill leachate in two stage vertical-flow wetland system with/without addition of carbon source. *Journal of Chemical Technology and Metallurgy*, *51*(2), 223–228.

Lavrova, S., & Koumanova, B. (2010). Influence of recirculation in a lab-scale vertical flow constructed wetland on the treatment efficiency of landfill leachate. *Bioresource Technology, 101*(6), 1756–1761. doi:10.1016/j.biortech.2009.10.028 PMID:19932959

Lin, C. Y., Chang, F. Y., & Chang, C. H. (2000). Co-digestion of leachate with septage using a UASB reactor. *Bioresource Technology, 73*(2), 175–178. doi:10.1016/S0960-8524(99)00166-2

Lopez, A., Pagano, M., Volpe, A., & Di Pinto, A. C. (2004). Fenton's pre-treatment of mature landfill leachate. *Chemosphere, 54*(7), 1005–1010. doi:10.1016/j.chemosphere.2003.09.015 PMID:14637358

Loukidou, M. X., & Zouboulis, A. I. (2001). Comparison of two biological treatment processes using attached-growth biomass for sanitary landfill leachate treatment. *Environmental Pollution, 111*(2), 273–281. doi:10.1016/S0269-7491(00)00069-5 PMID:11202731

Luederitz, V., Eckert, E., Lange-Weber, M., Lange, A., & Gersberg, R. M. (2001). Nutrient removal efficiency and resource economics of vertical flow and horizontal flow constructed wetlands. *Ecological Engineering, 18*(2), 157–171. doi:10.1016/S0925-8574(01)00075-1

Madera-Parra, C. A. (2016). Treatment of landfill leachate by polyculture constructed wetlands planted with native plants. *Ingeniería y competitividad, 18*(2), 183-192.

Madera-Parra, C. A. (2016). Treatment of landfill leachate by polyculture constructed wetlands planted with native plants. *Ingeniería y Competitividad, 18*(2), 183–192.

Madera-Parra, C. A., Peña-Salamanca, E. J., Peña, M. R., Rousseau, D. P. L., & Lens, P. N. L. (2015). Phytoremediation of landfill leachate with *Colocasia esculenta, Gynerum sagittatum* and *Heliconia psittacorum* in constructed wetlands. *International Journal of Phytoremediation, 17*(1), 16–24. doi:10.1080/15226514.2013.828014 PMID:25174421

Maehlum, T. (1995). Treatment of landfill leachate in on-site lagoons and constructed wetlands. *Water Science and Technology, 32*(3), 129–135. doi:10.2166/wst.1995.0134

Mathew, A., Dubey, A., & Mahindrakar, A. B. (2016). Study of Constructed Wetland for Treatment of Landfill Leachate. *International Journal of Chemtech Research, 9*(11), 87–95.

McCutcheon, S. C., & Schnoor, J. L. (2004). *Phytoremediation: transformation and control of contaminants* (Vol. 121). John Wiley & Sons.

Mohammed, O. N., & M-Ridha, M. J. (2019). Phytoremediation of organic pollutants in wastewater using native plants. *Association of Arab Universities Journal of Engineering Sci*ences, *26*(2), 54-65.

Mojiri, A., & Ziyang, L. (2015). Phenols and manganese removal from landfill leachate and municipal wastewater using the constructed wetland. *International Journal of Environmental Chemical Ecological Geological and Geophysical Engineering, 9*(8), 974–977.

Mojiri, A., Ziyang, L., Tajuddin, R. M., Farraji, H., & Alifar, N. (2016). Co-treatment of landfill leachate and municipal wastewater using the ZELIAC/zeolite constructed wetland system. *Journal of Environmental Management, 166*, 124–130. doi:10.1016/j.jenvman.2015.10.020 PMID:26496842

Mukhopadhyay, S., & Maiti, S. K. (2010). Phytoremediation of metal enriched mine waste: A review. *Global Journal of Environmental Research*, *4*(3), 135–150.

Ogata, Y., Ishigaki, T., Ebie, Y., Sutthasil, N., Chiemchaisri, C., & Yamada, M. (2015). Water reduction by constructed wetlands treating waste landfill leachate in a tropical region. *Waste Management (New York, N.Y.)*, *44*, 164–171. doi:10.1016/j.wasman.2015.07.019 PMID:26209341

Oloruntade, A. J., PA, A., & F Alao, F. (2013). Municipal solid waste collection and management strategies in Akure, South-Western Nigeria. *Caspian Journal of Environmental Sciences*, *11*(1), 1–10.

Prasad, M. N. V. (2003). Phytoremediation of metal-polluted ecosystems: Hype for commercialization. *Russian Journal of Plant Physiology: a Comprehensive Russian Journal on Modern Phytophysiology*, *50*(5), 686–701. doi:10.1023/A:1025604627496

Rana, V., & Maiti, S. K. (2018a). Differential distribution of metals in tree tissues growing on reclaimed coal mine overburden dumps, Jharia coal field (India). *Environmental Science and Pollution Research*, *25*(10), 9745–9758. doi:10.100711356-018-1254-5 PMID:29368202

Rana, V., & Maiti, S. K. (2018b). Municipal wastewater treatment potential and metal accumulation strategies of *Colocasia esculenta* (L.) Schott and *Typha latifolia* L. in a constructed wetland. *Environmental Monitoring and Assessment*, *190*(6), 328. doi:10.100710661-018-6705-4 PMID:29730705

Renou, S., Givaudan, J. G., Poulain, S., Dirassouyan, F., & Moulin, P. (2008). Landfill leachate treatment: Review and opportunity. *Journal of Hazardous Materials*, *150*(3), 468–493. doi:10.1016/j.jhazmat.2007.09.077 PMID:17997033

Saeed, T., Haque, I., & Khan, T. (2019). Organic matter and nutrients removal in hybrid constructed wetlands: Influence of saturation. *Chemical Engineering Journal*, *371*, 154–165. doi:10.1016/j.cej.2019.04.030

Sanchez-Galvan, G., & Bolanos-Santiago, Y. (2018). Phytofiltration of anaerobically digested sugarcane ethanol stillage using a macrophyte with high potential for biofuel production. *International Journal of Phytoremediation*, *20*(8), 805–812. doi:10.1080/15226514.2018.1438350 PMID:29775098

Sawaittayothin, V., & Polprasert, C. (2007). Nitrogen mass balance and microbial analysis of constructed wetlands treating municipal landfill leachate. *Bioresource Technology*, *98*(3), 565–570. doi:10.1016/j.biortech.2006.02.002 PMID:16546377

Saxena, G., Purchase, D., Mulla, S. I., Saratale, G. D., & Bharagava, R. N. (2020). Phytoremediation of heavy metal-contaminated sites: Eco-environmental concerns, field studies, sustainability issues, and future prospects. *Reviews of Environmental Contamination and Toxicology*, *249*, 71–131. doi:10.1007/398_2019_24 PMID:30806802

Schiopu, A. M., & Gavrilescu, M. (2010). Options for the treatment and management of municipal landfill leachate: Common and specific issues. *CLEAN–Soil, Air, Water (Basel)*, *38*(12), 1101–1110.

Singh, S., Kaushik, A., & Kaushik, C. (2016). Comparing efficacy of down-flow and up-flow vertical constructed wetlands for treatment of simulated dumpsite leachate. *Imperial Journal of Interdisciplinary Research*, *2*, 942–945.

Srivastava, P., Abbassi, R., Yadav, A. K., Garaniya, V., Kumar, N., Khan, S. J., & Lewis, T. (2020). Enhanced chromium (VI) treatment in electroactive constructed wetlands: Influence of conductive material. *Journal of Hazardous Materials*, *387*, 121722. doi:10.1016/j.jhazmat.2019.121722 PMID:31806439

Tang, X. Y., Yang, Y., McBride, M. B., Tao, R., Dai, Y. N., & Zhang, X. M. (2019). Removal of chlorpyrifos in recirculating vertical flow constructed wetlands with five wetland plant species. *Chemosphere*, *216*, 195–202. doi:10.1016/j.chemosphere.2018.10.150 PMID:30368084

Thakur, D., Ganguly, R., Gupta, A. K., & Ghali, V. (2020). Evaluation of Existing Solid Waste Management System in Una Town, India. In *Sustainable Waste Management: Policies and Case Studies* (pp. 367–381). Springer. doi:10.1007/978-981-13-7071-7_33

Townsend, T. G., Powell, J., Jain, P., Xu, Q., Tolaymat, T., & Reinhart, D. (2015). Leachate Control, Storage, and Treatment. In *Sustainable Practices for Landfill Design and Operation* (pp. 243–266). Springer. doi:10.1007/978-1-4939-2662-6_11

Vishnoi, S. R., & Srivastava, P. N. (2007). Phytoremediation–green for environmental clean. In *Proceedings of Taal2007: the 12th World lake conference* (Vol. 1016, p. 1021). Academic Press.

Vymazal, J. (2005). Horizontal sub-surface flow and hybrid constructed wetlands systems for wastewater treatment. *Ecological Engineering*, *25*(5), 478–490. doi:10.1016/j.ecoleng.2005.07.010

Vymazal, J. (2007). Removal of nutrients in various types of constructed wetlands. *The Science of the Total Environment*, *380*(1-3), 48–65. doi:10.1016/j.scitotenv.2006.09.014 PMID:17078997

Vymazal, J. (2010). Constructed wetlands for wastewater treatment. *Water (Basel)*, *2*(3), 530–549. doi:10.3390/w2030530 PMID:20795704

Vymazal, J. (2014). Constructed wetlands for treatment of industrial wastewaters: A review. *Ecological Engineering*, *73*, 724–751. doi:10.1016/j.ecoleng.2014.09.034

Vymazal, J., Brezinova, T. D., Kozeluh, M., & Kule, L. (2017). Occurrence and removal of pharmaceuticals in four full-scale constructed wetlands in the Czech Republic–the first year of monitoring. *Ecological Engineering*, *98*, 354–364. doi:10.1016/j.ecoleng.2016.08.010

Wang, K., Li, L., Tan, F., & Wu, D. (2018). Treatment of landfill leachate using activated sludge technology: A review. *Archaea (Vancouver, B.C.)*, *2018*, 1039453. Advance online publication. doi:10.1155/2018/1039453 PMID:30254508

Welander, U., & Henrysson, T. (1998). Physical and chemical treatment of a nitrified leachate from a municipal landfill. *Environmental Technology*, *19*(6), 591–599. doi:10.1080/09593331908616715

Wiszniowski, J., Robert, D., Surmacz-Gorska, J., Miksch, K., & Weber, J. V. (2006). Landfill leachate treatment methods: A review. *Environmental Chemistry Letters*, *4*(1), 51–61. doi:10.100710311-005-0016-z

Wojciechowska, E. (2017). Potential and limits of landfill leachate treatment in a multi-stage subsurface flow constructed wetland–evaluation of organics and nitrogen removal. *Bioresource Technology*, *236*, 146–154.

Xu, X., & Mills, G. L. (2018). Do constructed wetlands remove metals or increase metal bioavailability? *Journal of Environmental Management, 218*, 245–255. doi:10.1016/j.jenvman.2018.04.014 PMID:29680756

Yalcuk, A., & Ugurlu, A. (2009). Comparison of horizontal and vertical constructed wetland systems for landfill leachate treatment. *Bioresource Technology, 100*(9), 2521–2526. doi:10.1016/j.biortech.2008.11.029 PMID:19157867

Zhang, D. Q., Jinadasa, K. B. S. N., Gersberg, R. M., Liu, Y., Ng, W. J., & Tan, S. K. (2014). Application of constructed wetlands for wastewater treatment in developing countries–a review of recent developments (2000–2013). *Journal of Environmental Management, 141*, 116–131. doi:10.1016/j.jenvman.2014.03.015 PMID:24784754

Zhang, H., Choi, H. J., & Huang, C. P. (2005). Optimization of Fenton process for the treatment of landfill leachate. *Journal of Hazardous Materials, 125*(1-3), 166–174. doi:10.1016/j.jhazmat.2005.05.025 PMID:16011874

Zhang, X. B., Peng, L. I. U., Yue-suo, Y., & Chen, W. R. (2007). Phytoremediation of urban wastewater by model wetlands with ornamental hydrophytes. *Journal of Environmental Sciences (China), 19*(8), 902–909. doi:10.1016/S1001-0742(07)60150-8 PMID:17966845

Ziyang, L., Youcai, Z., Tao, Y., Yu, S., Huili, C., Nanwen, Z., & Renhua, H. (2009). Natural attenuation and characterization of contaminants composition in landfill leachate under different disposing ages. *The Science of the Total Environment, 407*(10), 3385–3391. doi:10.1016/j.scitotenv.2009.01.028 PMID:19217144

ADDITIONAL READING

Bakhshoodeh, R., Alavi, N., Oldham, C., Santos, R. M., Babaei, A. A., Vymazal, J., & Paydary, P. (2020). Constructed wetlands for landfill leachate treatment: A review. *Ecological Engineering, 146*, 105725. doi:10.1016/j.ecoleng.2020.105725

Batool, A., & Salch, T. A. (2020). Removal of toxic metals from wastewater in constructed wetlands as a green technology; catalyst role of substrates and chelators. *Ecotoxicology and Environmental Safety, 189*, 109924. doi:10.1016/j.ecoenv.2019.109924 PMID:31759740

Cano, V., Vich, D. V., Andrade, H. H., Salinas, D. T., & Nolasco, M. A. (2020). Nitrification in multistage horizontal flow treatment wetlands for landfill leachate treatment. *The Science of the Total Environment, 704*, 135376. doi:10.1016/j.scitotenv.2019.135376 PMID:31812428

Lekshmi, B., Sharma, S., Sutar, R. S., Parikh, Y. J., Ranade, D. R., & Asolekar, S. R. (2020). Circular Economy Approach to Women Empowerment Through Reusing Treated Rural Wastewater Using Constructed Wetlands. In *Waste Management as Economic Industry Towards Circular Economy* (pp. 1–10). Springer. doi:10.1007/978-981-15-1620-7_1

Rana, V., & Maiti, S. K. (2020). Municipal and Industrial Wastewater Treatment Using Constructed Wetlands. In *Phytoremediation* (pp. 329–367). Springer. doi:10.1007/978-3-030-00099-8_10

Wagner, T. V., Parsons, J. R., Rijnaarts, H. H., de Voogt, P., & Langenhoff, A. A. (2020). Benzotriazole removal mechanisms in pilot-scale constructed wetlands treating cooling tower water. *Journal of Hazardous Materials*, *384*, 121314. doi:10.1016/j.jhazmat.2019.121314 PMID:31581006

Wang, Y., Cai, Z., Sheng, S., Pan, F., Chen, F., & Fu, J. (2020). Comprehensive evaluation of substrate materials for contaminants removal in constructed wetlands. *The Science of the Total Environment*, *701*, 134736. doi:10.1016/j.scitotenv.2019.134736 PMID:31715485

Xu, Q., Renault, S., Goltz, D., & Yuan, Q. (2020). Phytoremediation of waste dumping site soil and landfill leachate by using cattail (*Typha latifolia*). *Environmental Technology*, *41*(9), 1101–1106. doi:10.1080/09593330.2018.1521474 PMID:30198835

KEY TERMS AND DEFINITIONS

Biochemical Oxygen Demand: Biochemical oxygen demand is the amount of dissolved oxygen needed (i.e., demanded) by aerobic biological organisms to break down organic material present in a given water sample at certain temperature over a specific time period.

Chemical Oxygen Demand: Chemical oxygen demand is a measure of the capacity of water to consume oxygen during the decomposition of organic matter and the oxidation of inorganic chemicals such as Ammonia and nitrite.

Constructed Wetland: A constructed wetland is an engineered wetland which encompasses natural processes such as sedimentation, filtration, etc. to treat different types of wastewaters.

Landfill: A landfill site (also known as a tip, dump, rubbish dump, garbage dump, or dumping ground) is a site for the disposal of waste materials.

Leachate: A leachate is any liquid that, in the course of passing through matter, extracts soluble or suspended solids, or any other component of the material through which it has passed.

Municipal Solid Waste: Municipal solid waste is a waste type consisting of everyday items that are discarded by the public.

Phytoremediation: Phytoremediation is a bioremediation process that uses various types of plants to remove, transfer, stabilize, and/or destroy contaminants in the different types of environments such as soil, water, and air.

Chapter 14
Safety and Efficacy of Pseudomonas Exopolymer in Sequestration of Iron From Aqueous Environments

Moushumi Ghosh
Thapar Institute of Engineering and Technology, India

Divya Sharma
Thapar Institute of Engineering and Technology, India

Taranpreet Kaur
Thapar Institute of Engineering and Technology, India

ABSTRACT

The present study reports the iron binding characteristics and safety of an exopolymer (EBP) of an environmental isolate of Pseudomonas *sp. The EBP was predominantly polysaccharide in composition with pyruvic and uronic acid residues. A prevalence of carboxyl and hydroxyl groups was observed in the Fourier-transform infrared spectroscopy (FTIR) results, while scanning electron microscopy (SEM) revealed a porous structure in a linear fashion with large number of grooves. The purified EBP was stable for over two months and exhibited rapid binding of iron (25mg/L) within 10 minutes at ambient temperature. X-ray diffraction (XRD) and energy-dispersive X-ray spectroscopy (EDAX) analysis of iron challenged EBP suggested the involvement of carboxyl groups in potentiating iron removal. Both Langmuir and Freundlich adsorption isotherms depicted high iron removal capacity in comparison to reported biomasses or biopolymers. Cytotoxic effects were not observed upon challenging various doses of EBP in RAW 264.7 cell lines implying a strong possibility of application of the EBP.*

DOI: 10.4018/978-1-7998-4888-2.ch014

INTRODUCTION

Metal ions, especially heavy metal ions, discharged into water bodies by human activities, persist and accumulates in the food chain leading to serious environmental problems. Additionally, the bioaccumulation and biomagnifications of heavy metals at each step of the food chain poses health issues in animals and humans due to their mutagenic and carcinogenic potential (Bawuro et al., 2018; Latif et al., 2018; Yen et al., 2018). As the second most abundant element in earth's crust, Iron is a crucial cofactor in enzymes involved in various metabolic processes and electron transport in plants and animals. Despite the physiological importance of iron in all genera of life, high levels of iron in water can exert untoward effects in aquatic ecosystems. An increase in iron levels beyond a threshold can impact species composition of lakes by stimulating growth of green algae and cyanobacteria while also adversely affecting growth of aquatic and submerged plants. Adverse health effects such as liver cancer, infertility and diabetes are common by overloading of iron (Staniek and Wojciak, 2018). The presence of iron beyond 0.3 mg/L imparts colour (yellow to reddish), odor and taste to drinking water. In view of this both World Health Organization (WHO) and the United States Environment Protection Agency (US EPA) have recommended levels of 0.3 mg/L iron in drinking water (WHO, 1996).

In recent years, biological methods of pollutant and toxicant removal have gained increasing attention as an easy and cost effective substitute to conventional chemical and physical methods of remediation (Wang et al., 2012; Xue et al., 2006; Yang et al., 2004). Although the bioremediation protocols have been successfully employed for non-specific removal of phosphate, metals and other pollutants/toxicants, iron removal in wastewater is generally achieved by conventional chemical and physical methods like precipitation, crystallisation, flocculation, ion exchange, ultrafiltration, reverse osmosis and others(Kurniawan et al., 2006). However, with the realization of environmental impact of iron in recent years, the research focus has now shifted to development of novel, bio-based systems for iron removal (Horzum et al., 2010; Kousalya et al., 2010; Ngah et al., 2005; Reiad et al., 2012; Tapia et al., 2011; Xue et al., 2006; Yang et al., 2004).

Microbial exoproducts, most notably exobiopolymers (EBPs), have found applications in removal of metals from industrial wastes, oil refining, waste water treatment and as thickeners and emulsifying agents (Hay et al., 2014; Paniagua-Michel Jde et al., 2014). EBPs are mainly composed of polysaccharide and proteins and exhibit metal ion sorption capability due to presence of carboxyl, amine and hydroxyl groups. Recent studies have demonstrated that dried biomass of activated sludge as well as exobiopolymers produced by several microbial species exhibit significant iron binding capacity (Aryal and Liakopoulou-Kyriakides, 2013; Emtiazi et al., 2004; Huang et al., 2011; Moppert et al., 2009; Shokoohi et al., 2009; Tapia et al., 2013; Tapia et al., 2011; Yu et al., 2009).

In an attempt to identify bacterial strains for iron bioremediation, an EBP-producing bacterium was isolated from industrial sludge. The major objectives of the study were identification of the microorganism, characterization of its iron sorption behaviour and *in vitro* cytotoxicity in cell line. This appears to be the first report describing exobiopolymer production by *Pseudomonas* sp. and its iron binding efficiency with an objective of removing iron from water.

METHODS AND MATERIALS

Chemicals, Microorganisms and Cultural Conditions

The chemicals used were of highest purity available and were procured from Merck (E.Merck, Darmstatd, Germany). Media components were purchased from Himedia Pvt Ltd, (Mumbai, India).

In this study, *Pseudomonas* sp. a bacterial isolate from activated sludge sample of industry, near Patiala was used. The strain was maintained in Luria Bertani (LB) broth (Hi Media, Mumbai, India) and grown in biopolymer producing (BP) medium for exobiopolymer production. The BP media contained ammonium sulphate (0.1%), dextrose (0.1%), calcium chloride (0.07%), magnesium sulphate (0.03%), peptone (0.5%), sodium chloride (0.01%) and agar (0.3%).

Purification of Exobiopolymer

The bacteria was cultured at 37°C for 48 h(Kaur etal,2013) and separated from the medium by centrifugation at 12,000 rpm at 4°C for 20 min. Biopolymer(EBP) was extracted from the supernatant by precipitating it with chilled ethanol for 24h and dialyzed extensively against deionised water (Ghosh et al., 2009). The stability of the biopolymer upon storage under ambient conditions was documented in an earlier study.

Compositional and Structural Analysis of Exobiopolymer

The content of sugar in the exobiopolymer was determined by phenol sulphuric acid method. Protein content was analyzed by Folin-Lowry method using BSA as standard. Pyruvic acid and uronic acid content was analyzed by the method of Friedemann and Haugen (1943) and carbazole-sulphate reaction, respectively. Powdered X-ray diffraction (XRD) analysis of exobiopolymer was carried out with X-ray diffractometer (Xpert Pro, Panalytical, US) using Cu- Kα radiation with 1.54Å wavelength. All samples were scanned between 5° and 85°. SEM analysis was preformed for surface characterization. Viscosity measurements of aqueous EBP solution (1 mg/mL) were done by Brooksfield viscometer.

Toxicity of the Exopolymer

Cytotoxic activity of the biopolymer and cell viability was evaluated using RAW 264.7 macrophages and MTT assay (Kaur and Ghosh, 2017). Briefly, RAW 264.7 cells (National Centre for Cell Sciences, Pune, India) were maintained in Dulbecco's modified Eagle's medium (DMEM) containing 10% fetal bovine serum (FBS), 1% penicillin-streptomycin solution, 1% L-glutamine and HEPES[4-(2-hydroxyethyl)-1-piperazineethanesulfonicacid]. The cells were grown in 5% carbon dioxide at 37°C throughout the experiment. Cells were added in plates with 96 wells (50,000 cells/0.2 mL/well), allowed to incubate overnight for attachment; the medium in wells was replaced with EBP-containing medium the following day. A stock solution of 1 mg/mL was prepared in sterile saline, suitably diluted in culture medium to obtain working solutions (10, 20, 30, 40 and 50 mg/L) and 0.2 mL was added. The cells were incubated for 24 or 48 hours and cell viability was determined by MTT assay as described earlier (Singh et al., 2013; Singh et al., 2008).

Standard Iron Solution

One hundred milligrams of ferrous ammonium sulphate was dissolved in one litre of deionized water and working concentrations (0.1 mg/L to 20 mg/L) were prepared by diluting standard solution with deionized water.

Iron Adsorption Studies and Optimization

For analyzing the effective binding of EBP with iron, both EBP and iron concentration was optimized in synthetic water containing varying amounts of iron. Batch experiments were carried out at various time intervals (0, 5, 10, 20, 30, 45, 60 and 90 min) to study the biosorption kinetics of iron. Optimum final concentration of EBP required for iron binding was evaluated by varying iron (0.1, 1, 5 and 20 mg/L) and EBP concentrations (1, 3, 5, 10 and 25 mg/L). EBP and iron working solutions were mixed, kept undisturbed at room temperature (25°C) for varied time intervals. The residual iron was determined by phenanthroline method by measuring the absorbance at 520 nm (Masawat et al., 2016).

The equilibria of adsorption of iron were modelled using adsorption-type isotherms. The Freundlich and Langmuir models were used to describe the adsorption equilibrium. Adsorption capacity of the exobiopolymer at equilibrium (q_e, mg/g) was calculated by using the following equation:

$$q_e = \frac{\left(C_o - C_e\right)V}{M}$$

where, C_0 and C_e are initial and final sample absorbance (O.D), respectively, V is the volume of the sample solution and 'M' is the weight of EBP added. Langmuir and Freundlich adsorption isotherms were used to analyze the binding efficiency of EBP.

The Langmuir model assumes adsorption on homogenous and equivalent binding sites and is expressed as:

$$q_e = \frac{q_m b C_e}{1 + bC_e}$$

where $\mathbf{q_m}$ is the maximum amount of metal adsorbed at saturation, per unit mass of adsorbent; b is an equilibrium constant (Langmuir constant), related to the energy of adsorption; and C_e is the equilibrium (final) concentration of the metal in the solution, expressed as mg/L.

Freundlich model assumes adsorption on a heterogeneous surface and is expressed as:

$$q_e = K_f C_e^{1/n}$$

where q_e is the amount of metal adsorbed per unit mass of adsorbent, expressed as mg/g; K_f is a constant, related to the adsorbent capacity, larger K_f value reflects a larger overall adsorption capacity. On the other hand; *n* is a constant related to the energy of sorption. The parameters of Langmuir and Freundlich isotherms were evaluated by employing computational software MATLab (Kaur and Ghosh, 2015).

Spectroscopic Analysis for Iron Adsorption

Energy dispersive X-ray spectroscopy (EDAX) was performed to evaluate the amount of iron bound by EBP. Samples of native EBP and iron-bound EBP were coated with a conductive layer of gold and then analyzed by scanning electron microscopy (SEM; 6510-LV, Joel, Japan) attached with an EDAX micro-analyzer (Oxford Instrument, UK). SEM analysis was performed at an accelerating voltage of 20.0 kV. Fourier transformed infrared spectroscopy (FTIR) was performed by potassium bromide pellet method on a Carry 660 FTIR spectrometer (Agilent Technologies, US).

Statistical Analysis

Data was converted to mean±standard deviation (SD) and was analysed by one-way analysis of variance (ANOVA), followed by Tukey's test or two-way ANOVA and Bonferroni's test using GrapPad Prism and considered significant for $p<0.05$.

Results and Discussion

The application of innovative processes for treating industrial wastewater as well as potable water especially those containing heavy metals have been strongly realized across the world. Bio-based approaches have been preferred for their safety and capability of many in serving as effective adsorbents. The presence of iron in water has remained a persistent health issue; for instance over 20 million people drink water containing iron and arsenic with levels exceeding those of national and international standards in many countries (Kumar and Puri, 2012; Oyem et al., 2015). High iron concentrations were reported very recently in district of Moradabad (Uttar Pradesh); over 50% of the samples were reported to be beyond permissible limit for iron. Highest iron concentration in groundwater sample was found to be 3820 ppb while 6294 ppb in surface water, against the permissible limit of 300 ppb (Kumar et al., 2017). Several reports likewise have indicated significant levels of iron in southern and eastern states of India. Few reports have described the potential of bacterial exopolymers as a bioadsorbent for iron especially with respect to its suitability in enhancing water quality.

Characterization of Exobiopolymer

The exobiopolymer(EBP) from *Pseudomonas* sp. was found to be off white in appearance. SEM micrographs indicated porous structure of the exobiopolymer and the presence of large number of linear grooves and projections on the surface of EBP indicated its high surface area. The irregular particle size with porous structure on the surface was also observed which is conducive for iron adsorption (Fig 1).

Approximately 2 mg of EBP was spread on carbon tape and visualized by SEM. The arrows show pores in EBP structure and the white circled area indicayes grooves and projections.

Viscosity of aqueous EBP solutions was elucidated for understanding the visco-elastic properties. Apparent viscosity of solution decreased with increasing shear stress which suggested a non-Newtonian behaviour of EBP solution. The flow index value was calculated to be 0.6. Similar non-Newtonian viscosity behaviour has also been reported for biopolymers produced by other microorganisms (Ismail and Nampoothiri, 2014; Kaur and Ghosh, 2015).

Figure 1. SEM micrograph of exobiopolymer produced by Pseudomonas sp.

Analysis of the EBP revealed presence of sugars, proteins, uronic acids and pyruvic acids (Table 1). The EBP contained higher sugars and amino sugars as compared to other strains such as *Acinetobacter* and *Klebsiella* previously reported (Ghosh et al., 2009; Kaur and Ghosh, 2015).

Table 1. **Composition of Exobiopolymer of *Pseudomonas* sp.**

Components	Composition (µg/mL)*
Sugar	87
Protein	35
Amino sugars	19
Pyruvic acid	25
Uronic acid	11

*EBP concentration used was 1 mg/mL.

Effect of Iron and Exobiopolymer Concentration and Contact Time Upon Sorption

The iron sorption on EBP was studied as a function of its initial concentration. It was observed that maximum iron binding to EBP was at 20 mg/L of iron concentration and 5 mg/L of EBP concentration. Further increase in iron concentration to 25 mg/L or EBP concentration to 10 and 25 mg/L showed no significant effect (p>0.05) on iron adsorption (Fig 2).

Figure 2. Effect of various iron concentration and EBP concentrations on its reduction (%)

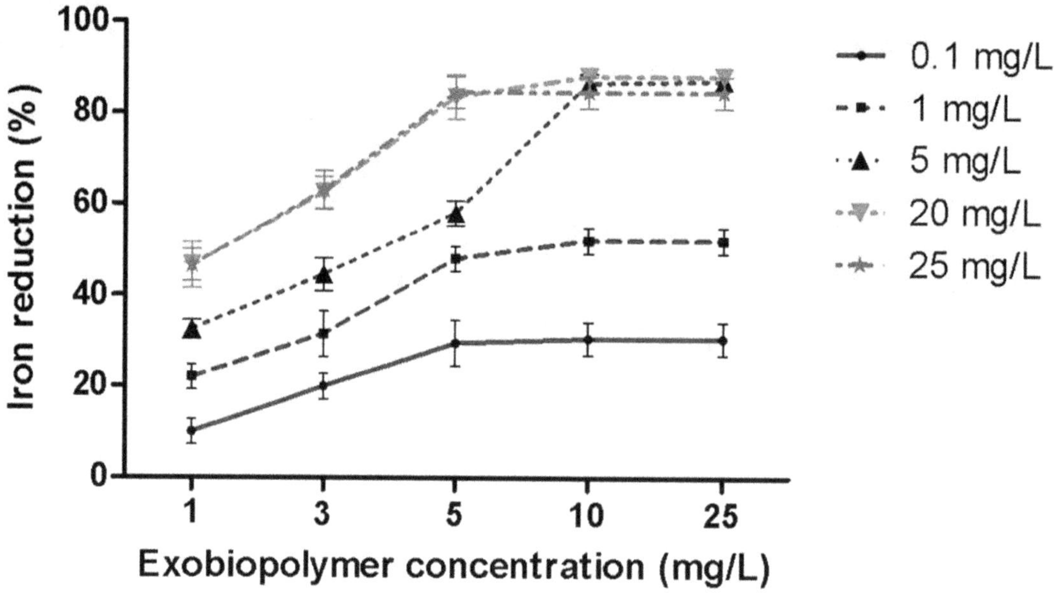

Iron solution and EBP solution were mixed to obtain varying concentrations (1-25 mg/L) and kept undisturbed at room temperature for 60 min. Unbound iron was separated by filtration and determined by phenanthroline method. The percentage of initial iron concentration that bound to EBP is referred to as % iron reduction.

The data is mean±SD of triplicate samples. The data was analysed by two-way ANOVA followed by Bonferroni's test to determine effect of iron concentration (F = 403; p<0.0001), EBP concentration (F = 229; p<0.0001) and interaction between iron and EBP (F = 8; p<0.0001).

Moreover, the amount of iron adsorbed, at each iron and EBP concentration, was maximum at 10 min; increase in contact time for up to 60 min led to no appreciable change in iron binding (Fig 3). A rapid process of iron binding by EBP is implicated, this results in saturation of binding sites at the concentrations studied. A similar rapid equilibration has also been demonstrated for metal ion binding in EBP produced by other microorganisms (Gupta and Diwan, 2017).

Figure 3. Effect of contact time of iron removal

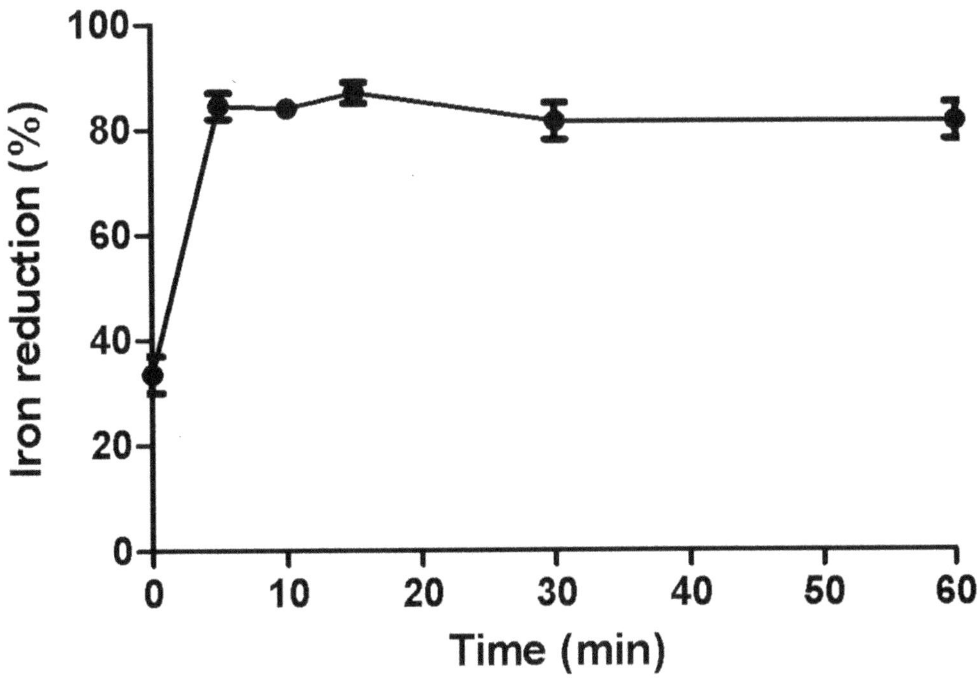

Iron solution and EBP solution were mixed to obtain 20 mg/L and 5 mg/L, respectively, and kept undisturbed at room temperature for up to 60 min. Unbound iron was separated by filtration and determined by phenanthroline method. The percentage of initial iron concentration bound to EBP is referred to as % iron reduction.

The data is mean±SD of triplicate samples. The data was analyzed by one-way ANOVA followed by Tukey's test to determine effect of time on iron reduction ($F = 53$). The difference was significant ($p<0.0001$) between 1 min and other time intervals (5, 10, 15, 30 and 60 min) but insignificant ($p>0.05$) at all other time points.

The equilibrium time of 10 min was selected to determine iron sorption isotherms. At this equilibrium time and 20 mg/L of iron concentration, it was observed that adsorption efficiency decreased from 8 g/g to 0.7 g/g and percentage removal of iron increased from 50% to 90% with increase in EBP concentration from 1 to 20 mg/L (Fig 4).

Iron solution (20 mg/L) and EBP solution (1-20 mg/L) were mixed and kept undisturbed at room temperature for 10 min. Unbound iron was separated by filtration and determined by phenanthroline method. The percentage of initial iron concentration that bound to EBP is referred to as % iron reduction. Adsorption capacity (qe) was determined using formula described in Materials and Methods. The data is mean±SD of triplicate samples. The data was analysed by one-way ANOVA followed by Tukey's test. The results were statistically different between 1 mg/L EBP and higher concentrations for qe and % iron reduction.*$p<0.05$, **$p<0.01$ compared to 1 mg/L EBP. The difference at 3 mg/L EBP and higher

Figure 4. Iron removal with adsorption efficiency with increasing EBP concentration

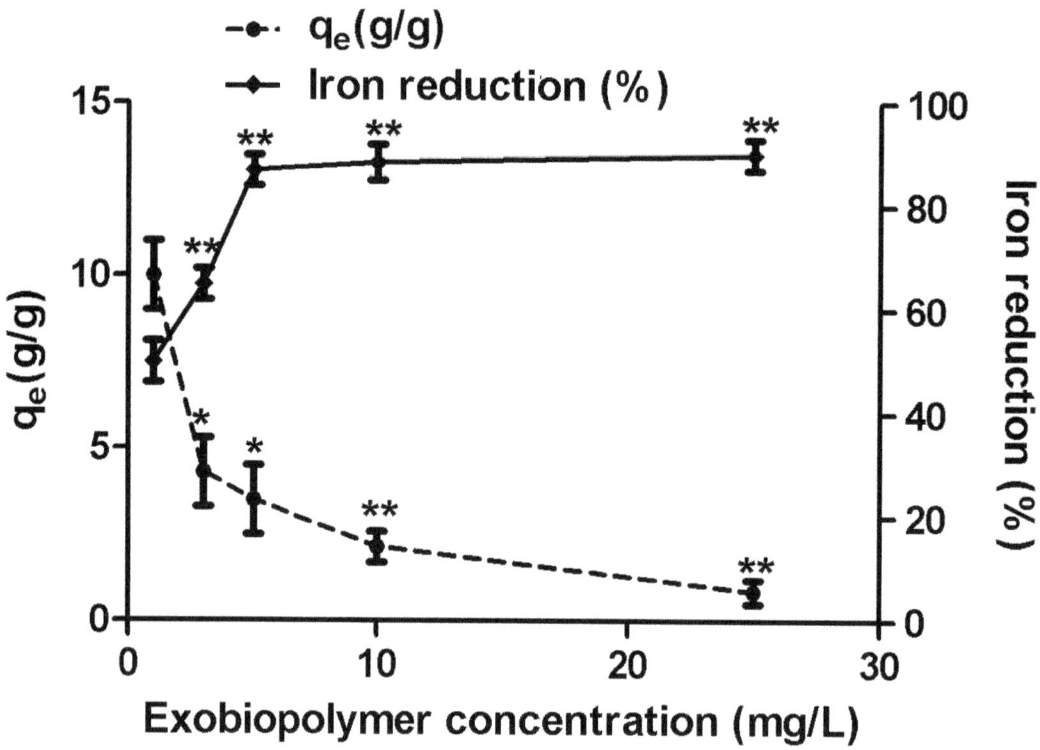

concentrations was insignificant for qe (p>0.05) but significant for % iron reduction (p<0.05). The results at other concentrations were statistically insignificant (p>0.05).

Langmuir and Freundlich Adsorption Approximation of Iron Biosorption

The Langmuir isotherm considers monolayer adsorption onto a surface containing a finite number of adsorption sites of uniform strategies with no migration of iron in the plane surface on the other hand, Freundlich adsorption model usually lays emphasis on heterogeneous or multilayer adsorption of iron onto exopolymer surface (Kaur and Ghosh, 2015).

The equilibrium uptake (q_e) of iron by EBP as biosorbant was plotted as a function of their equilibrium concentrations (C_e) as shown in Fig 5. Correlation coefficient and least sum of squares were used to select the best isotherm model. The iron adsorption data in Fig 4 fitted the reciprocal equation of the Freundlich adsorption isotherm, resulting in the regression coefficient (R^2) shown in Table 2.

The Freundlich isotherm parameters, K_f and n, were also calculated (Tables 2 and 3). The intercept K is a measure of the adsorption capacity of the adsorbent and the slope 1/ n, the intensity of adsorption, where n is a number greater than unity. The Freundlich adsorption isotherms (R^2 0.99) was found to represent the equilibrium adsorption data with better fitting than Langmuir isotherm (R^2 0.33). The results are presented based on mass (g) uptake per gram EBP. The value of q_e derived from the isotherm

Table 2. Langmuir and Freundlich isotherms constant of exobiopolymer at different concentrations

Exobiopolymer conc. (mg/L)	Langmuir $q_e = \frac{q_m bC}{1+bC_e}$				Freundlich $q_e = k_f C_e^{1/n}$		
	q_m	B	R_L	R^2	K_f	*N*	R^2
1	1.15×10^4	0.163	0.234	0.803	317.9	0.667	0.999
3	1.6×10^4	0.045	0.234	0.916	317.9	0.667	0.999
5	1.15×10^4	0.097	0.588	0.338	2.5	0.126	0.998
10	1.62×10^4	0.031	0.520	0.617	0.001	0.050	0.998
25	7675	0.026	0.428	0.651	173.1	0.847	0.425

*R_L - Separation factor which depends upon Langmuir constant and initial iron concentration

Figure 5. Fitting of adsorption efficiency with Langmuir and Freundlich models at 5 mg/L EBP and 20 mg/L iron after 10 min contact time. The data is mean±SD of triplicate samples. The data was analyzed by non-liner curve fitting using MATLAB.

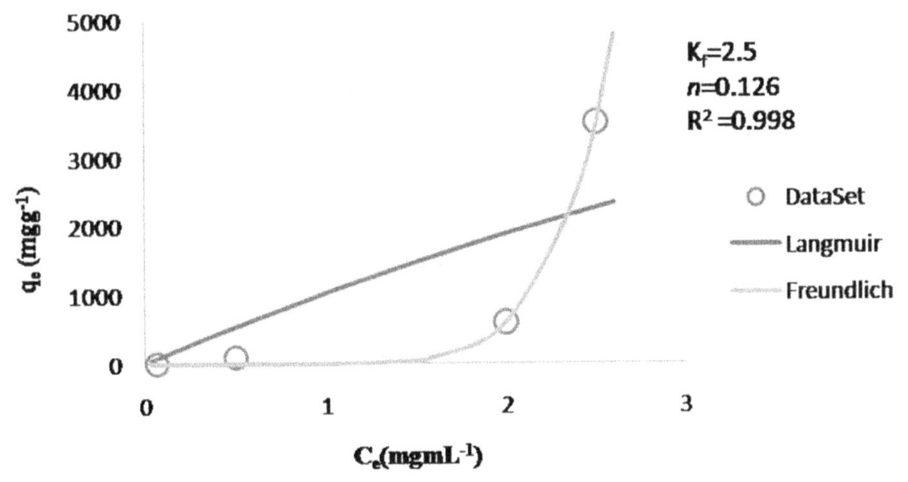

Table 3. Isotherm constants with their R^2 values

Isotherms	Equation	Constants	Value	R^2
Langmuir	$q_e = \frac{q_m bC_e}{1+bC_e}$	q_m	1.15×10^4	0.338
		b	0.097	
Freundlich	$q_e = k_f C_e^{1/n}$	K_f	2.5	0.998
		n	0.126	

equation was up to 3.5 g/g exobiopolymer and this value was much higher than the observed capacities or the estimated q_e values by other natural bacterial natural or biomasses and modified exopolysaccharides reported in literature (Wei et al., 2016; Zhang et al., 2017).

Spectroscopic Characterization of Iron Binding by EBP

EDAX analysis was performed to determine the quantity of elemental iron bound to EBP. The EDAX spectra of control samples (before iron adsorption) exhibited distinct peaks of carbon and oxygen. Following iron adsorption, the spectrum showed an additional, distinct peak for iron. EDAX confirmed the adsorption of 5% of iron by the exobiopolymer (Fig 6).

Figure 6. EDAX spectrum of iron bound EBP

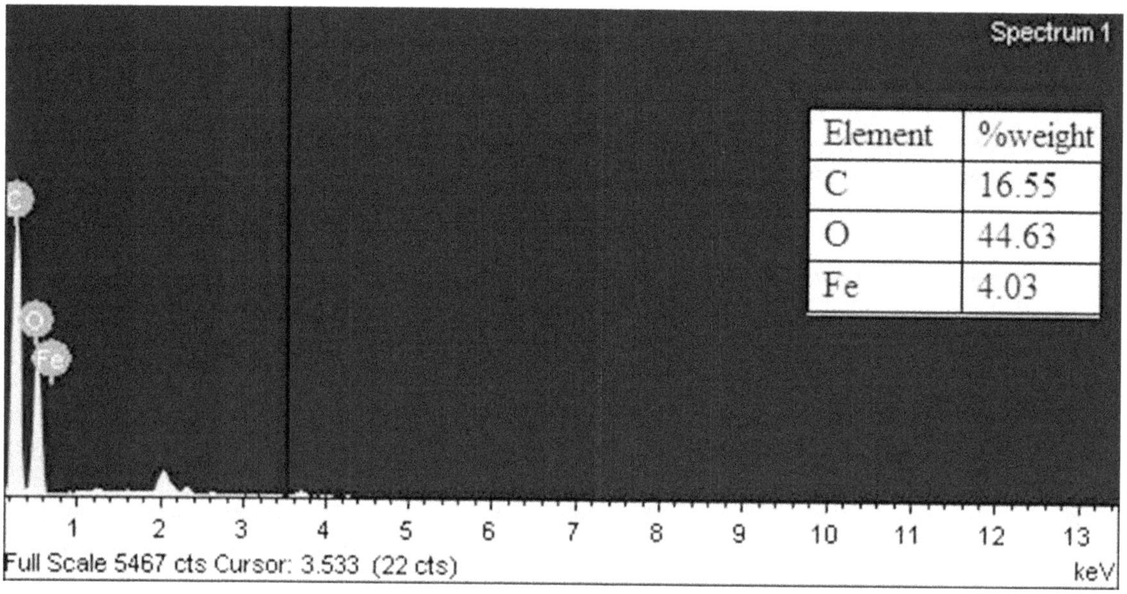

The FTIR spectra for iron-loaded (C_i=20 mg/L) EBP were obtained to elucidate the chemical groups involved in iron binding. This binding capacity of *Pseudomonas* sp. exobiopolymer is usually attributed to the high hydrophilicity, due to the presence of hydroxyl groups, sulphate, acetamido, primary amino and ester groups and also due to flexible structure of EBP chains (Jamil and Ahmed, 2008). The bands in spectrum of unloaded EBP, at 1,722cm^{-1} and 1,085 cm^{-1}, were shifted to the right in iron-loaded EBP (at 1,715cm^{-1} and 1,058 cm^{-1}, respectively). The bands in these ranges correspond to the asymmetric stretching vibration of C=O bond in the carboxylic group and to the stretching vibration of the hydroxyl group (O-H) (Omoike and Chorover, 2004). The peak intensity of the carboxyl group is higher in the spectrum of the native exobiopolymers than for exobiopolymer with iron. This suggests a preferential interaction of iron with carboxyl groups in solution (Fig 7 a and b). Hence, the role of carboxylic groups in iron binding can be suggested. The results obtained were supported by earlier reports (Aryal and Liakopoulou-Kyriakides, 2013).

Figure 7. FTIR spectra of EBP before and after binding of iron. The lower panel (a) depicts the spectrum of free EBP while the upper panel (b) shows spectrum of iron-bound EBP.

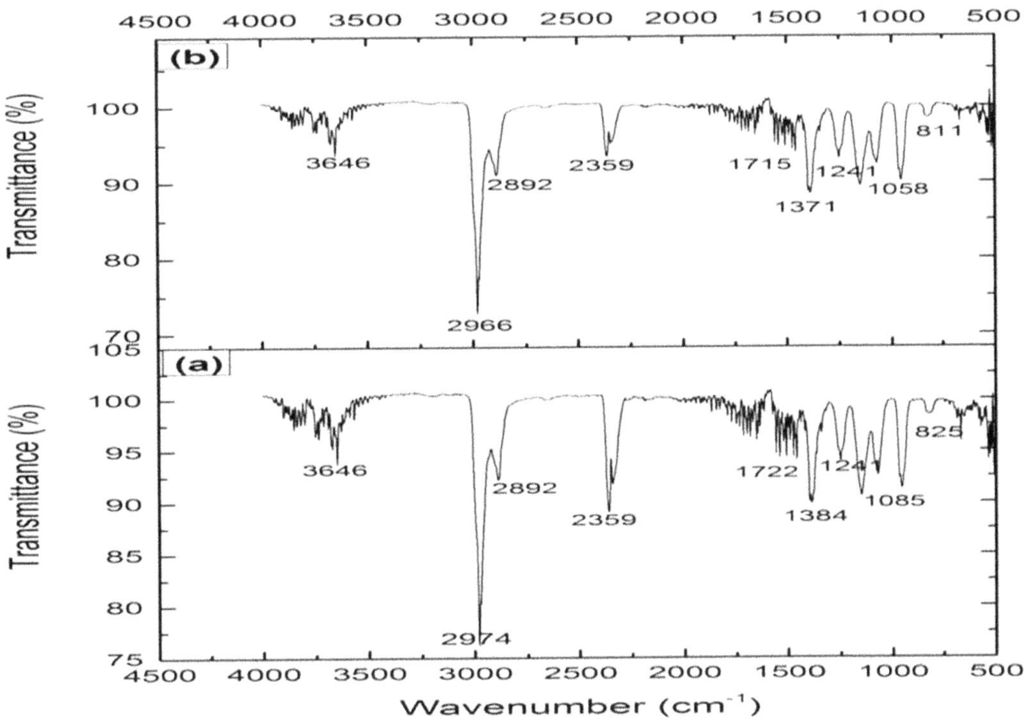

A comparison of XRD spectrum of native EBP with iron-challenged EBP revealed presence of iron in the latter samples. It is possible that the interaction of acidic –COOH groups with ferrous ions would result in precipitation of iron in the form of oxides. A similar mechanism of iron sol stabilization by organic acids, polyols and polysaccharides has also been reported whereby the acidic groups reduced the ferrous form to metallic iron and the acid cap stabilized the iron sol in solution (Carvalho et al., 2013; He and Zhao, 2007; Lodhia et al., 2009; Sahoo et al., 2005; Wu et al., 2008).

Cytotoxiciy

The cell viability of RAW 264.5 macrophages was not significantly ($p>0.05$) affected at 24 hours of incubation in the presence of the biopolymer (Table 4). Microscopic observations of the treated macrophages were also carried out. Cell morphology of macrophages in a monolayer culture following treatment with EBP (varying doses) displayed no distinct morphological changes at bactericidal doses (results not shown) even upon incubation for 48- 72 hours. These results indicated safety of the EBP.

There has been a profound interest, in recent times on extracellular microbial polymers. Mostly these EBPs are elaborated from the cell as a protection against harsh conditions of starvation, pH and temperature. In view of this the EBPs possess robust functional, rheological and physico-chemical properties. The net anionic makeup, a uniqueness of the Structure enables the biopolymer to sequester positively

Table 4. Impact of Pseudomonas EBP (10-50mg/L) on survival of RAW 264.7 cells after 24 hours incubation. Data represents mean±SD of two experiments performed in triplicate. No significant changes (p>0.05)(not shown) occurred after 48 hours.

Dose of biopolymer (mg/L)	Viability (%)
0	100% ±3.22
10	99.78% ± 4.13
20	98.97% ± 7.11
30	99.03% ± 5.34
40	99.64% ± 7.33
50	98.67% ± 6.77

charged heavy metal ions. Therefore bio-detoxification of terrestrial and aquatic systems contaminated with heavy metal is regarded as an ultimate benign and economic process.

CONCLUSION

The physicochemical characteristics of an exopolymer produced by *Pseudomonas* sp. and capable of binding ferrous ions were evaluated in the current study. The exopolymer, a polysaccharide could efficiently remove iron from water at ambient temperature. The adsorption was demonstrated by EDAX and XRD analysis and by Langmuir and Freundlich Isotherms. Carboxyl groups of the biopolymer play a major role for binding iron. In vitro toxicity studies established safety of the exobiopolymer. Overall, results of this study indicate that *Pseudomonas* sp. exobiopolymer can be further exploited as a good adsorbent for iron and has potential for treatment of water containing high levels of iron. However, further studies involving real time trials are necessary.

REFERENCES

Aryal, M., & Liakopoulou-Kyriakides, M. (2013). Binding Mechanism and Biosorption Characteristics of Fe(III) by *Pseudomonas* sp. cells. *Journal of Water Sustainability*, *3*(3), 117–131.

Bawuro, A. A., Voegborlo, R. B., & Adimado, A. A. (2018). Bioaccumulation of Heavy Metals in Some Tissues of Fish in Lake Geriyo, Adamawa State, Nigeria. *Journal of Environmental and Public Health*, *2018*, 7. doi:10.1155/2018/1854892 PMID:29887895

Carvalho, M. D., Henriques, F., Ferreira, L. P., Godinho, M., & Cruz, M. M. (2013). Iron oxide nanoparticles: The Influence of synthesis method and size on composition and magnetic properties. *Journal of Solid State Chemistry*, *201*(0), 144–152. doi:10.1016/j.jssc.2013.02.024

Emtiazi, G., Ethemadifar, Z., & Habibi, M. H. (2004). Production of extra-cellular polymer in *Azotobacter* and biosorption of metal by exopolymer. *African Journal of Biotechnology*, *3*(6), 330–333. doi:10.5897/AJB2004.000-2060

Ghosh, M., Ganguli, A., & Pathak, S. (2009). Application of a novel biopolymer for removal of Salmonella from poultry wastewater. *Environmental Technology, 30*(4), 337–344. doi:10.1080/09593330902732093 PMID:19492545

Gupta, P., & Diwan, B. (2017). Bacterial Exopolysaccharide mediated heavy metal removal: A Review on biosynthesis, mechanism and remediation strategies. *Biotechnology Reports (Amsterdam, Netherlands), 13*, 58–71. doi:10.1016/j.btre.2016.12.006 PMID:28352564

Hay, I. D., Wang, Y., Moradali, M. F., Rehman, Z. U., & Rehm, B. H. (2014). Genetics and regulation of bacterial alginate production. *Environmental Microbiology, 16*(10), 2997–3011. doi:10.1111/1462-2920.12389 PMID:24428834

He, F., & Zhao, D. (2007). Manipulating the size and dispersibility of zerovalent iron nanoparticles by use of carboxymethyl cellulose stabilizers. *Environmental Science & Technology, 41*(17), 6216–6221. doi:10.1021/es0705543 PMID:17937305

Horzum, N., Boyaci, E., Eroglu, A. E., Shahwan, T., & Demir, M. M. (2010). Sorption efficiency of chitosan nanofibers toward metal ions at low concentrations. *Biomacromolecules, 11*(12), 3301–3308. doi:10.1021/bm100755x PMID:21080700

Huang, J. H., Elzinga, E. J., Brechbuehl, Y., Voegelin, A., & Kretzschmar, R. (2011). Impacts of *Shewanella putrefaciens* strain CN-32 cells and extracellular polymeric substances on the sorption of As(V) and As(III) on Fe(III)-(hydr)oxides. *Environmental Science & Technology, 45*(7), 2804–2810. doi:10.1021/es103978r PMID:21375285

Ismail, B., & Nampoothiri, K. (2014). Molecular characterization of an exopolysaccharide from a probiotic *Lactobacillus plantarum* MTCC 9510 and its efficacy to improve the texture of starchy food. *Journal of Food Science and Technology, 51*(12), 4012–4018. doi:10.100713197-013-0928-8 PMID:25477674

Jamil, N., & Ahmed, N. (2008). Production of biopolymers by *Pseudomonas aeruginosa* isolated from marine source. *Brazilian Archives of Biology and Technology, 51*(3), 457–464. doi:10.1590/S1516-89132008000300004

Kaur, T., & Ghosh, M. (2015). *Acinetobacter haemolyticus* MG606 produces a novel, phosphate binding exobiopolymer. *Carbohydrate Polymers, 132*, 72–79. doi:10.1016/j.carbpol.2015.06.002 PMID:26256326

Kaur, T., & Ghosh, M. (2017). Characterization and toxicity of a phosphate-binding exobiopolymer produced by *Acinetobacter haemolyticus* MG606. *Journal of Water and Health, 15*(1), 103–111. doi:10.2166/wh.2016.176 PMID:28151444

Kaur, T., Sharma, J., Ganguli, A., & Ghosh, M. (2013). Application of biopolymer produced from metabolic engineered *Acinetobacter* sp. for the development of phosphate optoelectronic sensor. *Composite Interfaces, 21*(2), 143–151. doi:10.1080/15685543.2013.842116

Kousalya, G. N., Gandhi, M. R., Viswanathan, N., & Meenakshi, S. (2010). Preparation and metal uptake studies of modified forms of chitin. *International Journal of Biological Macromolecules, 47*(5), 583–589. doi:10.1016/j.ijbiomac.2010.07.014 PMID:20708027

Kumar, M., & Puri, A. (2012). A review of permissible limits of drinking water. *Indian Journal of Occupational and Environmental Medicine*, *16*(1), 40–44. doi:10.4103/0019-5278.99696 PMID:23112507

Kumar, V., Bharti, P. K., Talwar, M., Tyagi, A. K., & Kumar, P. (2017). Studies on high iron content in water resources of Moradabad district (UP), India. *Water Science*, *31*(1), 44–51. doi:10.1016/j.wsj.2017.02.003

Kurniawan, T. A., Chan, G. Y. S., Lo, W.-H., & Babel, S. (2006). Physico-chemical treatment techniques for wastewater laden with heavy metals. *Chemical Engineering Journal*, *118*(1-2), 83–98. doi:10.1016/j.cej.2006.01.015

Latif, A., Bilal, M., Asghar, W., Azeem, M., Ahmad, M. I., Abbas, A., & Shahzad, T. (2018). Heavy Metal Accumulation in Vegetables and Assessment of their Potential Health Risk. *Journal of Environmental and Analytical Chemistry*, *5*(1), 234. doi:10.4172/2380-2391.1000234

Lodhia, J., Mandarano, G., Ferris, N. J., Eu, P., & Cowell, S. F. (2009). Development and use of iron oxide nanoparticles (Part 1): Synthesis of iron oxide nano-particles for MRI. *Biomedical Imaging Intervention Journal*, *6*(2), e12. PMID:21611034

Masawat, P., Harfield, A., Srihirun, N., & Namwong, A. (2016). Green Determination of Total Iron in Water by Digital Image Colorimetry. *Analytical Letters*, *50*(1), 173–185. doi:10.1080/00032719.2016.1174869

Moppert, X., Le Costaouec, T., Raguenes, G., Courtois, A., Simon-Colin, C., Crassous, P., & Guezennec, J. (2009). Investigations into the uptake of copper, iron and selenium by a highly sulphated bacterial exopolysaccharide isolated from microbial mats. *Journal of Industrial Microbiology & Biotechnology*, *36*(4), 599–604. doi:10.100710295-009-0529-8 PMID:19198908

Ngah, W. S., Ab Ghani, S., & Kamari, A. (2005). Adsorption behaviour of Fe(II) and Fe(III) ions in aqueous solution on chitosan and cross-linked chitosan beads. *Bioresource Technology*, *96*(4), 443–450. doi:10.1016/j.biortech.2004.05.022 PMID:15491825

Omoike, A., & Chorover, J. (2004). Spectroscopic study of extracellular polymeric substances from *Bacillus subtilis*: Aqueous chemistry and adsorption effects. *Biomacromolecules*, *5*(4), 1219–1230. doi:10.1021/bm034461z PMID:15244434

Oyem, H. H., Oyem, I. M., & Usese, A. I. (2015). Iron, manganese, cadmium, chromium, zinc and arsenic groundwater contents of Agbor and Owa communities of Nigeria. *SpringerPlus*, *4*(1), 104. doi:10.118640064-015-0867-0 PMID:25853026

Paniagua-Michel, J. J., Olmos-Soto, J., & Morales-Guerrero, E. R. (2014). Algal and microbial exopolysaccharides: New insights as biosurfactants and bioemulsifiers. *Advances in Food and Nutraceutical Research*, *73*, 221–257. doi:10.1016/B978-0-12-800268-1.00011-1 PMID:25300549

Reiad, N. A., Abdel Salam, O. E., Abadir, E. F., & Harraz, F. A. (2012). Adsorptive removal of iron and manganese ions from aqueous solutions with microporous chitosan/polyethylene glycol blend membrane. [China]. *Journal of Environmental Sciences (China)*, *24*(8), 1425–1432. doi:10.1016/S1001-0742(11)60954-6 PMID:23513684

Sahoo, Y., Goodarzi, A., Swihart, M. T., Ohulchanskyy, T. Y., Kaur, N., Furlani, E. P., & Prasad, P. N. (2005). Aqueous ferrofluid of magnetite nanoparticles: Fluorescence labeling and magnetophoretic control. *The Journal of Physical Chemistry B, 109*(9), 3879–3885. doi:10.1021/jp045402y PMID:16851439

Shokoohi, R., Saghi, M. H., Ghafari, H. R., & Hadi, M. (2009). Biosorption of iron from aqueous solution by dried biomass of activated sludge. *Iranian Journal of Environmental Health Sciences & Engineering*, *6*(2), 107–114.

Singh, R. P., Jain, S., & Ramarao, P. (2013). Surfactant-assisted dispersion of carbon nanotubes: Mechanism of stabilization and biocompatibility of the surfactant. *Journal of Nanoparticle Research*, *15*(10), 1985. doi:10.100711051-013-1985-7

Singh, R. P., Jhamb, S. S., & Singh, P. P. (2008). Effects of morphine during *Mycobacterium tuberculosis* H37Rv infection in mice. *Life Sciences*, *82*(5-6), 308–314. doi:10.1016/j.lfs.2007.11.024 PMID:18191418

Staniek, H., & Wojciak, R. W. (2018). The Combined Effects of Iron Excess in the Diet and Chromium(III) Supplementation on the Iron and Chromium Status in Female Rats. *Biological Trace Element Research*, *184*(2), 398–408. doi:10.100712011-017-1203-z PMID:29164513

Tapia, J. M., Munoz, J., Gonzalez, F., Blazquez, M. L., & Ballester, A. (2013). Sorption of ferrous and ferric iron by extracellular polymeric substances (EPS) from acidophilic bacteria. *Preparative Biochemistry & Biotechnology*, *43*(8), 815–827. doi:10.1080/10826068.2013.805624 PMID:23876140

Tapia, J. M., Munoz, J. A., Gonzalez, F., Blazquez, M. L., & Ballester, A. (2011). Mechanism of adsorption of ferric iron by extracellular polymeric substances (EPS) from a bacterium *Acidiphilium* sp. *Water Science and Technology*, *64*(8), 1716–1722. doi:10.2166/wst.2011.649 PMID:22335116

Wang, W., Bu, W., Wang, L., Palo, P. E., Mallapragada, S., Nilsen-Hamilton, M., & Vaknin, D. (2012). Interfacial Properties and Iron Binding to Bacterial Proteins That Promote the Growth of Magnetite Nanocrystals: X-ray Reflectivity and Surface Spectroscopy Studies. *Langmuir*, *28*(9), 4274–4282. doi:10.1021/la205074n PMID:22316331

Wei, W., Wang, Q., Li, A., Yang, J., Ma, F., Pi, S., & Wu, D. (2016). Biosorption of Pb (II) from aqueous solution by extracellular polymeric substances extracted from *Klebsiella* sp. J1: Adsorption behavior and mechanism assessment. *Scientific Reports*, *6*(1), 31575. doi:10.1038rep31575 PMID:27514493

WHO. (1996). *World Health Organization & International Programme on Chemical Safety. Guidelines for drinking-water quality. Health criteria and other supporting information* (2nd ed.). Geneva: World Health Organization. https://www.who.int/iris/handle/10665/38551

Wu, W., He, Q., & Jiang, C. (2008). Magnetic Iron Oxide Nanoparticles: Synthesis and Surface Functionalization Strategies. *Nanoscale Research Letters*, *3*(11), 397–415. doi:10.100711671-008-9174-9 PMID:21749733

Xue, G., He, S. B., & Wang, X. Z. (2006). Factors affecting biological removal of iron and manganese in groundwater. *Huan Jing Ke Xue*, *27*(1), 95–100. PMID:16599128

Yang, H., Li, D., Zhang, J., Hao, R., & Li, B. (2004). Design of biological filter for iron and manganese removal from water. *Journal of Environmental Science and Health A Toxic Hazard Substance Environ Eng, 39*(6), 1447-1454.

Yen, Y., Cheng, B., Chan, C., Lin, C., & Chen, H. (2018). Heavy Metal Components in Blood and Urinary Stones of Urolithiasis Patients. *Biological Trace Element Research, 185*(2), 266–274. doi:10.100712011-018-1253-x PMID:29488068

Yu, G. H., He, P. J., & Shao, L. M. (2009). Characteristics of extracellular polymeric substances (EPS) fractions from excess sludges and their effects on bioflocculability. *Bioresource Technology, 100*(13), 3193–3198. doi:10.1016/j.biortech.2009.02.009 PMID:19269815

Zhang, Z., Cai, R., Zhang, W., Fu, Y., & Jiao, N. (2017). A Novel Exopolysaccharide with Metal Adsorption Capacity Produced by a Marine Bacterium *Alteromonas* sp. JL2810. *Marine Drugs, 15*(6), 175. doi:10.3390/md15060175 PMID:28604644

Chapter 15
Recent Advancements in Bioremediation of Metal Contaminants

Yuvarajan Ragunathan
Mahendra Arts and Science College (Autonomous), India

Kannan Nallakumar
Mahendra Arts and Science College (Autonomous), India

Thirumalaisamy Rathinavel
https://orcid.org/0000-0003-3137-5895
Mahendra Arts and Science College (Autonomous), India

Muthusamy Govarthanan
Kyungpook National University, South Korea

Selvankumar Thangaswamy
https://orcid.org/0000-0002-3500-8681
Mahendra Arts and Science College (Autonomous), India

ABSTRACT

Biofilms are an accumulation of single or various populations of microorganisms that are present on the surfaces through membrane-bound substances due to the gene expression, which differs from free-floating expression and leads to expressed genes regulating biofilm formation and development. In this regard, recent advances in microbial-based heavy metals have propelled bioremediation as a prospective alternative to conventional techniques. Adsorption and biodegradation of organic contaminants and the immobilization, mobilization, and/or transformation of metals are the main remediation processes that can be mediated by the action of several microorganisms surviving in hostile environments with high concentrations of pollutants. The chapter discussed the formation and regulation of biofilms to degrade the metal contaminant, the importance of gene transfer, and applications of biofilm-mediated bioremediation processes.

DOI: 10.4018/978-1-7998-4888-2.ch015

INTRODUCTION

Harmful materials are released into the environment as a result of human activities. Different pollution arising from the farming industry, transport and energy use, enter the air, soil, freshwater and the oceans. Pollutions effects can be small scale or global, gradual or dramatic and include threats to wildlife and health problems for the public, pollution is the present environmental issue. Pollutions are entering into the environment in the forms of Gases and smoke from industry and vehicles drift into the air household sewage, agricultural sprays and other liquids are released on to land and into the oceans and rivers. Solids too, such refuse and mining waste are dumped on the ground and into the sea.

The heavy metal consists of metal and metalloid elements that have a rather high density ranging from 3.5 to 7g cm^3 and is toxic or poisonous at low concentrations and includes mercury, cadmium, arsenic, chromium, thallium, zinc, nickel, copper and lead. It was widely documented and frequently applied to the widespread pollutants of soils and water bodies (Duffus, 2002). Refinement of heavy metals from wastewater has been a challenge for a while in that, most of the heavy metal salts are soluble in water very quickly and form aqueous solution and so, cannot easily be separated using ordinary physical means. However, several different conventional treatment processes are commonly employed to remove heavy metals from industrial wastewater before their discharge into the environment (Fomina and Gadd, 2014). Moreover, such treatments produce large amounts of sludge that are not environmentally friendly and need to be treated with great difficulties. Ion exchange membrane technologies and activated carbon adsorption processes are extremely expensive (Eccles, 1999).

In most natural environmental system, microbial like bacteria, fungal and algae are commonly found in close association with surfaces and interfaces in the form of multicellular aggregates glued together with the slime they secrete (Wimpenny, 2000). Biofilms are bacterial species in which cells are fixed in a matrix of extracellular polymeric compounds attached to a surface(Branda, 2005). Living in biofilms helps protect bacteria from toxic conditions and the formation of biofilms appears to be an important factor in the disease cycle of bacterial pathogens in both animals and plants. There is one chance for the increased resistance to environmental stresses observed in biofilm cells appears to be the increase in the outer layer surface of accumulation of persister cells within the biofilm (Davey and Toole, 2000). The molecular mechanisms of the biofilm formation differ from among the varies species, and even slight modification among the varies strains of the same species. Conversely, some features are recognized as general attributes of biofilm formation (Lewis, 2005). In this review, work discussed the recent advancement techniques of control of metal contaminant water using biofilms such as bacterial, fungal, algae and yeast.

Origin of Heavy Metals

Heavy metals are classified as different due to their chemical properties and are used broadly in electronics, machines and the artefacts of everyday life, as well as in high-tech applications. As a result, they can enter into the aquatic and edible systems of human and animals from the differentiation of anthropogenic sources as well as from the natural geochemical weathering of soil and rocks. The main reasons for contamination such as mining wastes, landfill leaches, municipal wastewater, urban reservoirs and industrial wastewaters, particularly from the electroplating, electronic and metal-finishing industries (Fig.1.). With the increasing generation of metals from technologies activities, the problem of waste disposal has become one of paramount importance. Many aquatic environments face metal concentrations that exceed water quality criteria designed to protect the environment, animals and human. The problems are exacerbated because metals tend to be transported with sediments, are persistent in the environment and can bioaccumulate in the food chain. The heavy metal contaminations were consisting of the oldest cases of environmental pollution in the world, for example, Cu, Hg and Pb mining, smelting and utilization by ancient peoples of civilizations in the developing countries, such as the Romans and the Phoenicians (Monds, O'Toole, 2009.).

Application of Heavy Metals

Cadmium is widely used in several industries to manufacturing the solders, batteries, electronic devices, ceramics, photography, insecticides, and metallurgical activities. It can be used in the modified form of metal-ore refining, cadmium containing pigments, alloys and electronic compounds, phosphate fertilizers, detergents and refined petroleum products. Copper is spreading through Mining, metallurgy and industrial applications are the major sources of copper exposure in the environment. Mining and metallurgical processing of zinc ores and burning of coal are the major sources of zinc in the air, soil and water. Nickel is one of the heavy metal is affected the people and its salts are used in several industrial applications such as in electroplating, automobile and aircraft parts, batteries, coins, spark plugs, cosmetics and stainless steel, and is used extensively in the production of both nickel-cadmium batteries on an industrial scale. It enters into the water bodies naturally by weathering of rocks and soils and through the leaching of the minerals (Ravindra et al, 2015; Dojlido and Best, 1993). Arsenic enters the environment through the natural weathering of rocks and anthropogenic activities, mining and smelting processes, and pesticide use and coal combustion. The toxicity of arsenic as a result of the contamination of groundwater bodies and surface waters is of great concern. Arsenic exists as arsenate and arsenite, in most of the groundwater (Volesky, 1990).

Heavy Metals in Water

There are one billion people cannot be utilised to clean drinking water and two million peoples were affected per year; due to the water contamination is one of the world's leading causes of death (Cullen and Reimer, 1989). This problem is only expected to degenerate as the World Health Organization (WHO) estimates that climate change will limit access to clean water for as much as half of the world's population (Gleick, 2003) and a recent United Nations report projects that the world could face a 40% water shortage in as few as 15 years (WWAP, 2015). The urgent needs of energy production and an exponential increase in heavy metal use in industrial processes have caused a rise in human exposure to toxic heavy metals in recent decades (Tchounwou,2012). The high toxicity and occurrence of cadmium, chromium, lead, arsenic, and mercury place them among the greatest concern. These metals, which play no role in human homeostasis, induce multiple organ damage, cause birth defects, and are classified carcinogens (Shannon, 2008). To courage the environmental and human well-being, we must find new solutions for the cheap, energy-efficient remediation of trace contaminants from water (Sholl,2016).

Treatment for Removal of Heavy Metals

Many researchers reported that the various physicochemical and biological processes are commonly used to remove heavy metals from industrial wastewater before delivered into the environment (Fomina

Table 1. The standard metal concentration in drinking water and the health effects

Metal	Effects	Standards level for drinking
Lead	· Toxic to humans, aquatic fauna and livestock · High doses cause metabolic poison · Tiredness, irritability anaemia and behavioural changes of children · Hypertension and brain damage · Phytotoxic	· By Environmental Protection Agency maximum concentration: 0.1 mg L^{-1} · European Community: 0.5 mg L^{-1} · Regulation of water quality (India) 0.1 mg L^{1}.
Nickel	· High concentration can cause DNA damage · Eczema of hands · High phytotoxicity · Damaging fauna	· By the Environmental Protection Agency maximum concentration: 0.1 mg L^{-1} · By European Community: 0.1 mg L^{-1} · Regulation of water quality (India) 0.1 mg L^{-1}
Chromium	· Necrosis nephritis and death in man (10 mg kg1 of body weight as hexavalent chromium) · Irritation of gastrointestinal mucosa	· By the Environmental Protection Agency, maximum concentration: (hexavalent and trivalent) total 0.1 mg L^{-1} · By European Community: 0.5 mg L^{-1} · Regulation of water quality (India) 0.1 mg L^{-1}
Copper	· Causes damage in a variety of aquatic fauna · Phytotoxic Mucosal irritation and corrosion Central nervous system · Irritation followed by depression	· By the Environmental Protection Agency maximum concentration: 1.0 mg L^{-1} · By European Community: 3 mg L^{-1} · Regulation of water quality (India) 0.01 mg L^{-1}
Zinc	· Phytotoxic · Anemia Lack of muscular coordination Abdominal pain etc.	· By the Environmental Protection Agency maximum concentration: 5 mg L-1 · By European Community: 5 mg L^{-1} · Regulation of water quality (India) 0.1 mg L^{-1}
Cadmium	· Cause serious damage to kidneys and bones in humans · Bronchitis, emphysema, anaemia Acute effects in children	· By the Environmental Protection Agency maximum concentration: 0.005 mg L^{-1} · By European Community: 0.2 mg L^{-1} Regulation of water quality (India) 0.001 mg L^{-1}
Mercury	· Poisonous Causes mutagenic effects Disturb the cholesterol	· By the Environmental Protection Agency maximum concentration: 0.002 mg L^{-1} · By European Community: 0.001 mg L^{-1} · Regulation of water quality (India) 0.004 mg L^{-1}
Arsenic	· Causes toxicological and carcinogenic effects cause melanosis, keratosis and hyperpigmentation in humans · Genotoxicity through the generation of reactive oxygen species and lipid peroxidation · Immunotoxic Modulation of co-receptor expression	· World Health Organization guideline of 10 mg L^{-1} By European Community: 0.01 mg L^{-1} · Regulation of water quality (India): 0.05 mg L^{-1}

and Gadd,2014). Conservative methods like electrochemical treatment, ion exchange, precipitation, osmosis, evaporation, and sorption were used for the water treatment because of low cost, and some of them are not environmentally friendly (Mulligan et al, 2001; Kadirvelu et al, 2002). On the other hand, bioremediation processes show promising results for the removal of metals, even when present in very low concentrations where physicochemical removal methods fail to operate. Furthermore, this is an eco-friendly and economically convenient option. The bioremediation strategy is based on the high metal binding capacity of biological agents, which can remove heavy metals from contaminated sites with high efficiency. The researchers can report that the microorganisms were considered as a biological tool for metal removal because it can be used to significant concentration, remove, and recover heavy metals from contaminated aquatic environments (Riggle and Kumamoto,2000). Many scientific peoples carried out these type of studies have been conducted by using microbial source for the uptake of heavy metals in polluted waters as an alternative strategy to conventional treatments (Tsezos and Volesky,1981; Gadd and White,1993, Texier,1999).

Advancement Water Treatment Methods

Biofilms are associated with many microbial especially bacterial species that are adherence on solid surfaces and moisture environment condition (Costerton et al, 1994). They occur nearly in every moist environment where required nutrient energy is available and surface attachment can be achieved. Biofilms can be formed by a single bacteria cell species, although they can also consist of many species of bacteria, fungi, algae and protozoa (Virginia Deibel, and Jean Schoeni,2003). Approximately 97% of the biofilm matrix is either water, which is removed to the capsules of microbial cells or solvent, the physical properties like viscosity are determined by the solutes dissolved in it (Sutherland,2001). The formation of Extracellular Polymeric Substances (EPS) enhances the ability of the cell to adhere to surfaces with the help of flagella, pili, fimbriae, or glycocalyx of the bacterial cell wall (Hans-Curt Flemming et al, 2007).

The diffusion processes that occur within the biofilm matrix are dependent on the water binding capacity and mobility of the biofilm. Heavy metals utilized by this microbial biomass is a new eco-compatible and economically feasible application that has been developed to remove heavy metals from wastewater (Gupta et al, 2016), and studies have exposed that interaction of a microbial substance with heavy metals reduced heavy metal ion concentrations in solution (Meylan et al, 2003, Ogbuagu et al 2011). This bioremediation option is based on the high concentration of metal-binding capacity of biological agents, which remove heavy metals from wastewater or contaminated sites with high efficiency. In these research findings has revealed that they act as metal bio-adsorbent as they have metal-sequestering properties (Sagar Aryal et al, 2015).

The microbial-based biofilm can also live in contaminated habitats because they are metabolically able to exploit contaminants as potential energy sources (Ilyina et al 2003; Zhang et al 2014). In biological treatment or removal of heavy metals, microorganisms with biological activity such as algae, bacteria, fungi and yeast can be used in their naturally occurring forms. The efficient removal of heavy metals from wastewater is dependent on several factors, including sludge concentration, the solubility of metal ions, pH, the metallic concentration and wastewater pollution load (Chipasa, 2003).

A similar study was concluded that the removal of heavy metals from static wastewater using biofilms; a bioremediation technology was considered as very important especially in developing countries such as Nigeria where wastewater discharge regulations are flouted and treatment does not have top priority due to high cost of treatment facilities. Meylan et al. (2006) and Ogbuagu et al. (2011) have conducted experiments on metal accumulation in algal biofilm in lotic streams and observed that biofilms are an efficient model for the removal of metals in solution. However, reports on the application of this biological technique in static environments which mimic industrial effluent reservoirs are lacking.

Kurniawan and Yamamoto (2013) reported that the characterization of the biofilm polymer is revealed by electrophoretic mobility measurements and acid-base titrations. To clarify the ability of biofilm for biosorption of pollutant ions, they study the adsorption kinetics of lithium-ion to biofilms formed in winter and spring. The followings are observed as the results of this study:

1) Both negative and positive charged sites exist on biofilm polymer;

2) The ion adsorption by biofilm polymer is a physicochemical process where the electrostatic interaction between the ion and the charged sites in biofilm polymers is a main driving force;
3) The adsorbing ion promotes ions desorption from biofilms through ion exchange mechanism;
4) The biofilms components change seasonally and appear to affect the ability of biofilms to adsorb ions. From all of the similar studies, this study is the first one investigating the utilization of natural biofilms formed in different seasons for biosorption of heavy metal ion that is lithium-ion. This study also reveals that biofilm may become a promising adsorbent for pollutant ions.

The basic treatment of regimes for remediation of heavy metals ions include methods like coagulation, chemical precipitation, electrodialysis, evaporative recovery, floatation, flocculation, ion exchange, nano-filtration, reverse osmosis, ultrafiltration etc. [Lakherwal, 2014]. Although effective, these methods are usually expensive due to high energy and reagent requirements. Moreover, they generate a large amount of toxic accumulation and its byproducts, which pollutes the environment. Many times the researchers may result in incomplete and unpredictable because of quantitative removal cannot be accurately estimated (Gavrilescu, 2004; Alluri,2007). Hence, there is an imperative need to devise effective, efficient, economic and environmentally safe strategies which can decrease the heavy metal ion concentration from toxic to safe limits in the environment. Sorption with specific reference to metal ions can be defined as any phenomenon that involves their association that is ranging from electrostatic to covalent with peripherally available one or more functional groups on sorbent material. When the sorbent involved in such a reaction is a biological agent, the phenomenon is defined as biosorption.

Prokaryotic as well as eukaryotic microbial biomasses both living and alive cells, like bacteria, fungi, yeast and few microalgae are such emerging candidates of biosorption which can uptake and reduce heavy metal ion concentration from contaminated water sources in an eco-friendly manner (Das et al,2008; Rani et al,2010; Comte et al 2008). When the suggested that specifically of bacterial cells, heavy metal ions in both particulate, as well as insoluble form, can potentially be accumulated by intact bacterial cells and their byproducts. Extracellular polymeric substances are such complex blend of high molecular weight microbial biopolymeric secretory byproducts. They contain proteins, polysaccharides, uronic acids, humic substances, lipids etc. Their most essential constituent, having ion sequestration capability, is extracellular polysaccharides or exopolysaccharide (EPS). Primarily it is composed of complex high molecular weight organic macro-molecules like polysaccharide along with smaller proportions of protein and uronic acid (Lau et al, 2005, Vandevivere and Kirchman,1993).

Immobilization Techniques

The recent advancements methods of water treatment by immobilization techniques have excelled the biological reaction kinetics exclusively in terms of reaction rate and specificity enhancement. Numerous reports also verify that attachment of bacterial cells to solid surfaces stimulates the exopolysaccharide production without altering the specific growth rate (Ozdemir et al, 2005(a)). This was demonstrated by *Chryseomonas luteola* immobilized in alginate bead along with its EPS for examining cadmium, cobalt, nickel and copper ion adsorption. Investigators make up *Chryseomonas luteola* and it's EPS with calcium salt of natural polymer – alginate in the form of beads in different combinations. Interestingly the combination of bacterial EPS immobilized in calcium alginate resulted in maximum cadmium and cobalt ion sequestration from aqueous solutions reportedly 64.10 mg/g and 55.25 mg/g of EPS beads respectively [Ozdemir et al,2005(b)). Langmuir isotherm also depicted an adsorption capacity of 1.989 mmol of

Cu_2^+/g and 1.224 mmol of Ni_2^+/g dry weight of alginate immobilized culture EPS. While alginate bead alone displayed comparatively low metal ion adsorption efficiency reportedly 1.505 mmol of Cu_2^+/g dry weight and 0.996 mmol $Ni_2\ ^+$/g dry weight. The reaction reached saturation within one hour of initiation at pH and temperature optima of 6 and 25°C respectively (Hassiba et al, 2014). This advancement studies in the area of immobilization have been reported with GRAS status and also in bacterial EPS. The utility of GRAS *Paenibacillus Polymyxa* EPS immobilized in agar beads was evaluated for the uptake of lead ions regulating solution pH, initial metal ion concentration etc. Maximum adsorption was found to be 111.11 mg/g immobilized EPS as per the Langmuir model and adsorption saturation was reached within 120 min of reaction commencement. The P^H optimum was found to be 5 (Ozdemir et al, 2003).

Bacterial Based Biofilm

Exopolysaccharides (EPS) of pure culture dead biomass, isolated from activated sludge has been one such example. *Ochrobactrum anthropi* removed cadmium ions along with other toxic metals under specific influence of pH and initial metal ion concentration (Rani et al, 2010). The organism not merely exhibited peripheral metal ion adsorption but also revealed high metal tolerance (up to 30 mg/L of cadmium ion concentration). Maximum uptake capacity of the biosorbent was found to be 29.5 mg/g at 100.6 ppm initial load. It depicted chromium ion tolerance up to a limit of 16 ppm and adsorption maxima of 57.8 mg/g of EPS from the maximum initial load of 280 ppm within two hours of reaction. The adsorption optimum was contrastingly observed at the acidic P^H of 2. Freundlich and Langmuir adsorption models established that dead biomass exopolysaccharides, chelated copper ions maximally at the acidic P^H of 2 and reached absorption equilibrium within two hours with a capacity of 26 mg/g EPS at the initial metal load of 91.6 ppm. However, a comparative metal ion adsorption analysis between dead and immobilized biomass associated EPS from electroplating effluent (heavy metal resistant) isolates, stated dead biomass to be little lagging than immobilized, for copper, cadmium and lead ion uptake. In a recent study life and dead biomass bound EPS of three different bacteria were evaluated for their chromium ion sorption capacity as a function of pH, temperature and initial metal ion concentration. Dead biomass EPS of *Bacillus cereus*, *Bacillus pumilus* and *Pantoea agglomerans* illustrated slightly superior uptake tendency at 37°C each showing more than 85% reduction in initial metal load. The pH optimum was reported to be again between 2 and 3 for the said species. The peculiar trend was repeatedly observed in several investigations where the pH range was lower for chromium and several other metal ion adsorption irrespective of the type of bacterial species. The reason (for chromium ions) being the repulsion between negatively charged chromate ions ($HCrO^{-4}$, CrO^{-7}, $Cr_4O_{13}^{-2}$, $Cr_4O_{10}^{-2}$) and negatively charged groups on EPS and cell surface. The lowering of pH creates an overall positive charge on account of functional group protonation and increased hydronium ion concentration (Sultan et al, 2012).

Water is might be germ-free before being distributed to the endpoint that is utilising the consumers and its microbial level before leaving the treatment plant should be within limits set by water authorities. Furthermore, though the process of sterilization significantly reduces the number of microorganisms, that means the germs are stable condition, allowed to surviving of microbes under favourable conditions. This decline in water quality may be moved to recovery and subsequent growth of sub mortally damaged bacteria due to system deficiencies such as cross-connections, broken water mains and contamination during bulk storage. Moreover, these bacterial cells can attach and form biofilms as accumulation on the surfaces of piping material from which cells may be released into the flow (Camper et al, 1998). The majority of bacteria in the drinking water system occur in biofilms rather than in the water phase

(Szewzyk et al, 2000). The organisms in biofilms tend to become more resistant to antibiotics and disinfectants thereby become a reservoir for the subsequent spread of pathogenic organisms. Besides, they offer increased virulence and resistance that potentially reduces the LD_{50} (Lethal Dose) by increasing the viable organisms to survive and pass through the human stomach and reach the intestine (Costerton et al, 1995). Moreover, biofilm can influence the taste and odour of the water & when developed on ferrous metal surfaces, they may cause corrosion of the pipes and also the release of iron particles into the water (Walch, 1992).

James *et al.*, 2000 reported that the microbial load of water from the Dental unit water system in general dental practices and the biofouling of DUWS tubing. Water and tube samples were taken from 55 dental surgeries in southwestern England. Contamination of water was determined by viable counts on various selective media, and biofouling was determined by using microscopic and image analysis techniques. Microbial loading ranged from 500 to 10^5 CFU. ml^{-1} in 95% of DUWS water samples, it exceeded European Union drinking water guidelines and in 83% it exceeded American Dental Association DUWS standards. Among visible bacteria, 68% were viable by *Bac*Light staining, but only 5% of this "viable by *Bac*Light" fraction produced colonies on agar plates. *Legionella pneumophila, Mycobacterium* spp., *Candida* spp., and *Pseudomonas* spp. were detected in one, five, two, and nine different surgeries, respectively. Presumptive oral streptococci and *Fusobacterium* spp. were detected in four and one surgeries, respectively, suggesting back siphonage and failure of entire detection devices. Hepatitis B virus was never detected. Decontamination strategies (5 of 55 surgeries) significantly reduced biofilm coverage but significantly increased microbial numbers in the water phase (in both cases, $P < 0.05$). Microbial loads were not significantly different in DUWS fed with soft, hard, deionized, or distilled water or in different DUWS (main, tank, or bottle-fed). Microbiologically, no DUWS can be considered cleaner than others. DUWS deliver water to patients with microbial levels exceeding those considered safe for drinking water (James et al, 2000).

Water quality in the drinking water system plays an important role in the general health and performance of broiler chickens. Conditions in the DWS of broilers are ideal for microbial biofilm formation. Since pathogens might reside within these biofilms, they serve as a potential source of waterborne transmission of pathogens to livestock and humans. Knowledge about the presence, importance and composition of biofilms in the DWS of broilers is largely missing. In this study, the researchers to monitor the occurrence of the bacterial in the way of chemicals and microbiological characterise based biofilms in the DWS of five broiler farms. The bacterial load after disinfection in DWSs was assessed by sampling with a flocked swab followed by enumerations of total aerobic flora and *Pseudomonas spp*. The most accumulation of the microbial species was identified to make biofilm-forming capacity was evaluated. For this method, some secondary metabolites like, proteins, carbohydrates and uronic acids were quantified to analyse the presence of extracellular polymeric substances of biofilms. Despite disinfection of the water and the DWS, average TAC was 6.03 ± 1.53 log CFU/20cm^2. Enumerations for *Pseudomonas spp*. were on average 0.88 log CFU/20cm^2 lower. The most identified dominant species from TAC were *Stenotrophomonas maltophilia*, *Pseudomonas geniculata* and *Pseudomonas aeruginosa*. However, in this species level, most of the identified microorganisms were farm unique characteristics. Almost all the isolates belonging to the three most abundant species were strong biofilm producers. Overall, nighty two percentages of all tested microorganisms were able to form biofilm under lab conditions. Furthermore, sixty-three percentages of the DWS surfaces appeared to be contaminated with microorganisms combined with at least one of the analysed chemical components, which is indicative for the presence of biofilm. *Stenotrophomonas maltophilia*, *Pseudomonas geniculata* and *Pseudomonas aeruginosa* are considered that the presence and involved pathogens and could consequently be a potential risk for animal health. Furthermore, the biofilm-forming capacity of these organisms could stimulate and supportive measures of the attachment of other bacterial pathogens such as *Campylobacter spp.* and *Salmonella spp.*

Fugal Biofilm

The heavy metal treatment using physiochemical processes with precipitation method produces undesirable effects due to its generation of hazards gas such as sulfide and secondary sludge wastes moreover its pollute nearby water sources. Bioremediation gains much attention due to efficient heavy metal remediation through its eco-friendly reactions. Also, there is growing hope scientific community especially molecular biology and bioengineering technology enlarge the microorganism efficiency at reasonable environmental cleanup bioremediation of heavy metals. Microbes world has already multiplied through the various mechanism for biologic chemical precipitation using H_2S producing. Bioremediation using one of the interesting group of microbes namely sulfate-reducing microorganisms (SRMs), utilizes sulfate as their terminal electron acceptor to produce H_2S leading to precipitation of nearby heavy metals. SRMs are obligate anaerobic microorganism requires optimal culture conditions and grow slowly.

Fungal culture mainly utilized for its growth by the biochemical and ecological capacity to degrade a wide range of environmental pollutants. The ability of fungi to form especially, mycelia network growth using their catabolic enzymes might be breaking their survival and it has utilized environmental pollutants as a substrate to make fungi and more advantageous over other biological agents for bioremediation processes. Also, metabolic and ecological features of fungi suited for use in various bioremediation and waste treatment processes. Yeast act as better biological alternatives to replace this traditional chemical precipitation method to overcome such ecosystem pollution. Yeast was naturally produced sulfides through yeast sulfate assimilation pathway, which will provide a solution to the handling of two important existing problems such as sulfide gas and heavy metal removal. Heavy metals are precipitated by H_2S generated through sulfate assimilation pathway of Yeast. These engineered yeasts provide a natural, environmentally responsible, low-cost H_2S source while also simplifying H_2S storage and transportation. Strategic knockouts and optimal culture conditions will helpful for efficient removal of heavy metal from various samples. Heavy metals such as mercury, lead and copper from numerous samples effectively removed by these sulfide-producing yeasts. Yeast displays biomineralization peptides on their surfaces which will eventually help to control the size distribution and crystallinity of precipitated heavy metal sulfide nanoparticles inside yeast cells. Yeast-based bioremediation removes heavy metals bioremediation and it also offers a viable platform for metal re-extraction, which is hopeful for other yeast-based bioremediation processes [Sharon Maes et al, 2019]. Future research investigation is required on biomineralization peptides to improve metal sulfide formation and capture. Further designing of biomineralization peptides have two advantages applications that are selective bioremediation of heavy metals from the polluted sample. Highly toxic elements such as cadmium and mercury in drinking waters can be selectively removed as less-toxic elements with engineered biomineralization peptides using known heavy-metal binding motifs found in fungal cells. Bioengineering of yeast cells to facilitate precipitation of multi-metal removal as their sulfides during effluent treatment. Genetically engineered fungal cells such as *Saccharomyces cerevisiae* (CP_2HP_3) can uptake cadmium and zinc simultaneously.

Fungi are used as bio-absorbents for the removal of toxic heavy metals with excellent capacities for metal uptake and recovery (Fu et al, 2012). Several researchers proved these kinds of studies showed that fungal cells play a significant role in the adhesion and removal of inorganic chemicals from wastewater samples. *Aspergillus sp.* used for the efficient removal (85%) of chromium in tannery wastewater in an optimal condition compared to a 65% chromium removal from the tannery effluent in non-optimal conditions (Srivastava and Thakur,2006). Organic pollutants may hamper the growth of the fungal organism in non-optimal conditions of tannery effluent. *Coprinopsis atramentaria* is bioaccumulated 76% of Cd_2^+ and 94.7% of Pb_2^+ effectively (Lakkireddy and Kües, 2017). Therefore, it has been documented as an effective accumulator of heavy metal ions for mycoremediation. Dead fungal biomass formed by the *Aspergillus niger*, *Rhizopus oryzae*, *Saccharomyces cerevisiae*, and *Penicillium chrysogenum* could effectively convert toxic chromium ions Cr (VI) to nontoxic chromium ions Cr (III) (Park et al, 2005). Some of the fungal cells produces biosurfactants to remove heavy metal ions from soil samples. For example, *Candida sphaerica* produces biosurfactants effectively for the removal iron, zinc, lead (95, 90, 75% respectively) through biosurfactant- metal ion complex formation causes detachment of metal from soil samples (Luna et al, 2016). *Candida spp.* accumulate a significant quantity of nickel (71%) and copper (68%). Several strains of yeast such as *Hansenula polymorpha*, *S. cerevisiae*, *Yarrowia lipolytica*, *Rhodotorula pilimanae*, *Pichia guilliermondii*, and *Rhodotorula mucilage* have been used to bio-convert Cr (VI) to Cr (III) (Dönmez and Aksu, 2001)

Algal Biofilm

Algal cells are autotrophic and require low nutrients and produce enormous biomass compared to other microbes. Biosorbents from algal cells have been used for efficient heavy metal removal with excellent biosorption capacity (Abbas et al, 2014). Algae biomass is used for bioremediation of heavy metal polluted effluent through adsorption or by integration into the cells. Phyco-remediation is the advantages of various types of algae and cyanobacteria for the removal of heavy metals by either removal or degradation of their toxic level concentration (Chabukdhara et al, 2017). Algae have various chemical moieties on their surface such as hydroxyl, carboxyl, phosphate, and amide, which act as metal-binding sites (He and Chen, 2014). Goher et al(2016) were reported that the used dead cells of *Chlorella vulgaris* which used to remove cadmium, copper, and lead ions from aqueous solution under various conditions of pH, biosorbent dosage, and contact time. In this research findings suggested that the biomass of *C. vulgaris* is an extremely efficient biosorbent for the removal of cadmium (Cd_2^+), copper (Cu_2^+) and lead (Pb_2^+) at the significant level of 95.5%, 97.7%, and 99.4%, respectively, from contaminated water in the concentration of 50 mg dm^{-3} of each metal ion.

CONCLUSION

In this review discussed the different methods for the remediation of contaminated water using physio-chemical methods and conventional techniques with the highest cost and low accuracy of the results. But the microbial-based biofilm showed the accurate and economically feasible then time-consuming. In this methods may apply without any hazards to the public and understanding and appropriate precaution for water contaminant in the environment in the way of minimal production to the small scale application and operation. The biofilm-based technique conducted by using bacteria, fungi and also algae. Bacterial based biofilm can applicable in the area of household water pipeline, the fungal may be used in the prevention of contamination in the food industry. Algal based bioremediation is applicable in the aquatic system especially prevent the water contaminant when the transport through the ship. Consequently, it can bring about a quantum leap in the direction of developing an extremely sustainable, economic and environmentally restorative method for the removal of heavy metals in aquatic systems. Herewith concluded that the removal of heavy metal ions also been done by the use of microbial sources effectively.

REFERENCES

Abbas, H. S., Ismail, M. I., Mostafa, M. T., & Sulaymon, H. A. (2014). Biosorption of heavy metals: A review. *Journal of Chemical Science and Technology*, *3*, 74–102.

Alluri, H. K., Ronda, S. R., Settalluri, V. S., Bondili, J. S., Suryanarayana, V., & Venkateshwar, P. (2007). Biosorption: An eco-friendly alternative for heavy metal removal. *African Journal of Biotechnology*, 6.

Aryal, S., Karki, G., & Pandey, S. (2015). Microbial Diversity in Freshwater and Marine Environment. *Nepal Journal of Biotechnology*, *3*(1), 68–70. doi:10.3126/njb.v3i1.14236

Barakat, M. (2011). New trends in removing heavy metals from industrial wastewater. *Arabian Journal of Chemistry*, *4*(4), 361–377. doi:10.1016/j.arabjc.2010.07.019

Branda, S. S., Vik, S., Friedman, L., & Kolter, R. (2005). Biofilms: The matrix revisited. *Trends in Microbiology*, *13*(1), 20–26. doi:10.1016/j.tim.2004.11.006 PMID:15639628

Camper, A., Burr, M., Ellis, B., Butterfield, P., & Abernathy, C. (1998). Development and structure [2] of drinking water biofilms and techniques for their study. *Journal of Applied Microbiology*, *85*(S1), 1–12. doi:10.1111/j.1365-2672.1998.tb05277.x PMID:21182687

Chabukdhara, M., Gupta, S. K., & Gogoi M. (2017). Phycoremediation of heavy metals coupled with the generation of bioenergy. *Algal Biofuels*, 163–188.

Chipasa, K. B. (2003). Accumulation and Fate of Selected Heavy Metals in a Biological Wastewater Treatment System. *Waste Management (New York, N.Y.)*, *23*(2), 135–143. doi:10.1016/S0956-053X(02)00065-X PMID:12623088

Comte, S., Guibaud, G., & Baudu, M. (2008). Biosorption properties of extracellular polymeric substances (EPS) towards Cd: Cu and Pb for different pH values. *Journal of Hazardous Materials*, *151*(1), 185–193. doi:10.1016/j.jhazmat.2007.05.070 PMID:17611021

Costerton, J. W., Lewandowski, Z., Caldwell, D. E., Korber, D. R., & Lappin-Scott, H. M. (1995). Microbial biofilms. *Annual Review of Microbiology*, *49*(1), 711–745. doi:10.1146/annurev.mi.49.100195.003431 PMID:8561477

Costerton, J. W., Lewandowski, Z., Debeer, D., Caldwell, D., Korber, D., & James, G. (1994). Biofilms, the Customized Microniche. *Journal of Bacteriology*, *176*(8), 2137–2142. doi:10.1128/JB.176.8.2137-2142.1994 PMID:8157581

Cullen, W. R., & Reimer, K. J. (1989). Arsenic speciation in the environment. *Chemical Reviews*, *89*(4), 713–764. doi:10.1021/cr00094a002

Das, N., Vimala, R., & Karthika, P. (2008). Biosorption of heavy metals—An overview. *Indian Journal of Biotechnology*, *7*, 159–169.

Davey, M. E., & O'Toole, G. A. (2000). Microbial biofilms: From ecology to molecular genetics. *Microbiology and Molecular Biology Reviews*, *64*(4), 847–867. doi:10.1128/MMBR.64.4.847-867.2000 PMID:11104821

Deibel & Schoeni. (2003). Biofilms: Forming a Defense Strategy for the Food Plant. *Food Safety Magazine*.

Dojlido, J., & Best, G. A. (1993). *Chemistry of Water and Water Pollution*. Ellis Horwood Ltd.

Dönmez, G., & Aksu, Z. (2001). Bioaccumulation of copper(ii) and nickel(ii) by the non-adapted and adapted growing Candida SP. *Water Research*, *35*(6), 1425–1434. doi:10.1016/S0043-1354(00)00394-8 PMID:11317889

Duffus, J. H. (2002). Heavy Metals – A meaningless Terms? (IUPAC Technical Report). *Pure and Applied Chemistry*, *74*(5), 793–807. doi:10.1351/pac200274050793

Eccles, H. (1999). Treatment of metal contamination waste. Why select a biological process? *Trends in Biotechnology*, *17*(12), 462–465. doi:10.1016/S0167-7799(99)01381-5 PMID:10557157

Flemming, H.-C., Neu, T. R., & Wozniak, D. J. (2007). The EPS Matrix: The "House of Biofilm Cells". *Journal of Bacteriology*, *189*(22), 7945–7947. doi:10.1128/JB.00858-07 PMID:17675377

Fomina, M., & Gadd, G. M. (2014). Biosorption: Current perspectives on concept, definition and application. *Bioresource Technology*, *160*, 3–14. doi:10.1016/j.biortech.2013.12.102 PMID:24468322

Fomina, M., & Gadd, G. M. (2014). Biosorption: Current perspectives on concept, definition and application. *Bioresource Technology*, *160*, 3–14. doi:10.1016/j.biortech.2013.12.102 PMID:24468322

Fu, Q. Y., Li, S., & Zhu, Y. H. (2012). Biosorption of copper (II) from aqueous solution by mycelial pellets of *Rhizopus oryzae*. *African Journal of Biotechnology*, *11*(6), 1403–1411.

Gadd, G. M., & White, C. (1993). Microbial treatment of metal pollution—Working biotechnologyy? *Trends in Biotechnology*, *11*(8), 353–359. doi:10.1016/0167-7799(93)90158-6 PMID:7764182

Gautam, Sharma, Mahiya, & Chattopadhyaya. (2015). Contamination of Heavy Metals in Aquatic Media: Transport, Toxicity and Technologies for Remediation. The Royal Society of Chemistry, 1, 1-25.

Gavrilescu, M. (2004). Removal of heavy metals from the environment by biosorption. *Chemistry (Weinheim an der Bergstrasse, Germany)*, *28*, 30.

Gleick, P. H. (2003). Global freshwater resources: Soft-path solutions for the 21st century. *Science*, *302*(5650), 1524–1528. doi:10.1126cience.1089967 PMID:14645837

Goher, M. E., El-Monem, A. M. A., Abdel-Satar, A. M., Ali, M. H., Hussian, A.-E. M., & Napiórkowska-Krzebietke, A. (2016). Biosorption of some toxic metals from aqueous solution using non-living algal cells of *Chlorella vulgaris*. *Journal of Elementology*, *21*(3), 703–714.

Gupta, P., Sarkar, S., Das, B., Bhattacharjee, S., & Tribedi, P. (2016). Biofilm, pathogenesis and prevention—a journey to break the wall: A review. *Archives of Microbiology*, *198*(1), 1–15. doi:10.100700203-015-1148-6 PMID:26377585

Hassiba, M., Naima, A., Yahia, K., & Zahra, S. (2014). Study of lead adsorption from aqueous solutions on agar beads with EPS produced from *Paenibacillus polymyxa*. *Chemical Engineering Transactions*, *38*, 31–36.

He, J., & Chen, J. P. (2014). A comprehensive review on biosorption of heavy metals by algal biomass: Materials, performances, chemistry, and modelling simulation tools. *Bioresource Technology*, *160*, 67–78. doi:10.1016/j.biortech.2014.01.068 PMID:24630371

Ilyina, A., Castillo, S.M.I., Villarreal, S.J.A., Ramirez, E.G., & Candelas, R.J. (2003). Isolation of soil bacteria for bioremediation of hydrocarbon contamination. *BECTII*, *44*(15), 88-91.

James, T. (2000). Marsh Microbial Biofilm Formation and Contamination of Dental-Unit Water Systems in General Dental Practice. *Applied and Environmental Microbiology*, *66*(8), 3363–3367. doi:10.1128/AEM.66.8.3363-3367.2000 PMID:10919792

Kadirvelu, K., Senthilkumar, P., Thamaraiselvi, K., & Subburam, V. (2002). Activated carbon prepared from biomass as adsorbent: Elimination of Ni (II) from aqueous solution. *Bioresource Technology*, *81*(1), 87–90. doi:10.1016/S0960-8524(01)00093-1 PMID:11708760

Kurniawan, A., & Yamamoto, T. (2013). Biofilm polymer for biosorption of pollutant ions. *Procedia Environmental Sciences*, *17*, 179–187. doi:10.1016/j.proenv.2013.02.027

Lakherwal, D. (2014). Adsorption of heavy metals: A review. *International Journal of Environmental Research and Development*, *4*, 41–48.

Lakkireddy, K., & Kües, U. (2017). Bulk isolation of basidiospores from wild mushrooms by electrostatic attraction with low risk of microbial contaminations. *AMB Express*, *7*(1), 28. doi:10.118613568-017-0326-0 PMID:28124290

Lau, T., Wu, X., Chua, H., Qian, P., & Wong, P. K. (2005). Wong Effect of exopolysaccharides on the adsorption of metal ions by Pseudomonas sp. CU-1. *Water Science and Technology*, *52*(7), 63–68. doi:10.2166/wst.2005.0182

Lewis, K. (2005). Persister cells and the riddle of biofilm survival. *Biochemistry (Moscow)*, *70*(2), 267–274. doi:10.100710541-005-0111-6 PMID:15807669

Luna, J. M., Rufino, R. D., & Sarubbo, L. A. (2016). Biosurfactant from *Candida sphaerica* UCP0995 exhibiting heavy metal remediation properties. *Process Safety and Environmental Protection*, *102*, 558–566. doi:10.1016/j.psep.2016.05.010

Maes, S., Vackier, T., Nguyen Huu, S., Heyndrickx, M., Steenackers, H., Sampers, I., Raes, K., Verplaetse, A., & De Reu, K. (2019). Occurrence and characterisation of biofilms in drinking water systems of broiler houses. *BMC Microbiology*, *19*(1), 77. doi:10.118612866-019-1451-5 PMID:30987581

Meylan, S., Behra, R., & Sigg, L. (2003). Accumulation of copper and zinc in periphyton in response to dynamic variations of metal speciation in freshwater. *Environmental Science & Technology*, *37*(22), 5204–5212. doi:10.1021/es034566+ PMID:14655709

Meylan, S., Behra, R., & Sigg, L. (2003). Accumulation of copper and zinc in periphyton in response to dynamic variations of metal speciation in freshwater. *Environmental Science & Technology*, *37*(22), 5204–5212. doi:10.1021/es034566+ PMID:14655709

Monds, R. D., & O'Toole, G. A. (2009). The developmental model of microbial biofilms: Ten years of a paradigm up for review. *Trends in Microbiology*, *17*(2), 73–87. doi:10.1016/j.tim.2008.11.001 PMID:19162483

Mulligan, C. N., Yong, R., & Gibbs, B. F. (2001). Remediation technologies for metal contaminated soils and groundwater: An evaluation. *Engineering Geology*, *60*(1–4), 193–207. doi:10.1016/S0013-7952(00)00101-0

Ogbuagu, D. H., Okoli, C. G., Emereibeole, E. I., Anyanwu, I. C., Onuoha, O., Ubah, N. O., Ndugbu, C. O., Okoroama, O. N., Okafor, A., Ewa, E., Ossai, R., & Ukah, F. (2011). Trace metals accumulation in biofilms of the upper and middle reaches of Otamiri River in Owerri, Nigeria. *Journal of Biodiversity and Environmental Sciences*, *1*(3), 19–26.

Ogbuagu, D. H., Okoli, C. G., Emereibeole, E. I., Anyanwu, I. C., Onuoha, O., Ubah, N. O., Ndugbu, C. O., Okoroama, O. N., Okafor, A., Ewa, E., Ossai, R., & Ukah, F. (2011). Trace metals accumulation in biofilms of the upper and middle reaches of Otamiri River in Owerri, Nigeria. *Journal of Biodiversity and Environmental Sciences*, *1*(3), 19–26.

Ozdemir, G., Ceyhan, N., & Manav, E. (2005). Utilization of an exopolysaccharide produced by *Chryseomonas luteola* TEM05 in alginate beads for adsorption of cadmium and cobalt ions. *Bioresource Technology*, *96*(15), 1677–1682. doi:10.1016/j.biortech.2004.12.031 PMID:16023570

Ozdemir, G., Ceyhan, N., & Manav, E. (2005). Utilization in alginate beads for Cu (II) and Ni (II) adsorption of an exopolysaccharide produced by *Chryseomonas luteola* TEM05. *World Journal of Microbiology & Biotechnology*, *21*(2), 163–167. doi:10.100711274-004-1563-3

Ozdemir, G., Ozturk, T., Ceyhan, N., Isler, R., & Cosar, T. (2003). Heavy metal biosorption by biomass of *Ochrobactrum anthropi* producing exopolysaccharide in activated sludge. *Bioresource Technology*, *90*(1), 71–74. doi:10.1016/S0960-8524(03)00088-9 PMID:12835060

Park, D., Yun, Y. S., Jo, J. H., & Park, J. M. (2005). Mechanism of hexavalent chromium removal by dead fungal biomass of Aspergillus niger. *Water Research*, *39*(4), 533–540. doi:10.1016/j.watres.2004.11.002 PMID:15707625

Rani, M. J., Hemambika, B., Hemapriya, J., & Kannan, V. R. (2010). Comparative assessment of heavy metal removal by immobilized and dead bacterial cells: A biosorption approach. *African Journal of Environmental Science and Technology*, 4.

Rani, M. J., Hemambika, B., Hemapriya, J., & Kannan, V. R. (2010). Comparative assessment of heavy metal removal by immobilized and dead bacterial cells: A biosorption approach. *African Journal of Environmental Science and Technology*, 4.

Riggle, P. J., & Kumamoto, C. A. (2000). Role of a Candida albicans P1-type ATPase in resistance to copper and silver ion toxicity. *Journal of Bacteriology*, *182*(17), 4899–4905. doi:10.1128/JB.182.17.4899-4905.2000 PMID:10940034

Shannon, M. A., Bohn, P. W., Elimelech, M., Georgiadis, J. G., Marinas, B. J., & Mayes, A. M. (2008). Science and technology for water purification in the coming decades. *Nature*, *452*(7185), 301–310. doi:10.1038/nature06599 PMID:18354474

Sholl, D. S., & Lively, R. P. (2016). Seven chemical separations to change the world. *Nature*, *532*(7600), 435–437. doi:10.1038/532435a PMID:27121824

Srivastava, S., & Thakur, I. S. (2006). Isolation and process parameter optimization of Aspergillus sp. for removal of chromium from tannery effluent. *Bioresource Technology*, *97*(10), 1167–1173. doi:10.1016/j.biortcch.2005.05.012 PMID:16023341

Sultan, S., Mubashar, K., & Faisal, M. (2012). Uptake of toxic Cr (VI) by biomass of exopolysaccharides producing bacterial strains. *African Journal of Microbiological Research*, *6*, 3329–3336.

Sutherland, I. W. (2001). Biofilm exopolysaccharides: A strong and sticky framework. *Microbiology*, *147*(1), 3–9. doi:10.1099/00221287-147-1-3 PMID:11160795

Szewzyk, U., Szewzyk, R., Manz, W., & Schleifer, K. H. (2000). Microbiological safety of drinking water. *Annual Review of Microbiology*, *54*(1), 81–127. doi:10.1146/annurev.micro.54.1.81 PMID:11018125

Tchounwou, P. B., Yedjou, C. G., Patlolla, A. K., & Sutton, D. J. (2012). Heavy Metals Toxicity and the Environment. *EXS*, *101*, 133–164. doi:10.1007/978-3-7643-8340-4_6 PMID:22945569

Texier, A. C., Andres, Y., & Le Cloirec, P. (1999). Selective biosorption of lanthanide (La, Eu, Yb) ions by Pseudomonas aeruginosa. *Environmental Science & Technology*, *33*(3), 89–495. doi:10.1021/es9807744

Tsezos, M., & Volesky, B. (1981). Biosorption of uranium and thorium. *Biotechnology and Bioengineering*, *23*(3), 583–604. doi:10.1002/bit.260230309

Vandevivere, P., & Kirchman, D. L. (1993). Attachment stimulates exopolysaccharide synthesis by a bacterium. *Applied and Environmental Microbiology*, *59*(10), 3280–3286. doi:10.1128/AEM.59.10.3280-3286.1993 PMID:16349064

Volesky, B. (1990). *Biosorption of Heavy Metals*. CRC Press.

Walch, M. (1992). *Microbial Corrosion.* In J. Lederberg (Ed.), *Encyclopedia of Microbiology* (Vol. 1, pp. 585–591). Academic Press.

Wimpenny, J., Manz, W., & Szewzyk, U. (2000). Heterogeneity in biofilms. *FEMS Microbiology Reviews*, *24*(5), 661–667. doi:10.1111/j.1574-6976.2000.tb00565.x PMID:11077157

WWAP (United Nations World Water Assessment Programme). (2015). *Water for a Sustainable World. The United Nations World Water Development Report*. UNESCO.

Zhang, C., & Zhiggang, Y. (2014). Effects of sediment geochemical properties on heavy metal bioavailability. *Environment International*, *73*, 270–281. doi:10.1016/j.envint.2014.08.010 PMID:25173943

Compilation of References

11. Gianfreda, L., & Rao, M. A. (2004). Potential of extra cellular enzymes in remediation of polluted soils: A review. *Enzyme and Microbial Technology*, *35*(4), 339–354. doi:10.1016/j.enzmictec.2004.05.006

18. Clement, R., Paredes, C., & Bernal, M. P. (2007). A field experiment investigating the effects of olive husk and cow manure on heavy metal availability in a contaminated calcareous soil from Murica (Spain). *Agriculture, Ecosystems & Environment*, *188*(1-4), 319–326. doi:10.1016/j.agee.2006.06.002

Abbas, Rafatullah, Ismail, & Lalung. (2014). Isolation, Identification, and Characterization of Cadmium Resistant *Pseudomonas* sp. M3 from Industrial Wastewater. *Journal of Waste Management*. . doi:10.1155/2014/160398

Abbas, H. S., Ismail, M. I., Mostafa, M. T., & Sulaymon, H. A. (2014). Biosorption of heavy metals: A review. *Journal of Chemical Science and Technology*, *3*, 74–102.

Abdollahi, M., Zeinali, S., Nasirimoghaddam, S., Sabbaghi, S. J. D., & Treatment, W. (2015). *Effective removal of As (III) from drinking water samples by chitosan-coated magnetic nanoparticles*. Academic Press.

Abou-Shanab, R. A. I., Angle, J. S., & Chaney, R. L. (2006). Bacterial inoculants affecting nickel uptake by Alyssum murale from low, moderate and high Ni soils. *Soil Biology & Biochemistry*, *38*(9), 2882–2889. doi:10.1016/j.soilbio.2006.04.045

Abou-Shanab, R. A., Angle, J. S., Delorme, T. A., Chaney, R. L., Van Berkum, P., Moawad, H., Ghanem, K., & Ghozlan, H. A. (2003). Rhizobacterial effects on nickel extraction from soil and uptake by Alyssum murale. *The New Phytologist*, *158*(1), 219–224. doi:10.1046/j.1469-8137.2003.00721.x

Abou-Shanab, R. I., Delorme, T. A., Angle, J. S., Chaney, R. L., Ghanem, K., Moawad, H., & Ghozlan, H. A. (2003). Phenotypic characterization of microbes in the rhizosphere of Alyssum murale. *International Journal of Phytoremediation*, *5*(4), 367–379. doi:10.1080/15226510309359043 PMID:14750563

Abyar, H., Safahieh, A., Zolgharnein, H., & Zamani, I. (2012). Isolation and identification of Achromobacter denitrificans and evaluation of its capacity in cadmium removal. *Polish Journal of Environmental Studies*, *21*(6), 1523–1527.

Ackerley, D. F., Gonzalez, C. F., Keyhan, M., Blake, I. I. R., & Matin, A. (2004b). Mechanism of chromate reduction by the *E. coli* protein, NfsA, and the role of different chromate reductases in minimizing oxidative stress during chromate reduction. *Environmental Microbiology*, *6*(8), 851–860. doi:10.1111/j.1462-2920.2004.00639.x PMID:15250887

Ackerley, D. F., Gonzalez, C. F., Park, C. H., Blake, R. II, Keyhan, M., & Matin, A. (2004a). Chromate reducing properties of soluble flavoproteins from *Pseudomonas putida* and *E. Coli*. *Applied and Environmental Microbiology*, *70*(2), 873–882. doi:10.1128/AEM.70.2.873-882.2004 PMID:14766567

Agency for Toxic Substances and Disease Registry (ATSDR). (2007). *Toxicological Profile for Arsenic*. U.S. Department of Health and Human Services, Public Health Services.

Agirtas, M. S., & Kilicel, F. (1999). Determination of Cu, Ni, Mn and Zn pollution in soil at the shore of Van Lake with Flame Atomic Spectrophotmetry. *Bulletin of Pure and Applied Science*, *18*, 45–47.

Agnihotri, R., Mahuli, S. K., Chauk, S. S., & Fan, L. S. (1999). Influenceof Surface Modifiers on the Structure of Precipitated Calcium Carbonate. *Industrial & Engineering Chemistry Research*, *38*(6), 2283–2291. doi:10.1021/ie9900521

Agrawal, P., Lee, C., Lim, H. C., & Ramkrishna, D. (1982). Theoretical Investigations of Dynamic Behavior of Isothermal Continuous Stirred Tank Biological Reactors. *Chemical Engineering Science*, *37*(3), 453–462. doi:10.1016/0009-2509(82)80098-5

Ahalya, N., Ramachandra, T. V., & Kanamadi, R. D. (2003). Biosorption of Heavy Metals. *Research Journal of Chemistry and Environment*, *7*, 71–79.

Ahluwalia, S. S., & Goyal, D. (2005). Removal of heavy metals by waste tea leaves from aqueous solution. *Engineering in Life Sciences*, *5*(2), 158–162. doi:10.1002/elsc.200420066

Ahluwalia, S. S., & Goyal, D. (2007). Microbial and plant derived biomass for removal of heavy metals from wastewater. *Bioresource Technology*, *98*(12), 2243–2257. doi:10.1016/j.biortech.2005.12.006 PMID:16427277

Ahmad, F., Ahmad, I., & Khan, M. S. (2008). Screening of free-living rhizospheric bacteria for their multiple plant growth promoting activities. *Microbiological Research*, *163*(2), 173–181. doi:10.1016/j.micres.2006.04.001 PMID:16735107

Ahmad, I., Akhtar, M. J., Asghar, H. N., Ghafoor, U., & Shahid, M. (2016). Differential effects of plant growth-promoting rhizobacteria on maize growth and cadmium uptake. *Journal of Plant Growth Regulation*, *35*(2), 303–315. doi:10.100700344-015-9534-5

Ahmed, D. E. (2007). Environmental pollution and its impact on human and animal health: A review. *The Sudan J Vet Res*, *22*, 37–46.

Ahmed, I., Zafar, S., & Ahmed, F. (2005). Heavy metal biosorption potential of *Aspergillus* and *Rhizopus* sp. isolated from waste water treated soil. *Journal of Applied Science & Environmental Management*, *9*, 123–126.

Akhavan, O., & Azimirad, R. (2009). Photocatalytic property of Fe2O3 nanograin chains coated by TiO_2 nanolayer in visible light irradiation. *Applied Catalysis A, General*, *369*(1-2), 77–82. doi:10.1016/j.apcata.2009.09.001

Akhtar, M. N., Sastry, K. S., & Mohan, P. M. (1996). Mechanism of metal ion absorption by fungal biomass. *Biometals*, *9*, 21–28.

Akinbile, C. O., Yusoff, M. S., & Zuki, A. A. (2012). Landfill leachate treatment using sub-surface flow constructed wetland by *Cyperus haspan*. *Waste Management (New York, N.Y.)*, *32*(7), 1387–1393. doi:10.1016/j.wasman.2012.03.002 PMID:22456086

Akratos, C. S., Tekerlekopoulou, A. G., & Vayenas, D. V. (2018). Treatment of wastewater from tanneries and the textile industry using Constructed Wetland systems. In Constructed Wetlands for Industrial Wastewater Treatment (pp. 327-341). John Wiley & Sons Ltd. doi:10.1002/9781119268376.ch16

Alavi, M., Antonic, V., Ravizee, A., Weina, P., Izadjoo, M., & Stojadinovic, A. (2011). An Enterobacter plasmid as a new genetic background for the transposon Tn1331. *Infection and Drug Resistance*, *4*, 209–213. doi:10.2147/IDR.S25408 PMID:22259249

Alcalde, M., Ferrer, M., Plou, F. J., & Ballesteros, A. (2006). Environmental biocatalysis: From remediation with enzymes to novel green processes. *Trends in Biotechnology*, *24*(6), 281–287. doi:10.1016/j.tibtech.2006.04.002 PMID:16647150

Ali, H., Khan, E., & Sajad, M. A. (2013). Phytoremediation of heavy metals-concepts and applications. *Chemosphere*, *91*(7), 869–881. doi:10.1016/j.chemosphere.2013.01.075 PMID:23466085

Ali, S. Z., Sandhya, V., Grover, M., Kishore, N., Rao, L. V., & Venkateswarlu, B. (2009). Pseudomonas sp. strain AKM-P6 enhances tolerance of sorghum seedlings to elevated temperatures. *Biology and Fertility of Soils*, *46*(1), 45–55. doi:10.100700374-009-0404-9

Alluri, H. K., Ronda, S. R., Settalluri, V. S., Bondii, J. S., Suryanarayana, V., & Venkateshwar, P. (2007). Biosorption: An eco-friendly alternative for heavy metal removal. *African Journal of Biotechnology*, *6*(25), 2924–2931. doi:10.5897/AJB2007.000-2461

Alluri, H. K., Ronda, S. R., Settalluri, V. S., Bondili, J. S., Suryanarayana, V., & Venkateshwar, P. (2007). Biosorption: An eco-friendly alternative for heavy metal removal. *African Journal of Biotechnology*, 6.

Aluko, O. O., & Sridhar, C. (2005). Application of constructed wetlands to the treatment of leachates from a municipal solid waste landfill in Ibadan, Nigeria. *Journal of Environmental Health*, *67*(10), 58–62. PMID:15991686

Aluko, O. O., Sridhar, M. K., & Oluwande, P. A. (2003). Characterization of leachates from a municipal solid waste landfill site in Ibadan, Nigeria. *Journal of Environmental Health Research*, *2*(1), 32–37.

Alvarez, A., Saez, J. M., Costa, J. S. D., Colin, V. L., Fuentes, M. S., Cuozzo, S. A., Benimeli, C. S., Polti, M. A., & Amoroso, M. J. (2017). Actinobacteria: Current research and perspectives for bioremediation of pesticides and heavy metals. *Chemosphere*, *166*, 41–62. doi:10.1016/j.chemosphere.2016.09.070 PMID:27684437

American Academy of Pediatric Dentistry. (1994). Special issue: Reference manual 1995. *Pediatric Dentistry*, *16*(special issue), 1–96.

American Academy of Pediatrics Committee on Nutrition. (1995). Fluoride supplementation for children: Interim policy recommendations. *Pediatrics*, *95*, 777. PMID:7724324

American Dental Association. (2003). *Fluoridation Facts. National and International Organizations that Recognize the Public Health Benefits of Community Water Fluoridation for Preventing Dental Decay*. Author.

Amoozegar, M. A., Ghasemi, A., Razavi, M. R., & Naddaf, S. (2007). Short Communication: Evaluation of hexavalent chromium reduction by chromate-resistant moderately halophile, *Nesterenkonia* sp. strain MF2. *Process Biochemistry*, *42*(10), 1475–1479. doi:10.1016/j.procbio.2007.07.001

Anjum, M. K., Mughal, S. M., Sayyed, U., Yaqub, A., Khaliqu, A., & Rashid, A. M. (2014). Influence of increasing fluoride dose rates on selected liver and kidney enzymes profile in domestic chicken (Gallus domesticus). *Journal of Animal & Plant Sciences*, *24*(1), 77–80.

Antunes, W. M., Luna, A. S., Henriques, C. A., & DaCosta, A. C. A. (2003). An evaluation of copper biosorption by a brown seaweed under optimized conditions. *Electronic Journal of Biotechnology*, *6*, 174–184.

Aoshima, K. (2016). Itai-itai disease: Renal tubular osteomalacia induced by environmental exposure to cadmium—historical review and perspectives. *Soil Science and Plant Nutrition*, *62*(4), 319–326. doi:10.1080/00380768.2016.1159116

Apambire, W. B., Boyle, D. R., & Michel, F. A. (2007). Geochemistry, genesis, and health implications of fluoriferous groundwaters in the upper regions of Ghana. *Environ Geol*, *33*(1), 13–24. doi:10.1007002540050221

Arao, T., Ishikawa, S., Murakami, M., Abe, K., Maejima, Y., & Makino, T. (2010). Heavy metal contamination of agricultural soil and countermeasures in Japan. *Paddy and Water Environment*, *8*(3), 247–257. doi:10.100710333-010-0205-7

Arévalo-Rangel, Cárdenas-González, Martínez-Juárez, & Acosta-Rodríguez. (2013). *Hexavalent Chromate Reductase Activity in Cell Free Extracts of Penicillium sp*. Hindawi Publishing Corporation. doi:10.1155/2013/909412

Arivoli, A., Mohanraj, R., & Seenivasan, R. (2015). Application of vertical flow constructed wetland in treatment of heavy metals from pulp and paper industry wastewater. *Environmental Science and Pollution Research International*, *22*(17), 13336–13343. doi:10.100711356-015-4594-4 PMID:25940487

Arora. (2012). Microbial Biotechnology: Energy and Environment. *CABI, 2012 - Science.*

Arora, P. K. (2010). Application of Monooxygenases in Dehalogenation, Desulphurization, Denitrification and Hydroxylation of Aromatic Compounds. *Journal of Bioremediation & Biodegradation*, *01*(03). Advance online publication. doi:10.4172/2155-6199.1000112

Aryal, M., & Liakopoulou-Kyriakides, M. (2013). Binding Mechanism and Biosorption Characteristics of Fe(III) by *Pseudomonas* sp. cells. *Journal of Water Sustainability*, *3*(3), 117–131.

Aryal, S., Karki, G., & Pandey, S. (2015). Microbial Diversity in Freshwater and Marine Environment. *Nepal Journal of Biotechnology*, *3*(1), 68–70. doi:10.3126/njb.v3i1.14236

Ashokkumar, P., Loashini, V. M., & Bhavya, V. (2017). Effect of pH, Temperature and biomass on biosorption of heavy metals by *Sphaerotilus natans*. *International Journal of Microbiology and Mycology*, *6*(1), 32–38.

Aussawasathien, D., Teerawattananon, C., & Vongachariya, A. (2008). Separation of micron to sub-micron particles from water: Electrospun nylon-6 nanofibrous membranes as pre-filters. *Journal of Membrane Science*, *315*(1-2), 11–19. doi:10.1016/j.memsci.2008.01.049

Ayangbenro, A. S., & Babalola, O. O. (2017). A new strategy for heavy metal polluted environments: A review of microbial biosorbents. *International Journal of Environmental Research and Public Health*, *14*(1), 94. doi:10.3390/ijerph14010094 PMID:28106848

Azad, M. A. K., Amin, L., & Sidik, N. M. (2014). Genetically engineered organisms for bioremediation of pollutants in contaminated sites. *Chinese Science Bulletin*, *59*(8), 703–714. doi:10.100711434-013-0058-8

Aziz, S. Q., Aziz, H. A., Yusoff, M. S., & Bashir, M. J. (2011). Landfill leachate treatment using powdered activated carbon augmented sequencing batch reactor (SBR) process: Optimization by response surface methodology. *Journal of Hazardous Materials*, *189*(1-2), 404–413. doi:10.1016/j.jhazmat.2011.02.052 PMID:21420786

Aziz, S. Q., Aziz, H. A., Yusoff, M. S., & Mohajeri, S. (2012). Removal of phenols and other pollutants from different landfill leachates using powdered activated carbon supplemented SBR technology. *Environmental Monitoring and Assessment*, *184*(10), 6147–6158. doi:10.100710661-011-2409-8 PMID:22068314

Azubuike, C. C., Chikere, C. B., & Okpokwasili, G. C. (2016). Bioremediation techniques- classification based on site of application: Principles, advantages, limitations and prospects. *World Journal of Microbiology & Biotechnology*, *32*(11), 4–18. doi:10.100711274-016-2137-x PMID:27638318

Babak, L., Šupinova, P., Zichova, M., Burdychova, R., & Vitova, E. (2012). Biosorption of Cu, Zn and Pb by thermophilic bacteria –effect of biomass concentration on biosorption capacity. *Acta Universitatis Agriculturae et Silviculturae Mendelianae Brunensis*, *40*(5), 2012. doi:10.11118/actaun201260050009

Baby, R., Saifullah, B., & Hussein, M. Z. (2019). Palm kernel shell as an effective adsorbent for the treatment of heavy metal contaminated water. *Scientific Reports*, *9*(1), 18955. doi:10.103841598-019-55099-6 PMID:31831850

Bae, W. C., Lee, H. K., Choe, Y. C., Jahng, D. J., Lee, S. H., & Kim, S. J. (2005). Purification and characterization of NADPH dependent Cr(VI) reductase from *Escherichia coli* ATCC 33456. *Journal of Microbiology (Seoul, Korea), 43*, 21–27. PMID:15765053

Bah, A. M., Dai, H., Zhao, J., Sun, H., Cao, F., Zhang, G., & Wu, F. (2011). Effect of cadmium, chromium and lead on growth, metal uptake and antioxidative capacity in *Typha angustifolia. Biological Trace Element Research, 142*(1), 77–92. doi:10.100712011-010-8746-6 PMID:20552296

Bajaj, A., Pathak, A., Mudiam, M. R., Mayilraj, S., & Manickam, N. (2010). Isolation and characterization of a Pseudomonas sp. strain IITR01 capable of degrading α-endosulfan and endosulfan sulfate: Endosulfan degrading Pseudomonas sp. *Journal of Applied Microbiology, 109*(6), 2135–2143. doi:10.1111/j.1365-2672.2010.04845.x PMID:20825519

Baker, A. J. M. (1981). Accumulators and excluders - strategies in the response of plants to heavy metals. *Journal of Plant Nutrition, 3*(1-4), 643–654. doi:10.1080/01904168109362867

Baker, A. J. M., & Brooks, R. R. (1989). Terrestrial higher plants which hyperaccumulate metallic element-A review of their distribution, ecology and phytochemistry. *Biorecovery, 1*(4), 811–826.

Baker, A. J. M., McGrath, S. P., Reeves, R. P., & Smith, J. A. C. (2000). Metal hyperaccumulator plants: a review of the ecology and physiology of a biochemical resource for phytoremediation of metal polluted soils. In N. Terry & G. Bauelos (Eds.), *Phytoremediation of contaminated soil and water* (pp. 85–107). Lewis Publishers.

Bakhshoodeh, R., Soltani-Mohammadi, A., Alavi, N., & Ghanavati, H. (2017). Treatment of high polluted leachate by subsurface flow constructed wetland with vetiver. *Civil Engineering (New York, N.Y.), 49*, 43–44.

Bandowe, M., Bigalke, M., Boamah, L., Nyarko, E., Saalia, F., & Wilcke, W. (2014). Polycyclic aromatic compounds (PAHs and oxygenated PAHs) and trace metals in fish species from Ghana (West Africa): Bioaccumulation and health risk assessment. *Environment International, 65*, 135–146. doi:10.1016/j.envint.2013.12.018 PMID:24486971

Bandyopadhyay, S., Ghanekar, M. D., & Pillai, H. K. (2006). Process water management. *Industrial & Engineering Chemistry Research, 45*(15), 5287–5297. doi:10.1021/ie060268k

Banerjee, P., Chakrabarti, S., Maitra, S., & Dutta, B. K. J. U. s. (2012). *Zinc oxide nano-particles–sonochemical synthesis, characterization and application for photo-remediation of heavy metal.* Academic Press.

Banerjee, A., Sharma, R., & Banerjee, U. C. (2002). The nitrile-degrading enzymes: Current status and future prospects. *Applied Microbiology and Biotechnology, 60*(1-2), 33–44. doi:10.100700253-002-1062-0 PMID:12382040

Barakat, M. (2011). New trends in removing heavy metals from industrial wastewater. *Arabian Journal of Chemistry, 4*(4), 361–377. doi:10.1016/j.arabjc.2010.07.019

Barak, Y., Ackerley, D. F., Dodge, C. J., Banwari, L., Alex, C., Francis, A. J., & Matin, A. (2006). Analysis of novel soluble chromate and uranyl reductases and generation of an improved enzyme by directed evolution. *Applied and Environmental Microbiology, 72*(11), 7074–7082. doi:10.1128/AEM.01334-06 PMID:17088379

Barałkiewicz, D., & Siepak, J. (1999). Chromium, Nickel and Cobalt in Environmental Samples and Existing Legal Norms. *Polish Journal of Environmental Studies, 8*(4), 201–208.

Barkay, T., & Schaefer, J. (2001). Metal and radionuclide bioremediation: Issues, considerations and potentials. *Current Opinion in Microbiology, 4*(3), 318–323. doi:10.1016/S1369-5274(00)00210-1 PMID:11378486

Barrow, N. J., & Ellis, A. S. (2006). Testing a mechanistic model. III The effect of pH on fluoride retention by a soil. *Journal of Soil Science, 37*(2), 287–293. doi:10.1111/j.1365-2389.1986.tb00030.x

Baryla, A., Carrier, P., Franck, F., Coulomb, C., Sahut, C., & Havaux, M. (2001). Leaf chlorosis in oilseed rape plants (*Brassica napus*) grown on cadmium polluted soil: Causes and consequences for photosynthesis and growth. *Planta*, *212*(5-6), 696–709. doi:10.1007004250000439 PMID:11346943

Bary, S. A. S., & Clark, C. (1978). Problems in interpreting the relationship between the amounts of Pb in plants and soil on metalliferous wastes. *The New Phytologist*, *81*(3), 773–783. doi:10.1111/j.1469-8137.1978.tb01652.x

Battarbee, R., Anderson, N., Appleby, P., Flower, R.G., Fritz, S., Haworth, E., Higgit, S., Jones, Reiser, A., Munro, M.A., Natkanski, J., Oldfield, F., Patrick, S.T., Richardson, N., Rippey, B., & Stevenson, A.C. (1988). *Lake Acidification in The United Kingdom*. ENSIS.

Baunthiyal, M., & Ranghar, S. (2013). Accumulation of fluoride by plants: Potential for phytoremediation. *Clean Soil Air Water*, *43*(1), 127–132. doi:10.1002/clen.201300353

Baunthiyal, M., & Sharma, V. (2012). Phytoremediation of fluoride contaminated water and soil: A search for fluoride hyperaccumulators. *J Agri Tech*, *8*, 1965–1978.

Baweja, M., Nain, L., Kawarabayasi, Y., & Shukla, P. (2016). Current Technological Improvements in Enzymes toward Their Biotechnological Applications. *Frontiers in Microbiology*, *7*. Advance online publication. doi:10.3389/fmicb.2016.00965 PMID:27379087

Bawuro, A. A., Voegborlo, R. B., & Adimado, A. A. (2018). Bioaccumulation of Heavy Metals in Some Tissues of Fish in Lake Geriyo, Adamawa State, Nigeria. *Journal of Environmental and Public Health*, *2018*, 7. doi:10.1155/2018/1854892 PMID:29887895

Beatriz, C. M., & Juan, A. (2011). Combination treatment of corn starch water by sedimentation, microfiltration and reverse osmosis. *Desalination*, *279*(1-3), 285–290. doi:10.1016/j.desal.2011.06.021

Beaver, J. R., Manis, E. E., Loftin, K. A., Graham, J. L., Pollard, A. I., & Mitchell, R. M. (2014). Land use patterns, ecoregion, and microcystin relationships in U.S. lakes and reservoirs: A preliminary evaluation. *Harmful Algae*, *36*, 57–62. doi:10.1016/j.hal.2014.03.005

Belchik, S. M., Kennedy, D. W., Dohnalkova, A. C., Wang, Y., Sevinc, P. C., Wu, H., Lin, Y., Lu, P. H., Fredrickson, J. K., & Shi, L. (2011). Extracellular reduction of hexavalent chromium by cytochromes MtrC and OmcA of *Shewanella oneidensis* MR-1. *Applied and Environmental Microbiology*, *77*(12), 4035–4041. doi:10.1128/AEM.02463-10 PMID:21498755

Benitez, E., Romero, E., Gomez, M., Gallardo-Lara, F., & Nogales, R. (2001). Biosolids and biosolids-ash as sources of heavy metals in a plant-soil system. *Water, Air, and Soil Pollution*, *132*(1-2), 75–87. doi:10.1023/A:1012012924151

Bernard, A. (2008). Biomarkers of metal toxicity in population studies:Research potential and interpretation issues. *Journal of Toxicology and Environmental Health*, *71*(18), 1259–1265. doi:10.1080/15287390802211885 PMID:18654897

Berti, W. R., & Cunningham, S. D. (2000). Phytostabilization of metals. In I. Raskin & B. D. Ensley (Eds.), *Phytoremediation of toxic metals: Using Plants to Clean up the environment* (pp. 71–88). Wiley and Sons Inc.

Berti, W. R., & Jacob, L. W. (1996). Chemistry and phytotoxicity of soil trace elements from repeated sewage sludge application. *Journal of Environmental Quality*, *25*(5), 1025–1032. doi:10.2134/jeq1996.00472425002500050014x

Bert, V., Meerts, P., & Laprade, P. S. (2003). Genetic basis of cadmium tolerance and hyperaccummulation in *Arabidopsis halleri. Plant and Soil*, *249*(1), 9–18. doi:10.1023/A:1022580325301

Beyersmann, D., & Hartwig, A. (2008). Carcinogenic metal compounds: Recent insight into molecular and cellular mechanisms. *Archives of Toxicology*, *82*(8), 493–512. doi:10.100700204-008-0313-y PMID:18496671

Bhandari, G. (2018). Environmental Nanotechnology: Applications of Nanoparticles for Bioremediation. In Approaches in Bioremediation (pp. 301-315). Springer.

Bharagava, R. N., & Mishra, S. (2018). Hexavalent chromium reduction potencial of *Cellulosimicrobium* sp. isolated from common effluent treatment plant of tannery industries. *Ecotoxicology and Environmental Safety*, *147*, 102–109. doi:10.1016/j.ecoenv.2017.08.040 PMID:28841524

Bhardwaj, N., & Kundu, S. C. (2010). Electrospinning: A fascinating fiber fabrication technique. *Biotechnology Advances*, *28*(3), 325–347. doi:10.1016/j.biotcchadv.2010.01.004 PMID:20100560

Bhardwaj, P., Chaturvedi, A., & Prasad, P. (2009). Effect of enhanced lead and cadmium in soil on physiological and biochemical attributes of *Phaseolus vulgaris* L. *Nature and Science*, *7*(8), 63–75.

Bhaskar, P. V., & Bhosle, N. B. (2006). Bacterial extracellular polymeric substance (EPS): A carrier of heavy metals in the marine food chain. *Environment International*, *32*(2), 191–198. doi:10.1016/j.envint.2005.08.010 PMID:16256198

Bhattacharyya, P., Chakrabarti, K., Chakraborty, A., Tripathy, S., & Powell, M. A. (2008). Fractionation and bioavailability of Pb in municipal solid waste compost and Pb uptake by rice straw and grain under submerged condition in amended soil. *Geosciences Journal*, *12*(1), 41–45. doi:10.100712303-008-0006-9

Bhunia, B., Basak, B., Mandal, T., Bhattacharya, P., & Dey, A. (2013). Effect of pH and temperature on stability and kinetics of novel extracellular serine alkaline protease (70 kDa). *International Journal of Biological Macromolecules*, *54*, 1–8. doi:10.1016/j.ijbiomac.2012.11.024 PMID:23219732

Bielicka, A., Bojanowska, I., & Wisniewski, A. (2005). Two faces of chromium pollutant and bioelement. *Polish Journal of Environmental Studies*, *14*(1), 5–10.

Bieliyakova, T. M. (2007). Fluorine in soils and plants as related to endemic fluorosis. *Почвоведение*, *8*, 55.

BIS. (1991). *Indian standard specification for drinking water. IS: 10500*. Bureau of Indian Standards.

Blake, R. C. II, Choate, D. M., Bardhan, S., Revis, N., Barton, L. L., & Zocco, T. C. (1993). Chemical transformation of toxic metals by *Pseudomonas* strain from a toxic waste site. *Environmental Toxicology and Chemistry*, *12*(8), 1365–1376. doi:10.1002/etc.5620120806

Blaylock, M. J., & Huang, J. W. (2000). Phytoextraction of metals. In I. Raskin & B. D. Ensley (Eds.), *Phytoremediation of toxic metals: using plants to clean-up the environment* (pp. 53–70). John Wiley & Sons, Inc.

BlaylockM.J.SaltD.E.DushenkovS.ZakharovaO.GussmanC.;

Bognitzki, M., Czado, W., Frese, T., Schaper, A., Hellwig, M., Steinhart, M., Greiner, A., & Wendorff, J. H. (2001). Nanostructured fibers via electrospinning. *Advanced Materials*, *13*(1), 70–72. doi:10.1002/1521-4095(200101)13:1<70::AID-ADMA70>3.0.CO;2-H

Boopathy, R. (2000). Factors limiting bioremediation technologies. *Bioresource Technology*, *74*(1), 63–67. doi:10.1016/S0960-8524(99)00144-3

Booth, S., Weljie, A., & Turner, J. (2015). Metabolomics reveals diferences of metal toxicity in cultures of Pseudomonas pseudoalcaligenes KF707 grown on different carbon sources. *Frontiers in Microbiology*, *6*, 827. doi:10.3389/fmicb.2015.00827 PMID:26347721

Bopp, L. H., & Ehrlich, H. L. (1988). Chromate resistance and reduction in *Pseudomonas fluorescens* strain LB300. *Archives of Microbiology*, *150*(5), 426–431. doi:10.1007/BF00422281

Bourbonnais, R., Paice, M. G., Reid, I. D., Lanthier, P., & Yaguchi, M. (1995). Lignin oxidation by laccase isozymes from *Trametes versicolor* and role of the mediator 2,2¢-azinobis(3-ethylbenzthiazoline-6-sulfonate) in kraft lignin depolymerization. *Applied and Environmental Microbiology, 61*(5), 1876–1880. doi:10.1128/AEM.61.5.1876-1880.1995 PMID:7646025

Bowen, H. J. M. (2006). *Trace elements in biochemistry*. Academic.

Boyd, R. S. (2010). Heavy metal pollutants and chemical ecology: Exploring new frontiers. *Journal of Chemical Ecology, 36*(1), 46–58. doi:10.100710886-009-9730-5 PMID:20108028

Branco, R., Chung, A. P., Johnston, T., Gurel, V., Morais, P., & Zhitkovich, A. (2008). The chromate inducible chr BACF operon from the transposable element TnOtChr confers resistance to chromium(VI) and superoxide. *Journal of Bacteriology, 190*(21), 6996–7003. doi:10.1128/JB.00289-08 PMID:18776016

Branda, S. S., Vik, S., Friedman, L., & Kolter, R. (2005). Biofilms: The matrix revisited. *Trends in Microbiology, 13*(1), 20–26. doi:10.1016/j.tim.2004.11.006 PMID:15639628

Braud, A., Jézéquel, K., Vieille, E., Tritter, A., & Lebeau, T. (2006). Changes in extractability of Cr and Pb in a polycontaminated soil after bioaugmentation with microbial producers of biosurfactants, organic acids and siderophores. *Water Air and Soil Pollution Focus, 6*(3-4), 261–279. doi:10.100711267-005-9022-1

Breil, C., Vian, M. A., Zemb, T., Kunz, W., & Chemat, F. (2017). "Bligh and Dyer" and Folch Methods for Solid–Liquid–Liquid Extraction of Lipids from Microorganisms. Comprehension of Solvatation Mechanisms and towards Substitution with Alternative Solvents International. *Journal of Molecular Sciences, 2017*(18), 708. doi:10.3390/ijms18040708 PMID:28346372

Brewer, R. F. (1966). Fluorine. In H. D. Chapman (Ed.), *Diagnostic criteria for plants and soils* (pp. 180–196). University of California.

Brooks, R. R. (1998). *Plants that hyperaccumulate heavy metals: their role in phytoremediation, microbiology, archaeology, mineral exploration and phytomining*. CAB International.

Browning, C. L., & Wise, J. P. Sr. (2017). Prolonged exposure to particulate chromate inhibits RAD51 nuclear import mediator proteins. *Toxicology and Applied Pharmacology, 331*, 101–107. doi:10.1016/j.taap.2017.05.030 PMID:28554658

Brown, S., Thompson, M., VerBerkmoes, N., Chourey, K., Shah, M., Zhou, J., Hettich, R., & Thompson, D. (2006). Molecular dynamics of the *Shewanella oneidensis* response to chromate stress. *Molecular & Cellular Proteomics, 5*(6), 1054–1071. doi:10.1074/mcp.M500394-MCP200 PMID:16524964

Bruins, M.R., Kapil, S., & Oehme, F.W. (2000). Microbial resistance to metals in the environment. *Ecotoxicology and Environmental Safety, 45*(3), 198-207.

Bruins, R., Kapil, S., & Oehme, F. W. (2000). Microbial Resistance to Metals in the Environment. *Ecotoxicology and Environmental Safety, 45*(3), 198–207. doi:10.1006/eesa.1999.1860 PMID:10702338

Bulc, T. G. (2006). Long term performance of a constructed wetland for landfill leachate treatment. *Ecological Engineering, 26*(4), 365–374. doi:10.1016/j.ecoleng.2006.01.003

Bulc, T., Vrhovsek, D., & Kukanja, V. (1997). The use of constructed wetland for landfill leachate treatment. *Water Science and Technology, 35*(5), 301–306. doi:10.2166/wst.1997.0222

Burd, G. I., Dixon, D. G., & Glick, B. R. (1998). A plant growth-promoting bacterium that decreases nickel toxicity in seedlings. *Applied and Environmental Microbiology, 64*(10), 3663–3668. doi:10.1128/AEM.64.10.3663-3668.1998 PMID:9758782

Burd, G. I., Dixon, D. G., & Glick, B. R. (2000). Plant growth-promoting bacteria that decrease heavy metal toxicity in plants. *Canadian Journal of Microbiology*, *46*(3), 237–245. doi:10.1139/w99-143 PMID:10749537

Burns, R. G., DeForest, J. L., Marxsen, J., Sinsabaugh, R. L., Stromberger, M. E., Wallenstein, M. D., Weintraub, M. N., & Zoppini, A. (2013). Soil enzymes in a changing environment: Current knowledge and future directions. *Soil Biology & Biochemistry*, *58*, 216–234. doi:10.1016/j.soilbio.2012.11.009

Cabuk, A., Ilhan, S., Filik, C., & Caliskan, F. (2005). Pb (2+) biosorption by pretreated fungal biomass. *Turkish Journal of Biology*, *29*, 23–28.

Calheiros, C. S., Rangel, A. O., & Castro, P. M. (2007). Constructed wetland systems vegetated with different plants applied to the treatment of tannery wastewater. *Water Research*, *41*(8), 1790–1798. doi:10.1016/j.watres.2007.01.012 PMID:17320926

Camargo, F. A. O., Okeke, B. C., Bento, F. M., & Frankenberger, W. T. (2003). In vitro reduction of hexavalent chromium by a cell-free extract of *Bacillus* sp. ES 29 stimulated by Cu^{2+}. *Applied Microbiology and Biotechnology*, *62*(5-6), 569–573. doi:10.100700253-003-1291-x PMID:12679851

Camargo, F. A. O., Okeke, B. C., Bento, F. M., & Frankenberger, W. T. (2004). Hexavalent chromium reduction by immobilized cells and cell-free extract of *Bacillus sp.* ES29. *Bioremediation Journal*, *8*(1-2), 23–30. doi:10.1080/10889860490453140

Camargo, F. A. O., Okeke, B. C., Bento, F. M., & Frankenberger, W. T. (2005). Diversity of chromium-resistant bacteria isolated from soils contaminated with dichromate. *Applied Soil Ecology*, *29*(2), 193–202. doi:10.1016/j.apsoil.2004.10.006

Campbell, P. G. C., Errecalde, O., Fortin, C., Hiriart-Baer, V. P., & Vigneault, B. (2002). Metal bioavailability to phytoplankton applicability of the biotic ligand model. *Comparative Biochemistry and Physiology. Toxicology & Pharmacology: CBP*, *133*(1-2), 189–206. doi:10.1016/S1532-0456(02)00104-7 PMID:12356527

Camper, A., Burr, M., Ellis, B., Butterfield, P., & Abernathy, C. (1998). Development and structure [2] of drinking water biofilms and techniques for their study. *Journal of Applied Microbiology*, *85*(S1), 1–12. doi:10.1111/j.1365-2672.1998.tb05277.x PMID:21182687

Campos, J., Martinez-Pacheco, M., & Cervantes, C. (1995). Hexavalent-chromate reduction by a chromate resistant *Bacillus* sp. *Antonie van Leeuwenhoek*, *68*(3), 203–208. doi:10.1007/BF00871816 PMID:8572677

Canovas, D., Cases, I., & de Lorenzo, V. (2003). Heavy metal tolerance and metal homeostasis in *Pseudomonas putida* as revealed by complete genome analysis. *Environmental Microbiology*, *5*(12), 1242–1256. doi:10.1111/j.1462-2920.2003.00463.x PMID:14641571

Cao, J., Zhao, Y., & Liu, J. (1997). Brick tea consumption as the cause of dental fluorosis among children from Mongol, Kazak and Yugu populations in China. *Food and Chemical Toxicology*, *35*(8), 827–833. doi:10.1016/S0278-6915(97)00049-5 PMID:9350228

Cargnelutti, D., Almeri Tabaldi, L., Spanevello, R. M., de Oliviera Jucoski, G., Battisti, V., Redin, M., Blanco Linares, C. E., Dressler, V. L., de Moraes Flores, E. M., Texeira Nicoloso, F., Morsch, V. M., & Chitolina, M. R. (2006). Mercury toxicity induces oxidative stress in growing cucumber seedlings. *Chemosphere*, *65*(6), 999–1006. doi:10.1016/j.chemosphere.2006.03.037 PMID:16674986

Carroll, C. C., Warnakulasuriyarachchi, D., Nokhbeh, M. R., & Lambert, I. B. (2002). *Salmonella typhimurium* mutagenicity tester strains that over express oxygen insensitive nitroreductases nfsA and nfsb. *Mutation Research*, *501*(1-2), 79–98. doi:10.1016/S0027-5107(02)00018-0 PMID:11934440

Carvalho, M. D., Henriques, F., Ferreira, L. P., Godinho, M., & Cruz, M. M. (2013). Iron oxide nanoparticles: The Influence of synthesis method and size on composition and magnetic properties. *Journal of Solid State Chemistry*, *201*(0), 144–152. doi:10.1016/j.jssc.2013.02.024

Cecchin, I., Reddy, K. R., Thomé, A., Tessaro, E. F., Schnaid, F. J. I. B., & Biodegradation. (2017). *Nanobioremediation: Integration of nanoparticles and bioremediation for sustainable remediation of chlorinated organic contaminants in soils*. Academic Press.

Cele, E. N., & Maboeta, M. (2016). Response of soil enzyme activities to synergistic effects of biosolids and plants in iron ore mine soils. *International Journal of Environmental Science and Technology*, *13*(9), 2117–2126. doi:10.100713762-016-1043-y

Centers for Disease Control and Prevention (CDC). (2001). Recommendations for Using Fluoride to Prevent and Control Dental Caries in the United States. *MMWR*, *50*(RR-14), 1–42. PMID:11521913

Ceribasi, I. H., & Yetis, U. (2001). Biosorption of Ni(II) and Pb(II) by *Phanerochaete chrysosporium* from a binary metal system- Kinetics. *Water S.A.*, *21*, 15–20.

Cervantes, C., Campos-Garc'ıa, J., Devars, S., Gutierrez-Corona, F., Loza-Tavera, H., Torres-Guzman, C., & Sanchez, R. (2001). Interactions of chromium with microorganisms and plants. *FEMS Microbiology Reviews*, *25*(3), 335–347. doi:10.1111/j.1574-6976.2001.tb00581.x PMID:11348688

Cervantes, C., & Campos-Garcia, J. (2007). *Reduction and efflux of chromate by bacteria. Molecular Microbiology of Heavy Metals*. Springer-Verlag.

Cervantes, C., Ohtake, H., Chu, L., Misra, T. K., & Silver, S. (1990). Cloning, nucleotide sequence and expression of the chromate resistance determinant of *Pseudomonas aeruginosa* plasmid pUM505. *Journal of Bacteriology*, *172*(1), 287–291. doi:10.1128/JB.172.1.287-291.1990 PMID:2152903

Chabukdhara, M., Gupta, S. K., & Gogoi M. (2017). Phycoremediation of heavy metals coupled with the generation of bioenergy. *Algal Biofuels*, 163–188.

Chabukdhara, M., Gupta, S. K., & Gogoi, M. (2017). Phycoremediation of heavy metals coupled with generation of bioenergy. In *Algal Biofuels* (pp. 163–188). Springer. doi:10.1007/978-3-319-51010-1_9

Chae, G. T., Yun, S. T., Mayer, B., Kim, K. H., Kim, S. Y., Kwon, J. S., Kim, K., & Koh, Y. K. (2017). Fluorine geochemistry in bedrock groundwater of South Korea. *The Science of the Total Environment*, *385*(1–3), 272–283. doi:10.1016/j.scitotenv.2007.06.038 PMID:17655916

Chai, L. Y., Huang, S. H., Yang, Z. H., Peng, B., Huang, Y., & Chen, Y. H. (2009). Hexavalent chromium reduction by *Pannonibacter phragmitetus* BB isolated from soil under chromium-containing slag heap. *Journal of Environmental Science and Health. Part A, Toxic/Hazardous Substances & Environmental Engineering*, *44*(6), 615–622. doi:10.1080/10934520902784690 PMID:19337925

Chaiyen, P., Fraaije, M. W., & Mattevi, A. (2012). The enigmatic reaction of flavins with oxygen. *Trends in Biochemical Sciences*, *37*(9), 373–380. doi:10.1016/j.tibs.2012.06.005 PMID:22819837

Chakraborty, D., Agarwal, V. K., Bhatia, S. K., & Bellare, J. (1994). Steadystate Transitions and Polymorph Transformations inContinuous Precipitation of Calcium Carbonate. *Industrial & Engineering Chemistry Research*, *33*(9), 2187–2197. doi:10.1021/ie00033a024

Chakraborty, D., & Bhatia, S. K. (1996). Formation and Aggregation ofPolymorphs in Continuous Precipitation. 2. Kinetics of CaCO3Precipitation. *Industrial & Engineering Chemistry Research*, *35*(6), 1995–2006. doi:10.1021/ie950402t

Chakroun, H. K., Souissi, F., Bouchardon, J., Moutte, J., Faure, O., Remon, E., & Abdeljaoued, S. (2010). Transfer and accumulation of lead, zinc, cadmium and copper in plants growing in abandoned mining district area. *African Journal of Environmental Science and Technology*, *4*(10), 651–659.

Chakroun, H., Mechichi, T., Martinez, M. J., Dhouib, A., & Sayadi, S. (2010). Purification and characterization of a novel laccase from the ascomycete Trichoderma atroviride: Application on bioremediation of phenolic compounds. *Process Biochemistry*, *45*(4), 507–513. doi:10.1016/j.procbio.2009.11.009

Chaney, R. L., Angle, J. S., Broadhurst, C. L., Peters, C. A., Tappero, R. V., & Sparks, D. L. (2007). Improved understanding of hyperaccumulation yields commercial phytoextraction and phytomining echnologies. *Journal of Environmental Quality*, *36*(5), 1429–1443. doi:10.2134/jeq2006.0514 PMID:17766822

Chang, H., & Chen, L. (1984). Bifurcation Characteristics of Nonlinear Systems under Conventional PID Control. *Chemical Engineering Science*, *39*(7-8), 1127–1142. doi:10.1016/0009-2509(84)85075-7

Chardin, B., Giudici-Orticoni, M. T., De Luca, G., Guigliarelli, B., & Bruschi, M. (2003). Hydrogenases in sulfate-reducing bacteria function as chromium reductase. *Applied Microbiology and Biotechnology*, *63*(3), 315–321. doi:10.100700253-003-1390-8 PMID:12861426

Chen, F., Wu, F. B., Dong, J., Vincze, E., Zhang, G., Wang, F., Huang, Y., & Wei, K. (2007). Cadmium translocation and accumulation in developing barley grains. *Planta*, *227*(1), 223–232. doi:10.100700425-007-0610-3 PMID:17713783

Cheng, G., & Li, X. (2009). Bioreduction of chromium (VI) by *Bacillus* sp. isolated from soils of iron mineral area. *European Journal of Soil Biology*, *45*(5-6), 483–487. doi:10.1016/j.ejsobi.2009.06.009

Chen, J. F., Wei, Y. H., & Guo, F. (2000). Synthesis of Nanoparticles With Novel Technology: High-Gravity Reactive Precipitation. *Industrial & Engineering Chemistry Research*, *39*(4), 948–954. doi:10.1021/ie990549a

Chen, J., Yan, L. G., Yu, H. Q., Li, S., Qin, L. L., Liu, G. Q., Li, Y., & Du, B. (2016). Efficient removal of phosphate by facile prepared magnetic diatomite and illite clay from aqueous solution. *Chemical Engineering Journal*, *287*, 162–172. doi:10.1016/j.cej.2015.11.028

Chen, K., Sun, Z., Fang, R., Shi, Y., Cheng, H. M., & Li, F. J. A. m. (2018). Metal–Organic Frameworks (MOFs)-. *Derived Nitrogen-Doped Porous Carbon Anchored on Graphene with Multifunctional Effects for Lithium–Sulfur Batteries.*, *28*(38), 1707592.

Chen, L. S., Ma, Y. W., & Cotterell, B. (1989). Impact Fracture Energyof Mineral-Filled Polypropylene. *Polymer Engineering and Science*, *29*(8), 505–512. doi:10.1002/pen.760290804

Chen, S., Yin, H., Ye, J., Peng, H., Liu, Z., Dang, Z., & Chang, J. (2014). Influence of co-existed benzo[a]pyrene and copper on the cellular characteristics of *Stenotrophomonas maltophilia* during biodegradation and transformation. *Bioresource Technology*, *158*, 181–187. doi:10.1016/j.biortech.2014.02.020 PMID:24603491

Chen, Y., Chao, Y., Li, Y., Lin, Q., Bai, J., Tang, L., Wang, S., Ying, R., & Qiu, R. (2016). Survival strategies of the plant-associated bacterium Enterobacter sp. strain EG16 under cadmium stress. *Applied and Environmental Microbiology*, *82*(6), 1734–1744. doi:10.1128/AEM.03689-15 PMID:26729719

Cheung, K. H., & Gu, J. D. (2007). Mechanism of hexavalent chromium detoxification by microorganisms and bioremediation application potential: A review. *International Biodeterioration & Biodegradation*, *59*(1), 8–15. doi:10.1016/j.ibiod.2006.05.002

Chhabra, R., Singh, A., & Abrol, I. P. (1999). Fluorine in sodic soils. *Soil Science Society of America Journal*, *44*(1), 33–36. doi:10.2136ssaj1980.03615995004400010008x

Chiemchaisri, C., Chiemchaisri, W., Junsod, J., Threedeach, S., & Wicranarachchi, P. N. (2009). Leachate treatment and greenhouse gas emission in subsurface horizontal flow constructed wetland. *Bioresource Technology*, *100*(16), 3808–3814. doi:10.1016/j.biortech.2008.12.028 PMID:19200716

Chipasa, K. B. (2003). Accumulation and Fate of Selected Heavy Metals in a Biological Wastewater Treatment System. *Waste Management (New York, N.Y.)*, *23*(2), 135–143. doi:10.1016/S0956-053X(02)00065-X PMID:12623088

Chirenje, T., Ma, L. Q., Reeves, M., & Szulczewski, M. (2004). Lead distribution in near-surface soils of two Florida cities: Gainesville and Miami. *Geoderma*, *119*(1-2), 113–120. doi:10.1016/S0016-7061(03)00244-1

Chong, M. N., Jin, B., Chow, C. W. K., & Saint, C. (2010). Recent developments in photocatalytic water treatment technology: A review. *Water Research*, *44*(10), 2997–3027. doi:10.1016/j.watres.2010.02.039 PMID:20378145

Cho, U. H., & Seo, N. H. (2005). Oxidative stress in *Arabidopsis thaliana* exposed to cadmium is due to hydrogen peroxide accumulation. *Plant Science*, *168*(1), 113–120. doi:10.1016/j.plantsci.2004.07.021

Choubisa, S. L. (2010). Natural amelioration of fluoride toxicity (fluorosis) in goats and sheep [scientific correspondence]. *Current Science*, *99*(10), 1331–1332.

Choubisa, S. L., & Choubisa, D. (2016). Status of industrial fluoride pollution and its diverse adverse health effects in man and domestic animals in India. *Environmental Science and Pollution Research International*, *23*(8), 7244–7254. doi:10.100711356-016-6319-8 PMID:26903127

Choudhury, S., & Panda, S. K. (2004). Toxic effects, oxidative stress and ultrastructural changes in moss *Taxithelium nepalense* (Schwaegr) Broth. under chromium and lead phytotoxicity. *Water, Air, and Soil Pollution*, *167*(1), 73–90.

Chouhan, S., & Flora, S. J. (2010). Arsenic and F: Two major ground water pollutants. *Indian Journal of Experimental Biology*, *48*, 666–678. PMID:20929051

Chourey, K., Thompson, M. R., Morrrell-Falvey, J., Verberkmoes, N. C., Brown, S. D., Shah, M., Zhou, J., Doktycz, M., Heitich, R. L., & Thompson, D. K. (2006). Global molecular and morphological effects of 24-hour chromium (VI) exposure on *Shewanella oneidensis* MR-1. *Applied and Environmental Microbiology*, *72*(9), 6331–6344. doi:10.1128/AEM.00813-06 PMID:16957260

Christensen, T. H., & Kjeldsen, P. (1989). Basic Biochemical Processes in Landfills. In *Sanitary Landfilling: Process, Technology and Environmental Impact*. Academic Press.

Clarissa, H. (2018). Brazilian fruit processing, wastes as a source of lipase and other biotechnological products: A review. *Annals of the Brazilian Academy of Sciences*, *90*(3), 2927–2943. doi:10.1590/0001-3765201820180095 PMID:30304225

Clemens, S. (2006). Toxic metal accumulation, responses to exposure and mechanisms of tolerance in plants. *Biochimie*, *88*(11), 1707–1719. doi:10.1016/j.biochi.2006.07.003 PMID:16914250

Clemente, R., Arco-Lázaro, E., Pardo, T., Martín, I., Sánchez-Guerrero, A., Sevilla, F., & Bernal, M. P. (2019). Combination of soil organic and inorganic amendments helps plants overcome trace element induced oxidative stress and allows phytostabilization. *Chemosphere*, *223*, 223–231. doi:10.1016/j.chemosphere.2019.02.056 PMID:30784729

Colberg, P. J. S., & Young, L. Y. (1995).Anaerobic Degradation of Nonhalogenated Homocyclic Aromatic Compounds Coupled with Nitrate, Iron, or Sulfate Reduction. Microbial Transformation and Degradation of Toxic Organic Chemicals, 307–330.

Comino, E., Riggio, V., & Rosso, M. (2013). Grey water treated by an hybrid constructed wetland pilot plant under several stress conditions. *Ecological Engineering*, *53*, 120–125. doi:10.1016/j.ecoleng.2012.11.014

Comte, S., Guibaud, G., & Baudu, M. (2008). Biosorption properties of extracellular polymeric substances (EPS) towards Cd: Cu and Pb for different pH values. *Journal of Hazardous Materials*, *151*(1), 185–193. doi:10.1016/j.jhazmat.2007.05.070 PMID:17611021

Congeevaram, S., Dhanarani, S., Park, J., Dexilin, M., & Thamaraiselvi, K. (2007). Biosorption of chromium and nickel by heavy metal resistant fungal and bacterial isolates. *Journal of Hazardous Materials*, *146*(1-2), 270–277. doi:10.1016/j.jhazmat.2006.12.017 PMID:17218056

Costerton, J. W., Lewandowski, Z., Caldwell, D. E., Korber, D. R., & Lappin-Scott, H. M. (1995). Microbial biofilms. *Annual Review of Microbiology*, *49*(1), 711–745. doi:10.1146/annurev.mi.49.100195.003431 PMID:8561477

Costerton, J. W., Lewandowski, Z., Debeer, D., Caldwell, D., Korber, D., & James, G. (1994). Biofilms, the Customized Microniche. *Journal of Bacteriology*, *176*(8), 2137–2142. doi:10.1128/JB.176.8.2137-2142.1994 PMID:8157581

Crnković, D., Ristić, M., & Antonović, D. (2006). Distribution of heavy metals and arsenic in soils of Belgrade (Serbia and Montenegro). *Soil & Sediment Contamination*, *15*(6), 581–589. doi:10.1080/15320380600959073

Cronin, S. J., Manoharan, V., Hedley, M. J., & Loganathan, P. (2000). Fluoride: A review of its fate, bioavailability, and risks of fluorosis in grazed-pasture systems in New Zealand. *New Zealand Journal of Agricultural Research*, *43*(3), 295–321. doi:10.1080/00288233.2000.9513430

Cullen, W. R., & Reimer, K. J. (1989). Arsenic speciation in the environment. *Chemical Reviews*, *89*(4), 713–764. doi:10.1021/cr00094a002

Cundy, A. B., Hopkinson, L., & Whitby, R. L. J. S. e. (2008). Use of iron-based technologies in contaminated land and groundwater remediation. *RE:view*, *400*(1-3), 42–51. PMID:18692221

Czarnowski, W., Wrzesniowska, K., & Krechniak, J. (2006). Fluoride in drinking water and human urine in Northern and Central Poland. *The Science of the Total Environment*, *191*(1-2), 177–184. doi:10.1016/0048-9697(96)05259-X PMID:8885430

Dan, A., Fujii, D., Soda, S., Machimura, T., & Ike, M. (2017a). Removal of phenol, bisphenol A, and 4-tert-butylphenol from synthetic landfill leachate by vertical flow constructed wetlands. *The Science of the Total Environment*, *578*, 566–576. doi:10.1016/j.scitotenv.2016.10.232 PMID:27836343

Dan, A., Oka, M., Fujii, Y., Soda, S., Ishigaki, T., Machimura, T., & Ike, M. (2017b). Removal of heavy metals from synthetic landfill leachate in lab-scale vertical flow constructed wetlands. *The Science of the Total Environment*, *584*, 742–750. PMID:28131455

Danh, L. T., Truong, P., Mammucari, R., & Foster, N. (2009). Vetiver grass, Vetiveria zizanioides: A choice plant for phytoremediation of heavy metals and organic wastes. *International Journal of Phytoremediation*, *11*(8), 664–691. doi:10.1080/15226510902787302 PMID:19810597

Dary, M., Chamber-Pérez, M. A., Palomares, A. J., & Pajuelo, E. (2010). "In situ" phytostabilisation of heavy metal polluted soils using Lupinus luteus inoculated with metal resistant plant-growth promoting rhizobacteria. *Journal of Hazardous Materials*, *177*(1-3), 323–330. doi:10.1016/j.jhazmat.2009.12.035 PMID:20056325

Das, A. K., Shenoy, U. V., & Bandyopadhyay, S. (2009). Evolution of resource allocation networks. *Industrial & Engineering Chemistry Research*, *48*(15), 7152–7167. doi:10.1021/ie9003392

Das, D. K., Burman, G. K., & Kidwai, A. L. (1981). Chemical composition of monsoon rainwater over Bhopal, Madhya Pradesh during 1977 and 1978. *Mausam (New Delhi)*, *32*, 221–228.

Das, N., Vimala, R., & Karthika, P. (2008). Biosorption of heavy metals- An overview. *Indian Journal of Biotechnology, 7*, 159–169.

Das, N., Vimala, R., & Karthika, P. (2008). Biosorption of heavy metals—An overview. *Indian Journal of Biotechnology, 7*, 159–169.

Das, S., & Chandra, A. L. (1990). Chromate reduction in *Streptomyces. Experientia, 46*(7), 731–733. doi:10.1007/BF01939949 PMID:2164957

Das, S., Dash, H. R., & Chakraborty, J. (2016). Genetic basis and importance of metal resistant genes in bacteria for bioremediation of contaminated environments with toxic metal pollutants. *Applied Microbiology and Biotechnology, 100*(7), 2967–2984. doi:10.100700253-016-7364-4 PMID:26860944

Das, S., Mishra, J., Das, S. K., Pandey, S., Rao, D. S., Chakraborty, A., Sudarshan, M., Das, N. N., & Thatoi, H. N. (2014). Investigation on mechanism of Cr(VI) reduction and removal by *Bacillus amyloliquefaciens*, a novel chromate tolerant bacterium isolated from chromite mine soil. *Chemosphere, 96*, 112–121. doi:10.1016/j.chemosphere.2013.08.080 PMID:24091247

Davey, M. E., & O'Toole, G. A. (2000). Microbial biofilms: From ecology to molecular genetics. *Microbiology and Molecular Biology Reviews, 64*(4), 847–867. doi:10.1128/MMBR.64.4.847-867.2000 PMID:11104821

Davies, F. T. Jr, Puryear, J. D., Newton, R. J., Egilla, J. N., & Saraiva Grossi, J. A. (2002). Mycorrhizal fungi increase Cr uptake by sunflower plants: Influence of tissue mineral concentration, growth and gas exchange. *Journal of Plant Nutrition, 25*(11), 2389–2407. doi:10.1081/PLN-120014702

Davies, L. C., Cabrita, G. J. M., Ferreira, R. A., Carias, C. C., Novais, J. M., & Martins-Dias, S. (2009). Integrated study of the role of *Phragmites australis* in azo-dye treatment in a constructed wetland: From pilot to molecular scale. *Ecological Engineering, 35*(6), 961–970. doi:10.1016/j.ecoleng.2008.08.001

De Jaysankar, N. R., & Vardanyan, L. (n.d.). Detoxification of toxic heavy metals by marine bacteria highly resistant to mercury. *Marine Biotechnology, 5*, 185–193.

de Mello-Farias, P. C., Chaves, A. L. S., & Lencina, C. L. (2011). Transgenic plants for enhanced phytoremediation–physiological studies. In *Genetic Transformation*. InTech. doi:10.5772/24355

De Silva, A., Carvalho, M., De Souza, S., Dias, P., Filho, R., Saramago, C., Bento, C., & Hofer, E. (2012). Heavy metal tolerance (Cr, Ag and Hg) in bacteria isolated from sewage. *Brazilian Journal of Microbiology, 43*(4), 1620–1631. doi:10.1590/S1517-83822012000400047 PMID:24031994

de Souza, J. T., Arnould, C., Deulvot, C., Lemanceau, P., Gianinazzi-Pearson, V., & Raaijmakers, J. M. (2003). Effect of 2, 4-diacetylphloroglucinol on Pythium: Cellular responses and variation in sensitivity among propagules and species. *Phytopathology, 93*(8), 966–975. doi:10.1094/PHYTO.2003.93.8.966 PMID:18943863

Deepa, K. K., Sathishkumar, M., Binupriya, A. R., Murugessan, G. S., Swaminathan, K., & Yun, S. E. (2006). Sorption of Cr(VI) from dilute solutions and wastewater by live and pretreated biomass of *Aspergillus flavus. Chemosphere, 62*(5), 833–840. doi:10.1016/j.chemosphere.2005.04.087 PMID:15982702

Deibel & Schoeni. (2003). Biofilms: Forming a Defense Strategy for the Food Plant. *Food Safety Magazine*.

Deitzel, J., Kleinmeyer, J., Harris, D., & Tan, N. B. (2001). The effect of processing variables on the morphology of electrospun nanofibers and textiles. *Polymer, 42*(1), 261–272. doi:10.1016/S0032-3861(00)00250-0

De, J., Ramaiah, N., & Vardanyan, L. (2008). Detoxification of toxic heavy metals by marine bacteria highly resistant to mercury. *Marine Biotechnology (New York, N.Y.), 10*(4), 471–477. doi:10.100710126-008-9083-z PMID:18288535

De, J., Sarkar, A., & Ramaiah, N. (2006). Bioremediation of toxic substances by mercury resistant marine bacteria. *Ecotoxicology (London, England), 15*(4), 385–389. Advance online publication. doi:10.100710646-006-0066-4 PMID:16673165

Delorme, T. A., Gagliardi, J. V., Angle, J. S., & Chaney, R. L. (2001). Influence of the zinc hyperaccumulatorThlaspicaerulescens J. & C. Presl. and the nonmetal accumulator Trifoliumpratense L. on soil microbial populations. *Canadian Journal of Microbiology, 47*(8), 773–776. doi:10.1139/w01-067 PMID:11575505

Demarche, P., Junghanns, C., Nair, R. R., & Agathos, S. N. (2012). Harnessing the power of enzymes for environmental stewardship. *Biotechnology Advances, 30*(5), 933–953. doi:10.1016/j.biotechadv.2011.05.013 PMID:21624453

Desai, C., Jain, K., & Madamwar, D. (2008). Evaluation of in vitro Cr(VI) reduction potential in cytosolic extracts. *Bioresource Technology, 99*(14), 6059–6069. doi:10.1016/j.biortech.2007.12.046 PMID:18255287

Desjardin, V., Bayard, R., Huck, N., Manceau, A., & Gourdon, R. (2002). Effect of microbial activity on the mobility of chromium in soils. *Waste Management (New York, N.Y.), 22*(2), 195–200. doi:10.1016/S0956-053X(01)00069-1 PMID:12003148

Dey, S., Dey, J., Patra, S., & Pothal, D. (2007). Changes in the antioxidative enzyme activities and lipid peroxidation in wheat seedlings exposed to cadmium and lead stress. *Brazilian Journal of Plant Physiology, 19*(1), 53–60. doi:10.1590/S1677-04202007000100006

Dey, S., & Paul, A. K. (2012). Optimization of cultural conditions for growth associated chromate reduction by *Arthrobacter* sp. SUK 1201 isolated from chromite mine overburden. *Journal of Hazardous Materials, 213–214*, 200–206. doi:10.1016/j.jhazmat.2012.01.078 PMID:22361630

Dey, S., & Paul, A. K. (2013). Hexavalent chromium reduction by aerobic heterotrophic bacteria indigenous to chromite mine overburden. *Brazilian Journal of Microbiology, 44*(1), 307–315. doi:10.1590/S1517-83822013000100045 PMID:24159321

Dey, S., & Paul, A. K. (2016). Evaluation of chromate reductase activity in the cell-free culture filtrate of *Arthrobacter* sp. SUK 1201 isolated from chromite mine overburden. *Chemosphere, 156*, 69–75. doi:10.1016/j.chemosphere.2016.04.101 PMID:27176938

Dhal, B., Thatoi, H., Das, N., & Pandey, B. D. (2010). Reduction of hexavalent chromium by *Bacillus* sp. isolated from chromite mine soils and characterization of reduced product. *Journal of Chemical Technology and Biotechnology (Oxford, Oxfordshire), 85*, 1471–1479. doi:10.1002/jctb.2451

Dhankar, R., & Hooda, A. (2011). Fungal biosorption- an alternative to meet the challenges of heavy metal pollution in aqueous solutions. *Environmental Technology, 3*(5), 467–491. doi:10.1080/09593330.2011.572922 PMID:21877528

Dhir, B., Sharmila, P., Saradhi, P. P., & Nasim, S. A. (2009). *Physiological and antioxidant responses of Salvinia natans exposed to Chromium-rich wastewater*. Ecotoxicol. doi:10.1016/j.ecoenv.2009.03.015

Di Gregorio, S., Barbafieri, M., Lampis, S., Sanangelantoni, A. M., Tassi, E., & Vallini, G. (2006). Combined application of Triton X-100 and Sinorhizobium sp. Pb002 inoculum for the improvement of lead phytoextraction by Brassica juncea in EDTA amended soil. *Chemosphere, 63*(2), 293–299. doi:10.1016/j.chemosphere.2005.07.020 PMID:16153689

Dien, V. T., Morris, S. E., Karadeema, R. J., & Romesberg, F. E. (2018). Expansion of the genetic code via expansion of the genetic alphabet. *Current Opinion in Chemical Biology, 46*, 196–202. doi:10.1016/j.cbpa.2018.08.009 PMID:30205312

Diep, P., Mahadevan, R., & Yakunin, A. (2018). Heavy metal removal by bioaccumulation using genetically engineered microorganisms. *Frontiers in Bioengineering and Biotechnology, 6*, 1–20. doi:10.3389/fbioe.2018.00157 PMID:30420950

Dinakar, N., Nagajyothi, P. C., Suresh, S., Damodharam, T., & Suresh, C. (2009). Cd induced changes on proline, antioxidant enzymes, nitrate and nitrite reductases in *Arachis hypogea* L. *J. Environ. Biol.* 30(2): 289-294. *Environ. Safe.*, *72*, 1790–1797.

Ding, H., Wu, Y., Zou, B., Lou, Q., Zhang, W., Zhong, J., Lu, L., & Dai, G. (2016). Simultaneous removal and degradation characteristics of sulfonamide, tetracycline, and quinolone antibiotics by laccase-mediated oxidation coupled with soil adsorption. *Journal of Hazardous Materials*, *307*, 350–358. doi:10.1016/j.jhazmat.2015.12.062 PMID:26826938

Divyasree, C., & Prabhakaran Subramanian, S. (2016). Studies on the bioremediation of hromium from aqueous solutions using *Corynebacterium paurometabolum*. *Transactions of the Indian Institute of Metals*, *70*, 497–509.

Dobbelaere, S., Vanderleyden, J., & Okon, Y. (2003). Plant growth-promoting effects of diazotrophs in the rhizosphere. *Critical Reviews in Plant Sciences*, *22*(2), 107–149. doi:10.1080/713610853

Doble, M., & Kumar, A. (2005). *Biotreatment of industrial effluents*. Elsevier, Butterworth-Heinemann.

Dodd, D., & Cann, I. O. (2010). Enzymatic deconstruction of xylan for biofuel production. *Biology and Bioenergy*, *1*(1), 2–17. PMID:20431716

Dojlido, J., & Best, G. A. (1993). *Chemistry of Water and Water Pollution*. Ellis Horwood Ltd.

Domingos, M., Klumpp, A., Rinaldi, M. C. S., Modesto, I. F., Klumpp, G., & Delitti, W. B. C. (2003). Combined effects of air and soil pollution by fluoride emissions on *Tibouchina pulchra* Cogn., at Cubatão, SE Brazil, and their relations with aluminium. *Plant and Soil*, *249*(2), 297–308. doi:10.1023/A:1022800225753

Dönmez, G., & Aksu, Z. (2001). Bioaccumulation of copper(ii) and nickel(ii) by the non-adapted and adapted growing Candida SP. *Water Research*, *35*(6), 1425–1434. doi:10.1016/S0043-1354(00)00394-8 PMID:11317889

Doran, J. W., & Zeiss, M. R. (2000). Soil health and sustainability: Managing the biotic component of soil quality. *Applied Soil Ecology*, *15*(1), 3–11. doi:10.1016/S0929-1393(00)00067-6

Douglas, G. B., Lurling, M., & Spears, B. M. (2016). Assessment of changes in potential nutrient limitation in an impounded river after application of lanthanum-modified bentonite. *Water Research*, *97*, 47–54. doi:10.1016/j.watres.2016.02.005 PMID:26879191

Duffus, J. H. (2002). Heavy Metals – A meaningless Terms? (IUPAC Technical Report). *Pure and Applied Chemistry*, *74*(5), 793–807. doi:10.1351/pac200274050793

Dunbar, W. S. (2017). Biotechnology and the Mine of Tomorrow. *Trends in Biotechnology*, *35*(1), 79–89. doi:10.1016/j.tibtech.2016.07.004 PMID:27612568

Dunne, C., Crowley, J. J., Moënne-Loccoz, Y., Dowling, D. N., & O'Gara, F. (1997). Biological control of Pythiumultimum by Stenotrophomonasmaltophilia W81 is mediated by an extracellular proteolytic activity. *Microbiology*, *143*(12), 3921–3931. doi:10.1099/00221287-143-12-3921

Duran, N., & Esposito, E. (2000). Potential applications of oxidative ′ enzymes and phenoloxidase-like compounds in wastewater and soil treatment: A review. *Applied Catalysis B: Environmental*, *28*(2), 83–99. doi:10.1016/S0926-3373(00)00168-5

Dushenkov, V., Kumar, P. N., Motto, H., & Raskin, I. (1995). Rhizofiltration: The use of plants to remove heavy metals from aqueous streams. *Environmental Science & Technology*, *29*(5), 1239–1245. doi:10.1021/es00005a015 PMID:22192017

Eccles, H. (1999). Treatment of metal contamination waste. Why select a biological process? *Trends in Biotechnology*, *17*(12), 462–465. doi:10.1016/S0167-7799(99)01381-5 PMID:10557157

Edmunds, W. M., & Smedley, P. L. (2005). Fluoride in natural waters. In O. Selinus (Ed.), *Essentials of medical geology* (pp. 301–329). Elsevier.

Elangovan, R., Abhipsa, S., Rohit, B., Ligy, P., & Chandraraj, K. (2006). Reduction of Cr(VI) by a *Bacillus* sp. *Biotechnology Letters, 28*(4), 247–252. doi:10.100710529-005-5526-z PMID:16555008

Elangovan, R., Philip, L., & Chandraraj, K. (2010). Hexavalent chromium reduction by free and immobilized cell-free extract of *Arthrobacter rhombi*-RE. *Applied Biochemistry and Biotechnology, 160*(1), 81–97. doi:10.100712010-008-8515-6 PMID:19165627

Eldyasti, A., Andalib, M., Hafez, H., Nakhla, G., & Zhu, J. (2011). Comparative modeling of biological nutrient removal from landfill leachate using a circulating fluidized bed bioreactor (CFBBR). *Journal of Hazardous Materials, 187*(1-3), 140–149. doi:10.1016/j.jhazmat.2010.12.115 PMID:21255923

El-morsy, E. M., Nour El-Dein, M. M., & El-Diamoney, S. M. M. (2013). *Mucor recemosus* as a biosorbant of metal ions from polluted water in northern delta of Egypt. *Mycosphere, 4*(6), 1118–1131. doi:10.5943/mycosphere/4/6/9

Emtiazi, G., Ethemadifar, Z., & Habibi, M. H. (2004). Production of extra-cellular polymer in *Azotobacter* and biosorption of metal by exopolymer. *African Journal of Biotechnology, 3*(6), 330–333. doi:10.5897/AJB2004.000-2060

Ergul-Ulger, Z., Ozkan, A. D., Tunca, E., Atasagun, S., & Tekinay, T. (2014). Chromium(VI) biosorption and bioaccumulation by live and acid modified biomass of a novel *Morganella morganii* isolate. *Separation Science and Technology, 49*(6), 907–914. doi:10.1080/01496395.2013.866681

Essalhi, M., & Khayet, M. (2013). Self-sustained webs of polyvinylidene fluoride electrospun nanofibers at different electrospinning times: 1. Desalination by direct contact membrane distillation. *Journal of Membrane Science, 433*, 167–179. doi:10.1016/j.memsci.2013.01.023

European Union. (2002). *Heavy Metals in Wastes*. European Commission on Environment. https://ec.europa.eu/environment/waste/ studies/pdf/heavy_metalsreport.pdf

Fadel, M., Naziha, M. H., Elshafei, M. M., Mostafa, A. H., Ahmed, M. A., & Khater, H. M. (2015). Biosorption of manganese from groundwater by biomass of *Sacchromyces cerevisae. Housing and Building National Research Centre Journal, 13*, 106–113.

Faisal, M., & Hasnain, S. (2006). Growth stimulatory effect of Ochrobactrum intermedium and Bacillus cereus on Vigna radiata plants. *Letters in Applied Microbiology, 43*(4), 461–466. doi:10.1111/j.1472-765X.2006.01977.x PMID:16965380

Fakayode, S. O. (2005). Impact assessment of industrial effluent on water quality of the receiving Alaro river in Ibadan Nigeria. *AJEAM-RAGEE*, 1-13

Falcaro, P., Ricco, R., Yazdi, A., Imaz, I., Furukawa, S., Maspoch, D., . . . Doonan, C. J. J. C. C. R. (2016). *Application of metal and metal oxide nanoparticles@ MOFs*. Academic Press.

Farag, S., & Zaki, S. (2010). Identification of bacterial strains from tannery effluent and reduction of hexavalent chromium. *Journal of Environmental Biology, 31*, 877–882.

Farhan, S. N., & Khadom, A. A. (2015). Biosorption of heavy metals fromaqueous solutions by *Saccharaomyces cerevisae. International Journal of Industrial Chemistry, 6*(2), 119–130. doi:10.100740090-015-0038-8

Fashola, M. O., Ngole-Jeme, V. M., & Babalola, O. O. (2016). Heavy metal pollution from gold mines: Environmental effects and bacterial strategies for resistance. *International Journal of Environmental Research and Public Health, 13*(11), 1047. doi:10.3390/ijerph13111047 PMID:27792205

Feng, Y. Z., Grogan, P., Caporaso, J. G., Zhang, H. Y., Lin, X. G., Knight, R., & Chu, H. (2014). pH is a good predictor of the distribution of anoxygenic purple phototrophic bacteria in Arctic soils. *Soil Biology & Biochemistry*, *74*, 193–200. doi:10.1016/j.soilbio.2014.03.014

Feng, Y. Z., Lin, X. G., Zhang, J., Mao, T. T., & Zhu, J. G. (2011). Soil purple phototrophic bacterial diversity under double cropping (rice-wheat) with free-air CO2 enrichment (FACE). *European Journal of Soil Science*, *62*(4), 533–540. doi:10.1111/j.1365-2389.2011.01357.x

Fereidoun, H., Nourddin, M. S., Rreza, N. A., Mohsen, A., Ahmad, R., & Pouria, H. (2007). The Effect of Long-Term Exposure to Particulate Pollution on the Lung Function of Teheranian and Zanjanian Students, Pakistan. *The Journal of Physiology*, *3*(2), 1–5.

Ferner, D. J. (2001). *Toxicity, heavy metals. eMed J, 2(5), 1.* Google Scholar.

Ferroudi, N., Nzimoto, J., Davidson, A., Talbot, D., Briot, E., Dupuis, V., & Abramson, S. (2013). Maghemite nanoparticles and maghemite/silica nanocomposite microspheres as magnetic fenton catalysts for the removal of water pollutants. *Applied Catalysis B: Environmental*, *136*, 9–18. doi:10.1016/j.apcatb.2013.01.046

Field, E. K., Blaskovich, J. P., Peyton, B. M., & Gerlach, R. (2018). Carbon-dependent chromate toxicity mechanism in an environmental *Arthrobacter* isolate. *Journal of Hazardous Materials*, *355*, 162–169. doi:10.1016/j.jhazmat.2018.05.020 PMID:29800910

Filipovic-Kovacevic, Z., Sipos, L., & Briski, F. (2000). Biosorption of chromium, copper, nickel and zinc ions onto fungal pellets of *Aspergillus niger* 405 from aqueous solutions. *Food Technology and Biotechnology*, *38*, 211–216.

Fiset, J. F., Blais, J. F., & Riverso, P. A. (2008). Review on the removal of metal ions from effluents using seaweeds, alginate derivatives and other sorbents. *Revue des Sciences de l'Eau*, *21*(3), 283–308. doi:10.7202/018776ar

Flemming, H.-C., Neu, T. R., & Wozniak, D. J. (2007). The EPS Matrix: The "House of Biofilm Cells". *Journal of Bacteriology*, *189*(22), 7945–7947. doi:10.1128/JB.00858-07 PMID:17675377

Flores, L., García, J., Pena, R., & Garfí, M. (2019). Constructed wetlands for winery wastewater treatment: A comparative Life Cycle Assessment. *The Science of the Total Environment*, *659*, 1567–1576. doi:10.1016/j.scitotenv.2018.12.348 PMID:31096366

Fomina, M., & Gadd, G. M. (2014). Biosorption: Current perspectives on concept, definition and application. *Bioresource Technology*, *160*, 3–14. doi:10.1016/j.biortech.2013.12.102 PMID:24468322

Fourest, E., & Roux, J. C. (1992). Heavy metal biosorption by fungal mycelial by-product, mechanisms and influence of pH. *Applied Microbiology and Biotechnology*, *37*(3), 399–403. doi:10.1007/BF00211001

Fox, B. G., Borneman, J. G., Wackett, L. P., & Lipscomb, J. D. (1990). Haloalkene oxidation by the soluble methane monooxygenase from *Methylosinus trichosporium* OB3b: Mechanistic and environmental implications. *Biochemistry*, *29*(27), 6419–6427. doi:10.1021/bi00479a013 PMID:2207083

Frederick, T. M., Taylor, E. A., Willis, J. L., Shultz, M. S., & Woodruff, P. J. (2013). Chromate reduction is expedited by bacteria engineered to produce the compatible solute trehalose. *Biotechnology Letters*, *35*(8), 1291–1296. doi:10.100710529-013-1200-z PMID:23563698

Fredrickson, J. K., Kostandarithes, H. M., Li, S. W., Plymale, A. E., & Daly, M. J. (2000). Reduction of Fe (III), Cr (VI), U (VI), and Tc (VII) by *Deinococcus radiodurans* R1. *Applied and Environmental Microbiology*, *66*(5), 2006–2011. doi:10.1128/AEM.66.5.2006-2011.2000 PMID:10788374

Fuge, R., & Andrews, M. J. (1988). Fluorine in U.K. environment. *Environmental Geochemistry and Health, 10*(3–4), 96–104. doi:10.1007/BF01758677 PMID:24213649

Fung, K. F., Zhang, Z. Q., Wong, J. W. C., & Wong, M. H. (1999). Fluoride contents in tea and soil from tea plantations and the release of fluoride into tea liquor during infusion. *Environmental Pollution, 104*(2), 197–205. doi:10.1016/S0269-7491(98)00187-0

Fu, Q. Y., Li, S., & Zhu, Y. H. (2012). Biosorption of copper (II) from aqueous solution by mycelial pellets of *Rhizopus oryzae*. *African Journal of Biotechnology, 11*(6), 1403–1411.

Gadd, G. M. (2009). Heavy Metal Pollutants: Environmental and Biotechnological Aspects. Encyclopedia of Microbiology, 321-334. doi:10.1016/B978-012373944-5.00149-8

Gadd, G. M. (1990). Heavy metal accumulation by bacteria and other microorganisms. *Experientia, 46*(8), 834–840. doi:10.1007/BF01935534

Gadd, G. M. (1992). Metals and microorganisms: A problem of definition. *FEMS Microbiology Letters, 100*(1-3), 197–203. doi:10.1111/j.1574-6968.1992.tb05703.x PMID:1478456

Gadd, G. M. (1993). Interactions of fungi with toxic metals. *The New Phytologist, 124*(1), 25–60. doi:10.1111/j.1469-8137.1993.tb03796.x

Gadd, G. M., & White, C. (1985). Copper uptake by *Penicillium ochro-chloron*: Influence of pH on toxicity and demonstration of energy dependent copper influx using protoplasts. *Journal of General Microbiology, 131*, 1875–1879.

Gadd, G. M., & White, C. (1993). Microbial treatment of metal pollution—Working biotechnologyy? *Trends in Biotechnology, 11*(8), 353–359. doi:10.1016/0167-7799(93)90158-6 PMID:7764182

Gadhe, A., Sonawane, S. S., & Varma, M. N. (2015). Enhancement effect of hematite and nickel nanoparticles on biohydrogen production from dairy wastewater. *International Journal of Hydrogen Energy, 40*(13), 4502–4511. doi:10.1016/j.ijhydene.2015.02.046

Ganesan, V. (2008). Rhizoremediation of cadmium soil using a cadmium-resistant plant growth-promoting rhizopseudomonad. *Current Microbiology, 56*(4), 403–407. doi:10.100700284-008-9099-7 PMID:18239844

Gao, C. Z., Dong, Y., Zhang, H. J., & Zhang, J. M. (2007). Utilisation of Distiller Waste and Residual Mother Liquor to Prepare Precipitated Calcium Carbonate. *Journal of Cleaner Production, 15*(15), 1419–1425. doi:10.1016/j.jclepro.2006.06.024

Gao, L., & Yan, X. (2016). Nanozymes: An emerging field bridging nanotechnology and biology. *Science China. Life Sciences, 59*(4), 400–402. doi:10.100711427-016-5044-3 PMID:27002958

Garbarino, J. (1995). *Raising Children in a Socially Toxic Environment*. Jossey-Bass Publishers.

Garcia, M. A., Alonso, J., & Melgar, M. J. (2005). *Agaricus macrosporus* as a potential bioremediation agent for substrates contaminated with heavy metals. *Journal of Chemical Technology and Biotechnology (Oxford, Oxfordshire), 80*(3), 325–330. doi:10.1002/jctb.1203

Garg, A., Visht, S., Sharma, P. K., & Kumar, N. J. D. P. S. (2011). *Formulation, characterization and application on nanoparticle: a review*. Academic Press.

Garrido, S., Del Campo, G. M., Esteller, M. V., Vaca, R., & Lugo, J. (2005). Heavy metals in soil treated with sewage sludge composting, their effect on yield and uptake of broad bean seeds (Viciafaba L.). *Water, Air, and Soil Pollution, 166*(1-4), 303–319. doi:10.100711270-005-5269-4

Gasik, K., & Korban, S. S. (2007). Expression of *Arabidopsis* phytochelatin synthase in Indian mustard (*Brassica juncea*) plants enhance tolerance for cadmium and zinc. *Planta*, *225*(5), 1277–1285. doi:10.100700425-006-0421-y PMID:17086401

Gaur, N., Flora, G., Yadav, M., Tiwari, A. J. E. S. P., & Impacts. (2014). *A review with recent advancements on bioremediation-based abolition of heavy metals*. Academic Press.

Gautam, Sharma, Mahiya, & Chattopadhyaya. (2015). Contamination of Heavy Metals in Aquatic Media: Transport, Toxicity and Technologies for Remediation. The Royal Society of Chemistry, 1, 1-25.

Gauthier, T., Norwood, P., Prepas, E., & Pyle, G. (2014). Metal-PAH mixtures in the aquatic environment: A review of co-toxic mechanisms leading to more-than-additive outcomes. *Aquatic Toxicology (Amsterdam, Netherlands)*, *154*, 253–269. doi:10.1016/j.aquatox.2014.05.026 PMID:24929353

Gavrilescu, M. (2004). Removal of heavy metals from the environment by biosorption. *Chemistry (Weinheim an der Bergstrasse, Germany)*, *28*, 30.

Geva, P., Kahta, R., Nakonechny, F., Aronov, S., & Nisnevitch, M. (2016). Increased copper bioremediation ability of new transgenic and adapted Saccharomyces cerevisiae strains. *Environmental Science and Pollution Research International*, *23*(19), 19613–19625. doi:10.100711356-016-7157-4 PMID:27392627

Ghosh, M., Ganguli, A., & Pathak, S. (2009). Application of a novel biopolymer for removal of Salmonella from poultry wastewater. *Environmental Technology*, *30*(4), 337–344. doi:10.1080/09593330902732093 PMID:19492545

Ghosh, M., & Singh, S. P. (2005). Comparative uptake and phytoextraction study of soil induced chromium by accumulator and high biomass weed species. *Applied Ecology and Environmental Research*, *3*(2), 67–79. doi:10.15666/aeer/0302_067079

Gianfreda, L., & Bollag, J.-M. (2002). Isolated enzymes for the transformation and detoxification of organic pollutants. In R. G. Burns & R. Dick (Eds.), *Enzymes in the Environment: Activity, Ecology and Applications* (pp. 491–538). Marcel Dekker. doi:10.1201/9780203904039.ch19

Gianfreda, L., & Rao, M. A. (2008). Bonifica di suoli contaminati e depurazione dell'acqua. In M. Gennari & M. Trevisan (Eds.), *Agrofarmaci. Conoscenze per un Uso Sostenibile* (pp. 521–564). Oasi Alberto Perdisa.

Gianfreda, L., Xu, F., & Bollag, J. M. (1999). *Laccases: a useful group of deligniðcation and possible industrial applications*. In A. Messerschmidt (Ed.), *Multi-Copper Oxidases* (pp. 201–224). World Scientiðc.

Giardina, P., Faraco, V., Pezzella, C., Piscitelli, A., Vanhulle, S., & Sannia, G. (2010). Laccases: A never-ending story. *Cellular and Molecular Life Sciences*, *67*(3), 369–385. doi:10.100700018-009-0169-1 PMID:19844659

Gilotra, U., & Srivastava, S. (1997). Plasmid-encoded sequestration of copper by *Pseudomonas pickettii* strain US321. *Current Microbiology*, *34*(6), 378–381. doi:10.1007002849900199 PMID:9142746

Gilpin, L., & Johnson, A. H. (2014). Fluorine in agricultural soils of Southern Pennsylvania. *Soil Science Society of America Journal*, *44*(2), 255–258. doi:10.2136ssaj1980.03615995004400020010x

Gleick, P. H. (2003). Global freshwater resources: Soft-path solutions for the 21st century. *Science*, *302*(5650), 1524–1528. doi:10.1126cience.1089967 PMID:14645837

Glick, B. R. (2001). Phytoremediation: Synergistic use of plants and bacteria to clean up the environment. *Biotechnology Advances*, *21*(5), 383–393. doi:10.1016/S0734-9750(03)00055-7 PMID:14499121

Godheja, J., Sk, S., Siddiqui, S. A., & Dr, M. (2016). Xenobiotic Compounds Present in Soil and Water: A Review on Remediation Strategies. *Journal of Environmental & Analytical Toxicology*, *6*(5). Advance online publication. doi:10.4172/2161-0525.1000392

Goher, M. E., El-Monem, A. M. A A., Abdel-Satar, M., Ali, M. H., Hussian, A.-E. M., & Napiórkowska-Krzebietke, A. (2016). Biosorption of some toxic metals from aqueous solution using non-living algal cells of *Chlorella vulgaris*. *Journal of Elementology*, *21*(3), 703–714.

Goher, M. E., El-Monem, A. M. A., Abdel-Satar, A. M., Ali, M. H., Hussian, A.-E. M., & Napiórkowska-Krzebietke, A. (2016). Biosorption of some toxic metals from aqueous solution using non-living algal cells of *Chlorella vulgaris*. *Journal of Elementology*, *21*(3), 703–714.

Gomathy, M., & Sbarinathan, K. (2010). Microbial mechanisms of heavy metal tolerance- A review. *Agricultural Research Communication Centre*, *31*(2), 133–138.

Gomez, L. D., Steele-King, C. G., & McQueen-Mason, S. J. (2008). Sustainable liquid biofuels from biomass: The writing's on the walls. *The New Phytologist*, *178*(3), 473–485. doi:10.1111/j.1469-8137.2008.02422.x PMID:18373653

Gong, Y., Chen, J., & Pu, R. (2019). The enhanced removal and phytodegradation of sodium dodecyl sulfate (SDS) in wastewater using controllable water hyacinth. *International Journal of Phytoremediation*, *21*(11), 1080–1089. doi:10.1080/15226514.2019.1606779 PMID:31044608

Gonzalez, C. F., Ackerley, D. F., Lynch, S. V., & Matin, A. (2005). ChrR a soluble quinine reductase of *Pseudomonas putida* that defends against H2O2. *The Journal of Biological Chemistry*, *280*(24), 22590–22595. doi:10.1074/jbc.M501654200 PMID:15840577

Gopakumar, D.A., Vishnu, A., Daniel, P., Leu, S.Y., Abdul, K.H.P.S., & Sabu, T. 2019. Nanocellulose based membrane for water purification. *Nanoscale Materials in Water Purification*, *4*, 59-85.

Gopal, R., & Khurana, N. (2011). Effect of heavy metal pollutants on sunflower. *African Journal of Plant Science*, *5*(9), 531–536.

Goulhen, F., Gloter, A., Guyot, F., & Bruschi, M. (2006). Cr(VI) detoxification by Desulfovibrio vulgaris strain Hildenborough: Microbe-metal interactions studies. *Applied and Environmental Microbiology*, *71*(6), 892–897. PMID:16896506

Gratao, P. L., Prasad, M. N. V., Cardoso, P. F., Lea, P. J., & Azevedo, R. A. (2005). Phytoremediation: Green technology for the clean up of toxic metals in the environment. *Brazilian Journal of Plant Physiology*, *17*(1), 53–64. doi:10.1590/S1677-04202005000100005

Grisey, E., Laffray, X., Contoz, O., Cavalli, E., Mudry, J., & Aleya, L. (2012). The bioaccumulation performance of reeds and cattails in a constructed treatment wetland for removal of heavy metals in landfill leachate treatment (Etueffont, France). *Water, Air, and Soil Pollution*, *223*(4), 1723–1741. doi:10.100711270-011-0978-3

Grosse, S., Laramee, L., Wendlandt, K.-D., McDonald, I. R., Miguez, C. B., & Kleber, H.-P. (1999). Purification and characterization of the soluble methane monooxygenase of the type II methanotrophic bacterium Methylocystis sp. strain WI 14. *Applied and Environmental Microbiology*, *65*(9), 3929–3935. doi:10.1128/AEM.65.9.3929-3935.1999 PMID:10473397

Groudev, S. N., & Groudeva, V. I. (1988). Microbiological solubilization of gold. *Symposium of Precious and Rare Metals*.

Gruji, C. S., Vasic, S., Radojevic, I., Comic, L., & Ostojic, A. (2017). Comparison of the *Rhodotorula mucilaginosa* biofilm and planktonic culture on heavy metal susceptibility and removal potential. *Water, Air, and Soil Pollution*, *228*(2), 73. doi:10.100711270-017-3259-y

Guo, Y. (2016). Arsenic methylation by an arsenite S-adenosylmethionine methyltransferase fromSpirulinaplatensis. *Journal of Environmental Sciences (China)*, 34–35.

Gupta, Schmoll, Maki, Tuohy, & Mazutti. (n.d.). *Applications of Microbial Engineering*. CRC Press Taylor & Francis Group.

Gupta, Tuohy, O'Donovan, & Lohani. (n.d.). *Biotechnology of Bioactive Compounds: Sources and Applications*. Wiley & Blackwell.

Gupta, A., Joia, J., Sood, A., Sood, R., & Sidhu, C. (2016). Microbes as Potential Tool for Remediation of Heavy Metals: A Review. *Journal of Microbial & Biochemical Technology*, *8*(4), 364–372. doi:10.4172/1948-5948.1000310

Gupta, A., Meyer, J. M., & Goel, R. (2002). Development of heavy metal-resistant mutants of phosphate solubilizing Pseudomonas sp. NBRI 4014 and their characterization. *Current Microbiology*, *45*(5), 323–327. doi:10.100700284-002-3762-1 PMID:12232661

Gupta, N., Sabat, J., Parida, R., & Kerkatta, D. (2007). Solubilization of tricalcium phosphate and rock phosphate by microbes isolated from chromite, iron and manganese mines. *Acta Botanica Croatica*, *66*, 197–204.

Gupta, P., & Diwan, B. (2017). Bacterial Exopolysaccharide mediated heavy metal removal: A Review on biosynthesis, mechanism and remediation strategies. *Biotechnology Reports (Amsterdam, Netherlands)*, *13*, 58–71. doi:10.1016/j.btre.2016.12.006 PMID:28352564

Gupta, P., Sarkar, S., Das, B., Bhattacharjee, S., & Tribedi, P. (2016). Biofilm, pathogenesis and prevention—a journey to break the wall: A review. *Archives of Microbiology*, *198*(1), 1–15. doi:10.100700203-015-1148-6 PMID:26377585

Gupta, R., Sundarrajan, S., & Priya, M. (2015, January-March). Isolation and Characterization of Heavy Metal Tolerant Bacterial Isolates VITNJ12 and VITNJ13 from Paper Mill Effluent, Erode District, Tamilnadu, India. *Int. J. Drug Dev. & Res.*, *7*(1), 145–149.

Gupta, S. K., & Shukla, P. (2016). Advanced technologies for improved expression of recombinant proteins in bacteria: Perspectives and applications. *Critical Reviews in Biotechnology*, *36*(6), 1089–1098. doi:10.3109/07388551.2015.1084264 PMID:26384140

Gupta, S., Nayek, S., Saha, R. N., & Satpati, S. (2008). Assessment of heavy metal accumulation in macrophyte, agricultural soil and crop plants adjacent to discharge zone of sponge iron factory. *Environ. Geol*, *55*(4), 731–739. doi:10.100700254-007-1025-y

Gurave, Korde, Dhas, & Disale. (2015). Isolation and identification of heavy metal resistant bacteria from petroleum soil of Loni, Ahmednagar, Pelagia Research Library. *European Journal of Experimental Biology*, *5*(12), 6-11.

Gurunathan, S., Kalishwaralal, K., Vaidyanathan, R., Venkataraman, D., Pandian, S. R. K., Muniyandi, J., . . . Biointerfaces, S. B. (2009). *Biosynthesis, purification and characterization of silver nanoparticles using Escherichia coli*. Academic Press.

Gu, S., Cai, J., Wang, J., Yuan, Y., Chang, D., & Chikhotkin, V. F. (2015). Compound Method to Disperse $CaCO_3$ Nanoparticles to Nano-Size in Water. *Journal of Nanoscience and Nanotechnology*, *15*(12), 9488–9493. doi:10.1166/jnn.2015.10332 PMID:26682370

Guzik, U., Hupert-Kocurek, K., & Wojcieszysk, D. (2013). Intradiol Dioxygenases — The Key Enzymes in Xenobiotics Degradation. In R. Chamy (Ed.), *Biodegradation of Hazardous and Special Products*. doi:10.5772/56205

Gyuricza, V., Fodor, F., & Szigeti, Z. (2010). Phytotoxic Effects of Heavy Metal contaminated Soil Reveal Limitations of Extract-Based Ecotoxicological Tests. *Water, Air, and Soil Pollution*, *210*(1-4), 113–122. doi:10.100711270-009-0228-0

Haidouti, C. (2001). Fluoride distribution in soils in the vicinity of a point emission source in Greece. *Geoderma, 49*(1-2), 129–138. doi:10.1016/0016-7061(91)90096-C

Hall, C., Brachat, S., & Dietrich, F. S. (2005). Contribution of horizontal gene transfer to the evolution of Saccharomyces cerevisiae. *Eukaryotic Cell, 4*(6), 1102–1115. doi:10.1128/EC.4.6.1102-1115.2005 PMID:15947202

Handa, B. K. (1975). Geochemistry and genesis of fluoride containing ground waters in India. *Ground Water, 13*(3), 275–281. doi:10.1111/j.1745-6584.1975.tb03086.x

Handique, G. K., & Handique, A. K. (2009). Proline accumulation in lemongrass (*Cymbopogon flexuosus* Stapf.) due to heavy metal stress. *Journal of Environmental Biology, 30*(2), 299–302. PMID:20121035

Haruta, S., Hanafusa, T., Fukase, H., Miyajima, H., & Oki, T. (2003). An Effective Absorption Behaviour of Insulin for DiabeticTreatment Following Intranasal Delivery using PorousSpherical Calcium Carbonate in Monkey and Healthy HumanVolunteers. *Diabetes Technology & Therapeutics, 5*(1), 1–9. doi:10.1089/152091503763816409 PMID:12725701

Hashmat, A. J., Afzal, M., Fatima, K., Anwar-ul-Haq, M., Khan, Q. M., Arias, C. A., & Brix, H. (2019). Characterization of hydrocarbon-degrading bacteria in constructed wetland microcosms used to treat crude oil polluted water. *Bulletin of Environmental Contamination and Toxicology, 102*(3), 358–364. doi:10.100700128-018-2518-y PMID:30542756

Hassiba, M., Naima, A., Yahia, K., & Zahra, S. (2014). Study of lead adsorption from aqueous solutions on agar beads with EPS produced from *Paenibacillus polymyxa. Chemical Engineering Transactions, 38*, 31–36.

Hauser, L. J., Miriam, L., Brown, S. D., Larimer, F., Keller, K. L., Rapp-Giles, B. J., & … . (2011). Complete genome sequence and updated annotation of *Desulfovibrio alaskensis* G20. *Journal of Bacteriology, 193*(16), 4268–4269. doi:10.1128/JB.05400-11 PMID:21685289

Hay, I. D., Wang, Y., Moradali, M. F., Rehman, Z. U., & Rehm, B. H. (2014). Genetics and regulation of bacterial alginate production. *Environmental Microbiology, 16*(10), 2997–3011. doi:10.1111/1462-2920.12389 PMID:24428834

He, F., & Zhao, D. (2007). Manipulating the size and dispersibility of zerovalent iron nanoparticles by use of carboxymethyl cellulose stabilizers. *Environmental Science & Technology, 41*(17), 6216–6221. doi:10.1021/es0705543 PMID:17937305

He, J., & Chen, J. P. (2014). A comprehensive review on biosorption of heavy metals by algal biomass: Materials, performances, chemistry, and modelling simulation tools. *Bioresource Technology, 160*, 67–78. doi:10.1016/j.biortech.2014.01.068 PMID:24630371

Hellström, L., Elinder, C. G., Dahlberg, B., Lundberg, M., Järup, L., Persson, B., & Axelson, O. (2001). Cadmium exposure and end-stage renal disease. *American Journal of Kidney Diseases, 38*(5), 1001–1008. doi:10.1053/ajkd.2001.28589 PMID:11684553

Hem, J. D. (2015). U.S. Geological Survey Water-Supply, Paper: Vol. 2254. The study and interpretation of the chemical characteristics of natural water (3rd ed.). Academic Press.

He, M., Li, X., Guo, L., Miller, S. J., Rensing, C., & Wang, G. (2010). Characterization and genomic analysis of chromate resistant and reducing *Bacillus cereus* Strain SJ1. *BMC Microbiology, 10*(221), 1–10. doi:10.1186/1471-2180-10-221 PMID:20723231

He, S. Y., Feng, Y. Z., Ni, J., Sun, Y. F., Xue, L. H., Feng, Y. F., Yu, Y., Lin, X., & Yang, L. (2016). Different responses of soil microbial metabolic activity to silver and iron oxide nanoparticles. *Chemosphere, 147*, 195–202. doi:10.1016/j.chemosphere.2015.12.055 PMID:26766356

Heyrman, J., Mergaert, J., Denys, R., & Swings, J. (1999). *The use of fatty acid methyl ester analysis (FAME) for the identification of heterotrophic bacteria present on three mural paintings showing severe damage by microorganisms Federation of European Microbiological Societies*. Elsevier Science.

He, Z., Gao, F., Sha, T., Hu, Y., & He, C. (2009). Isolation and characterization of a Cr(VI)-reduction *Ochrobactrum* sp. strain CSCr-3 from chromium landfill. *Journal of Hazardous Materials*, *163*(2-3), 869–873. doi:10.1016/j.jhazmat.2008.07.041 PMID:18722054

Hildebrandt, U., Regvar, M., & Bothe, H. (2007). Arbuscular mycorrhiza and heavy metal tolerance. *Phytochemistry*, *68*(1), 139–146. doi:10.1016/j.phytochem.2006.09.023 PMID:17078985

Holum, J. R. (1983). *Elements of General and Biological Chemistry* (6th ed.). John Wiley and Sons.

Homaeigohar, S., & Elbahri, M. (2014). Nanocomposite electrospun nanofiber membranes for environmental remediation. *Materials (Basel)*, *7*(2), 1017–1045. doi:10.3390/ma7021017 PMID:28788497

Horcsik, Z., Kovacs, L., Laposi, R., Meszaros, I., Lakatos, G., & Garab, G. (2007). Effect of Cr on photosystem 2 in the unicellular alga, *Chlorella pyrenoidosa. Photosynthetica*, *45*(1), 65–69. doi:10.100711099-007-0010-8

Horton, R. N., Apel, W. A., Thompson, V. S., & Sheridan, P. P. (2006). Low temperature reduction of hexavalent chromium by a microbial enrichment consortium and a novel strain of *Arthrobacter aurescens. BMC Microbiology*, 6. PMID:16436214

Horzum, N., Boyaci, E., Eroglu, A. E., Shahwan, T., & Demir, M. M. (2010). Sorption efficiency of chitosan nanofibers toward metal ions at low concentrations. *Biomacromolecules*, *11*(12), 3301–3308. doi:10.1021/bm100755x PMID:21080700

Ho, Y. S., & McKay, G. (1999). Pseudo-second order models for adsorption systems. *Journal of Hazardous Materials*, *136*(3), 681–689. doi:10.1016/j.jhazmat.2005.12.043 PMID:16460877

Hristovski, K. D., Nguyen, H., Westerhoff, P. K. J. J. o. E. S., & A, H. P. (2009). *Removal of arsenate and 17α-ethinyl estradiol (EE2) by iron (hydr) oxide modified activated carbon fibers*. Academic Press.

Huang, J., Cao, Y., Liu, Z., Deng, Z., Tang, F., & Wang, W. J. C. E. J. (2012). *Efficient removal of heavy metal ions from water system by titanate nanoflowers*. Academic Press.

Huang, C. P., Westman, D., Quirk, K., & Huang, J. P. (1988). The removal of cadmium (II) from dilute solutions by fungal biomass. *Water Science and Technology*, *20*(11-12), 369–376. doi:10.2166/wst.1988.0308

Huang, F., Ge, L., Zhang, B., Wang, Y., Tian, H., Zhao, L. P., He, Y., & Zhang, X. (2014). A fullerene colloidal suspension stimulates the growth and denitrification ability of wastewater treatment sludge-derived bacteria. *Chemosphere*, *108*, 411–417. doi:10.1016/j.chemosphere.2014.02.042 PMID:24630252

Huang, J. H., Elzinga, E. J., Brechbuehl, Y., Voegelin, A., & Kretzschmar, R. (2011). Impacts of *Shewanella putrefaciens* strain CN-32 cells and extracellular polymeric substances on the sorption of As(V) and As(III) on Fe(III)-(hydr)oxides. *Environmental Science & Technology*, *45*(7), 2804–2810. doi:10.1021/es103978r PMID:21375285

Huang, J. W., & Cunningham, S. D. (1996). Lead phytoextraction: Species variation in lead uptake and translocation. *The New Phytologist*, *134*(1), 75–84. doi:10.1111/j.1469-8137.1996.tb01147.x

Huang, K., Chen, C., Shen, Q., Rosen, B. P., & Zhao, F.-J. (2015). Genetically engineering *Bacillus subtilis*with a heat-resistantarsenitemethyltransferase for bioremediation of arsenic contaminated organic waste. *Applied and Environmental Microbiology*, *81*(19), 6718–6724. doi:10.1128/AEM.01535-15 PMID:26187966

Huang, K., Chen, C., Zhang, J., Tang, Z., Shen, Q., Rosen, B. P., & Zhao, F.-J. (2016). Efficient Arsenic Methylation and Volatilization Mediated by a Novel Bacterium From an Arsenic-Contaminated Paddy Soil. *Environmental Science & Technology, 50*(12), 6389–6396. doi:10.1021/acs.est.6b01974 PMID:27258163

Huang, Y., Xi, Y., Gan, L., Johnson, D., Wu, Y., Ren, D., & Liu, H. (2019). Effect of lead and cadmium on photosynthesis in Amaranthus spinosus and assessment of phytoremediation potential. *International Journal of Phytoremediation, 21*(10), 1041–1049. doi:10.1080/15226514.2019.1594686 PMID:31020865

Hudak, P. F., & Sanmanee, S. (2003). Spatial patterns of nitrate, chloride, sulfate, and fluoride concentrations in the woodbine aquifer of north-central Texas. *Environmental Monitoring and Assessment, 82*(3), 311–320. doi:10.1023/A:1021946402095 PMID:12602633

Hu, P., Brodie, E. L., Suzuki, Y., McAdams, H. H., & Andersen, G. L. (2005). Whole-genome transcriptional analysis of heavy metal stresses in *Caulobacter crescentus. Journal of Bacteriology, 187*(24), 8437–8449. doi:10.1128/JB.187.24.8437-8449.2005 PMID:16321948

Hussain, D., Haydon, M. J., Wang, Y., Wong, W., Sherson, S. M., Young, J., Camakaris, J., Harper, J. F., & Cobbett, C. S. (2004). P-Type ATPase heavy metal transporters with roles in essential zinc homeostasis in *Arabidopsis. The Plant Cell, 16*(5), 1327–1339. doi:10.1105/tpc.020487 PMID:15100400

Hu, Z. S., & Deng, Y. L. (2003). Supersaturation Control in AragoniteSynthesis Using Sparingly Soluble Calcium Sulfate as Reactants. *Journal of Colloid and Interface Science, 266*(2), 359–365. doi:10.1016/S0021-9797(03)00699-4 PMID:14527459

Hu, Z. S., Deng, Y. L., & Sun, Q. H. (2004). Synthesis of Precipitated Calcium Carbonate Nanoparticles using a Two-MembraneSystem. *Colloid Journal, 66*(6), 745–750. doi:10.100710595-005-0017-4

Hyndman, D. W. (1985). *Petrology of igneous rocks* (2nd ed.). McGraw-Hill.

Hynninen, A., Touzé, T., Pitkänen, L., Mengin-Lecreulx, D., & Virta, M. (2009). An efflux transporter PbrA and a phosphatase PbrB cooperate in a lead-resistance mechanism in bacteria. *Molecular Microbiology, 74*(2), 384–394. doi:10.1111/j.1365-2958.2009.06868.x PMID:19737357

Ianieva, O. (2009). *Heavy metal resistance mechanisms in bacteria.* Retrieved from: https://www.ncbi.nlm.nih.gov/pubmed/20455433

Ibrahim, A. S. S., El-Tayeb, M. A., Elbadawi, Y. B., & Al-Salamah, A. A. (2011). Isolation and characterization of novel potent Cr(VI) reducing alkaliphilic *Amphibacillus* sp. KSUCr3 from hypersaline soda lakes. *Electronic Journal of Biotechnology, 14*(4). Advance online publication. doi:10.2225/vol14-issue4-fulltext-4

Ibrahim, Z., Ahmad, W. A., & Baba, A. B. (2001). Bioaccumulation of silver and the isolation of metal-binding protein from *P. diminuta. Brazilian Archives of Biology and Technology, 44*(3), 223–225. doi:10.1590/S1516-89132001000300001

Idi, A., Nor, M. H. M., Wahab, M. F. A., & Ibrahim, Z. (2015). Photosynthetic bacteria: An eco-friendly and cheap tool for bioremediation. *Reviews in Environmental Science and Biotechnology, 14*(2), 271–285. doi:10.100711157-014-9355-1

Idris, R., Trifonova, R., Puschenreiter, M., Wenzel, W. W., & Sessitsch, A. (2004). Bacterial communities associated with flowering plants of the Ni hyperaccumulator Thlaspi goesingense. *Applied and Environmental Microbiology, 70*(5), 2667–2677. doi:10.1128/AEM.70.5.2667-2677.2004 PMID:15128517

Igiri, B., Okoduwa, S., Idoko, G., Akabuogu, E., Adeyi, A., & Ejiogu, I. (2018). Toxicity and bioremediation of heavy metals contaminated ecosystem from tannery wastewater: A review. *Journal of Toxicology, 2018*, 1–16. doi:10.1155/2018/2568038 PMID:30363677

Ilyina, A., Castillo, S.M.I., Villarreal, S.J.A., Ramirez, E.G., & Candelas, R.J. (2003). Isolation of soil bacteria for bioremediation of hydrocarbon contamination. *BECTH, 44*(15), 88-91.

Ilhan, S., Cabuk, A., Filik, C., & Calishan, F. (2004). Effect of pretreatment of biosorption of heavy metals by fungal biomass. *Trakya University Journal of Natural Sciences, 5*, 11–17.

Imperato, M., Adamo, P., Naimo, D., Arienzo, M., Stanzione, D., & Violante, P. (2003). Spatial distribution of heavy metals in urban soils of Naples city (Italy). *Environmental Pollution, 124*(2), 247–256. doi:10.1016/S0269-7491(02)00478-5 PMID:12713924

Indu, T. S., & Shaili, S. (2011). Bioremediation and bioconversion of chromium and pentachlorophenol in tannery effluent by microorganisms. *International Journal of Technology, 3*, 224–233.

Institute of Environmental Conservation and Research (INECAR). (2000). *Position Paper Against Mining in Rapu-Rapu.* INECAR, Ateneo de Naga University. www.adnu.cdu.ph/Institutes/Inecar/pospaper1.asp

Iram, M., Guo, C., Guan, Y. P., Ishfaq, A., & Liu, H. Z. (2010). Adsorption and magnetic removal of neutral red dye from aqueous solution using Fe3O4 hollow nanospheres. *Journal of Hazardous Materials, 181*(1-3), 1039–1050. doi:10.1016/j.jhazmat.2010.05.119 PMID:20566240

Ishibashi, Y., Cervantes, C., & Silver, S. (1990). Chromium reduction in *Pseudomonas putida. Applied and Environmental Microbiology, 56*(7), 2268–2270. doi:10.1128/AEM.56.7.2268-2270.1990 PMID:2389940

Isin, E. M., & Guengerich, F. P. (2007). Complex reactions catalyzed by cytochrome P450 enzymes. *Biochimica et Biophysica Acta, 1770*(3), 314–329. doi:10.1016/j.bbagen.2006.07.003 PMID:17239540

Ismail, B., & Nampoothiri, K. (2014). Molecular characterization of an exopolysaccharide from a probiotic *Lactobacillus plantarum* MTCC 9510 and its efficacy to improve the texture of starchy food. *Journal of Food Science and Technology, 51*(12), 4012–4018. doi:10.100713197-013-0928-8 PMID:25477674

Jafari, S. A., Cheraghi, S., Mirbakhsh, M., Mirza, R., & Maryamabadi, A. (2015). Employing response surface methodology for optimization of mercury bioremediation by *Vibrio parahaemolyticus* PG02 in coastal sediments of Bushehr, Iran. *Clean - Soil, Air. Water (Basel), 43*(1), 118–126.

Jagaba, A. H., Sule, A., Ibrahim, M. L., Ab, A. A. L., & Umaru, I. (2018). Wastewater Treatment Using Alum, the Combinations of Alum-Ferric Chloride, Alum-Chitosan, Alum-Zeolite and Alum- Moringa Oleifera as Adsorbent and Coagulant. *International J of Eng Management, 3*, 67–75.

James, T. (2000). Marsh Microbial Biofilm Formation and Contamination of Dental-Unit Water Systems in General Dental Practice. *Applied and Environmental Microbiology, 66*(8), 3363–3367. doi:10.1128/AEM.66.8.3363-3367.2000 PMID:10919792

Jamil, M., Zeb, S., Anees, M., Roohi, A., Ahmed, I., ur Rehman, S., & Rha, E. (2014). Role of Bacillus licheniformis in phytoremediation of nickel contaminated soil cultivated with rice. *International Journal of Phytoremediation, 16*(6), 554–571. doi:10.1080/15226514.2013.798621 PMID:24912242

Jamil, N., & Ahmed, N. (2008). Production of biopolymers by *Pseudomonas aeruginosa* isolated from marine source. *Brazilian Archives of Biology and Technology, 51*(3), 457–464. doi:10.1590/S1516-89132008000300004

Jarosławiecka & Piotrowska-Seget. (2014). Lead resistance in micro-organisms. *Microbiology, 160*, 12–25.

Järup, L. J. B. m. b. (2003). *Hazards of heavy metal contamination. Academic* Press.

Javaid, A., Bajwa, R., & Manzoor, T. (2011). Biosorption of heavy metals by pretreated biomass of *Aspergillus niger. Pakistan Journal of Botany*, *43*, 419–425.

Javaid, N. M. A., Bajwa, K. U. R., & Manzoor, R. A. (2011). Biosorption of heavy metals by pre-treatment of biomass of *Aspergillus niger. Pakistan Journal of Botany*, *43*(1), 419–425.

Jha, S. K., Nayak, A. K., Sharma, Y. K., Mishra, V. K., & Sharma, D. K. (2014). Fluoride accumulation in soil and vegetation in the vicinity of brick fields. *Bulletin of Environmental Contamination and Toxicology*, *80*(4), 369–373. doi:10.100700128-008-9391-z PMID:18345473

Jha, S., Dikshit, S., & Pandy, G. (2011). Comparative study of agitation rate and stationary phase for the removal of Cu^{2+} by *A. lentulus. International Journal of Pharma and Bio Sciences*, *2*, 208–211.

Jiang, S., Park, S., Yoon, Y., Lee, J. H., Wu, W. M., Phuoc Dan, N., Sadowsky, M. J., & Hur, H.-G. (2013). Methanogenesis facilitated by geobiochemical iron cycle in a novel syntrophic methanogenic microbial community. *Environmental Science & Technology*, *47*(17), 10078–10084. doi:10.1021/es402412c PMID:23919295

Ji, M., Hu, Z., Hou, C., Liu, H., Ngo, H. H., Guo, W., Shaoyong, L., & Zhang, J. (2020). New insights for enhancing the performance of constructed wetlands at low temperatures. *Bioresource Technology*, *301*, 122722. doi:10.1016/j.biortech.2019.122722 PMID:31983576

Jordao, C. P., Nascentes, C. C., Cecon, P. R., Fontes, R. L. F., & Pereira, J. L. (2006). Heavy metal availability in soil amended with composted urban solid wastes. *Environmental Monitoring and Assessment*, *112*(1-3), 309–326. doi:10.100710661-006-1072-y PMID:16404548

Joshi. (2018). *Biosorption a green approach for heavy metal remedial from water and waste waters*. doi:10.26479/2018.0401.05

Joshi, B. H., & Modi, K. G. J. (2013, April-June). Soil screening and characterization of heavy metal resistant bacteria for its prospects in bioremediation of contaminated. *Environ. Res. Develop. Journal of Environmental Research and Development*, *7*(4A).

Juhnke, S., Peitzsch, N., Hubener, N., Große, C., & Nies, D. H. (2002). New genes involved in chromate resistance in *Ralstonia metallidurans* strain CH34. *Archives of Microbiology*, *179*(1), 15–25. doi:10.100700203-002-0492-5 PMID:12471500

Ju, K.-S., & Parales, R. E. (2006). Control of Substrate Specificity by Active-Site Residues in Nitrobenzene Dioxygenase. *Applied and Environmental Microbiology*, *72*(3), 1817–1824. doi:10.1128/AEM.72.3.1817-1824.2006 PMID:16517627

Jung, M. C., & Thornton, I. (1996). Heavy metal contamination of soils and plants in the vicinity of a lead–zinc mine Korea. *Applied Geochemistry*, *11*(1-2), 53–59. doi:10.1016/0883-2927(95)00075-5

Juturu, V., & Wu, J. C. (2011). Microbial xylanases: Engineering, production and industrial appli- cations. *Biotechnology Advances*. PMID:22138412

Kabata-Pendias, A., & Pendias, H. (2004). Elements of group VII. In Trace elements in soils and plants. CRC.

Kabata-Pendias, A. (1992). Trace metals in soils in Poland occurrence and behaviour. *Soil Science*, *140*, 53–70.

Kadirvelu, K., Senthilkumar, P., Thamaraiselvi, K., & Subburam, V. (2002). Activated carbon prepared from biomass as adsorbent: Elimination of Ni (II) from aqueous solution. *Bioresource Technology*, *81*(1), 87–90. doi:10.1016/S0960-8524(01)00093-1 PMID:11708760

Kamaludeen, S. P. B., Megharaj, M., Juhasz, A. L., Sethunathan, N., & Naidu, R. (2003). Chromium microorganism interactions in soils: Remediation implications. *Reviews of Environmental Contamination and Toxicology*, *178*, 93–164. doi:10.1007/0-387-21728-2_4 PMID:12868782

Kamarudzaman, A. N., Ismail, N. S., Aziz, R. A., & Jalil, M. F. (2011). Removal of nutrients from landfill leachate using subsurface flow constructed wetland planted with Limnocharis flava and Scirpus atrovirens. *2011 International Conference on Environmental and Computer Science.*

Kanel, S. R., Greneche, J.-M., Choi, H. J. E. s., & Technology. (2006). *Arsenic (V) removal from groundwater using nano scale zero-valent iron as a colloidal reactive barrier material.* Academic Press.

Kang, G. D., & Cao, Y. M. (2014). Application and modification of poly (vinylidene fluoride)(PVDF) membranes –a review. *Journal of Membrane Science*, *463*, 145–165. doi:10.1016/j.memsci.2014.03.055

Kanmani, Aravind, & Preston. (2012). Remediation of chromium contaminants using bacteria. *International Journal of Environmental Science and Technology, 9*(1), 183-193.

Kanmani, P., Aravind, J., & Preston, D. (2012). Remediation of chromium contaminants using bacteria. *International Journal of Environmental Science and Technology*, *9*(1), 183–193. doi:10.100713762-011-0013-7

Kanu, I., & Achi, O.K. (2011). Industrial Effluents and their Impact on Water Quality of Receiving Rivers in Nigeria. *Journal of Applied Technology in Environmental Sanitation, 1*(1), 75-86.

Kapulnik, Y., Ensley, B. D., & Raskin, I. (1997). Enhanced accumulation of Pb in Indian mustard by soil-applied chelating agents. *Environmental Science & Technology*, *31*(3), 860–865. doi:10.1021/es960552a

Karigar, C. S., & Rao, S. S. (2011). Role of Microbial Enzymes in the Bioremediation of Pollutants: A Review. *Enzyme Research*, *2011*, 1–11. doi:10.4061/2011/805187 PMID:21912739

Karimi, N., Ghadederian, S. M., Raab, A., Fledmann, J., & Meharg, A. A. (2009). An arsenic accumulating, hypertolerant *Brassica isatis*, Cappadocica. *The New Phytologist*, *184*(1), 41–47. doi:10.1111/j.1469-8137.2009.02982.x PMID:19656300

Karlen, D. L., Doran, J. W., Weinhold, B. J., & Andrews, S. S. (2003). Soil quality Humankind's foundation for survival. *Journal of Soil and Water Conservation*, 58.

Karn, B., Kuiken, T., & Otto, M. J. E. h. p. (2009). *Nanotechnology and in situ remediation: a review of the benefits and potential risks.* Academic Press.

Kashefi, K., & Loveley, D. R. (2000). Reduction of Fe(III), Mn(IV), and toxic metals at 100 °C by *Pyrobaculum islandicum. Applied and Environmental Microbiology*, *66*(3), 1050–1056. doi:10.1128/AEM.66.3.1050-1056.2000 PMID:10698770

Katiyar, P., Pandey, N., & Sahu, K. K. (2020). Biological approaches of fluoride remediation: Potential for environmental clean-up. *Environmental Science and Pollution Research International*, *27*(12), 1–12. doi:10.100711356-020-08224-2 PMID:32146673

Kato, S., Hashimoto, K., & Watanabe, K. (2012). Methanogenesis facilitated by electric syntrophy via (semi)conductive iron-oxide minerals. *Environmental Microbiology*, *14*(7), 1646–1654. doi:10.1111/j.1462-2920.2011.02611.x PMID:22004041

Kaur, T., & Ghosh, M. (2015). *Acinetobacter haemolyticus* MG606 produces a novel, phosphate binding exobiopolymer. *Carbohydrate Polymers*, *132*, 72–79. doi:10.1016/j.carbpol.2015.06.002 PMID:26256326

Kaur, T., & Ghosh, M. (2017). Characterization and toxicity of a phosphate-binding exobiopolymer produced by *Acinetobacter haemolyticus* MG606. *Journal of Water and Health*, *15*(1), 103–111. doi:10.2166/wh.2016.176 PMID:28151444

Kaur, T., Sharma, J., Ganguli, A., & Ghosh, M. (2013). Application of biopolymer produced from metabolic engineered *Acinetobacter* sp. for the development of phosphate optoelectronic sensor. *Composite Interfaces*, *21*(2), 143–151. doi:10.1080/15685543.2013.842116

Keerthisinghe, G., McLaughlin, M. J., & Freney, J. R. (2001). Use of gypsum, phosphogypsum and fluoride to ameliorate subsurface acidity in a pasture soil. In R. J. Wright, V. C. Aligar, & R. P. Murrmann (Eds.), Plant soil interactions at low pH (pp. 509–517). Academic Press.

Kelishadi, R., & Poursafa, P. (2010). Air pollution and non-respiratory health hazards for children. *Archives of Medical Science*, *6*(4), 483–495. doi:10.5114/aoms.2010.14458 PMID:22371790

Keller, E. A. (1979). *Environmental geology*. Charles & Merril.

Kennedy, K. J., & Lentz, E. M. (2000). Treatment of landfill leachate using sequencing batch and continuous flow upflow anaerobic sludge blanket (UASB) reactors. *Water Research*, *34*(14), 3640–3656. doi:10.1016/S0043-1354(00)00114-7

Khan, I., Saeed, K., & Khan, I. J. A. J. o. C. (2019). *Nanoparticles: Properties, applications and toxicities*. Academic Press.

Khan, S. B., Marwani, H. M., Asiri, A. M., Bakhsh, E. M. J. D., & Treatment, W. (2016). *Exploration of calcium doped zinc oxide nanoparticles as selective adsorbent for extraction of lead ion*. Academic Press.

Khan, A. L., & Lee, I. J. (2013). Endophytic Penicillium funiculosum LHL06 secretes gibberellin that reprograms Glycine max L. growth during copper stress. *BMC Plant Biology*, *13*(1), 86. doi:10.1186/1471-2229-13-86 PMID:23721090

Khandelwal, H., Dhar, H., Thalla, A. K., & Kumar, S. (2019). Application of life cycle assessment in municipal solid waste management: A worldwide critical review. *Journal of Cleaner Production*, *209*, 630–654. doi:10.1016/j.jclepro.2018.10.233

Khan, M. H. (2007). Induction of oxidative stress and antioxidant metabolism in *Calamus tenius* leaves under chromium and zinc toxicity. *Indian Journal of Plant Physiology / Official Publication of the Indian Society for Plant Physiology*, *12*, 353–359.

Khusnutdinova, A. N., Ovchenkova, E. P., Khristova, A. P., Laurinavichene, T. V., Shastik, E. S., Liu, J., & Tsygankov, A. A. (2012). New tolerant strains of purple nonsulfur bacteria for hydrogen production in a two-stage integrated system. *International Journal of Hydrogen Energy*, *37*(10), 8820–8827. doi:10.1016/j.ijhydene.2012.02.003

Kilice, F. (1999). Investigation of Toxic Heavy Metals Pollution in the Road Dust a the Centre of Van Turkey. *Bulletin of Pure and Applied Science*, *18c*, 1–4.

Kim, B. C., Nair, S., Kim, J., Kwak, J. H., Grate, J. W., Kim, S. H., & Gu, M. B. J. N. (2005). *Preparation of biocatalytic nanofibres with high activity and stability via enzyme aggregate coating on polymer nanofibres*. Academic Press.

Kimani, N. G. (2007). *Environmental Pollution and Impacts on Public Health: Implications of the Dandora Dumping Site Municipal in Nairobi*. United Nations Environment Programme.

Kim, J.-S., Park, J.-W., Lee, S.-E., & Kim, J.-E. (2002). Formation of Bound Residues of 8-Hydroxybentazon by Oxidoreductive Catalysts in Soil. *Journal of Agricultural and Food Chemistry*, *50*(12), 3507–3511. doi:10.1021/jf011504z PMID:12033819

Kim, K., & Jeong, G. Y. (2005). Factors influencing natural occurrence of fluoride-rich ground waters: A case study in the southeastern part of the Korean Peninsula. *Chemos*, *58*(10), 1399–1408. doi:10.1016/j.chemosphere.2004.10.002 PMID:15686758

Kirk-Othmer. (1996). *Encyclopedia of chemical technology* (3rd ed., Vol. 10). Wiley.

Kirkpatrick, B., Kohler, K., Byrne, M., Fleming, L. E., Scheller, K., Reich, A., Hitchcock, G., Kirkpatrick, G., Ullmann, S., & Hoagland, P. (2014). Human responses to Florida red tides: Policy awareness and adherence to local fertilizer ordinances. *The Science of the Total Environment*, *493*, 898–909. doi:10.1016/j.scitotenv.2014.06.083 PMID:25003583

Kitamura, M., Konno, H., Yasui, A., & Masuoka, H. (2002). ControllingFactors and Mechanism of Reactive Crystallisation of CalciumCarbonate Polymorphs from Calcium Hydroxide Suspensions. *Journal of Crystal Growth*, *236*(1-3), 323–332. doi:10.1016/S0022-0248(01)02082-6

Kjeldsen, P., Barlaz, M. A., Rooker, A. P., Baun, A., Ledin, A., & Christensen, T. H. (2002). Present and long-term composition of MSW landfill leachate: A review. *Critical Reviews in Environmental Science and Technology*, *32*(4), 297–336. doi:10.1080/10643380290813462

Kloepper, J. W., Lifshitz, R., & Zablotowicz, R. M. (1989). Free-living bacterial inocula for enhancing crop productivity. *Trends in Biotechnology*, *7*(2), 39–44. doi:10.1016/0167-7799(89)90057-7

Klonowska, A., Heulin, T., & Vermeqlio, A. (2005). Selenite and tellurite reduction by *Shewanella oneidensis. Applied and Environmental Microbiology*, *71*(9), 5607–5609. doi:10.1128/AEM.71.9.5607-5609.2005 PMID:16151159

Kloos, H., Tekle-Haimanot, R., Kloos, H., & Zein, A. H. (Eds.). (1993). *Fluorosis. The ecology of health and disease in Ethiopia*. West View Press.

Koblar, A., Tavcar, G., & Ponikvar-Svet, M. (2011). Effects of airborne F on soil and vegetation. *Journal of Fluorine Chemistry*, *132*(10), 755–775. doi:10.1016/j.jfluchem.2011.05.022

Kobori, H., Harrison-Bernard, L. M., & Navar, L. G. (2001). Expression of angiotensinogen mRNA and protein in angiotensin II-dependent hypertension. *Journal of the American Society of Nephrology*, *12*, 431–439. PMID:11181790

Koepenick, M. (1976). *Papeti'ereduQu ́ebec*. Groupe d'informationsd'Affaires.

Kohler, J., Hernández, J. A., Caravaca, F., & Roldán, A. (2009). Induction of antioxidant enzymes is involved in the greater effectiveness of a PGPR versus AM fungi with respect to increasing the tolerance of lettuce to severe salt stress. *Environmental and Experimental Botany*, *65*(2-3), 245–252. doi:10.1016/j.envexpbot.2008.09.008

Komori, K., Wang, P. C., Toda, K., & Ohtake, H. (1989). Factors affecting chromate reduction in *Enterobacter cloacae* strain HO1. *Applied Microbiology and Biotechnology*, *31*(31), 567–570. doi:10.1007/BF00270796

Koppel, L. B. (1982). Input Multiplicities in Nonlinear Multivariable Control Systems. *AIChE Journal. American Institute of Chemical Engineers*, *28*(6), 935–945. doi:10.1002/aic.690280608

Kornhauser, C., Wrobel, K., Wrobel, K., Malacara, J. M., Nava, L. E., Gómez, L., & González, R. (2002). Possible adverse effect of chromium in occupational exposure of tannery workers. *Industrial Health*, *40*(2), 207–213. doi:10.2486/indhealth.40.207 PMID:12064563

Kotrba, P., Najmanova, J., Macek, T., Ruml, T., & Mackova, M. (2009). Genetically modified plants in phytoremediation of heavy metal and metalloid soil and sediment pollution. *Biotechnology Advances*, *27*(6), 799–810. doi:10.1016/j.biotechadv.2009.06.003 PMID:19567265

Kotrotsiou, O., & Costas, K. (2019). Water treatment by molecularly imprinted materials. *Nanoscale Materials in Water Purification*, *7*, 179-230.

Koul, B., & Taak, P. (2018). Chemical Methods of Soil Remediation. In *Biotechnological Strategies for Effective Remediation of Polluted Soils* (pp. 77–84). Springer. doi:10.1007/978-981-13-2420-8_4

Kousalya, G. N., Gandhi, M. R., Viswanathan, N., & Meenakshi, S. (2010). Preparation and metal uptake studies of modified forms of chitin. *International Journal of Biological Macromolecules, 47*(5), 583–589. doi:10.1016/j.ijbiomac.2010.07.014 PMID:20708027

Kowshik, M., Ashtaputre, S., Kharrazi, S., Vogel, W., Urban, J., Kulkarni, S. K., & Paknikar, K. J. N. (2002). *Extracellular synthesis of silver nanoparticles by a silver-tolerant yeast strain MKY3*. Academic Press.

Kramar, U. (2010). Metal hyperaccumulation in plants. *Annual Review of Plant Biology, 61*(1), 517–534. doi:10.1146/annurev-arplant-042809-112156 PMID:20192749

Kramer, U. (2005). Phytoremediation: Novel approaches to cleaning up polluted soils. *Current Opinion in Biotechnology, 16*(2), 133–141. doi:10.1016/j.copbio.2005.02.006 PMID:15831377

Krantev, A., Yordanova, R., & Popova, L. (2006). Salicylic acid decreases Cd toxicity in maize plants. *General and Applied Plant Physiology, 1*, 45–52.

Kristin, B. R., Love, S., Nicholas, R. R., McAdory, D., Halder, S. S., Kern, N., Kent, J., & Strickland, J. (2010). Chromium is not an essential trace element for mammals. Effect of a low chromium diet. *Journal of Biological Inorganic Chemistry, 16*(3), 3–81.

Kudzin, Y. K., & Pashova, V. T. (2009). Fluorine content of soils and plants after prolonged application of fertilisers. *Soils and Fertilizers, 33*, 451.

Kumar, V., Yadav, S. K. J. J. o. C. T., Biotechnology: International Research in Process, E., & Technology, C. (2009). *Plant-mediated synthesis of silver and gold nanoparticles and their applications*. Academic Press.

Kumaran, N. S., Sundaramanicam, A., & Bragadeeswaran, S. (n.d.). Adsorption studies on heavy metals by isolated cyanobacterial strain (nostoc sp.) from uppanar estuarine water, southeast coast of India. *Journal of Applied Sciences Research, 7*(11), 1609–1615.

Kumar, B., Naaz, A., Shukla, K., Narayan, C., Singh, G., Kumar, A., Ramanathan, A. L., & Anshumali. (2016). Spatial variability of fluorine in agricultural soils around Sidhi District, Central India. *Journal of the Geological Society of India, 87*(2), 227–235. doi:10.100712594-016-0391-z

Kumar, I. P., & Choudhary, A. K. (2018). Constructed Wetlands-A Sustainable Solution for Landfill Leachate Treatment. International J*ournal of Latest Technology in Engineering, Management &. Applied Sciences (Basel, Switzerland), 7*(6), 101–106.

Kumar, M., & Puri, A. (2012). A review of permissible limits of drinking water. *Indian Journal of Occupational and Environmental Medicine, 16*(1), 40–44. doi:10.4103/0019-5278.99696 PMID:23112507

Kumar, R., Bhatia, D., Singh, R., Rani, S., & Bishnoi, N. R. (2011). Sorption of heavy metals from electroplating effluent using immobilized biomass *Trichoderma viride* in a continuous packed-bed column. *International Biodeterioration & Biodegradation, 65*(8), 1133–1139. doi:10.1016/j.ibiod.2011.09.003

Kumar, S. R., & Gopinath, P. (2017). Nano-bioremediation applications of nanotechnology for bioremediation. In *Handbook of Advanced Industrial and Hazardous Wastes Management* (pp. 27–48). CRC Press. doi:10.1201/9781315117423-2

Kumar, V., Bharti, P. K., Talwar, M., Tyagi, A. K., & Kumar, P. (2017). Studies on high iron content in water resources of Moradabad district (UP), India. *Water Science, 31*(1), 44–51. doi:10.1016/j.wsj.2017.02.003

Kuo, W. C. J., & Smith, R. (1998). Designing for the interactions between water use and effluent treatment. *Transactions of the Institution of Chemical Engineers, 76*(3), 287–288. doi:10.1205/026387698524938

Kupper, H., & Kochain, L. V. (2010). Transcriptional regulation of metal transport genes and mineral nutrition during acclimatization to cadmium and zinc in the Cd/Zn hyperaccumulator, *Thalspi caerulescens* (Ganges population). *The New Phytologist*, *185*(1), 114–129. doi:10.1111/j.1469-8137.2009.03051.x PMID:19843304

Kupper, H., Lombi, E., Zhao, F. J., & McGrath, S. P. (2000). Cellular compartmentation of cadmium and zinc in relation to other elements in the hyperaccumulator *Arabidopsis halleri. Planta*, *212*(1), 75–84. doi:10.1007004250000366 PMID:11219586

Kurniawan, A., & Yamamoto, T. (2013). Biofilm polymer for biosorption of pollutant ions. *Procedia Environmental Sciences*, *17*, 179–187. doi:10.1016/j.proenv.2013.02.027

Kurniawan, T. A., Chan, G. Y. S., Lo, W.-H., & Babel, S. (2006). Physico-chemical treatment techniques for wastewater laden with heavy metals. *Chemical Engineering Journal*, *118*(1-2), 83–98. doi:10.1016/j.cej.2006.01.015

Kvande & Wahbeck. (2006). Theory for the Determination ofVapour Pressures by the Transpiration Method. *Acta Chem. Scand. Ser. A*, *30*, 297–302.

Kwak, Y. H., Lee, D. S., & Kim, H. B. (2003). *Vibrio harveyi* nitroreductase is also a chromate reductase. *Applied Microbiology and Biotechnology*, *69*, 4390–4395. PMID:12902220

L'opez-Luna, J., Gonz'alez-Ch'avez, M. C., Esparza-Garc'ıa, F. J., & Rodr'ıguez-V'azquez, R. (2009). Toxicity assessment of soil amended with tannery sludge, trivalent chromium and hexavalent chromium, using wheat, oat and sorghum plants. *Journal of Hazardous Materials*, *163*(2-3), 829–834. doi:10.1016/j.jhazmat.2008.07.034 PMID:18814962

Lagergren, S. (1898). About the theory of so called adsorption of souble substances. *Kungliga Svenska Vetenkapsakademiens Handlingar*, *24*, 1–39.

Lahermo, P., Ilmasti, M., Juntunen, R., & Taka, M. (1990). *The geochemical atlas of Finland. Part 1: the hydro-geochemical mapping of Finnish groundwater*. Geological Survey of Finland.

Lakherwal, D. (2014). Adsorption of heavy metals: A review. *International Journal of Environmental Research and Development*, *4*, 41–48.

Lakkireddy, K., & Kües, U. (2017). Bulk isolation of basidiospores from wild mushrooms by electrostatic attraction with low risk of microbial contaminations. *AMB Express*, *7*(1), 28. doi:10.118613568-017-0326-0 PMID:28124290

Lalley, J., Han, C., Li, X., Dionysiou, D. D., & Nadagouda, M. N. (2016). Phosphate adsorption using modified iron oxide-based sorbents in lake water: Kinetics, equilibrium, and column tests. *Chemical Engineering Journal*, *284*, 1386–1396. doi:10.1016/j.cej.2015.08.114

Langmuir, I. (1918). The adsorption of gases on plane surfaces of glass, mica and platinum. *Journal of the American Chemical Society*, *40*(9), 1361–1403. doi:10.1021/ja02242a004

Larsson, D. G. (2014). Pollution from drug manufacturing: Review and perspectives. *Philosophical Transactions of the Royal Society of London. Series B, Biological Sciences*, *369*(1656), 369. doi:10.1098/rstb.2013.0571 PMID:25405961

Lasat, M. M. (2002). Reviews and Analyses: Phytoextraction of Toxic Metals- A Review of Biological Mechanisms. *Journal of Environmental Quality*, *31*, 109–120. PMID:11837415

Lasat, M. M., Fuhrmann, M., Ebbs, S. D., Cornish, J. E., & Kochian, L. V. (1998). Phytoremediation of a radiocesium contaminated soil: Evaluation of cesium- 137bioaccumulation in the shoots of three plant species. *Journal of Environmental Quality*, *27*(1), 165–168. doi:10.2134/jeq1998.00472425002700010023x

Latif, A., Bilal, M., Asghar, W., Azeem, M., Ahmad, M. I., Abbas, A., & Shahzad, T. (2018). Heavy Metal Accumulation in Vegetables and Assessment of their Potential Health Risk. *Journal of Environmental and Analytical Chemistry*, *5*(1), 234. doi:10.4172/2380-2391.1000234

Lau, T., Wu, X., Chua, H., Qian, P., & Wong, P. K. (2005). Wong Effect of exopolysaccharides on the adsorption of metal ions by Pseudomonas sp. CU-1. *Water Science and Technology*, *52*(7), 63–68. doi:10.2166/wst.2005.0182

Lavrova, S. (2016). Treatment of landfill leachate in two stage vertical-flow wetland system with/without addition of carbon source. *Journal of Chemical Technology and Metallurgy*, *51*(2), 223–228.

Lavrova, S., & Koumanova, B. (2010). Influence of recirculation in a lab-scale vertical flow constructed wetland on the treatment efficiency of landfill leachate. *Bioresource Technology*, *101*(6), 1756–1761. doi:10.1016/j.biortech.2009.10.028 PMID:19932959

Lee, A. H., Kang, C.-M., Lee, Y. M., Lee, H., Yun, C.-W., Kim, G.-H., & Kim, J.-J. (2016). Heterologous expression of a new manganese-dependent peroxidase gene from Peniophora incarnata KUC8836 and its ability to remove anthracene in Saccharomyces cerevisiae. *Journal of Bioscience and Bioengineering*, *122*(6), 716–721. doi:10.1016/j.jbiosc.2016.06.006 PMID:27353859

Lee, S. W., Glickmann, E., & Cooksey, D. A. (2001). Chromosomal locus for cadmium resistance in Pseudomonas putida consisting of a cadmium-transporting ATPase and a MerR family response regulator. *Applied and Environmental Microbiology*, *67*(4), 1437–1444. doi:10.1128/AEM.67.4.1437-1444.2001 PMID:11282588

Lee, S., Moon, J. S., Ko, T. S., Petros, D., Goldbrough, P. B., & Schuyler, S. K. (2003). Overexpression of *Arabidopsis* phytochelatin synthase paradoxically leads to hypersensitivity to cadmium stress. *Plant Physiology*, *131*(2), 656–663. doi:10.1104/pp.014118 PMID:12586889

Legerská, B., Chmelová, D., & Ondrejovič, M. (2016). Degradation of Synthetic Dyes by Laccases – A Mini-Review. *Nova Biotechnologica et Chimica*, *15*(1), 90–106. doi:10.1515/nbec-2016-0010

Leisola, M., Jokela, J., Pastinen, O., & Turunen, O. (2006). *Industrial use of Enzymes—Essay*. Laboratory of Bioprocess Engineering, Helsinki University of Technology.

Lenz, M., Enright, A. M., O'aherty, V., van Aelst, A. C., & Lens, P. N. L. (2009). Bioaugmentation of UASB reactors with immobilized *Sulfurospirillum barnesii* for simultaneous selenate and nitrate removal. *Applied Microbiology and Biotechnology*, *83*(2), 377–388. doi:10.100700253-009-1915-x PMID:19305993

Lewis, K. (2005). Persister cells and the riddle of biofilm survival. *Biochemistry (Moscow)*, *70*(2), 267–274. doi:10.100710541-005-0111-6 PMID:15807669

Li, Y.-H., Wang, S., Luan, Z., Ding, J., Xu, C., & Wu, D. J. C. (2003). *Adsorption of cadmium (II) from aqueous solution by surface oxidized carbon nanotubes*. Academic Press.

Liang, Y., Wong, J. W. C., & Wei, L. (2005). Silicon-mediated enhancement of cadmium tolerance in maize (Zea mays L.) grown in cadmium contaminated soil. *Chemosphere*, *58*(4), 475–483. doi:10.1016/j.chemosphere.2004.09.034 PMID:15620739

Lima, A. I. G., Corticeiro, S. C., & Figueira, E. M. A. P. (2006). Glutathione-mediated cadmium sequestration in *Rhizobium leguminosarum*. *Enzyme and Microbial Technology*, *39*(4), 763–769. doi:10.1016/j.enzmictec.2005.12.009

Lin, C. Y., Chang, F. Y., & Chang, C. H. (2000). Co-digestion of leachate with septage using a UASB reactor. *Bioresource Technology*, *73*(2), 175–178. doi:10.1016/S0960-8524(99)00166-2

Lin, T., Wang, H., Wang, H., & Wang, X. (2004). The charge effect of cationic surfactants on the elimination of fibre beads in the electrospinning of polystyrene. *Nanotechnology, 15*(9), 1375–1381. doi:10.1088/0957-4484/15/9/044

Liu, S., Zhang, F., Chen, J., & Sun, G. (2011). Arsenic removal from contaminated soil via biovolatilization by genetically engineered bacteria under laboratory conditions. *Journal of Environmental Sciences (China), 23*(9), 1544–1550. doi:10.1016/S1001-0742(10)60570-0 PMID:22432292

Liu, W., Graham, M., Evans, E. A., & Reneker, D. H. (2002). Poly (meta-phenylene isophthalamide) nanofibers: Coating and post processing. *Journal of Materials Research, 17*(12), 3206–3212. doi:10.1557/JMR.2002.0464

Liwarska-Bizukojc, E. (2019). Nano- and microparticles-induced effect on activated sludge properties. *International Journal of Environmental Science and Technology, 16*(12), 8663–8670. doi:10.100713762-019-02496-z

Llagostera, M., Garido, S., Guerrero, R., & Barbe, J. (1986). Induction of SOS genes of *Escherichia coli* by chromium compounds. *Environmental Mutagenesis, 8*(4), 571–577. doi:10.1002/em.2860080408 PMID:3525136

Lodhia, J., Mandarano, G., Ferris, N. J., Eu, P., & Cowell, S. F. (2009). Development and use of iron oxide nanoparticles (Part 1): Synthesis of iron oxide nano-particles for MRI. *Biomedical Imaging Intervention Journal, 6*(2), e12. PMID:21611034

Logan, T. J., & Chaney, R. L. (1983). Utilization of municipal wastewater and sludge on land-metals. In *Proceedings of the Workshop on Utilization of Municipal Wastewater and Sludge on Land* (pp. 235-326). University of California Riverside.

Lopez, A., Pagano, M., Volpe, A., & Di Pinto, A. C. (2004). Fenton's pre-treatment of mature landfill leachate. *Chemosphere, 54*(7), 1005–1010. doi:10.1016/j.chemosphere.2003.09.015 PMID:14637358

Loukidou, M. X., & Zouboulis, A. I. (2001). Comparison of two biological treatment processes using attached-growth biomass for sanitary landfill leachate treatment. *Environmental Pollution, 111*(2), 273–281. doi:10.1016/S0269-7491(00)00069-5 PMID:11202731

Loux, N. T., Su, Y. S., & Hassan, S. M. (2011). Issues in assessing environmental exposures to manufactured nanomaterials. *International Journal of Environmental Research and Public Health, 8*(9), 3562–3578. doi:10.3390/ijerph8093562 PMID:22016703

Lovley, D. R., & Phillips, E. J. P. (1994). Reduction of Chromate by *Desulfovibrio vulgaris* and its C_3 Cytochrome. *Applied and Environmental Microbiology, 60*(2), 726–728. doi:10.1128/AEM.60.2.726-728.1994 PMID:16349200

Ludwig, R. D., Su, C., Lee, T. R., Wilkin, R. T., Acree, S. D., Ross, R. R., . . . technology. (2007). *In situ chemical reduction of Cr (VI) in groundwater using a combination of ferrous sulfate and sodium dithionite: a field investigation.* Academic Press.

Luederitz, V., Eckert, E., Lange-Weber, M., Lange, A., & Gersberg, R. M. (2001). Nutrient removal efficiency and resource economics of vertical flow and horizontal flow constructed wetlands. *Ecological Engineering, 18*(2), 157–171. doi:10.1016/S0925-8574(01)00075-1

Luna, J. M., Rufino, R. D., & Sarubbo, L. A. (2016). Biosurfactant from *Candida sphaerica* UCP0995 exhibiting heavy metal remediation properties. *Process Safety and Environmental Protection, 102*, 558–566. doi:10.1016/j.psep.2016.05.010

Macaskie, L., Creamer, N., Essa, A., Brown, N. J. B., & bioengineering. (2007). *A new approach for the recovery of precious metals from solution and from leachates derived from electronic scrap.* Academic Press.

Madera-Parra, C. A. (2016). Treatment of landfill leachate by polyculture constructed wetlands planted with native plants. *Ingeniería y competitividad, 18*(2), 183-192.

Madera-Parra, C. A. (2016). Treatment of landfill leachate by polyculture constructed wetlands planted with native plants. *Ingeniería y Competitividad, 18*(2), 183–192.

Madera-Parra, C. A., Peña-Salamanca, E. J., Peña, M. R., Rousseau, D. P. L., & Lens, P. N. L. (2015). Phytoremediation of landfill leachate with *Colocasia esculenta, Gynerum sagittatum* and *Heliconia psittacorum* in constructed wetlands. *International Journal of Phytoremediation, 17*(1), 16–24. doi:10.1080/15226514.2013.828014 PMID:25174421

Madrid, L., Díaz-Barrientos, E., Reinoso, R., & Madrid, F. (2004). Metals in urban soils of Sevilla: Seasonal changes and relations with other soil components and plant contents. *European Journal of Soil Science, 55*(2), 209–217. doi:10.1046/j.1365-2389.2004.00589.x

Madrid, Y., & Camara, C. (1997). Biological substrates for metal preconcentration and speciation. *Trends in Analytical Chemistry, 16*(1), 36–44. doi:10.1016/S0165-9936(96)00075-1

Maehlum, T. (1995). Treatment of landfill leachate in on-site lagoons and constructed wetlands. *Water Science and Technology, 32*(3), 129–135. doi:10.2166/wst.1995.0134

Maes, S., Vackier, T., Nguyen Huu, S., Heyndrickx, M., Steenackers, H., Sampers, I., Raes, K., Verplaetse, A., & De Reu, K. (2019). Occurrence and characterisation of biofilms in drinking water systems of broiler houses. *BMC Microbiology, 19*(1), 77. doi:10.118612866-019-1451-5 PMID:30987581

Magnuson, T. S., Swenson, M. W., Paszczynski, A. J., Deobald, L. A., Kerk, D., & Cummings, D. E. (2010). Proteogenomic and functional analysis of chromate reduction in *Acidiphilium cryptum* JF-5, an Fe(III)-respiring acidophile. *Biometals, 23*(6), 1129–1138. doi:10.100710534-010-9360-y PMID:20593301

Magyarosy, A., Laidlaw, R., Kilaas, R., Echer, C., Clark, D., & Keasling, J. (2002). Nickel accumulation and nickel oxalate precipitation by *Aspergillus niger. Applied Microbiology and Biotechnology, 59*(2-3), 382–388. doi:10.100700253-002-1020-x PMID:12111174

Mahmood, S., Khalid, A., Arshad, M., Mahmood, T., & Crowley, D. E. (2015). Detoxification of azo dyes by bacterial oxidoreductase enzymes. *Critical Reviews in Biotechnology*, •••, 1–13. doi:10.3109/07388551.2015.1004518 PMID:25665634

Mahmud, H., Huq, A., & Yahya, R. (2017). *Polymer-based adsorbent for heavy metals removal from aqueous solution.* Paper presented at the IOP Conference Series: Materials Science and Engineering. 10.1088/1757-899X/206/1/012100

Ma, J. F., Ueno, D., Zhao, F. J., & McGrath, S. P. (2005). Subcellular localization of cadmium and zinc in the leaves of a Cd- hyperaccumulating ecotype of *Thlaspi caerulescens. Planta, 220*(5), 736. doi:10.100700425-004-1392-5 PMID:15517354

Majid, N. M., Islam, M. M., Justin, V., Abdu, A., & Ahmadpour, P. (2011). Evaluation of heavy metal uptake and translocation by *Acacia mangium* as a phytoremediator of copper contaminated soil. *African Journal of Biotechnology, 10*(42), 8373–8379. doi:10.5897/AJB11.392

Mala, J. G. S., Nair, B. U., & Puvanakrishnan, R. (2006). Bioaccumulation and biosorption of chromium by *Aspergillus niger* MTCC 2594. *The Journal of General and Applied Microbiology, 52*(3), 179–186. doi:10.2323/jgam.52.179 PMID:16960334

Malik, A. (2004). Metal bioremediation through growing cells. *Environment International, 30*(2), 261–278. doi:10.1016/j.envint.2003.08.001 PMID:14749114

Malkoc, E., & Nuhoglu, Y. (2005). Investigations of nickel(II) removal from aqueous solutions using tea factory waste. *Journal of Hazardous Materials, 127*(1-3), 120–128. doi:10.1016/j.jhazmat.2005.06.030 PMID:16125314

Mambali, S. S. (1982). Necessity and efforts made to map the occurrence of fluoride within the country, Tanzania Mainland. A report the workshop on domestic water health standards with emphasis on fluoride Arusha, Tanzania Ministry of Water and Energy.

Manasi, M., Rajesh, N., & Rajesh, V. (2016). Evaluation of the genetic basis of heavy metal resistance in an isolate from electronic industry effluent. *Journal of Genetic Engineering and Biotechnology*, *14*(1), 177–180. doi:10.1016/j.jgeb.2016.02.002 PMID:30647612

Mane, P. C., & Bhosle, A. B. (2012). Bioremoval of some metals by living algae *spirogyra* sp. and *spirullina* sp. from aqueous solution. *International Journal of Environmental of Research*, *6*(2), 571–576.

Mani, D., & Kumar, C. (2014). Biotechnological advances in bioremediation of heavy metals contaminated ecosystems: An overview with special reference to phytoremediation. *International Journal of Environmental Science and Technology*, *11*(3), 843–872. doi:10.100713762-013-0299-8

Manoharan, V., Loganathan, P., Parfltt, R. L., & Tillman, R. W. (2006). Changes in soil solution composition and aluminium speciation under legume-based pastures in response to long-term phosphate fertiliser applications. *Australian Journal of Soil Research*, *34*(6), 985–998. doi:10.1071/SR9960985

Marandi, R. (2011). Biosorption of hexavalent chromium from aqueous solution by dead fungal biomass of *Phanerochaete chrysosporium*: Batch and fixed bed studies. *Canadian Journal of Chemical Engineering and Technology*, *2*, 8–22.

Margesin, R., Zimmerbauer, A., & Schinner, F. (1999). Soil lipase activity – a useful indicator of oil biodegradation. *Biotechnology Techniques*, *13*(12), 859–863. doi:10.1023/A:1008928308695

Marschner, H. (1995). Adaptation of plants to adverse chemical soil conditions. *Mineral Nutrition of Higher Plants*.

Martinez-Juarez, V. M., Cardenas-Gonzalez, J. F., Torre-Bouscoulet, M. E., & Acosta-Rodriguez, I. (2012). Biosorption of mercury (II) from aqueous solutions on to fungal biomass. *Bioinorganic Chemistry and Applications*, *1*, 1–5. doi:10.1155/2012/156190 PMID:23028382

Martinkova, L., Vejvoda, V., Kaplan, O., Kubac, D., Malandra, A., Cantarella, M., Bezouska, K., & Kfen, V. (2009). Fungal nitrilases as biocatalysts: Recent developments. *Biotechnology Advances*, *27*(6), 661–670. doi:10.1016/j.biotechadv.2009.04.027 PMID:19427375

Martins, L. L., Mourato, M. P., Cardoso, A. I., Pinto, A. P., & Mota, A. M. (2011). Oxidative stress induced by Cd in *Nicotiana tabaccum* L.: Effects on growth parameters, oxidative damage and antioxidant responses in different plant parts. *Acta Physiologiae Plantarum*, *33*(4), 1375–1383. doi:10.100711738-010-0671-y

Masawat, P., Harfield, A., Srihirun, N., & Namwong, A. (2016). Green Determination of Total Iron in Water by Digital Image Colorimetry. *Analytical Letters*, *50*(1), 173–185. doi:10.1080/00032719.2016.1174869

Mash, H. E., Chin, Y. P., Sigg, L., Hari, R., & Xue, H. B. (2003). Complexation of copper by zwitterionic aminosulfonic (good) buffers. *Analytical Chemistry*, *75*(3), 671–677. doi:10.1021/ac0261101 PMID:12585500

Masood, F., & Malik, A. (2011). Hexavalent chromium reduction by *Bacillus* sp. Strain FM1 isolated from heavy-metal contaminated soil. *Bulletin of Environmental Contamination and Toxicology*, *86*(1), 114–119. doi:10.100700128-010-0181-z PMID:21181113

Mathew, A., Dubey, A., & Mahindrakar, A. B. (2016). Study of Constructed Wetland for Treatment of Landfill Leachate. *International Journal of Chemtech Research*, *9*(11), 87–95.

Mauter, M. S., Elimelech, M. J. E. S., & Technology. (2008). *Environmental applications of carbon-based nanomaterials*. Academic Press.

Ma, X., Zhang, L., Xia, M., Li, S., Zhang, X., & Zhang, Y. (2017). Mimicking the Active Sites of Organophosphorus Hydrolase on the Backbone of Graphene Oxide to Destroy Nerve Agent Simulants. *ACS Applied Materials & Interfaces*, *9*(25), 21089–21093. doi:10.1021/acsami.7b07770 PMID:28621926

Maximous, N., Nakhla, G., & Wan, W. (2009). Comparative assessment of hydrophobic and hydrophilic membrane fouling in wastewater applications. *Journal of Membrane Science*, *339*(1), 93–99. doi:10.1016/j.memsci.2009.04.034

Ma, Y., Prasad, M. N. V., Rajkumar, M., & Freitas, H. (2011). Plant growth promoting rhizobacteria and endophytes accelerate phytoremediation of metalliferous soils. *Biotechnology Advances*, *29*(2), 248–258. doi:10.1016/j.biotechadv.2010.12.001 PMID:21147211

Ma, Y., Rajkumar, M., & Freitas, H. (2009). Improvement of plant growth and nickel uptake by nickel resistant-plant-growth promoting bacteria. *Journal of Hazardous Materials*, *166*(2-3), 1154–1161. doi:10.1016/j.jhazmat.2008.12.018 PMID:19147283

Mayo, J., Yavuz, C., Yean, S., Cong, L., Shipley, H., Yu, W., . . . Materials, T. o. A. (2007). *The effect of nanocrystalline magnetite size on arsenic removal*. Academic Press.

Ma, Z., Zhu, W., Long, H., Chai, L., & Wang, Q. (2007). Chromate reduction by resting cells of *Achromobacter* sp. Ch-1 under aerobic conditions. *Process Biochemistry*, *42*(6), 1028–1032. doi:10.1016/j.procbio.2007.03.007

Mazotto, A. M., Coelho, R. R. R., Cedrola, S. M. L., de Lima, M. F., Couri, S., Paraguai de Souza, E., & Vermelho, A. B. (2011). Keratinase Production by Three *Bacillus* spp. Using Feather Meal and Whole Feather as Substrate in a Submerged Fermentation. *Enzyme Research*, *2011*, 1–7. doi:10.4061/2011/523780 PMID:21822479

McCluggage, D. (1991). Heavy metal poisoning. NCS Magazine.

McCutcheon, S. C., & Schnoor, J. L. (2004). *Phytoremediation: transformation and control of contaminants* (Vol. 121). John Wiley & Sons.

McEldowney, S. (2000). The impact of surface attachment on cadmium accumulation by *Pseudomonas fluorescens* H2. *FEMS Microbiology Ecology*, *33*(2), 121–128. doi:10.1111/j.1574-6941.2000.tb00734.x PMID:10967211

McEldowney, S., Hardman, D. J., & Waite, S. (1993). Treatment Technologies. In S. McEldowney, D. J. Hardman, & S. Waite (Eds.), *Pollution Ecology and Biotreatment*. Longman Scientific & Technical.

McLaughlin, M. J., Simpson, P. G., Fleming, N., Stevens, D. P., Cozens, G., & Smart, M. K. (2006). Effect of fertiliser type on cadmium and fluorine concentrations in clover herbage. *Australian Journal of Experimental Agriculture*, *37*(8), 1019–1026. doi:10.1071/EA96117

Megharaj, M., Avudainayagam, S., & Naidu, R. (2003). Toxicity of hexavalent chromium and its reduction by bacteria isolated from soil contaminated with tannery waste. *Current Microbiology*, *47*(1), 51–54. doi:10.100700284-002-3889-0 PMID:12783193

Mehrizad, A., Zare, K., Dashti Khavidaki, H., Dastmalchi, S., Aghaie, H., & Gharbani, P. (2011). Kinetic and thermodynamic studies of adsorption of 4-chloro-2-nitrophenol on nano-TiO2. *J. Phys. Theor. Chem*, *8*, 33–38.

Mendham, J., Denney, R. C., Barnes, J. D., Thomas, M. J. K., Denney, R. C., & Thomas, M. J. K. (2000). *Vogel's Quantitative ChemicalAnalysis* (6th ed.). Prentice Hall.

Mengoni, A., Barzanti, R., Gonnelli, C., Gabbrielli, R., & Bazzicalupo, M. (2001). Characterization of nickel-resistant bacteria isolated from serpentine soil. *Environmental Microbiology*, *3*(11), 691–698. doi:10.1046/j.1462-2920.2001.00243.x PMID:11846759

Meryam Sardar, R. A. (2015). Enzyme Immobilization: An Overview on Nanoparticles as Immobilization Matrix. *Biochemistry and Analytical Biochemistry*, *04*(02). Advance online publication. doi:10.4172/2161-1009.1000178

Meylan, S., Behra, R., & Sigg, L. (2003). Accumulation of copper and zinc in periphyton in response to dynamic variations of metal speciation in freshwater. *Environmental Science & Technology*, *37*(22), 5204–5212. doi:10.1021/es034566+ PMID:14655709

Michel, C., Brugna, M., Aubert, C., Bernadac, A., & Bruschi, M. (2001). Enzymatic reduction of chromate: Comparative studies using sulphate reducing bacteria. *Applied Microbiology and Biotechnology*, *55*(1), 95–100. doi:10.1007002530000467 PMID:11234966

Miranda, A. T., González, M. V., González, G., Vargas, E., Campos-García, J., & Cervantes, C. (2005). Involvement of DNA helicases in chromate resistance by *Pseudomonas aeruginosa* PAO1. *Mutation Research*, *578*(1-2), 202–209. doi:10.1016/j.mrfmmm.2005.05.018 PMID:16105671

Mishra, A., & Malik, A. (2013). Recent advances in microbial metal bioaccumulation. *Critical Reviews in Environmental Science and Technology*, *43*(11), 1162–1222. doi:10.1080/10934529.2011.627044

Mistry, K., Desai, C., La, S., Patel, K., & Patel, B. (2010). Hexavalent chromium reduction by *Staphylococcus sp.* isolated from Cr (VI) contaminated land fill. International. *Journal of Biotechnology and Biochemistry*, *6*, 117–129.

Mistry, K., Desai, C., & Patel, K. (2009). Reduction of chromium (VI) by bacterial strain KK15 isolated from contaminated soil. *Journal of Cell and Tissue Research*, *9*, 1821–1826.

Mjengera, H., & Mkongo, G. (2003). appropriate defluoridation technology for use in fluorotic areas in Tanzania. *Physics and Chemistry of the Earth*, *28*(20-27), 1097–1104. doi:10.1016/j.pce.2003.08.030

Modak, J. M., & Natarajan, K. A. (1995). Biosorption of metals using nonliving biomass-a review. *Minerals & Metallurgical Processing*, *12*, 189–196.

Mohammed, O. N., & M-Ridha, M. J. (2019). Phytoremediation of organic pollutants in wastewater using native plants. *Association of Arab Universities Journal of Engineering Sci*ences, *26*(2), 54-65.

Mohanpuria, P., Rana, N. K., & Yadav, S. K. J. J. o. n. r. (2008). *Biosynthesis of nanoparticles: technological concepts and future applications*. Academic Press.

Mojiri, A., & Ziyang, L. (2015). Phenols and manganese removal from landfill leachate and municipal wastewater using the constructed wetland. *International Journal of Environmental Chemical Ecological Geological and Geophysical Engineering*, *9*(8), 974–977.

Mojiri, A., Ziyang, L., Tajuddin, R. M., Farraji, H., & Alifar, N. (2016). Co-treatment of landfill leachate and municipal wastewater using the ZELIAC/zeolite constructed wetland system. *Journal of Environmental Management*, *166*, 124–130. doi:10.1016/j.jenvman.2015.10.020 PMID:26496842

Monds, R. D., & O'Toole, G. A. (2009). The developmental model of microbial biofilms: Ten years of a paradigm up for review. *Trends in Microbiology*, *17*(2), 73–87. doi:10.1016/j.tim.2008.11.001 PMID:19162483

Monteiro, M. S., Rodriguez, E., Loureiro, J., Mann, R. M., Soares, A. M. V. M., & Santos, C. (2010). Flow cytometric assessment of Cd genotoxicity in three plants with different metal accumulation and detoxification capacities. *Ecotoxicology and Environmental Safety*, *73*(6), 1231–1237. doi:10.1016/j.ecoenv.2010.06.020 PMID:20663557

Moppert, X., Le Costaouec, T., Raguenes, G., Courtois, A., Simon-Colin, C., Crassous, P., & Guezennec, J. (2009). Investigations into the uptake of copper, iron and selenium by a highly sulphated bacterial exopolysaccharide isolated from microbial mats. *Journal of Industrial Microbiology & Biotechnology*, *36*(4), 599–604. doi:10.100710295-009-0529-8 PMID:19198908

Moreno, J. L., Hernández, T., & Garcia, C. (1999). Effects of a cadmium-contaminated sewage sludge compost on dynamics of organic matter and microbial activity in an arid soil. *Biology and Fertility of Soils*, *28*(3), 230–237. doi:10.1007003740050487

Moskvina, M. I., Brekhovskikh, A. A., & Nikandrov, V. V. (2003). Role of the heterotrophic bacteria associated with the cyanobacterium *Nostoc muscorum* in cadmium sulfide synthesis. [In Russian]. *Mikrobiologiia*, *72*(2), 284–285. PMID:12751257

Motesharezadeh, B. (2008). *Study of the possibility of increasing phytoremediation efficiency in heavy metal-contaminated soil by biological factors* (Doctoral dissertation). Univ College Agr Nat Resour, Univ of Tehran, Iran. (in Persian)

Mu, Y., Ai, Z., Zhang, L., Song, F. J. A. a. m., & interfaces. (2015). *Insight into core–shell dependent anoxic Cr (VI) removal with Fe@ Fe2O3 nanowires: indispensable role of surface bound Fe (II)*. Academic Press.

MubarakAli, D., Arunkumar, J., Nag, K. H., SheikSyedIshack, K., Baldev, E., Pandiaraj, D., . . . Biointerfaces, S. B. (2013). *Gold nanoparticles from Pro and eukaryotic photosynthetic microorganisms—Comparative studies on synthesis and its application on biolabelling*. Academic Press.

Mueller, N. C., Braun, J., Bruns, J., Černík, M., Rissing, P., Rickerby, D., . . . Research, P. (2012). *Application of nanoscale zero valent iron (NZVI) for groundwater remediation in Europe*. Academic Press.

Mukherjee, P. (2012). Cloning, Sequencing and Expression of Novel Trichloroethylene Degradation Genes from Stenotrophomonas maltophilia PM102: A Case of Gene Duplication. *Journal of Bioremediation & Biodegradation*, *04*(02). Advance online publication. doi:10.4172/2155-6199.1000177

Mukherjee, S., & Halder, G. (2016). Assessment of fluoride uptake performance of raw biomass and activated biochar of Colocasia esculenta stem: Optimization through response surface methodology. *Environmental Progress & Sustainable Energy*, *35*(5), 1305–1316. doi:10.1002/ep.12346

Mukherjee, S., Sahu, P., & Halder, G. (2017). Microbial remediation of fluoridecontaminated water via a novel bacterium *Providencia vermicola* (KX926492). *Journal of Environmental Management*, *204*, 413–423. doi:10.1016/j.jenvman.2017.08.051 PMID:28915476

Mukhopadhyay, S., & Maiti, S. K. (2010). Phytoremediation of metal enriched mine waste: A review. *Global Journal of Environmental Research*, *4*(3), 135–150.

Mulligan, C. N., Yong, R., & Gibbs, B. F. (2001). Remediation technologies for metal contaminated soils and groundwater: An evaluation. *Engineering Geology*, *60*(1–4), 193–207. doi:10.1016/S0013-7952(00)00101-0

Muneer, B., Iqbal, M. J., Shakoori, F. R., & Shakoori, A. R. (2013). Tolerance and biosorption of mercury by microbial consortia: Potential use in bioremediation of wastewater. *Pakistan Journal of Zoology*, *45*(1), 247–254.

Muneer, S., & Qadri, T. N., & Mahmooduzaffar, S. T. O. (2011). Cytogenetic and biochemical investigations to study the response of *Vigna radiata* to cadmium stress. *African Journal of Plant Science*, *5*(3), 183–192.

Mushtaq, F., Guerrero, M., Sakar, M. S., Hoop, M., Lindo, A. M., Sort, J., Chen, X., Nelson, B. J., Pellicer, E., & Pane, S. (2015). Magnetically driven bi $2O_3$/ BiOCl-based hybrid microrobots for photocatalytic water remediation. *Journal of Materials Chemistry. A, Materials for Energy and Sustainability*, *3*(47), 23670–23676. doi:10.1039/C5TA05825B

Muthukamalam, S., Sivagangavathi, S., Dhrishya, D., & Sudha Rani, S. (2017). Characterization of dioxygenases and biosurfactants produced by crude oil degrading soil bacteria. *Brazilian Journal of Microbiology*, *48*(4), 637–647. doi:10.1016/j.bjm.2017.02.007 PMID:28629972

Myers, C. R., Carstens, B. P., Antholine, W. E., & Myers, J. M. (2001). Chromium(VI) reductase activity is associated with the cytoplasmic membrane of anaerobically grown *Shewanella putrefaciens* MR01. *Journal of Applied Microbiology*, *88*(1), 98–106. doi:10.1046/j.1365-2672.2000.00910.x PMID:10735248

Naik, M., Shamim, K., & Dubey, S. (2012). Biological characterization of lead resistant bacteria to explore role of bacterial metallothionein in lead resistance. *Current Science*, *103*, 426–429.

Nancharaiah, Y., Mohan, S. V., & Lens, P. J. B. t. (2016). *Recent advances in nutrient removal and recovery in biological and bioelectrochemical systems*. Academic Press.

Narain, S., Ojha, C. S. P., Mishra, S. K., Chaube, U. C., & Sharma, P. K. (2011). Cadmium and Chromium removal by aquatic plants. *International Journal of Environmental Sciences*, *1*(6), 1297–1304.

Nasreen, S. A. A. N., Sundarrajan, S., Nizar, S. A. S., Balamurugan, R., & Ramakrishna, S. (2013). Advancement in electrospun nanofibrous membranes modification and their application in water treatment. *Membranes*, *3*(4), 266–284. doi:10.3390/membranes3040266 PMID:24957057

National Cancer Institute. (2001). *Cancer Facts: Fluoridated Water*. http://fluoride.oralhealth.org/ papers/2001/ncicancerfacts101000.htm

Nepple, B. B., Kessi, J., & Buchofen, R. (2000). Chromate reduction by *Rhodobacter sphaeroides*. *Journal of Industrial Microbiology & Biotechnology*, *25*(4), 198–203. doi:10.1038j.jim.7000049

Neuhold, J. M., & Sigler, W. F. (1960). Effects of sodium fluoride on carp and rainbow trout. *Transactions of the American Fisheries Society*, *89*(4), 358–370. doi:10.1577/1548-8659(1960)89[358:EOSFOC]2.0.CO;2

Neumuller, O. A. (1981). Rommps Chemie Lexicon (8th ed., vol. 2). Franck'sche Verlagshandlung.

Newman, L. A., Doty, S. L., Gery, K. L., Heilman, P. E., Muiznieks, I., Shang, T. Q., Siemieniec, S. T., Strand, S. E., Wang, X., Wilson, A. M., & Gordon, M. P. (1998). Phytoremediation of organic contaminants: A review of phytoremediation research at the University of Washington. *Soil & Sediment Contamination*, *7*(4), 531–542. doi:10.1080/10588339891334366

Ngah, W. S., Ab Ghani, S., & Kamari, A. (2005). Adsorption behaviour of Fe(II) and Fe(III) ions in aqueous solution on chitosan and cross-linked chitosan beads. *Bioresource Technology*, *96*(4), 443–450. doi:10.1016/j.biortech.2004.05.022 PMID:15491825

Ng, D. K. S., Foo, D. C. Y., & Tan, R. R. (2009). Automated targeting technique for single-impurity resource conservation networks. Part 2: Single-pass and partitioning waste-interception systems. *Industrial & Engineering Chemistry Research*, *48*(16), 7647–7661. doi:10.1021/ie900127r

Ngiam, M., Ramakrishna, S., Raghunath, M., & Chan, C. K. (2007). Nanofiber patent landscape. *Recent Patents on Nanotechnology*, *2*(2), 137–144. doi:10.2174/187221007780859645 PMID:19076028

Nies, A., Nies, D. H., & Silver, S. (1990). Nucleotide sequence and expression of a plasmid encoded chromate resistance determinant from *Alcaligenes eutrophus*. *The Journal of Biological Chemistry*, *265*, 5648–5653. PMID:2180932

Nies, D. H. (2000). Heavy metal-resistant bacteria as extremophiles: Molecular physiology and biotechnological use of *Ralstonia* sp. CH34. *Extremophiles*, *4*(2), 77–82. doi:10.1007007920050140 PMID:10805561

Nies, H. D. (2003). Efflux-mediated heavy metal resistance in prokaryotes. *FEMS Microbiology Reviews*, *27*(2-3), 313–339. doi:10.1016/S0168-6445(03)00048-2 PMID:12829273

Nnaji, J., Omotugba, S. K., Ampitan, A., Babatunde, K., & Oluwole, E. J. F. J. O. S.-I. (2019). Article. *Review of Nanomaterials for Remediation of Petroleum Impacted Soil and Water*, *3*(4), 276–284.

Noctor, G., Gomez, L., Vanacker, H., & Foyer, C. H. (2002). Interactions between biosynthesis, compartmentation and transport in the control of glutathione homeostasis and signaling. *Journal of Experimental Botany*, *53*(372), 1283–1304. doi:10.1093/jexbot/53.372.1283 PMID:11997376

Nolan, K. R. (1983). Copper toxicity syndrome. *Journal of Orthomolecular Psychiatry*, *12*(4), 270–282.

Noumavo, P. A., Agbodjato, N. A., Baba-Moussa, F., Adjanohoun, A., & Baba-Moussa, L. (2016). Plant growth promoting rhizobacteria: Beneficial effects for healthy and sustainable agriculture. *African Journal of Biotechnology*, *15*(27), 1452–1463. doi:10.5897/AJB2016.15397

Novotný, Č., Svobodová, K., Erbanová, P., Cajthaml, T., Kasinath, A., Lang, E., & Šašek, V. (2004). Ligninolytic fungi in bioremediation: Extracellular enzyme production and degradation rate. *Soil Biology & Biochemistry*, *36*(10), 1545–1551. doi:10.1016/j.soilbio.2004.07.019

Nriagu' & Pacyna. (1988). Quantitative assessment of worldwide contamination of air, water and soils by trace metals. *Nature, 333*.

Nriagu, J. O. (1989). A global Assessment of Natural Sources of Atmospheric Trace Metals. *Nature*, *338*(6210), 47–49. doi:10.1038/338047a0

Nyangababo, J. T., & Hamya, J. W. (1986). The deposition of lead, cadmium, zinc and copper from motor traffic on *Brachiariaenimi* and soil along a major Bombo road in Kampala city. *The International Journal of Environmental Studies*, *27*(1-2), 115–119. doi:10.1080/00207238608710280

O'Donnell, T. A. (1975). Fluorine. In *Comprehensive inorganic chemistry* (Vol. 2). Pergamon Press.

O'Reilly, M. M., Featherstone, J. D. (1987). Demineralization and remineralization around orthodontic appliances: an in vivo study. *Am J Orthod Dentofacial Orthop.*, *92*(1), 33-40.

Ogata, Y., Ishigaki, T., Ebie, Y., Sutthasil, N., Chiemchaisri, C., & Yamada, M. (2015). Water reduction by constructed wetlands treating waste landfill leachate in a tropical region. *Waste Management (New York, N.Y.)*, *44*, 164–171. doi:10.1016/j.wasman.2015.07.019 PMID:26209341

Ogbuagu, D. H., Okoli, C. G., Emereibeole, E. I., Anyanwu, I. C., Onuoha, O., Ubah, N. O., Ndugbu, C. O., Okoroama, O. N., Okafor, A., Ewa, E., Ossai, R., & Ukah, F. (2011). Trace metals accumulation in biofilms of the upper and middle reaches of Otamiri River in Owerri, Nigeria. *Journal of Biodiversity and Environmental Sciences*, *1*(3), 19–26.

Ogino, T., Suzuki, T., & Sawada, K. (1987). The Formation andTransformation Mechanism of Calcium-Carbonate in Water. *Geochimica et Cosmochimica Acta*, *51*(10), 2757–2767. doi:10.1016/0016-7037(87)90155-4

Ogunsola, O. J., Oluwole, A. F., Obioh, I. B., Asubiojo, O. I., Akeredolu, F. A., Akanle, O. A., & Spyrou, N. M. (1993). Analysis of suspended air particulates along some motorways in Nigeria by PIXE and EDXRF. *Nuclear Instruments & Methods in Physics Research. Section B, Beam Interactions with Materials and Atoms*, *79*(1-4), 404–407. doi:10.1016/0168-583X(93)95373-D

Ogwuegbu, M. O. C., & Muhanga, W. (2005). Investigation of lead concentration in the blood of people in the copper belt province of Zambia. *J Environ, 1*, 66–75.

Ohtake, H., Fujii, E., & Toda, K. (1990). Reduction of toxic chromate in an industrial effluent by use of a chromate-reducing srain of *Enterobacter cloacae*. *Environmental Technology*, *11*(7), 663–668. doi:10.1080/09593339009384909

Ojedokun, A. T., & Bello, O. S. (2016). Sequestering heavy metals from wastewater using cow dung. *Water Resources and Industry*, *13*, 7–13. doi:10.1016/j.wri.2016.02.002

Olaniran, A., Balgobind, A., & Pillay, B. (2013). Bioavailability of heavy metals in soil: Impact on microbial biodegradation of organic compounds and possible improvement strategies. *International Journal of Molecular Sciences*, *14*(5), 10197–10228. doi:10.3390/ijms140510197 PMID:23676353

Oloruntade, A. J., PA, A., & F Alao, F. (2013). Municipal solid waste collection and management strategies in Akure, South-Western Nigeria. *Caspian Journal of Environmental Sciences*, *11*(1), 1–10.

Ommati, M. M., Jamshidzadeh, A., Heidari, R., Sun, Z., Zamiri, M. J., Khodaei, F., ... Yeganeh, B. S. (2019). Carnosine and histidine supplementation blunt lead-induced reproductive toxicity through antioxidative and mitochondria-dependent mechanisms. *Biological Trace Element Research*, *187*(1), 151–162. doi:10.100712011-018-1358-2 PMID:29767280

Omoike, A., & Chorover, J. (2004). Spectroscopic study of extracellular polymeric substances from *Bacillus subtilis*: Aqueous chemistry and adsorption effects. *Biomacromolecules*, *5*(4), 1219–1230. doi:10.1021/bm034461z PMID:15244434

Omueti, J. A. I., & Jones, R. L. (1977). Fluoride adsorption by Illinois soil. *Journal of Soil Science*, *28*(4), 564–572. doi:10.1111/j.1365-2389.1977.tb02264.x

Omueti, J. A. I., & Jones, R. L. (2012). Fluorine distribution with depth in relation to profile development in Illinois. *Soil Science Society of America Journal*, *44*(2), 247–249. doi:10.2136ssaj1980.03615995004400020007x

Oniyama, E., & Walhbeck, P. G. (1995). Application of TranspirationTheory to TGA Data: Calcium Carbonate and Zinc Chloride. *Thermochimica Acta*, *250*(1), 41–53. doi:10.1016/0040-6031(94)01935-A

Ontañon, O. M., González, P. S., & Agostini, E. (2015). Biochemical and molecular mechanisms involved in simultaneous phenol and Cr(VI) removal by *Acinetobacter guillouiae* SFC 500-1A. *Environmental Science and Pollution Research International*, *22*(17), 13014–13023. doi:10.100711356-015-4571-y PMID:25916475

Opperman, D. J., Piater, L. A., & van Heerden, E. (2008). A novel chromate reductase from *Thermus scotoductus* SA- 01 related to old yellow enzyme. *Journal of Bacteriology*, *190*(8), 3076–3082. doi:10.1128/JB.01766-07 PMID:18263719

Ortiz-Bernad, I., Anderson, R. T., Vrionis, H. A., & Lovley, D. R. (2004). Vanadium respiration by *Geobacter metallireducens*: Novel strategy for in situ removal of vanadium from groundwater. *Applied and Environmental Microbiology*, *70*(5), 3091–3095. doi:10.1128/AEM.70.5.3091-3095.2004 PMID:15128571

Osman, D., & Cavet, J. (2008). Copper Homeostasis in Bacteria. *Advances in Applied Microbiology*, *65*, 217–247. doi:10.1016/S0065-2164(08)00608-4 PMID:19026867

Oyem, H. H., Oyem, I. M., & Usese, A. I. (2015). Iron, manganese, cadmium, chromium, zinc and arsenic groundwater contents of Agbor and Owa communities of Nigeria. *SpringerPlus*, *4*(1), 104. doi:10.118640064-015-0867-0 PMID:25853026

Ozdemir, G., Ceyhan, N., & Manav, E. (2005). Utilization in alginate beads for Cu (II) and Ni (II) adsorption of an exopolysaccharide produced by *Chryseomonas luteola* TEM05. *World Journal of Microbiology & Biotechnology*, *21*(2), 163–167. doi:10.100711274-004-1563-3

Ozdemir, G., Ceyhan, N., & Manav, E. (2005). Utilization of an exopolysaccharide produced by *Chryseomonas luteola* TEM05 in alginate beads for adsorption of cadmium and cobalt ions. *Bioresource Technology*, *96*(15), 1677–1682. doi:10.1016/j.biortech.2004.12.031 PMID:16023570

Ozdemir, G., Ozturk, T., Ceyhan, N., Isler, R., & Cosar, T. (2003). Heavy metal biosorption by biomass of *Ochrobactrum anthropi* producing exopolysaccharide in activated sludge. *Bioresource Technology*, *90*(1), 71–74. doi:10.1016/S0960-8524(03)00088-9 PMID:12835060

Ozsvath, D. L. (2006). Fluoride concentrations in a crystalline bedrock aquifer, marathon county, Wisconsin. *Environ Geol*, *50*(10), 132–138. doi:10.100700254-006-0192-6

Ozturk, A. (2007). Removal of nickel from aqueous solution by the bacterium *Bacillus thuringiensis*. *Journal of Hazardous Materials*, *147*(1-2), 518–523. doi:10.1016/j.jhazmat.2007.01.047 PMID:17320284

Pabst, M. W., Miller, C. D., Dimkpa, C. O., Anderson, A. J., & McLean, J. E. (2010). Defining the surface adsorption and internalization of copper and cadmium in a soil bacterium, *Pseudomonas putida*. *Chemosphere*, *81*(7), 904–910. doi:10.1016/j.chemosphere.2010.07.069 PMID:20797767

Pal & Kundu. (2014). Study of metal resistance potential of the Cd, Cr tolerant Alligator Weed. *Journal of Stress Physiology & Biochemistry*, *10*(1), 244–261.

Pal and Kundu. (2011). A Field Study of Chromium and Cadmium uptake, accumulation and tolerance in Alligator Weed [*Alternanthera philoxeroides* (Mart.) Griseb]. *Proceeding of National Academy of Sciences, Section B: Biological Science, India*, *81*(4), 404–410.

Pal. (2018). Comparative Estimation of Cadmium, Chromium Resistance Capacity of *Amaranthus viridis, Tianthema portulacastrum, Alternanthera philoxeroides* collected from contaminated site and Assessment of phytoremedial potential. *Environment and Ecology*, *36*(2), 409-417.

Pal, A. K., & Sengupta, C. (2019). Isolation of Cadmium and Lead Tolerant Plant Growth Promoting Rhizobacteria: Lysinibacillus varians and Pseudomonas putida from Indian Agricultural Soil. *Soil and Sediment Contamination: An International Journal*, *28*(7), 601–629. doi:10.1080/15320383.2019.1637398

Pal, A., Dutta, S., & Paul, A. K. (2005). Reduction of hexavalent chromium by cell free extract of *Bacillus sphaericus* AND 303 isolated from serpentine soil. *Current Microbiology*, *51*(5), 327–330. doi:10.100700284-005-0048-4 PMID:16163455

Pal, A., Ghosh, S., & Paul, A. K. (2006). Biosorption of cobalt by fungi from serpentine soil of Andaman. *Bioresource Technology*, *97*(10), 1253–1258. doi:10.1016/j.biortech.2005.01.043 PMID:16023340

Pal, A., & Paul, A. K. (2004). Aerobic chromate reduction by chromium-resistant bacteria isolated from serpentine soil. *Microbiological Research*, *159*(4), 347–354. doi:10.1016/j.micres.2004.08.001 PMID:15646381

Pan, B., Zhang, W., Lv, L., Zhang, Q., & Zheng, S. (2009). Development of polymeric and polymer-based hybrid adsorbents for pollutants removal from waters. *Chemical Engineering Journal*, *151*(1-3), 19–29. doi:10.1016/j.cej.2009.02.036

Pandey, A., & Banerjee, D. (2012). Biosorption of Cd(II) using discarded biomass of *Aspergillus aculeatus* DBF9. *Terestrial and Aquatic Environmental Toxicology*, *6*, 8–13.

Pandey, N., & Singh, G. K. (2012). Studies on antioxidative enzymes induced by cadmium in pea plants (*Pisum sativum*). *Journal of Environmental Biology*, *33*, 201–206. PMID:23033681

Pandey, S. N. (2006). Accumulation of heavy metal (Cd, Cr, Cu, Ni, Zn) in *Raphanus sativus* L. and *Spinacea oleracea* L. plants irrigated with industrial effluents. *Journal of Environmental Biology*, *27*(2), 381–384. PMID:17436528

Pandey, V., Dixit, V., & Shyam, R. (2005). Antioxidative responses in relation to growth of mustard (*Brassica juncea* cv. Pusa Jai Kishan) plants exposed to hexavalent chromium. *Chemosphere*, *61*(1), 40–47. doi:10.1016/j.chemosphere.2005.03.026 PMID:16157168

Pandey, V., Dixit, V., & Shyam, R. (2009). Chromium effect on ROS generation and detoxification in pea (*Pisum sativum*) leaf chloroplasts. *Protoplasma, 236*(1-4), 85–95. doi:10.100700709-009-0061-8 PMID:19582547

Paniagua-Michel, J. J., Olmos-Soto, J., & Morales-Guerrero, E. R. (2014). Algal and microbial exopolysaccharides: New insights as biosurfactants and bioemulsifiers. *Advances in Food and Nutraceutical Research, 73,* 221–257. doi:10.1016/B978-0-12-800268-1.00011-1 PMID:25300549

Pant, P. P., Tripathi, A. K., & Dwivedi, V. (2011). Effects of some heavy metals on some biochemical parameters of Sal (*Shorea robusta*) seedlings at nursery level, Doon valley. *Indian Journal of Agricultural Sciences, 66*(2), 45–51. doi:10.1080/09766898.2011.11884667

Parameswari, E., Lakshmanan, A., & Thilagavathi, T. (2009). Chromate resistance and reduction by bacterial isolates. *Australian Journal of Basic and Applied Sciences, 3,* 1363–1368.

Pardo, R., Herguedas, M., Barrado, E., & Vega, M. (2003). Biosorption of cadmium, copper, lead and zinc by inactive biomass of *Pseudomonas putida*. *Analytical and Bioanalytical Chemistry, 376*(1), 26–32. doi:10.100700216-003-1843-z PMID:12734614

Park, C. H., Gonzalez, C. F., Ackerley, D. F., Keyhan, M., & Matin, A. (2002). Molecular engineering of soluble bacterial proteins with chromate reductase activity. *Proceedings of the 1st International Conference on Remediation of Contaminated Sediments.*

Park, C. H., Keyhan, B., Wielinga, B., Fendorf, S., & Matin, A. (2000). Purification to homogeneity and characterization of a novel *Pseudomonas putida* chromate reductase. *Applied and Environmental Microbiology, 66*(5), 1788–1795. doi:10.1128/AEM.66.5.1788-1795.2000 PMID:10788340

Park, D., Yun, Y. S., Jo, J. H., & Park, J. M. (2005). Mechanism of hexavalent chromium removal by dead fungal biomass of Aspergillus niger. *Water Research, 39*(4), 533–540. doi:10.1016/j.watres.2004.11.002 PMID:15707625

Park, J.-W., Park, B.-K., & Kim, J.-E. (2006). Remediation of soil contaminated with 2,4-dichlorophenol by treatment of minced shepherd's purse roots. *Archives of Environmental Contamination and Toxicology, 50*(2), 191–195. doi:10.100700244-004-0119-8 PMID:16392021

Parthasarathy, G., & El-Halwagi, M. M. (2000). Optimum mass integration strategies for condensation and allocation of multi component VOCs. *Chemical Engineering Science, 55,* 881–895. doi:10.1016/S0009-2509(99)00382-6

Pathak, R. K., & Dikshit, A. K. (2012, June). Effect of Various Environmental Parameters on Biosorptive Removal of Atrazine from Water Environment. *International Journal of Environmental Sciences and Development, 3*(3).

Patra, R. C., Malik, S., Beer, M., Meghraj, M., & Naidu, R. (2010). Molecular characterization of chromium (VI) reducing potential in Gram positive isolated from contaminated sites. *Soil Biology & Biochemistry, 42*(10), 1857–1863. doi:10.1016/j.soilbio.2010.07.005

Pattanapipitpaisal, P., Brown, N. L., & Macaskie, L. E. (2001). Chromate reduction by *Microbacterium liquefaciens* immobilized in polyvinyl alcohol. *Biotechnology Letters, 23*(1), 61–65. doi:10.1023/A:1026750810580

Pendergast, M. M., & Hoek, E. M. (2011). A review of water treatment membrane nanotechnologies. *Energy & Environmental Science, 4*(6), 1946–1971. doi:10.1039/c0ee00541j

Peplow, D. (1999). *Environmental Impacts of Mining in Eastern Washington, Centre for Water and Watershed Studies Fact Sheet.* University of Washington.

Pesti, M., Gazdag, Z., Emri, T., Farkas, N., Koosz, Z., Belagy, J., & Pocsi, I. (2002). Chromate sensitivity in fission yeast is caused by increased glutathione reductase activity and peroxide overproduction. *Journal of Basic Microbiology*, *42*(6), 408–419. doi:10.1002/1521-4028(200212)42:6<408::AID-JOBM408>3.0.CO;2-8 PMID:12442303

Petrinic, I., Andersen, N. P. R., Sostar-Turk, S., & Le Marechal, A. M. (2007). The removal of reactive dye printing compounds using nanofiltration. *Dyes Pigm.*, *74*(3), 512–518. doi:10.1016/j.dyepig.2006.11.003

Pi, Y., Li, X., Xia, Q., Wu, J., Li, Y., Xiao, J., & Li, Z. J. C. E. J. (2018). *Adsorptive and photocatalytic removal of Persistent Organic Pollutants (POPs) in water by metal-organic frameworks (MOFs)*. Academic Press.

Pickering, W. F. (2015). The mobility of soluble fluoride in soils. *Environ Pollut Ser B*, *9*(44), 281–308.

Piontek, K., Smith, A. T., & Blodig, W. (2001). Lignin peroxidase structure and function. *Biochemical Society Transactions*, *29*(2), 111–116. doi:10.1042/bst0290111 PMID:11356137

Piotrowska-Długosz, A. (2017). The Use of Enzymes in Bioremediation of Soil Xenobiotics. *Xenobiotics in the Soil Environment, 49, 243–265.*, doi:10.1007/978-3-319-47744-2_17

Pointing, S. B., & Vrijmoed, L. L. P. (2000). Decolorization of azo and triphenylmethane dyes by Pycnoporus sanguineus producing laccase as the sole phenoloxidase. *World Journal of Microbiology & Biotechnology*, *16*(3), 317–318. doi:10.1023/A:1008959600680

Poljsak, B., Pocsi, I., Raspor, P., & Pesti, M. (2010). Interference of chromium with biological systems in yeasts and fungi: A review. *Journal of Basic Microbiology*, *50*(1), 21–36. doi:10.1002/jobm.200900170 PMID:19810050

Pollard, A. J., Reeves, R. D., & Baker, A. J. M. (2014). Facultative hyperaccumulation of heavy metals and metalloids. *Plant Science*, *218*, 8–17. doi:10.1016/j.plantsci.2013.11.011 PMID:24467891

Polomski, J., Fluhler, H., & Blaser, P. (2012). Fluoride induced mobilization and leaching of organic matter, iron and aluminium. *Journal of Environmental Quality*, *11*(3), 452–456. doi:10.2134/jeq1982.00472425001100030027x

Polti, M. A., Amoroso, M. J., & Abate, C. M. (2010). Chromate reductase activity in *Streptomyces* sp. MC1. *The Journal of General and Applied Microbiology*, *56*, 11–18. doi:10.2323/jgam.56.11 PMID:20339215

Popescu, M., Velea, A., Lőrinczi, A. J. D. j. o. n., & biostructures. (2010). *Biogenic production of nanoparticles*. Academic Press.

Popova, L. P., Maslenkova, L. T., Yordanova, R. Y., Ivanova, A. P., Krantev, A. P., Szalai, G., & Janda, T. (2009). Exogenous treatment with salicylic acid attenuates cadmium toxicity in pea seedlings. *Plant Physiology and Biochemistry*, *47*(3), 224–231. doi:10.1016/j.plaphy.2008.11.007 PMID:19091585

Prabhu, S., & Poulose, E. K. (2012). Silver nanoparticles: Mechanism of antimicrobial action, synthesis, medical applications, and toxicity effects. *International Nano Letters*, *2*(32), 1–10. doi:10.1186/2228-5326-2-32

Prado, F. E., Prado, C., Rodriguez-Montelongo, L., Gonzalez, J. A., Pagano, E. A., & Hilal, M. (2010). Uptake of Chromium by *Salvinia minima*: Effect on plant growth, leaf respiration and carbohydrate metabolism. *Journal of Hazardous Materials*, *177*(1-3), 546–553. doi:10.1016/j.jhazmat.2009.12.067 PMID:20080338

Pramanik, K., Ghosh, P. K., Ghosh, A., Sarkar, A., & Maiti, T. K. (2016). Characterization of PGP traits of a hexavalent chromium resistant Raoultella sp. isolated from the rice field near industrial sewage of Burdwan District, WB, India. *Soil and Sediment Contamination: An International Journal*, *25*(3), 313–331. doi:10.1080/15320383.2016.1137861

Prasad, M. N. V. (2003). Phytoremediation of metal-polluted ecosystems: Hype for commercialization. *Russian Journal of Plant Physiology: a Comprehensive Russian Journal on Modern Phytophysiology*, *50*(5), 686–701. doi:10.1023/A:1025604627496

Prasad, M. N. V., & Freitas, H. M. O. (2003). Metal hyperaccumulation in plants- biodiversity prospecting for phytoremediation technology. *Electronic Journal of Biotechnology*, *6*(3), 275–321. doi:10.2225/vol6-issue3-fulltext-6

Prasenjit, B., & Sumathi, S. (2005). Uptake of chromium by *Aspergillus foetidus. Journal of Material Cycles and Waste Management*, *7*(2), 88–92. doi:10.100710163-005-0131-8

Priester, J. H., Olson, S. G., Webb, S. M., Neu, M. P., Hersman, L. E., & Holden, P. A. (2006). Enhanced exopolymer production and chromium stabilization in *Pseudomonas putida* unsaturated biofilms. *Applied and Environmental Microbiology*, *72*(3), 1988–1996. doi:10.1128/AEM.72.3.1988-1996.2006 PMID:16517647

Prud'homme, M. (1990). Fluorspar. In: Canadian mineral year book, 1989 energy, mines, and resources. Academic Press.

Purakayastha, R. P. P., & Mitra, A. K. (1992). Metal uptake by mycelia during submerged Biology growth and by sporocarps of an edible fungus *Volvariella volvacea. Indian Journal of Experimental*, *302*, 1184–1187.

Puzon, G. J., Petersen, J. N., Roberts, A. G., Kramer, D. B., & Xun, L. (2002). A bacterial flavin reductase system reduces chromate to a soluble chromium (III)–NAD(+) complex. *Journal of Biochemical and Biophysical Research Communication*, *294*(1), 76–81. doi:10.1016/S0006-291X(02)00438-2 PMID:12054743

Qiao, J. T., Liu, T. X., Wang, X. Q., Li, F. B., Lv, Y. H., & Liu, C. P. (2017). Simultaneous alleviation of cadmium and arsenic accumulation in rice by applying zero-valent iron and biochar to contaminated paddy soils. *Chemosphere*, *195*, 260–271. doi:10.1016/j.chemosphere.2017.12.081 PMID:29272795

Qin, J., Rosen, B. P., Zhang, Y., Wang, G., Franke, S., & Rensing, C. (2006). Arsenic detoxification and evolution of trimethylarsine gas by a microbial arsenite S-adenosylmethionine methyltransferase. *Proceedings of the National Academy of Sciences of the United States of America*, *103*(7), 2075–2080. doi:10.1073/pnas.0506836103 PMID:16452170

Qiu, R. L., Fang, X. H., Tang, Y. T., Du, S. J., Zeng, X. W., & Brewer, E. (2006). Zinc hyperaccumulation and uptake by *Potentilla griffithi* hook. *International Journal of Phytoremediation*, *8*(4), 299–310. doi:10.1080/15226510600992865 PMID:17305304

Rafati, M., Khorasani, N., Moattar, F., Shirvany, A., Moraghebi, F., & Hosseinzadeh, S. (2011). Phytoremediation potential of *Populus alba* and *Morus alba* for cadmium, chromuim and nickel absorption from polluted soil. *International Journal of Environmental of Research*, *5*(4), 961–970.

Raghunandan, K., Kumar, A., Kumar, S., Permaul, K., & Singh, S. (2018). Production of gellan gum, an exopolysaccharide, from biodiesel-derived waste glycerol by sphingomonas spp. *Biotech*, *8*(71), 1–10. doi:10.100713205-018-1096-3 PMID:29354382

RajaRao & Bhargavi. (2013). Studies on biosorption of heavy metals using pretreated biomass of fungal species. *International Journal of Chemistry and Chemical Engineering*, *3*, 171–180.

Rajkumar, M., Ma, Y., & Freitas, H. (2008). Characterization of metal-resistant plant-growth promoting Bacillus weihenstephanensis isolated from serpentine soil in Portugal. *Journal of Basic Microbiology*, *48*(6), 500–508. doi:10.1002/jobm.200800073 PMID:18785659

Ramakrishna, S., & Shirazi, M. M. A. (2013). Electrospun membranes: Next generation membranes for desalination and water/wastewater treatment. *Desalination*, *308*, 198–208.

Rana, V., & Maiti, S. K. (2018a). Differential distribution of metals in tree tissues growing on reclaimed coal mine overburden dumps, Jharia coal field (India). *Environmental Science and Pollution Research*, *25*(10), 9745–9758. doi:10.100711356-018-1254-5 PMID:29368202

Rana, V., & Maiti, S. K. (2018b). Municipal wastewater treatment potential and metal accumulation strategies of *Colocasia esculenta* (L.) Schott and *Typha latifolia* L. in a constructed wetland. *Environmental Monitoring and Assessment*, *190*(6), 328. doi:10.100710661-018-6705-4 PMID:29730705

Rani, M. J., Hemambika, B., Hemapriya, J., & Kannan, V. R. (2010). Comparative assessment of heavy metal removal by immobilized and dead bacterial cells: A biosorption approach. *African Journal of Environmental Science and Technology*, 4.

Ranjan, R., & Ranjan, A. (2015). *Fluoride toxicity in animals. Springer Briefs in Animal Sciences (part of an E-book collection)*. Springer International Publishing.

Rao, G. P., Lu, C., Su, F. J. S., & Technology, P. (2007). *Sorption of divalent metal ions from aqueous solution by carbon nanotubes: a review*. Academic Press.

Rao, M. A., Scelza, R., Scotti, R., & Gianfreda, L. (2010). Role of enzymes in the remediation of polluted environments. *Journal of Soil Science and Plant Nutrition*, *10*(3). Advance online publication. doi:10.4067/S0718-95162010000100008

Rao, M. B., Tanksale, A. M., Ghatge, M. S., & Deshpande, V. V. (1998). Molecular and biotechnological aspects of microbial proteases. *Microbiology and Molecular Biology Reviews*, *62*(3), 597–635. doi:10.1128/MMBR.62.3.597-635.1998 PMID:9729602

Rascio, N., & Navari-Izzo, F. (2011). Heavy metal hyperaccumulating plants: How and why do they do it? And what makes them so interesting? *Plant Science*, *180*(2), 169–181. doi:10.1016/j.plantsci.2010.08.016 PMID:21421358

Rashidi, H. R., Sulaiman, N. N., & Hashim, N. A. (2012). Batik industry synthetic wastewater treatment using nanofiltration membrane. *Proc. Eng.*, *44*, 2010–2012. doi:10.1016/j.proeng.2012.09.025

Raskin & Ensley (Eds.). (n.d.). *Phytoremediation of toxic metals: using plants to clean-up the environment*. New York: John Wiley & Sons, Inc.

Raunkiaer, C. (1934). *The life forms of plants and statistical plant geography*. Clarendon Press.

Raven, P. H., Berg, L. R., & Johnson, G. B. (1998). *Environment* (2nd ed.). Saunders College Publishing.

Ray, S.S., Chen, S.S., Nguyen, N.C., & Hau, T.N. (2019). Electrospinning: A versatile fabrication technique for nanofibrous membranes for use in desalination. *Nanoscale Materials in Water Purification*, *9*, 247- 273.

Ray, S. S., Chen, S. S., Hsu, H. T., Cao, D. T., Nguyen, H. T., & Nguyen, N. C. (2017). Uniform hydrophobic electrospun nanofibrous layer composed of polysulfone and sodium dodecyl sulfate for improved desalination perfor-mance. *Separation and Purification Technology*, *186*, 352–365. doi:10.1016/j.seppur.2017.06.032

Ray, S. S., Chen, S. S., Li, C. W., Nguyen, N. C., & Nguyen, H. T. (2016). A comprehensive review: Electrospinning technique for fabrication and surface modification of membranes for water treatment application. *RSC Advances*, *6*(88), 85495–85514. doi:10.1039/C6RA14952A

Ray, S. S., Chen, S. S., Nguyen, N. C., Hsu, H. T., Nguyen, H. T., & Chang, C. T. (2014). Poly (vinyl alcohol) incorporated with surfactant based electrospun nanofibrous layer onto polypropylene mat for improved desalination by using membrane distillation. *Desalination*, *414*, 18–27. doi:10.1016/j.desal.2017.03.032

Reardon, C. L., Dohnalkova, A. C., Nachimuthu, P., Kennedy, D. W., Saffarini, D. A., Arey, B. W., Shi, L., Wang, Z., Moore, D., McLean, J. S., Moyles, D., Marshall, M. J., Zachara, J. M., Fredrickson, J. K., & Beliaev, A. S. (2010). Role of outer-membrane cytochromes MtrC and OmcA in the biomineralization of ferrihydrite by *Shewanella oneidensis* MR-1. *Geobiology*, *8*(1), 56–68. doi:10.1111/j.1472-4669.2009.00226.x PMID:20002197

Reddy, M. M., & Nancollas, G. H. (1971). The Crystallisation ofCalcium Carbonate: Isotopic Exchange and Kinetics. *Journal of Colloid and Interface Science*, *37*, 166–172. doi:10.1016/0021-9797(71)90161-5

Redjala, T., Sterckeman, T., & Morel, J. L. (2009). Cadmium uptake by roots: Contribution of apoplast and of high and low affinity membrane transport systems. *Environmental and Experimental Botany*, *67*(1), 235–242. doi:10.1016/j.envexpbot.2009.05.012

Reeves, R. D. (2003). Tropical hyperaccumulators of metals and their potential for phytoextraction. *Plant and Soil*, *249*(1), 57–65. doi:10.1023/A:1022572517197

Reeves, R. D., & Baker, J. M. (2000). Metal accumulating plants. In H. Raskin & B. D. Ensley (Eds.), *Phytoremediation of toxic metals: Using plants to clean up the environment* (pp. 193–230). John Wiley & Sons Inc.

Reeves, R. D., & Brooks, R. R. (1984). Studies on metal uptake by plants from serpentine and non-serpentine populations of Thlaspi goesingense H'al'acsy (Cruciferae). *The New Phytologist*, *98*(1), 191–204. doi:10.1111/j.1469-8137.1984.tb06108.x PMID:29681122

Rehman, A., Zahoor, A., Muneer, B., & Hasnain, S. (2008). Chromium tolerance and reduction potential of a *Bacillus* sp.ev3 isolated from metal contaminated wastewater. *Bulletin of Environmental Contamination and Toxicology*, *81*(1), 25–29. doi:10.100700128-008-9442-5 PMID:18498008

Rehman, Y., Rizvi, F. Z., Faisal, M., & Hasnain, S. (2013). Arsenic and chromium reduction in co-cultures of bacteria isolated from industrial sites of Pakistan. *Microbiology*, *82*(4), 458–433. doi:10.1134/S0026261713040188

Reiad, N. A., Abdel Salam, O. E., Abadir, E. F., & Harraz, F. A. (2012). Adsorptive removal of iron and manganese ions from aqueous solutions with microporous chitosan/polyethylene glycol blend membrane. [China]. *Journal of Environmental Sciences (China)*, *24*(8), 1425–1432. doi:10.1016/S1001-0742(11)60954-6 PMID:23513684

Reimann, C., & Decaritat, P. (Eds.). (1998). *In: Chemical elements in environment. Fact sheets for the geochemist and environmental scientist*. Springer.

Reisinger, S., Schiavon, M., Terry, N., Elizabeth, A. H., & Pilon, S. (2008). Heavy metal tolerance and accumulation in Indian mustard (*Brassica juncea*) expressing bacterial γ- glutamyl cysteine synthetase or glutathione synthetase. *International Journal of Phytoremediation*, *10*(5), 1–5. doi:10.1080/15226510802100630 PMID:19260225

Reneker, D. H., Yarin, A. L., Fong, H., & Koombhongse, S. (2000). Bending instability of electrically charged liquid jets of polymer solutions in electrospinning. *Journal of Applied Physics*, *87*(9), 4531–4547. doi:10.1063/1.373532

Renou, S., Givaudan, J. G., Poulain, S., Dirassouyan, F., & Moulin, P. (2008). Landfill leachate treatment: Review and opportunity. *Journal of Hazardous Materials*, *150*(3), 468–493. doi:10.1016/j.jhazmat.2007.09.077 PMID:17997033

Ren, X., Chena, C., Nagatsu, M., & Wang, X. (2011). Carbon nanotubes as adsorbents in environmental pollution management: A review. *Chemical Engineering Journal*, *170*(2-3), 395–410. doi:10.1016/j.cej.2010.08.045

Reports of UNICEF. (2017) *Thirsting for a Future: Water and children in a changing climate.* UNICEF.

Reuss, J. O., Dooley, H. L., & Griffis, W. (1978). Uptake of Cadmium from Phosphate Fertilizers by Peas, Radishes, and Lettuce 1. *Journal of Environmental Quality*, *7*(1), 128–133. doi:10.2134/jeq1978.00472425000700010026x

Rezvani, M., & Zaefarian, F. (2011). Bioaccumulation and translocation factor of cadmium and lead in *Aeluropus littorali. A.J.A.E, 2*(4), 114–119.

Rial, Vázquez, & Murado. (n.d.). *Effects of three heavy metals on the bacteria growth kinetics. A bivariate model for toxicological assessment.* Academic Press.

Riffaldi, R., Levi-Minzi, R., Cardelli, R., Palumbo, S., & Saviozzi, A. (2006). Soil biological activities in monitoring the bioremediation of diesel oil-contaminated soil. *Water, Air, and Soil Pollution, 170*(1–4), 3–15. doi:10.100711270-006-6328-1

Riggle, P. J., & Kumamoto, C. A. (2000). Role of a Candida albicans P1-type ATPase in resistance to copper and silver ion toxicity. *Journal of Bacteriology, 182*(17), 4899–4905. doi:10.1128/JB.182.17.4899-4905.2000 PMID:10940034

Ripp, S., & Henry, T. B. (2011). *Biotechnology and nanotechnology risk assessment: minding and managing the potential threats around us.* ACS Publications.

Rixon, E., Ferreira, L. M. A., Durrant, A. J., Laurie, J. I., Hazlewood, G. P., & Gilbert, H. J. (1992). Characterization of the gene celD and its encoded product 1,4-β-D-glucan glucohydrolase D from Pseudomonas fluorescens subsp. cellulosa. *The Biochemical Journal, 285*(3), 947–955. doi:10.1042/bj2850947 PMID:1497631

Rizwan, M., Singh, M., Mitra, C. K., & Morve, R. K. J. J. o. N. (2014). *Ecofriendly application of nanomaterials: nanobioremediation.* Academic Press.

Rizwan, M., Ali, S., Ali, B., Adrees, M., Arshad, M., Hussain, A., Zia ur Rehman, M., & Waris, A. A. (2019). Zinc and iron oxide nanoparticles improved the plant growth and reduced the oxidative stress and cadmium concentration in wheat. *Chemosphere, 214*, 269–277. doi:10.1016/j.chemosphere.2018.09.120 PMID:30265934

Robins, K. J., Hooks, D. O., Rehm, B. H. A., & Ackerley, D. F. (2013). Escherichia coli NemA is an efficient chromate reductase that can be biologically immobilized to provide a cell free system for remediation of hexavalent chromium. *PLoS One, 8*(3), 1–8. doi:10.1371/journal.pone.0059200 PMID:23527133

Robinson, W. D., & Edington, G. (1946). Fluorine in soils. *Soil Science, 61*(5), 341–353. doi:10.1097/00010694-194605000-00001

Roccatano, D. (2015). Structure, dynamics, and function of the monooxygenase P450 BM-3: Insights from computer simulations studies. *Journal of Physics Condensed Matter, 27*(27), 273102. doi:10.1088/0953-8984/27/27/273102 PMID:26061496

Rodríguez Couto, S., & Toca Herrera, J. L. (2006). Industrial and biotechnological applications of laccases: A review. *Biotechnology Advances, 24*(5), 500–513. doi:10.1016/j.biotechadv.2006.04.003 PMID:16716556

Rodriguez, E., Santos, C., Lucas, E., & Pereira, M. J. (2011). Evaluation of Chromium (VI) toxicity to *Chlorella vulgaris* Beijerinck cultures. *Fresenius Environmental Bulletin, 20*, 334–339.

Roh, Y., Liu, S. V., Li, G., Huang, H., Phelps, T. J., & Zhou, J. (2002). Isolation and characterization of metal-reducing thermoanaerobacter strains from deep subsurface environment of the Piceane basin, Colorado. *Applied and Environmental Microbiology, 68*(12), 6013–6020. doi:10.1128/AEM.68.12.6013-6020.2002 PMID:12450823

Romanenko, V. I., & Korenkov, V. N. (1977). A pure culture of bacteria utilising chromates and bichromates as hydrogen acceptors in growth under anaerobic conditions. *Microbiology, 46*, 329–332.

Romero, M. C., Reinoso, E. H., Urrutia, M. I., & Kiernan, A. M. (2006). Biosorption of heavy metals by *Talaromyces helicus*: A trained fungus for copper and biphenyl detoxification. *Electronic Journal of Biotechnology, 9*, 221–226. doi:10.2225/vol9-issue3-fulltext-11

Romuald, B. K., Hamoudi, S., Faical, L., & Khaled, B. (2012). Synthesis of $CaCO_3$ nanoparticles by controlled precipitation of saturated carbonate and calcium nitrate aqueous solutions. *Canadian Journal of Chemical Engineering*, *90*(1), 26–33. doi:10.1002/cjce.20673

Rosselli, W., Keller, C., & Boschi, K. (2003). Phytoextraction capacity of tree growing on a metal contaminated soil. *Plant and Soil*, *256*(2), 265–272. doi:10.1023/A:1026100707797

Rubilar, O., Diez, M. C., & Gianfreda, L. (2008). Transformation of chlorinated phenolic compounds by white rot fungi. *Critical Reviews in Environmental Science and Technology*, *38*(4), 227–268. doi:10.1080/10643380701413351

Rude, R. K., & Gruber, H. E. (2004). Magnesium deficiency and osteoporosis: Animal and human observations. *The Journal of Nutritional Biochemistry*, *15*(12), 710–716. doi:10.1016/j.jnutbio.2004.08.001 PMID:15607643

Ruiz-Duenas, F. J., Morales, M., Pérez-Boada, M., Choinowski, T., Martínez, M. J., Piontek, K., & Martínez, Á. T. (2007). Manganese oxidation site in Pleurotus cryngii versatile peroxidase: A site-directed mutagenesis, kinetic, and crystallographic study. *Biochemistry*, *46*(1), 66–77. doi:10.1021/bi061542h PMID:17198376

Rutledge, G. C., & Fridrikh, S. V. (2007). Formation of fibers by electrospinning. *Advanced Drug Delivery Reviews*, *59*(14), 1384–1391. doi:10.1016/j.addr.2007.04.020 PMID:17889398

Sadhu, S., & Maiti, K. (2013). Cellulase production by bacteria: A review. *British Microbiology Research Journal*, *3*(3), 235–258. doi:10.9734/BMRJ/2013/2367

Saeed, T., Haque, I., & Khan, T. (2019). Organic matter and nutrients removal in hybrid constructed wetlands: Influence of saturation. *Chemical Engineering Journal*, *371*, 154–165. doi:10.1016/j.cej.2019.04.030

Sag, Y., & Kutsal, T. (1996). Fully competitive biosorption of chromium(VI) and iron(III) ions from a binary metal mixture by *Rhizopus arrhizus*: Use of the competitive Langmuir model. *Process Biochemistry*, *31*(6), 573–585. doi:10.1016/S0032-9592(96)00003-9

Sag, Y., & Kutsal, T. (2001). Recent trends in biosorption of heavy metals. *Biotechnology and Bioprocess Engineering; BBE*, *6*(6), 376–385. doi:10.1007/BF02932318

Sag, Y., Tatar, B., & Kutsal, T. (2003). Biosorption of Pb(II) and Cu(II) by activated sludge in batch and continuous-flow stirred reactors. *Bioresource Technology*, *87*(1), 27–33. doi:10.1016/S0960-8524(02)00210-9 PMID:12733571

Sahi, S.V., Bryant, N.L, Sharma, N.C., & Singh, S.R. (2002). Characterization of a Lead hyperaccumulator shrub, *Sesbania drummondii. Environ. Sci. Technol., 36* 4676-4680.

Sahi, C., Agarwal, M., Reddy, M. K., Sopory, S. K., & Grover, A. (2003). Isolation and expression analysis of salt stress-associated ESTs from contrasting rice cultivars using a PCR-based subtraction method. *Theoretical and Applied Genetics*, *106*(4), 620–628. doi:10.100700122-002-1089-8 PMID:12595990

Sahin, Y., & Ozturk, A. (2005). Biosorption of Cr(VI) ions from aqueous solution by the bacterium *Bacillus thuringiensis. Process Biochemistry*, *40*(5), 1895–1901. doi:10.1016/j.procbio.2004.07.002

Sahoo, D. K., Das, A., Thatoi, H., Mondal, K. C., & Mohapatra, P. K. (2012). Keratinase production and biodegradation of whole chicken feather keratin by a newly isolated bacterium under submerged fermentation. *Applied Biochemistry and Biotechnology*, *167*(5), 1040–1051. doi:10.100712010-011-9527-1 PMID:22246728

Sahoo, Y., Goodarzi, A., Swihart, M. T., Ohulchanskyy, T. Y., Kaur, N., Furlani, E. P., & Prasad, P. N. (2005). Aqueous ferrofluid of magnetite nanoparticles: Fluorescence labeling and magnetophoretic control. *The Journal of Physical Chemistry B*, *109*(9), 3879–3885. doi:10.1021/jp045402y PMID:16851439

Saier, M. H. Jr. (2003). Tracing pathways of transport protein evolution. *Molecular Microbiology, 48*(5), 1145–1156. doi:10.1046/j.1365-2958.2003.03499.x PMID:12787345

Salnikow, K., & Zhitkovich, A. (2007). Genetic and epigenetic mechanisms in metal carcinogenesis and cocarcinogenesis: Nickel, Arsenic, and Chromium. *Chemical Research in Toxicology, 21*(1), 28–44. doi:10.1021/tx700198a PMID:17970581

Samantaray, D., Mohapatra, S., & Mishra, B. (2014). Bioremediation of industrial effluents. In *Microbial Biodegradation and Bioremediation* (pp. 325–339). Elsevier. doi:10.1016/B978-0-12-800021-2.00014-5

Sanchez-Galvan, G., & Bolanos-Santiago, Y. (2018). Phytofiltration of anaerobically digested sugarcane ethanol stillage using a macrophyte with high potential for biofuel production. *International Journal of Phytoremediation, 20*(8), 805–812. doi:10.1080/15226514.2018.1438350 PMID:29775098

Sandalio, L. M., Dalurzo, H. C., Gomez, M., Romero-Puertas, M. C., & del Rio, L. A. (2001). Cadmium induced changes in the induced change in the growth and oxidative metabolism of pea plants. *Journal of Experimental Botany, 52*(364), 2116–2126. doi:10.1093/jexbot/52.364.2115

Sandrin, T., & Maier, R. M. (2002). Effect of pH on cadmium toxicity, speciation, and accumulation during naphthalene biodegradation. *Environmental Toxicology and Chemistry, 21*(10), 2075–2079. doi:10.1002/etc.5620211010 PMID:12371483

Sankarammal, M., Thatheyus, A., & Ramya, D. (2014). Bioremoval of Cadmium Using *Pseudomonas fluorescens. Open Journal of Water Pollution and Treatment, 2*(2), 92–100. doi:10.15764/WPT.2014.02010

Santhosh, C., Velmurugan, V., Jacob, G., Jeong, S. K., Grace, A. N., & Bhatnagar, A. (2016). Role of nanomaterials in water treatment applications: A review. *Chemical Engineering Journal, 306*, 1116–1137. doi:10.1016/j.cej.2016.08.053

Santos-Díaz, M. S., & Zamora-Pedrazaa, C. (2010). Fluoride removal from water by plant species that are tolerant and highly tolerant to hydrogen fluoride. *Fluoride, 43*, 150–156.

Sarangi, A., & Krishnan, C. (2008). Comparison of in vitro Cr(VI) reduction by CFEs of chromate resistant bacteria isolated from chromate contaminated soil. *Bioresource Technology, 99*(10), 4130–4137. doi:10.1016/j.biortech.2007.08.059 PMID:17920879

Saraswat, S., & Rai, J. P. N. (2009). Phytoextraction potential of six plant species grown in multimetal contaminated soil. *Chemistry and Ecology, 25*(1), 1–11. doi:10.1080/02757540802657185

Sarathambal, C., Khankhane, P. J., Gharde, Y., Kumar, B., Varun, M., & Arun, S. (2017). The effect of plant growth-promoting rhizobacteria on the growth, physiology, and Cd uptake of Arundo donax L. *International Journal of Phytoremediation, 19*(4), 360–370. doi:10.1080/15226514.2016.1225289 PMID:27592507

Sarkar, M., Banerjee, A., Pramanick, P. P., & Sarkar, A. R. (2006). Use of laterite for the removal of fluoride from contaminated drinking water. *Journal of Colloid and Interface Science, 302*(2), 432–441. doi:10.1016/j.jcis.2006.07.001 PMID:16899254

Sasser. (1990, May). *Identification of Bacteria by Gas Chromatography of Cellular Fatty Acids Revised February 2001*. Academic Press.

Sathish Kumar, K., Murugan, K., Benelli, G., Higuchi, A., & Rajasekar, A. (2016). Bioreduction of hexavalent chromium by *Pseudomonas stutzeri* L1 and *Acinetobacter baumannii* L2. *Annals of Microbiology, 67*, 91-98.

Satyanarayana, T. (2005, July). Extremophilic microbes. *Diversity and Perspectives Current Science, 89*(1), 10.

Sawada, K., & Ohtaki, H. (1998). *Crystallisation Process*. John Wiley & Sons, Inc.

Sawaittayothin, V., & Polprasert, C. (2007). Nitrogen mass balance and microbial analysis of constructed wetlands treating municipal landfill leachate. *Bioresource Technology*, *98*(3), 565–570. doi:10.1016/j.biortech.2006.02.002 PMID:16546377

Saxena, J., Sharma, M. M., Gupta, S., & Singh, A. J. W. J. P. S. (2014). *Emerging role of fungi in nanoparticle synthesis and their applications*. Academic Press.

Saxena, D., Joshi, N., & Srivastava, S. (2002). Mechanism of copper resistance in a copper mine isolate *Pseudomonas putida* strain S4. *Current Microbiology*, *45*(6), 410–414. doi:10.100700284-002-3787-5 PMID:12402081

Saxena, G., Purchase, D., Mulla, S. I., Saratale, G. D., & Bharagava, R. N. (2020). Phytoremediation of heavy metal-contaminated sites: Eco-environmental concerns, field studies, sustainability issues, and future prospects. *Reviews of Environmental Contamination and Toxicology*, *249*, 71–131. doi:10.1007/398_2019_24 PMID:30806802

Saxena, V. K., & Ahmed, S. (2001). Dissolution of fluoride in groundwater: A water reaction study. *Environ Geol*, *40*(9), 1084–1087. doi:10.1007002540100290

Saxena, V. K., & Ahmed, S. (2003). Inferring the chemical parameters for the dissolution of fluoride in ground water. *Environ Geol*, *43*(6), 731–736. doi:10.100700254-002-0672-2

Schiopu, A. M., & Gavrilescu, M. (2010). Options for the treatment and management of municipal landfill leachate: Common and specific issues. *CLEAN–Soil, Air. Water (Basel)*, *38*(12), 1101–1110.

Scientific Committee on Health and Environmental Risks (SCHER). (2011 May 16). *Opinion of critical review of any new evidence on the hazard profile, health effects, and human exposure to fluoride and the fluoridating agents of drinking water*. Brussels, Belgium: Directorate General for Health and Consumers, European Commission.

Scott, C., Begley, C., Taylor, M. J., Pandey, G., Momiroski, V., & French, N. … Russell, R. J. (2011). Free-Enzyme Bioremediation of Pesticides: A Case Study for the Enzymatic Remediation of Organophosphorous Insecticide Residues. In K. S. Goh, B. L. Bret, T. L. Potter, & J. Gan (Eds.), Pesticide Mitigation Strategies for Surface Water Quality (Vol. 1075, pp. 155–174). doi:10.1021/bk-2011-1075.ch011

Sedláček, V., & Kučera, I. (2010). Chromate reductase activity of the *Paracoccus denitrificans* ferric reductase B (FerB) protein and its physiological relevance. *Archives of Microbiology*, *192*(11), 919–926. doi:10.100700203-010-0622-4 PMID:20821194

Sehn, P. (2008). Fluoride removal with extra low energy reverse osmosis membranes: Three years of large scale field experience in Finland. *Desalination*, *223*(1-3), 73–84. doi:10.1016/j.desal.2007.02.077

Sekora, Lawrence, Agudelo, Van Santen, & McInroy (2009). Using FAME Analysis to Compare, Differentiate, and Identify Multiple Nematode Species. *Journal of Nematology*, *41*(3).

Sengupta, D., Mitra, A. K., Choudhury, S. S., & Chandra, A. (2014). Isotherm Study in Arsenic Tolerant Bacteria Isolated from Arsenic Affected Area in West-Bengal, India. *IOSR Journal of Environmental Science, Toxicology and Food Technology*, *8*(1), 8-19.

Sen, M., & Ghosh Dastidar, M. (2007). Biosorption of Cr(VI) by resting cells of *Aspergillus* sp. *Iranian Journal of Environmental Health Sciences & Engineering*, *4*, 9–12.

Seppä, L., Pöllänen, L., & Hausen, H. (1994). Caries-preventive effect of fluoride varnish with different fluoride concentrations. *Caries Research*, *28*(1), 64–67. doi:10.1159/000261623 PMID:8124700

Sessitsch, A., Kuffner, M., Kidd, P., Vangronsveld, J., Wenzel, W. W., Fallmann, K., & Puschenreiter, M. (2013). The role of plant-associated bacteria in the mobilization and phytoextraction of trace elements in contaminated soils. *Soil Biology & Biochemistry*, *60*, 182–194. doi:10.1016/j.soilbio.2013.01.012 PMID:23645938

Sethi, P. K., & Khandeiwal, D. J. (2006). Cadmium exposure: Health hazards in developing countries. *Medical Toxicology*, (2), 14–15. doi:10.1007/BF03161007 PMID:18072106

Seymour, M. B., Su, C., Gao, Y., Lu, Y., & Li, Y. J. J. o. N. R. (2012). *Characterization of carbon nano-onions for heavy metal ion remediation.* Academic Press.

Shahi, S. K., & Patra, M. J. R. o. A. M. S. (2003). *Microbially synthesized bioactive nanoparticles and their formulation active against human pathogenic fungi.* Academic Press.

Shahid, M., Shamshad, S., Rafiq, M., Khalid, S., Bibib, I., & Khan, N. (2017). Chromium speciation, bioavailability, uptake, toxicity and detoxification in soil-plant system: A review. *Chemosphere*, *178*, 513–533. doi:10.1016/j.chemosphere.2017.03.074 PMID:28347915

Shahzad, Z., Gosti, F., Frerot, H., Lacombe, E., & Rossens, N. (2010). The five AhMTP1 zinc transporters undergo different evolutionary fates towards adaptive evolution to Zn tolerance in *Arabidopsis halleri. PLOS Genetics*, *6*(4), e1000911. doi:10.1371/journal.pgen.1000911 PMID:20419142

Shanker, A. K., Cervanter, C., & Loza-Tavera, H. (2005). Chromium toxicity in plants. *Environment International*, *31*(5), 739–753. doi:10.1016/j.envint.2005.02.003 PMID:15878200

Shannon, M. A., Bohn, P. W., Elimelech, M., Georgiadis, J. G., Marinas, B. J., & Mayes, A. M. (2008). Science and technology for water purification in the coming decades. *Nature*, *452*(7185), 301–310. doi:10.1038/nature06599 PMID:18354474

Shao, C., Guan, H., Liu, Y., Gong, J., Yu, N., & Yang, X. (2004). A novel method for making ZrO_2 nano fibres via an electrospinning technique. *Journal of Crystal Growth*, *267*(1-2), 380–384. doi:10.1016/j.jcrysgro.2004.03.065

Sharma, S., Tiwari, S., Hasan, A., Saxena, V., & Pandey, L. M. (2018). Recent advances in conventional and contemporary methods for remediation of heavy metal-contaminated soils. *3 Biotech*, *8*(4), 216. doi:10.100713205-018-1237-8

Sharma, T., Sharma, A., & Kanwar, S. S. (2017). An overview on esterases: structure, classification, sources and their application. In Recent advances in biotechnology (vol. 2, pp. 216-229). Shree Publishers.

Sharma, B., Dangi, A. K., & Shukla, P. (2018). Contemporary enzyme based technologies for bioremediation: A review. *Journal of Environmental Management*, *210*, 10–22. doi:10.1016/j.jenvman.2017.12.075 PMID:29329004

Sharma, D., Sharma, B., & Shukla, A. K. (2011). Biotechnological approach of microbial lipase: A review. *Biotechnology (Faisalabad)*, *10*(1), 23–40. doi:10.3923/biotech.2011.23.40

Sharmili, A. S., & Ramasamy, P. E. (2016). Fatty Acid Methyl Ester (FAME) Analysis of Moderately Thermophilic Bacteria Isolated from the Coramandal Coast, Chennai, Tamilnadu. *European Journal of Experimental Biology*, *6*(5).

Shazia, I. U., Sadia, G. R., & Talat, A. (2013). Bioremediation of heavy metals using isolates of filamentous fungus *Aspergillus fumigatus* collected from polluted soil of Kasur, Pakistan. *International Research Journal of Biological Sciences*, *2*, 66–73.

Shekhar, C., Sammaiah, D., Rambabu, M., & Reddy, K. J. (2011). Effect of cadmium on tomato growth and yield attributes. *J. Microbiol. Biotech. Res*, *1*(3), 109–112.

Sheng, X. F., & Xia, J. J. (2006). Improvement of rape (Brassica napus) plant growth and cadmium uptake by cadmium-resistant bacteria. *Chemosphere*, *64*(6), 1036–1042. doi:10.1016/j.chemosphere.2006.01.051 PMID:16516946

Shen, H., & Wang, Y. T. (1993). Characterization of enzymatic reduction of hexavalent chromium by *Escherichia coli* ATCC 33456. *Applied and Environmental Microbiology*, *59*(11), 3771–3777. doi:10.1128/AEM.59.11.3771-3777.1993 PMID:8285683

Shen, H., & Wang, Y. T. (1994). Biological reduction of chromium by *E. coli. Journal of Environmental Engineering*, *120*(3), 560–572. doi:10.1061/(ASCE)0733-9372(1994)120:3(560)

Shin, H. Y., Park, T. J., & Kim, M. I. (2015). Recent Research Trends and Future Prospects in Nanozymes. *Journal of Nanomaterials*, *2015*, 1–11. doi:10.1155/2015/756278

Shokoohi, R., Saghi, M. H., Ghafari, H. R., & Hadi, M. (2009). Biosorption of iron from aqueous solution by dried biomass of activated sludge. *Iranian Journal of Environmental Health Sciences & Engineering*, *6*(2), 107–114.

Sholl, D. S., & Lively, R. P. (2016). Seven chemical separations to change the world. *Nature*, *532*(7600), 435–437. doi:10.1038/532435a PMID:27121824

Short, H. E., McRobert, T. W., Bernard, A. S., & Mannadinayer, A. S. (1937). Endemic fluorosis in the Madras Presidency. *The Indian Journal of Medical Research*, *25*, 553–561.

Shupe, J. L., & Sharma, R. P. (2006). Fluoride distribution in natural ecosystem and related effects on wild animals. In D. D. Hemphill (Ed.), *Trace substances in environmental health* (Vol. 10, p. 137). University of Missouri.

Sieghardt, H. (1987). Schwermetall- und N¨ahrelementgehalte von Pflanzen und Bodenproben schwermetallhaltiger Halden im Raum Bleiberg in Ka¨rnten (O¨ sterreich): I. krautige Pflanzen. Z. Pflanzener¨ahr. *Bodenk*, *150*(3), 129–134. doi:10.1002/jpln.19871500302

Silver, S. (1997). The bacterial view of the periodic table: specific functions for all elements. In J. F. Banfield & K. H. Nealson (Eds.), *Geomicrobiology: Integration between Microbes and Minerals. Mineralogical society of America* (pp. 345–360). doi:10.1515/9781501509247-012

Singh, R., Behera, M., & Kumar, S. (2020). Nano-bioremediation: An innovative remediation technology for treatment and management of contaminated sites. In Bioremediation of Industrial Waste for Environmental Safety (pp. 165-182). Springer.

Singh, A. K., Srivastava, M., Rajneesh, N. K., & Shukla, S. (2017). Corncobs as low cost bio adsorbent for water and waste water treatment. *Int J of Inn Res Sci Eng Tech.*, *6*, 20487–20491.

Singh, A. L., & Sarma, P. N. (2010). Removal of arsenic(III) from waste water using *Lactobacillus acidophilus. Bioremediation Journal*, *14*(2), 92–97. doi:10.1080/10889861003767050

Singh, K. P., Mohan, D., Sinha, S., & Dalwani, R. (2004). Impact assessment of treated/untreated wastewater toxicants discharged by sewage treatment plants on health, agricultural, and environmental quality in the wastewater disposal area. *Chemosphere*, *55*(2), 227–255. doi:10.1016/j.chemosphere.2003.10.050 PMID:14761695

Singh, R. P., Jain, S., & Ramarao, P. (2013). Surfactant-assisted dispersion of carbon nanotubes: Mechanism of stabilization and biocompatibility of the surfactant. *Journal of Nanoparticle Research*, *15*(10), 1985. doi:10.100711051-013-1985-7

Singh, R. P., Jhamb, S. S., & Singh, P. P. (2008). Effects of morphine during *Mycobacterium tuberculosis* H37Rv infection in mice. *Life Sciences*, *82*(5-6), 308–314. doi:10.1016/j.lfs.2007.11.024 PMID:18191418

Singh, R., Singh, D. P., Kumar, N., Bhargav, S. K., & Barman, S. C. (2010). Accumulation and translocation of heavy metal in soil and plant from fly ash contaminated areas. *Journal of Environmental Biology*, *31*, 421–430. PMID:21186714

Singh, S. P., Khare, P., Satsangi, G. S., Lakhani, A., Kumari, S. S., & Srivastava, S. S. (2011). Rainwater composition at a regional representative site of a semi-arid region of India. *Water, Air, and Soil Pollution*, *127*(1/4), 93–108. doi:10.1023/A:1005295215338

Singh, S., Kang, S. H., Mulchandani, A., & Chen, W. (2008). Bioremediation: Environmental clean-up through pathway engineering. *Current Opinion in Biotechnology*, *19*(5), 437–444. doi:10.1016/j.copbio.2008.07.012 PMID:18760355

Singh, S., Kaushik, A., & Kaushik, C. (2016). Comparing efficacy of down-flow and up-flow vertical constructed wetlands for treatment of simulated dumpsite leachate. *Imperial Journal of Interdisciplinary Research*, *2*, 942–945.

Singh, S., Parihar, P., Singh, R., Singh, V., & Prasad, S. (2015). Heavy metal tolerance in plants: Role of transcriptomics, proteomics, metabolomics and Ionomics. *Frontiers in Plant Science*, *6*, 1143. PMID:26904030

Sirisha, V. L., Jain, A., & Jain, A. (2016). Enzyme Immobilization. Advances in Food and Nutrition Research, 79, 179–211. doi:10.1016/bs.afnr.2016.07.004

Sirotkina, M., Lyagin, I., & Efremenko, E. (2012). Hydrolysis of organophosphorus pesticides in soil: New opportunities with ecocompatible immobilized His6-OPH. *International Biodeterioration & Biodegradation*, *68*, 18–23. doi:10.1016/j.ibiod.2011.12.004

Skjelkvale, B. L. (1994). Factors influencing fluoride concentrations in Norwegian Lakes. *Water, Air, and Soil Pollution*, *77*, 151–167.

Slavek, J., Farrah, H., & Pickering, W. F. (1996). Interaction of clays with dilute fluoride solutions. *Water, Air, and Soil Pollution*, *23*(2), 209–220. doi:10.1007/BF00206977

Slobodkina, G. B., Bonch-Osmolovskaia, E. A., & Slobodkin, A. I. (2007). Reduction of chromate, selenite, tellurite, and iron (III) by the moderately thermophilic bacterium *Bacillus thermoamylovorans* SKC1. [in Russian]. *Mikrobiologiia*, *76*(5), 602–607. PMID:18069319

Smiri, M., Chaoui, A., Rouhier, N., Gelhaye, E., Jacquot, J. P., & El Ferjani, E. (2010). *RETRACTED: Effect of cadmium on resumption of respiration in cotyledons of germinating pea seeds*. Academic Press.

Sobrino-Plata, J., Ortega-Villasante, C., Laura Flores-Cáceres, M., Escobar, C., Del Campo, F. F., & Hernández, L. E. (2009). Differential alteration of antioxidant defences as bioindicators of mercury and Cd toxicity in alfa alfa. *Chemosphere*, *77*(7), 946–954. doi:10.1016/j.chemosphere.2009.08.007 PMID:19732935

Song, H. P., Li, X. G., Sun, J. S., Xu, S. M., & Han, X. (2008). Application of magnetotactic bacterium, *Strenotrophomonas* sp. to the removal of Au (III) from contaminated wastewater with a magnetic separator. *Chemosphere*, *72*(4), 616–621. doi:10.1016/j.chemosphere.2008.02.064 PMID:18439649

Spivak, A., Dzenis, Y., & Reneker, D. (2000). A model of steady state jet in the electrospinning process. *Mechanics Research Communications*, *27*(1), 37–42. doi:10.1016/S0093-6413(00)00060-4

Srinath, T., Khare, S., & Ramteke, P. W. (2001). Isolation of hexavalent chromium-reducing Cr-tolerant facultative anaerobes from tannery effluent. *The Journal of General and Applied Microbiology*, *47*(6), 307–312. doi:10.2323/jgam.47.307 PMID:12483605

Srivastava,, R., Khan,, R., & Manzoor,, N., & Mahmooduzzafar. (2011). Responses of cadmium exposures on growth, physio-biochemical characteristics and the antioxidative defence system of soybean (*Glycine max* L.). *Journal of Phytology*, *3*(10), 20–25.

Srivastava, P., Abbassi, R., Yadav, A. K., Garaniya, V., Kumar, N., Khan, S. J., & Lewis, T. (2020). Enhanced chromium (VI) treatment in electroactive constructed wetlands: Influence of conductive material. *Journal of Hazardous Materials*, *387*, 121722. doi:10.1016/j.jhazmat.2019.121722 PMID:31806439

Srivastava, S., & Thakur, I. S. (2006). Isolation and process parameter optimization of *Aspergillus* sp. for removal of chromium from tannery effluent. *Bioresource Technology, 97*(10), 1167–1173. doi:10.1016/j.biortech.2005.05.012 PMID:16023341

Staniek, H., & Wojciak, R. W. (2018). The Combined Effects of Iron Excess in the Diet and Chromium(III) Supplementation on the Iron and Chromium Status in Female Rats. *Biological Trace Element Research, 184*(2), 398–408. doi:10.100712011-017-1203-z PMID:29164513

Statyukha, G., Kvitka, O., Dzhygyrey, I., & Jezowski, J. (2008). A simple sequential approach for designing industrial wastewater treatment networks. *Journal of Cleaner Production, 16*(2), 215–224. doi:10.1016/j.jclepro.2006.09.002

Stevens, D. P., McLaughlin, M. J., & Alston, A. M. (1997). Phyto-toxicity of aluminium–fluoride complexes and their uptake from solution culture by *Avena sativa* and *Lycopersicon esculentum. Plant and Soil, 192*(1), 81–93. doi:10.1023/A:1004224526067

Strong, P. J., & Claus, H. (2011). Laccase: A Review of Its Past and Its Future in Bioremediation. *Critical Reviews in Environmental Science and Technology, 41*(4), 373–434. doi:10.1080/10643380902945706

Subrahmanyam, D. (2008). Effect of Cr toxicity on leaf photosynthetic characteristics and oxidative changes in wheat. *Photosynthetica, 46*(3), 339–340. doi:10.100711099-008-0062-4

Subramanian, S., & Seeram, R. (2013). New directions in nanofiltration applications—Are nanofibers the right mate-rials as membranes in desalination. *Desalination, 308*, 198–208. doi:10.1016/j.desal.2012.08.014

Sukigara, S., Gandhi, M., Ayutsede, J., Micklus, M., & Ko, F. (2003). Regeneration of Bombyx mori silk by electrospinning—part 1: Processing parameters and geometric properties. *Polymer, 44*(19), 5721–5727. doi:10.1016/S0032-3861(03)00532-9

Sultan, S., & Hasnain, S. (2007). Reduction of toxic hexavalent chromium by *Ochrobactrum intermedium* strain SDCr-5 stimulated by heavy metals. *Bioresource Technology, 98*(2), 340–344. doi:10.1016/j.biortech.2005.12.025 PMID:16488604

Sultan, S., Mubashar, K., & Faisal, M. (2012). Uptake of toxic Cr (VI) by biomass of exopolysaccharides producing bacterial strains. *African Journal of Microbiological Research, 6*, 3329–3336.

Sunqing, Q., Junxiu, D., & Guoxu, C. (2000). Wear and Friction Behaviour of CaCO3Nanoparticles used as Additives in Lubricating Oils. *Lubrication Science, 12*(2), 205–212. doi:10.1002/ls.3010120207

Sun, R., Jin, C., & Zhou, Q. (2010). Characteristics of cadmium accumulation and tolerance in *Rorippa globosa* (Turcz.) Thell., a species with some characteristics of cadmium hyperaccumulation. *Plant Growth Regulation, 61*(1), 67–74. doi:10.100710725-010-9451-3

Sun, Y. B., Zhou, Q. X., Sun, F. H., & Jin, C. X. (2007). Antioxidative defense and proline/ phytochelatin accumulation in a newly discovered Cd- hyperaccumulator, *Solanum nigrum* L. *Environmental and Experimental Botany, 60*(3), 468–476. doi:10.1016/j.envexpbot.2007.01.004

Suseem, S. R., & Mary Saral, A. (2014). Biosorption of heavy metals using mushroom *Pleurotus eous. Journal of Chemical and Pharmaceutical Research, 6*, 2163–2168.

Susheela, A. K. (2003). *Treatise on fluorosis* (2nd ed.). Fluorosis Research & Rural Development Foundation.

Susheela, A. K., Kumar, A., Bhatnagar, M., & Bahadur, R. (1993). Prevalence of endemic fluorosis with gastrointestinal manifestations in people living in some North-Indian villages. *Fluoride, 26*(2), 97–104.

Sutherland, I. W. (2001). Biofilm exopolysaccharides: A strong and sticky framework. *Microbiology*, *147*(1), 3–9. doi:10.1099/00221287-147-1-3 PMID:11160795

Suzuki, T., Miyata, N., Horitsu, H., Kawai, K., Takamizawa, K., Tai, Y., & Okazaki, M. (1992). NAD(P)H-Dependent Chromium(VI) Reductase of *Pseudomonas ambigua* G-1: A Cr(V) intermediate is formed during the reduction of Cr(VI) to Cr(III). *Journal of Bacteriology*, *174*(16), 5340–5345. doi:10.1128/JB.174.16.5340-5345.1992 PMID:1322884

Szewzyk, U., Szewzyk, R., Manz, W., & Schleifer, K. H. (2000). Microbiological safety of drinking water. *Annual Review of Microbiology*, *54*(1), 81–127. doi:10.1146/annurev.micro.54.1.81 PMID:11018125

Tahir, A., Lateef, Z., Abdel-Mageed, A., Sholkamy, E. N., & Mostafa, A. (2017). In vitro compatibility of fungi for the biosorption of zinc(II) and copper(II) from electroplating effluent. *Current Science*, *112*(04), 839–844. doi:10.18520/cs/v112/i04/839-844

Tai, C. Y., & Chen, P. C. (1995). Nucleation, Agglomeration andCrystal Morphpology of Calcium Carbonate. *AIChE Journal. American Institute of Chemical Engineers*, *41*(1), 68–77. doi:10.1002/aic.690410108

Taiz, L., & Zeiger, E. (2006). *Plant Physiology* (4th ed.). Sinauer Associates.

Takami, H., Kudo, T., & Horikoshi, K. (1997). Isolation of extradiol dioxygenasegenes that is phylogenetically distant from other meta cleavage dioxygenase genes. *Bioscience, Biotechnology, and Biochemistry*, *61*(3), 530–532. doi:10.1271/bbb.61.530 PMID:9095556

Takeuchi, M., Kawahata, H., Gupta, L. P., Kita, N., Morishita, Y., Yoshiro, O., & Komai, T. (2007). Arsenic resistance and removal by marine and non-marine bacteria. *Journal of Biotechnology*, *127*(3), 434–442. doi:10.1016/j.jbiotec.2006.07.018 PMID:16934903

Tang, C. Y., Criddle, Q. S., Fu, C. S., & Leckie, J. O. (2007). Effect of flux (trans membrane pressure) and membranes properties on fouling and rejection of reverse osmosis and nano filtration membranes treating perfluoro octane sulfonate containing waste water. *Journal of Environmental Science and Technology*, *41*(6), 2008–2014. doi:10.1021/es062052f PMID:17410798

Tang, X. Y., Yang, Y., McBride, M. B., Tao, R., Dai, Y. N., & Zhang, X. M. (2019). Removal of chlorpyrifos in recirculating vertical flow constructed wetlands with five wetland plant species. *Chemosphere*, *216*, 195–202. doi:10.1016/j.chemosphere.2018.10.150 PMID:30368084

Tanhan, P., Kruatrachu, M., & Pokethitiyook, P. (2007). Uptake and accumulation of Cadmium, Lead and Zinc by Siam weed [Chromolaena odoruta King and Robinson]. *Chemosphere*, *68*(2), 323–329. doi:10.1016/j.chemosphere.2006.12.064 PMID:17280700

Tantrey, M. S., & Agnihotri, R. K. (2010). Chlorophyll and proline content of gram (*Cicer arietinum* L.) under cadmium and mercury treatments. *Research Journal of Agricultural Science*, *1*(2), 119–122.

Tapia, J. M., Munoz, J. A., Gonzalez, F., Blazquez, M. L., & Ballester, A. (2011). Mechanism of adsorption of ferric iron by extracellular polymeric substances (EPS) from a bacterium *Acidiphilium* sp. *Water Science and Technology*, *64*(8), 1716–1722. doi:10.2166/wst.2011.649 PMID:22335116

Tapia, J. M., Munoz, J., Gonzalez, F., Blazquez, M. L., & Ballester, A. (2013). Sorption of ferrous and ferric iron by extracellular polymeric substances (EPS) from acidophilic bacteria. *Preparative Biochemistry & Biotechnology*, *43*(8), 815–827. doi:10.1080/10826068.2013.805624 PMID:23876140

Tarangini, K., & Gyana, R. (2009, January-March). Optimization of heavy metal biosorption using attenuated cultures of *Bacillus subtilis* and *Pseudomonas aeruginosa*. *Journal of Environmental Research and Development*, *3*(3).

Tastan, B. E., & Donmez, G. (2010). Single and binary biosorption of heavy metal ions and a reactive dye by *Aspergillus versicolor. Communications de la Faculty of Sciences University Ankara Series B*, *56*, 13–25. doi:10.1501/Commub_0000000549

Tastan, B. E., Ertugrul, S., & Dönmez, G. (2010). Effective bioremoval of reactive dye and heavy metals by *Aspergillus versicolor. Bioresource Technology*, *10*(3), 870–876. doi:10.1016/j.biortech.2009.08.099 PMID:19773159

Tausz, M., Wonisch, A., Grill, D., Morales, D., & Jimenez, M. S. (2003). Measuring antioxidants in tree species in the natural environment: From sampling to data evaluation. *Journal of Experimental Botany*, *54*(387), 1505–1510. doi:10.1093/jxb/erg175 PMID:12759368

Tawabini, B. S. (2015). Removal of methyl tertiary butyl ether (MTBE) from contaminated water using UV-assisted nano composite materials. *Desalination and Water Treatment*, *55*(2), 549–554. doi:10.1080/19443994.2014.924035

Tchounwou, P. B., Yedjou, C. G., Patlolla, A. K., & Sutton, D. J. (2012). Heavy Metals Toxicity and the Environment. *EXS*, *101*, 133–164. doi:10.1007/978-3-7643-8340-4_6 PMID:22945569

Ten Have, R., & Teunissen, P. J. M. (2001). Oxidative mechanisms involved in lignin degradation by white-rot fungi. *Chemical Reviews*, *101*(11), 3397–3413. doi:10.1021/cr000115l PMID:11749405

Texier, A. C., Andres, Y., & Le Cloirec, P. (1999). Selective biosorption of lanthanide (La, Eu, Yb) ions by Pseudomonas aeruginosa. *Environmental Science & Technology*, *33*(3), 89–495. doi:10.1021/es9807744

Thacker, U., Parikh, R., Shouche, Y., & Madamwar, D. (2007). Reduction of chromate by cell-free Extract of *Brucella* sp. isolated from Cr(VI) contaminated sites. *Bioresource Technology*, *98*(8), 1541–1547. doi:10.1016/j.biortech.2006.06.011 PMID:16931000

Thakur, D., Ganguly, R., Gupta, A. K., & Ghali, V. (2020). Evaluation of Existing Solid Waste Management System in Una Town, India. In *Sustainable Waste Management: Policies and Case Studies* (pp. 367–381). Springer. doi:10.1007/978-981-13-7071-7_33

Theriappan, P., Gupta, K. A., & Dhasarathan, P. (2011). Accumulation of proline under salinity and heavy metal stress in Cauliflower seedlings. J. *Appl. Sci. Environ. Manage*, *15*(2), 251–255. doi:10.4314/jasem.v15i2.68497

Theron, S., Yarin, A., Zussman, E., & Kroll, E. (2005). Multiple jets in electrospinning: Experiment and modeling. *Polymer*, *46*(9), 2889–2899. doi:10.1016/j.polymer.2005.01.054

Theron, S., Zussman, E., & Yarin, A. (2004). Experimental investigation of the governing parameters in the electrospinning of polymer solutions. *Polymer*, *45*(6), 2017–2030. doi:10.1016/j.polymer.2004.01.024

Tijhuis, L., Brattli, B., & Sæther, O. M. (2002). A geochemical survey of topsoil in the city of Oslo, Norway. *Environmental Geochemistry and Health*, *24*(1), 67–94. doi:10.1023/A:1013979700212

Tiwari, K. K., Dwivedi, S., Singh, N. K., Rai, U. N., & Tripathi, R. D. (2009). Chromium (VI) induced phytotoxicity and oxidative stress in pea (*Pisum sativum* L.): Biochemical changes and translocation of essential nutrients. *Journal of Environmental Biology*, *30*, 389–394. PMID:20120464

Tofighy, M. A., & Mohammadi, T. J. J. o. h. m. (2011). *Adsorption of divalent heavy metal ions from water using carbon nanotube sheets*. Academic Press.

Torres, E., Bustos-Jaimes, I., & Le Borgne, S. (2003). Potential use of oxidative enzymes for the detoxification of organic pollutants. *Applied Catalysis B: Environmental*, *46*(1), 1–15. doi:10.1016/S0926-3373(03)00228-5

Tosco, T., Papini, M. P., Viggi, C. C., & Sethi, R. J. J. o. c. p. (2014). *Nanoscale zerovalent iron particles for groundwater remediation: a review*. Academic Press.

Townsend, T. G., Powell, J., Jain, P., Xu, Q., Tolaymat, T., & Reinhart, D. (2015). Leachate Control, Storage, and Treatment. In *Sustainable Practices for Landfill Design and Operation* (pp. 243–266). Springer. doi:10.1007/978-1-4939-2662-6_11

Treesubsuntorn, C., Dhurakit, P., Khaksar, G., & Thiravetyan, P. (2018). Effect of microorganisms on reducing cadmium uptake and toxicity in rice (Oryza sativa L.). *Environmental Science and Pollution Research International, 25*(26), 25690–25701. doi:10.100711356-017-9058-6 PMID:28480489

Tripathi, S., Sanjeevi, R., Anuradha, J., Chauhan, D. S., & Rathoure, A. K. (2018). Nano-Bioremediation: Nanotechnology and Bioremediation. In Biostimulation Remediation Technologies for Groundwater Contaminants (pp. 202-219). IGI Global.

Tripathi, M., Munot, H. P., Shouche, Y., Meyer, J. M., & Goel, R. (2005). Isolation and functional characterization of siderophore-producing lead-and cadmium-resistant Pseudomonas putida KNP9. *Current Microbiology, 50*(5), 233–237. doi:10.100700284-004-4459-4 PMID:15886913

Trotta, A., Falaschi, P., Cornara, L., Minganti, V., Fusconi, A., Drava, G., & Berta, G. (2006). Arbuscular mycorrhizae increase the arsenic translocation factor in the As hyperaccumulating fern *Pteris vitata* L. *Chemosphere, 65*(1), 74–81. doi:10.1016/j.chemosphere.2006.02.048 PMID:16603227

Tsekova, K., Christova, D., & Ianis, M. (2006). Heavy metal biosorption sites in *Penicillium cyclopium. Journal of Applied Science & Environmental Management, 10*(3), 117–121. doi:10.4314/jasem.v10i3.17341

Tsezos, M., & Volesky, B. (1981). Biosorption of uranium and thorium. *Biotechnology and Bioengineering, 23*(3), 583–604. doi:10.1002/bit.260230309

Tsiros, T. X., Haidouti, C., & Chronopoulou, A. (1998). Airborne F contamination of soil and olive trees near an aluminium plant: Measurement and simulation. *J Environ Sci Health, 33*(7), 1309–1324. doi:10.1080/10934529809376790

Tsukihara, T., Honda, Y., Sakai, R., Watanabe, T., & Watanabe, T. (2006). Exclusive overproduction of recombinant versatile peroxidase MnP2 by genetically modified white rot fungus, Pleurotus ostreatus. *Journal of Biotechnology, 126*(4), 431–439. doi:10.1016/j.jbiotec.2006.05.013 PMID:16820241

Tucker, N., Stanger, J., Staiger, M., Razzaq, H., & Hofman, K. (2012). The history of the science and technology of electrospinning from 1600 to 1995. *Journal of Engineered Fibers and Fabrics, 7*(2_suppl), 63–73. doi:10.1177/155892501200702S10

Tuomela, M., & Hatakka, A. (2011). Oxidative Fungal Enzymes for Bioremediation. Comprehensive Biotechnology, 183–196. doi:10.1016/B978-0-08-088504-9.00370-6

Tyagi, M., da Fonseca, M. M. R., & de Carvalho, C. C. C. R. (2011). Bioaugmentation and biostimulation strategies to improve the effectiveness of bioremediation processes. *Biodegradation, 22*(2), 231–241. doi:10.100710532-010-9394-4 PMID:20680666

Umrania, V. V. (2006). Bioremediation of toxic heavy metals using acidothermophilic autotrophes. *Bioresource Technology, 97*(10), 1237–1242. doi:10.1016/j.biortech.2005.04.048 PMID:16324838

UNICEF. (1999). States of the art report on the extent of fluoride in drinking water and the resulting endemicity in India. Report by Fluorosis and Rural Development Foundation for UNICEF.

United Nations Environmental Protection/Global Program of Action. (2004). *Why the marine Environment needs Protection from Heavy metals, Heavy metal 2004.* UNEP/GPA Coordination office. http:// www.oceansatlas.org/unatlas/uses/uneptextsph/wasteph/2602gpa

Upadhyay, N., Vishwakarma, K., Singh, J., Mishra, M., Kumar, V., Rani, R., Mishra, R. K., Chauhan, D. K., Tripathi, D. K., & Sharma, S. (2017). Tolerance and reduction of chromium(VI) by *Bacillus* sp. MNU16 isolated from contaminated coal mining soil. *Frontiers in Plant Sciences*, *8*, 778. doi:10.3389/fpls.2017.00778 PMID:28588589

Uraguchi, S., Watanabu, I., Kiyono, M., & Kuno, K. (2006). Characteristics of Cadmium accumulation and tolerance in novel Cd- accumulating crops, *Avena strigosa* and *Crotalaria juncea*. *Exp. Botany*, *57*(12), 2955–2965. doi:10.1093/jxb/erl056 PMID:16873452

US Food and Drug Administration. (1999). Anticaries drug products for over-the-counter human use. *Code of Federal Regulations, 21*(CFR Part 355), 280-5.

USEPA. (2001). Microwave assisted digestion of sediments.Sludge,soils and oils. Official Methods/USEPA Methods, 3051, 846.

Vahabi, K., Mansoori, G. A., & Karimi, S. J. I. J. (2011). *Biosynthesis of silver nanoparticles by fungus Trichoderma reesei (a route for large-scale production of AgNPs)*. Academic Press.

Vajpayee, P., Rai, U. N., Ali, M. B., Tripathi, R. D., Yadav, V., & Sinha, S. N. (2001). Chromium induced physiological changes in *Vallisneria spiralis* L. and its role in phytoremediation of tannery effluents. *Bulletin of Environmental Contamination and Toxicology*, *67*, 246–265. PMID:11429683

Vandevivere, P., & Kirchman, D. L. (1993). Attachment stimulates exopolysaccharide synthesis by a bacterium. *Applied and Environmental Microbiology*, *59*(10), 3280–3286. doi:10.1128/AEM.59.10.3280-3286.1993 PMID:16349064

Varma, S., Sarode, D., Wakale, S., Bhanvase, B. A., & Deosarkar, M. P. (2013). Removal of nickel from waste water using grapheme nanocomposite. *Int. J. Chem. Phys. Sci.*, *2*, 132–139.

Vassilev, A., & Lidon, F. (2011). Cadmium induced membrane damage and changes in soluble protein and free amino acid contents in young barley plants. *Emer. Journal of Food & Agriculture*, *23*(2), 130–136.

Verbruggen, N., Hermans, C., & Schat, H. (2009). Molecular mechanisms of metal hyperaccumulation in plants. *The New Phytologist*, *181*(4), 759–776. doi:10.1111/j.1469-8137.2008.02748.x PMID:19192189

Verma, S., & Kuila, A. (2019). Bioremediation of heavy metals by microbial process. *Environmental Technology and Innovation*, *14*, 100369. doi:10.1016/j.eti.2019.100369

Verma, S., Saxena, J., Prasanna, R., Sharma, V., & Nain, L. (2012). Medium optimization for a novel crude-oil degrading lipase from Pseudomonas aeruginosa SL-72 using statistical approaches for bioremediation of crude-oil. *Biocatalysis and Agricultural Biotechnology*, *1*(4), 321–329. doi:10.1016/j.bcab.2012.07.002

Verma, S., Verma, P. K., Meher, A. K., Dwivedi, S., Bansiwal, A. K., Pande, V., Srivastava, P. K., Verma, P. C., Tripathi, R. D., & Chakrabarty, D. (2016). A novel arsenic methyltransferase gene of *Westerdykell aaurantiaca* isolated 1 from arsenic contaminated soil: Phylogenetic, physiological, biochemical studies and its role in arsenic bioremediation. *Metallomics*, *8*(3), 344–353. doi:10.1039/C5MT00277J PMID:26776948

Vernay, P., Gauthier-Moussard, C., Jean, L., Bordas, F., Faure, O., Ledoigt, G., & Hitmi, A. (2008). Effect of chromium species on phytochemical and physiological parameters in *Datura innoxia*. *Chemosphere*, *72*(5), 763–771. doi:10.1016/j.chemosphere.2008.03.018 PMID:18472140

Vijay Kumar, E., Srijana, M., Kiran Kumar, K., Harikrishna, N., & Reddy, G. (2011). A novel serine alkaline protease from *Bacillus altitudinis* GVC11 and its application as a dehairing agent. *Bioprocess and Biosystems Engineering*, *34*(4), 403–409. doi:10.100700449-010-0483-x PMID:21080200

Vinay, S.P., Udayabhanu Nagaraju, G., Chandrappa, C.P., & Chandrasekhar, N. (2019). Novel Gomutra (Cow urine) mediated synthesis of silver oxide nanoparticles and their enhanced photocatalytic, photoluminescence and antibacterial studies. *Journal of Science: Advanced Materials and Devices*, *4*, 392-399.

Visa, M. (2016). Synthesis and characterization of new zeolite materials obtained from fly ash for heavy metals removal in advanced wastewater treatment. *Powder Technology*, *294*, 338–347. doi:10.1016/j.powtec.2016.02.019

Vishnoi, S. R., & Srivastava, P. N. (2007). Phytoremediation–green for environmental clean. In *Proceedings of Taal2007: the 12th World lake conference* (Vol. 1016, p. 1021). Academic Press.

Vivas, A., Biro, B., Ruiz-Lozano, J. M., Barea, J. M., & Azcon, R. (2006). Two bacterial strains isolated from a Zn-polluted soil enhance plant growth and mycorrhizal efficiency under Zn-toxicity. *Chemosphere*, *62*(9), 1523–1533. doi:10.1016/j.chemosphere.2005.06.053 PMID:16098559

Vogel-Mikus, K., Drobne, D., & Regvar, M. (2005). Zn, Cd and Pb accumulation and arbuscular mycorrhizal colonisation of pennycress Thlaspi praecox Wulf. (Brassicaceae) from the vicinity of a lead mine and smelter in Slovenia. *Environmental Pollution*, *133*(2), 233–242. doi:10.1016/j.envpol.2004.06.021 PMID:15519454

Volesky, B. (1990). *Biosorption of heavy metals*. CRC Press.

Volesky, B. (1990). *Biosorption of Heavy Metals*. CRC Press.

Volesky, B., & Holan, Z. R. (1995). Biosorption of heavy metals. *Biotechnology Progress*, *11*(3), 235–250. doi:10.1021/bp00033a001 PMID:7619394

Vormann, J. (2003). Magnesium: Nutrition and metabolism. *Molecular Aspects of Medicine*, *24*(1-3), 27–37. doi:10.1016/S0098-2997(02)00089-4 PMID:12537987

Vos, C. M., Tesfahun, A. N., Panis, B., De Waele, D., & Elsen, A. (2012). Arbuscular mycorrhizal fungi induce systemic resistance in tomato against the sedentary nematode Meloidogyne incognita and the migratory nematode Pratylenchus penetrans. *Applied Soil Ecology*, *61*, 1–6. doi:10.1016/j.apsoil.2012.04.007

Vucak, M., Pons, M. N., Peric, J., & Vivier, H. (1998). Effect of Precipitation Conditions on the Morphology of CalciumCarbonate: Quantification of Crystal Shapes using ImageAnalysis. *Powder Technology*, *97*(1), 1–5. doi:10.1016/S0032-5910(97)03375-5

Vymazal, J. (2005). Horizontal sub-surface flow and hybrid constructed wetlands systems for wastewater treatment. *Ecological Engineering*, *25*(5), 478–490. doi:10.1016/j.ecoleng.2005.07.010

Vymazal, J. (2007). Removal of nutrients in various types of constructed wetlands. *The Science of the Total Environment*, *380*(1-3), 48–65. doi:10.1016/j.scitotenv.2006.09.014 PMID:17078997

Vymazal, J. (2010). Constructed wetlands for wastewater treatment. *Water (Basel)*, *2*(3), 530–549. doi:10.3390/w2030530 PMID:20795704

Vymazal, J. (2014). Constructed wetlands for treatment of industrial wastewaters: A review. *Ecological Engineering*, *73*, 724–751. doi:10.1016/j.ecoleng.2014.09.034

Vymazal, J., Brezinova, T. D., Kozeluh, M., & Kule, L. (2017). Occurrence and removal of pharmaceuticals in four full-scale constructed wetlands in the Czech Republic–the first year of monitoring. *Ecological Engineering*, *98*, 354–364. doi:10.1016/j.ecoleng.2016.08.010

Wagner, G. J. (1993). Accumulation of cadmium in crop plants and its consequences to human health. *Advances in Agronomy*, *51*, 173–212. doi:10.1016/S0065-2113(08)60593-3

Wahlbeck, P. G. (1992). TGA Experimental Data – Analysis Providedby Transpiration Theory. *Thermochimica Acta*, *197*(2), 469–478. doi:10.1016/0040-6031(92)85048-Z

Walch, M. (1992). *Microbial Corrosion.* In J. Lederberg (Ed.), *Encyclopedia of Microbiology* (Vol. 1, pp. 585–591). Academic Press.

Wanekaya, A. K. J. A. (2011). *Applications of nanoscale carbon-based materials in heavy metal sensing and detection.* Academic Press.

Wang, Tay, Tay, & Hung. (n.d.). *Environmental bio engineering*. HUMANA Press.

Wang, G., & Fowler, B. A. (2008). Roles of biomarkers in evaluating interactions amoug mixtures of lead, cadmium and arsenic. *Toxicology and Applied Pharmacology*, *233*(1), 92–99. doi:10.1016/j.taap.2008.01.017 PMID:18325558

Wang, H., Yuan, X., Wu, Y., Zeng, G., Chen, X., Leng, L., & Li, H. (2015). Facile synthesis of amino-functionalized titanium metal-organic frameworks and their superior visible-light photocatalytic activity for Cr (VI) reduction. *Journal of Hazardous Materials*, *286*, 187–194. doi:10.1016/j.jhazmat.2014.11.039 PMID:25585267

Wang, J. L., & Chen, C. (2006). Biosorption of heavy metals by *Saccharomyces cerevisiae*: A review. *Biotechnology Advances*, *24*(5), 427–451. doi:10.1016/j.biotechadv.2006.03.001 PMID:16737792

Wang, J. L., Han, Y. J., & Qian, Y. (2000). Progress in metal biosorption by microorganisms. *Journal of Microbiology (Seoul, Korea)*, *27*, 449–452.

Wang, J., Weihua, L., Zhang, C., & Shisheng, K. (2010). Physiological responses and detoxific mechanisms to Pb, Zn, Cu and Cd in young seedlings of *Paulownia fortunei. Journal of Environmental Sciences (China)*, *22*(2), 1916–1922. doi:10.1016/S1001-0742(09)60339-9 PMID:21462710

Wang, K., Li, L., Tan, F., & Wu, D. (2018). Treatment of landfill leachate using activated sludge technology: A review. *Archaea (Vancouver, B.C.)*, *2018*, 1039453. Advance online publication. doi:10.1155/2018/1039453 PMID:30254508

Wang, L. F., & Huang, J. Z. (1995). Outline of control practice of endemic fluorosis in China. *Social Science & Medicine*, *41*(8), 1191–1195. doi:10.1016/0277-9536(94)00429-W PMID:8578341

Wang, P. C., Mori, T., Komori, K., Sasatsu, M., Toda, K., & Ohtake, H. (1989). Isolation and characterization of an *Enterobacter cloacae* strain that reduces hexavalent chromium under anaerobic conditions. *Applied and Environmental Microbiology*, *55*(7), 1665–1669. doi:10.1128/AEM.55.7.1665-1669.1989 PMID:16347962

Wang, P. C., Toda, K., Ohtake, H., Kusaka, I., & Yale, I. (1991). Membrane bound respiratory system of *Enterobacter cloacae* strain HO1 grown anaerobically with chromate. *FEMS Microbiology Letters*, *76*(1), 11–16. doi:10.1111/j.1574-6968.1991.tb04408.x PMID:1851711

Wang, W., Bu, W., Wang, L., Palo, P. E., Mallapragada, S., Nilsen-Hamilton, M., & Vaknin, D. (2012). Interfacial Properties and Iron Binding to Bacterial Proteins That Promote the Growth of Magnetite Nanocrystals: X-ray Reflectivity and Surface Spectroscopy Studies. *Langmuir*, *28*(9), 4274–4282. doi:10.1021/la205074n PMID:22316331

Wang, Y. T., & Shen, H. (1995). Bacterial reduction of hexavalent chromium. *Journal of Industrial Microbiology*, *14*(2), 159–163. doi:10.1007/BF01569898 PMID:7766208

Wani, P. A., & Khan, M. S. (2010). Bacillus species enhance growth parameters of chickpea (Cicer arietinum L.) in chromium stressed soils. *Food and Chemical Toxicology*, *48*(11), 3262–3267. doi:10.1016/j.fct.2010.08.035 PMID:20813149

Wani, P. A., Khan, M. S., & Zaidi, A. (2008). Chromium-reducing and plant growth-promoting Mesorhizobium improves chickpea growth in chromium-amended soil. *Biotechnology Letters*, *30*(1), 159–163. doi:10.100710529-007-9515-2 PMID:17849087

Wani, R., Kodam, K. M., Gawai, K. R., & Dhakephalkar, P. K. (2007). Chromate reduction by *Burkholderia cepacia* MCMB-821 isolated from the pristine habitat of alkaline crater lake. *Applied Microbiology and Biotechnology*, *75*(3), 627–632. doi:10.100700253-007-0862-7 PMID:17361433

Warren, C., Burgess, W. G., & Garcia, M. G. (2005). Hydro-chemical associations and depth profiles of arsenic and fluoride in quaternary loess aquifers of Northern Argentina. *Mining Magazine (London)*, *69*(5), 877–886. doi:10.1180/0026461056950295

Wei, S. H., Mahuli, S. K., Agnihotri, R., & Fan, L. S. (1997). HighSurface Area Calcium Carbonate: Pore Structurual Propertiesand Sulfation Characteristics. *Industrial & Engineering Chemistry Research*, *36*(6), 2141–2148. doi:10.1021/ie9607681

Wei, S. H., Zhou, Q. X., & Wang, X. (2004). A newly discovered Cd- hyperaccumulator *Solanum nigrum* L. *Chinese Science Bulletin*, *49*(24), 2568–2573.

Wei, S., Zhou, Q., & Saha, U. K. (2008). Hyperaccumulative characteristic of weed species to heavy metals. *Water, Air, and Soil Pollution*, *192*(1-4), 173–181. doi:10.100711270-008-9644-9

Wei, W., Wang, Q., Li, A., Yang, J., Ma, F., Pi, S., & Wu, D. (2016). Biosorption of Pb (II) from aqueous solution by extracellular polymeric substances extracted from *Klebsiella* sp. J1: Adsorption behavior and mechanism assessment. *Scientific Reports*, *6*(1), 31575. doi:10.1038rep31575 PMID:27514493

Welander, U., & Henrysson, T. (1998). Physical and chemical treatment of a nitrified leachate from a municipal landfill. *Environmental Technology*, *19*(6), 591–599. doi:10.1080/09593331908616715

Wenzel, W. W., & Blum, W. E. H. (2002). Fluorine speciation and mobility in contaminated soils. *Soil Science*, *153*(5), 357–364. doi:10.1097/00010694-199205000-00003

Whiteway, J., Koziarz, P., Veall, J., Sandhu, N., Kumar, P., Hoecher, B., & Lambert, I. B. (1998). Oxygen-insensitive nitroreductases: Analysis of the roles of *nfsA* and *nfsB* in development of resistance to 5-nitrofuran derivatives in *Escherichia coli*. *Journal of Bacteriology*, *180*(21), 5529–5539. doi:10.1128/JB.180.21.5529-5539.1998 PMID:9791100

Whittemore, D. O., MacFarlane, P. A., Doveton, J. H., Butler, J. J., Chu, T., Bassler, R., Smith, M., Mitchell, J., & Wade, A. (2013). The Dakota aquifer program annual report, FY92. Kansas. *Geological Survey Open-File Report (United States)*, 93–1.

WHO. (1996). *Guidelines for drinking water quality. Recommendations* (Vol. 1). World Health Organisation.

WHO. (1996). *World Health Organization & International Programme on Chemical Safety. Guidelines for drinking-water quality. Health criteria and other supporting information* (2nd ed.). Geneva: World Health Organization. https://www.who.int/iris/handle/10665/38551

WHO. (2002). *Fluorides: Environmental health criteria 227*. Geneva: World Health Organization. Available from: http://www.inchem.org/documents/ehc/ehc/ehc227.htm#1.5

WHO. (2004). *Guidelines for drinking water quality* (3rd ed.). World Health Organisation.

WHO. (2005). *Water-related diseases – fluorosis, the disease and how it affects people*. World Health Organization.

Wielinga, B., Mizuba, M. M., Hansel, C. M., & Fendorf, S. (2001). Iron promoted reduction of chromate by dissimilatory iron-reducing bacteria. *Environmental Science & Technology*, *35*(3), 522–527. doi:10.1021/es001457b PMID:11351723

Wildung, R. E., Gorby, Y. A., Krupka, K. M., Hess, N. J., Li, S. W., Plymale, A. E., McKinley, J. P., & Fredrickson, J. K. (2000). Effect of electron donor and solution chemistry on products of dissimilatory reduction of technetium by *Shewanella putrefaciens*. *Applied and Environmental Microbiology*, *66*(6), 2451–2460. doi:10.1128/AEM.66.6.2451-2460.2000 PMID:10831424

Williams, P. P. (1977). Metabolism of synthetic organic pesticides by anaerobic microorganisms. *Residue Reviews*, *66*, 63–135. doi:10.1007/978-1-4612-6352-4_3 PMID:335464

Wimpenny, J., Manz, W., & Szewzyk, U. (2000). Heterogeneity in biofilms. *FEMS Microbiology Reviews*, *24*(5), 661–667. doi:10.1111/j.1574-6976.2000.tb00565.x PMID:11077157

Wiszniowski, J., Robert, D., Surmacz-Gorska, J., Miksch, K., & Weber, J. V. (2006). Landfill leachate treatment methods: A review. *Environmental Chemistry Letters*, *4*(1), 51–61. doi:10.100710311-005-0016-z

Wojciechowska, E. (2017). Potential and limits of landfill leachate treatment in a multi-stage subsurface flow constructed wetland–evaluation of organics and nitrogen removal. *Bioresource Technology*, *236*, 146–154.

Wong, S. C., Li, X. D., Zhang, G., Qi, S. H., & Min, Y. S. (2002). Heavy metals in agricultural soils of the Pearl River Delta, South China. *Environmental Pollution*, *119*(1), 33–44. doi:10.1016/S0269-7491(01)00325-6 PMID:12125727

Wray, J. L., & Farrington, D. (1957). Precipitation of Calcite and Aragonite. *Journal of the American Chemical Society*, *79*(9), 2031–2034. doi:10.1021/ja01566a001

Wu, Y., Pang, H., Liu, Y., Wang, X., Yu, S., Fu, D., . . . Wang, X. J. E. p. (2019). *Environmental remediation of heavy metal ions by novel-nanomaterials: a review*. Academic Press.

Wu, C. H., Wood, T. K., Mulchandani, A., & Chen, W. (2006). Engineering plant-microbe symbiosis for rhizoremediation of heavy metals. *Applied and Environmental Microbiology*, *72*(2), 1129–1134. doi:10.1128/AEM.72.2.1129-1134.2006 PMID:16461658

Wu, W., He, Q., & Jiang, C. (2008). Magnetic Iron Oxide Nanoparticles: Synthesis and Surface Functionalization Strategies. *Nanoscale Research Letters*, *3*(11), 397–415. doi:10.100711671-008-9174-9 PMID:21749733

WWAP (United Nations World Water Assessment Programme). (2015). *Water for a Sustainable World. The United Nations World Water Development Report*. UNESCO.

Xia, L., Xu, X., Zhu, W., Huang, Q., & Chen, W. (2015). A comparative study on the biosorption of Cd^{2+} onto *Paecilomyces lilacinus* XLA and *Mucoromycote* sp. XLC. *International Journal of Molecular Sciences*, *16*(7), 15670–15687. doi:10.3390/ijms160715670 PMID:26184169

Xiang, L., Xiang, Y., Wen, Y., & Wei, F. (2004). Formation of CaCO3Nano-Particles in the Presence of Terpineol. *Materials Letters*, *58*(6), 959–965. doi:10.1016/j.matlet.2003.07.034

Xintaras, C. (1992). *Analysis Paper: Impact of Lead-Contaminated Soil on Public Health*. U.S. Department of Health and Human Services. Public Health Service Agency for Toxic Substances and Disease Registry.

Xue, G., He, S. B., & Wang, X. Z. (2006). Factors affecting biological removal of iron and manganese in groundwater. *Huan Jing Ke Xue*, *27*(1), 95–100. PMID:16599128

Xu, F. (1996). Catalysis of novel enzymatic iodide oxidation by fungal laccase. *Applied Biochemistry and Biotechnology*, *59*(3), 221–230. doi:10.1007/BF02783566

Xu, P., Liu, L., Zeng, G., Huang, D., Lai, C., Zhao, M., Huang, C., Li, N., Wei, Z., Wu, H., Zhang, C., Lai, M., & He, Y. (2014). Heavy metal-induced glutathione accumulation and its role in heavy metal detoxification in Phanerochaete chrysosporium. *Applied Microbiology and Biotechnology*, *98*(14), 6409–6418. doi:10.100700253-014-5667-x PMID:24723291

Xu, P., Zeng, G. M., Huang, D. L., Feng, C. L., Hu, S., Zhao, M. H., Lai, C., Wei, Z., Huang, C., Xie, G. X., & Liu, Z. F. (2012). Use of iron oxide nanomaterials in wastewater treatment: A review. *The Science of the Total Environment*, *424*, 1–10. doi:10.1016/j.scitotenv.2012.02.023 PMID:22391097

Xu, X., & Mills, G. L. (2018). Do constructed wetlands remove metals or increase metal bioavailability? *Journal of Environmental Management*, *218*, 245–255. doi:10.1016/j.jenvman.2018.04.014 PMID:29680756

Yadav, K., Singh, J., Gupta, N., & Kumar, V. J. J. M. E. S. (2017). *A review of nanobioremediation technologies for environmental cleanup: a novel biological approach*. Academic Press.

Yalcuk, A., & Ugurlu, A. (2009). Comparison of horizontal and vertical constructed wetland systems for landfill leachate treatment. *Bioresource Technology*, *100*(9), 2521–2526. doi:10.1016/j.biortech.2008.11.029 PMID:19157867

Yanez, L., Ortiz, D., Calderon, J., Batres, L., Carrizales, L., Mejia, J., Martínez, L., García-Nieto, E., & Díaz-Barriga, F. (2002). Overview of human health and chemical mixtures: Problems facing developing countries. *Environmental Health Perspectives*, *110*(s6, Suppl. 6), 901–909. doi:10.1289/ehp.02110s6901 PMID:12634117

Yang, H., Li, D., Zhang, J., Hao, R., & Li, B. (2004). Design of biological filter for iron and manganese removal from water. *Journal of Environmental Science and Health A Toxic Hazard Substance Environ Eng, 39*(6), 1447-1454.

Yang, Y., Dong, H., Wang, Y., Wang, Y., Liu, N., Wang, D., & Zhang, X. J. I. C. C. (2017). *A facile synthesis for porous CuO/Cu2O composites derived from MOFs and their superior catalytic performance for CO oxidation*. Academic Press.

Yan, G., & Viraraghavan, T. (2000). Effect of pretreatment on the biosorption of heavy metals on *Mucor rouxii. Water S.A.*, *26*, 119–123.

Yang, Q., Agouri, S., Tyrrell, J., & Walsh, T. (2018). Heavy metal resistance genes are associated with blaNDM-1 and blaCTX-M15 carrying enterobecteriaceae. *Antimicrobial Agents and Chemotherapy*, *62*(5), e02614–e02642. doi:10.1128/AAC.02642-17 PMID:29507071

Yang, R. T. (2003). *Adsorbents: fundamentals and applications*. John Wiley & Sons. doi:10.1002/047144409X

Yang, X. E., Long, X. X., Ye, H. B., He, Z. L., Calvert, D. V., & Stoffella, P. J. (2004). Cadmium tolerance and hyperaccumulation in a new zinc hyperaccumulating plant species (*Sedum alfredii Hance*). *Plant and Soil*, *259*(1-2), 181–189. doi:10.1023/B:PLSO.0000020956.24027.f2

Yannarelli, G. G., Fernandez-Alvarez, A. J., Santa-Cruz, D. M., & Tomaro, M. L. (2006). Glutathione reductase activity and isomorph in leaves and roots of wheat plants subjected to cadmium stress. *Phytochemistry*, *68*(4), 505–512. doi:10.1016/j.phytochem.2006.11.016 PMID:17174990

Ybarra, G. R., & Webb, R. (1999). Effects of divalent metal cations and resistance mechanisms of the cyanobacterium *Synechococcus* sp. strain PCC 7942. *Journal of Hazardous Substance Research*, *2*(1), 1–9. doi:10.4148/1090-7025.1011

Yen, Y., Cheng, B., Chan, C., Lin, C., & Chen, H. (2018). Heavy Metal Components in Blood and Urinary Stones of Urolithiasis Patients. *Biological Trace Element Research*, *185*(2), 266–274. doi:10.100712011-018-1253-x PMID:29488068

Yildiz, M., & Terzi, H. (2012). Effect of hexavalent chromium stress in roots of Cr- tolerant and Cr- sensitive barley cultivars. *Acta Agronomica Hungarica*, *60*(1), 29–36. doi:10.1556/AAgr.60.2012.1.4

Yim, S.-K., Jung, H.-C., Pan, J.-G., Kang, H.-S., Ahn, T., & Yun, C.-H. (2006). Functional expression of mammalian NADPH–cytochrome P450 oxidoreductase on the cell surface of Escherichia coli. *Protein Expression and Purification, 49*(2), 292–298. doi:10.1016/j.pep.2006.05.013 PMID:16829136

Yin, K., Lv, M., Wang, Q., Wu, Y., Liao, C., Zhang, W., & Chen, L. (2016). Simultaneous bioremediation and biodetection of mercury ion through surface display of carboxylesterase E2 from Pseudomonas aeruginosa PA1. *Water Research, 103*, 383–390. doi:10.1016/j.watres.2016.07.053 PMID:27486950

Yin, P., Yu, Q., Jin, B., & Ling, Z. (1999). Biosorption removal of cadmium from aqueous solution by using pretreated fungal biomass cultured from starch waste water. *Water Research, 33*(8), 1960–1963. doi:10.1016/S0043-1354(98)00400-X

Yin, X. X., Wang, L. H., Bai, R., Huang, H., & Sun, G. X. (2012). Accumulation and transformation of arsenic in the blue-green alga Synechocysis sp. PCC6803. *Water, Air, and Soil Pollution, 223*(3), 1183–1190. doi:10.100711270-011-0936-0

Yoshida, S., Chatani, A., Honda, Y., Watanabe, T., & Kuwahara, M. (1998). Reaction of manganese peroxidase of Bjerkandera adusta with synthetic lignin in acetone solution. *Journal of Wood Science, 44*(6), 486–490. doi:10.1007/BF00833415

Young, R. A. (2005). *Toxicity profiles: toxicity summary for cadmium, risk assessment information system. RAIS*. University of Tennessee.

Yuan, X., Huang, H., Zeng, G., Li, H., Wang, J., Zhou, C., Zhu, H., Pei, X., Liu, Z., & Liu, Z. (2011). Total concentrations and chemical speciation of heavy metals in liquefaction residues of sewage sludge. *Bioresource Technology, 102*(5), 4104–4110. doi:10.1016/j.biortech.2010.12.055 PMID:21211964

Yuebing, S., Qising, Z., Shuhe, W., & Liping, R. (2007). Growth responses of newly discovered Cd- hyperaccumulator *Rorippa globosa* and its accumulation characteristics of cadmium and arsenic under joint stress of cadmium and arsenic. *Environmental Sciences, 1*(1), 107–113.

Yu, G. H., He, P. J., & Shao, L. M. (2009). Characteristics of extracellular polymeric substances (EPS) fractions from excess sludges and their effects on bioflocculability. *Bioresource Technology, 100*(13), 3193–3198. doi:10.1016/j.biortech.2009.02.009 PMID:19269815

Yu, J., Lei, M., Cheng, B., & Zhao, X. (2004). Effects of PAA Additiveand Temperature on Morphology of Calcium CarbonateParticles. *Journal of Solid State Chemistry, 177*(3), 681–689. doi:10.1016/j.jssc.2003.08.017

Zaidi, S., Usmani, S., Singh, B. R., & Musarrat, J. (2006). Significance of Bacillus subtilis strain SJ-101 as a bioinoculant for concurrent plant growth promotion and nickel accumulation in Brassica juncea. *Chemosphere, 64*(6), 991–997. doi:10.1016/j.chemosphere.2005.12.057 PMID:16487570

Zakaria, Z. A., Zakaria, Z., Surif, S., & Ahmad, W. A. (2007). Hexavalent chromium reduction by *Acinetobacter haemolyticus* isolated from heavy metal-contaminated wastewater. *Journal of Hazardous Materials, 146*(1-2), 30–38. doi:10.1016/j.jhazmat.2006.11.052 PMID:17188812

Zayed, A. M., & Terry, N.Agency for Toxic Substances and Disease Registry. (2003). Chromium in the environment: Factors affecting biological remediation. *Plant and Soil, 249*(1), 139–156. doi:10.1023/A:1022504826342

Zdarta, J., Meyer, A. S., Jesionowski, T., & Pinelo, M. (2018). Developments in support materials for immobilization of oxidoreductases: A comprehensive review. *Advances in Colloid and Interface Science, 258*, 1–20. doi:10.1016/j.cis.2018.07.004 PMID:30075852

Zemel, M. B., Richards, J., Milstead, A., & Campbell, P. (2005). Effects of calcium and dairy on body composition and weight loss in African- American adults. *Obesity Research, 13*(7), 1213–1225. doi:10.1038/oby.2005.144 PMID:16076991

Zeng, F., Qiu, B., Wu, X., Niu, S., & Zhang, G. (2012). *Glutathione mediated alleviation of Cr toxicity in rice plants*. Researchgate.

Zengin, F. K., & Munzuroglue, O. (2005). Effect of some heavy metals on content of chlorophyll, proline and some antioxidant chemicals in bean (*P. vulgaris* L.) seedlings. *Acta Biologica Cracoviensia, 47*(2), 157–164.

Zhang, D., Wei, S., Kaila, C., Su, X., Wu, J., Karki, A. B., . . . Guo, Z. J. N. (2010). *Carbon-stabilized iron nanoparticles for environmental remediation*. Academic Press.

Zhang, S., Cheng, F., Tao, Z., Gao, F., Chen, J. J. J. o. A., & Compounds. (2006). *Removal of nickel ions from wastewater by Mg (OH) 2/MgO nanostructures embedded in Al2O3 membranes*. Academic Press.

Zhang, Xu, Cao, Chen, Rosen, & Zhao. (2017). Arsenic methylation by a genetically engineered Rhizobium-legume symbiont.Plant. *World Journal of Microbiology & Biotechnology, 16*(3), 317–318.

Zhang, C., & Zhiggang, Y. (2014). Effects of sediment geochemical properties on heavy metal bioavailability. *Environment International, 73*, 270–281. doi:10.1016/j.envint.2014.08.010 PMID:25173943

Zhang, D. Q., Jinadasa, K. B. S. N., Gersberg, R. M., Liu, Y., Ng, W. J., & Tan, S. K. (2014). Application of constructed wetlands for wastewater treatment in developing countries–a review of recent developments (2000–2013). *Journal of Environmental Management, 141*, 116–131. doi:10.1016/j.jenvman.2014.03.015 PMID:24784754

Zhang, F., Ge, Z., Grimaud, J., Hurst, J., & He, Z. (2013). Long-term performance of liter-scale microbial fuel cells treating primary effluent installed in a municipal wastewater treatment facility. *Environmental Science & Technology, 47*(9), 4941–4948. doi:10.1021/es400631r PMID:23517192

Zhang, H., Choi, H. J., & Huang, C. P. (2005). Optimization of Fenton process for the treatment of landfill leachate. *Journal of Hazardous Materials, 125*(1-3), 166–174. doi:10.1016/j.jhazmat.2005.05.025 PMID:16011874

Zhang, H., Zhang, S., He, F., Qin, X., Zhang, X., & Yang, Y. (2016). Characterization of a manganese peroxidase from white-rot fungus Trametes sp.48424 with strong ability of degrading different types of dyes and polycyclic aromatic hydrocarbons. *Journal of Hazardous Materials, 320*, 265–277. doi:10.1016/j.jhazmat.2016.07.065 PMID:27551986

Zhang, S., Wang, R., Zhang, S., Li, G., & Zhang, Y. (2013). Development of phosphorylated silica nanotubes (PSNTs)/ polyvinylidene fluoride (PVDF) composite membranes for wastew-ater treatment. *Chemical Engineering Journal, 230*, 260–271. doi:10.1016/j.cej.2013.06.098

Zhang, W., & Zhang, X. (2007). A forecast analysis on fertilizers consumption worldwide. *Environmental Monitoring and Assessment, 133*(1-3), 427–434. doi:10.100710661-006-9597-7 PMID:17286167

Zhang, X. B., Peng, L. I. U., Yue-suo, Y., & Chen, W. R. (2007). Phytoremediation of urban wastewater by model wetlands with ornamental hydrophytes. *Journal of Environmental Sciences (China), 19*(8), 902–909. doi:10.1016/S1001-0742(07)60150-8 PMID:17966845

Zhang, X. H., Liu, J., Huang, H. T., Chen, Z., & Wang, D. Q. (2007). Chromium accumulation by the hyperaccumulator plant *Leersia hexandra* Swartz. *Chemosphere, 67*(6), 1138–1143. doi:10.1016/j.chemosphere.2006.11.014 PMID:17207838

Zhang, Y. F., He, L. Y., Chen, Z. J., Zhang, W. H., Wang, Q. Y., Qian, M., & Sheng, X. F. (2011). Characterization of lead-resistant and ACC deaminase-producing endophytic bacteria and their potential in promoting lead accumulation of rape. *Journal of Hazardous Materials, 186*(2-3), 1720–1725. doi:10.1016/j.jhazmat.2010.12.069 PMID:21227577

Zhang, Z., Cai, R., Zhang, W., Fu, Y., & Jiao, N. (2017). A Novel Exopolysaccharide with Metal Adsorption Capacity Produced by a Marine Bacterium *Alteromonas* sp. JL2810. *Marine Drugs, 15*(6), 175. doi:10.3390/md15060175 PMID:28604644

Zhelev, T. K., & Bhaw, N. (2000). Combined water–oxygen pinch analysis for better waste water treatment management. *Waste Management (New York, N.Y.)*, *20*(8), 665–670. doi:10.1016/S0956-053X(00)00050-7

Zhong, L. S., Hu, J. S., Liang, H. P., Cao, A. M., Song, W. G., & Wan, L. J. J. A. M. (2006). *Self-Assembled 3D flower-like iron oxide nanostructures and their application in water treatment*. Academic Press.

Zhou, F. L., Gong, R. H., & Porat, I. (2009). Mass production of nanofibre assemblies by electrostatic spinning. *Polymer International*, *58*(4), 331–342. doi:10.1002/pi.2521

Zhuang, J., & Gentry, R. W. (2011). Environmental application and risks of nanotechnology: a balanced view. In *Biotechnology and Nanotechnology Risk Assessment: Minding and Managing the Potential Threats around Us* (pp. 41–67). ACS Publications. doi:10.1021/bk-2011-1079.ch003

Ziyang, L., Youcai, Z., Tao, Y., Yu, S., Huili, C., Nanwen, Z., & Renhua, H. (2009). Natural attenuation and characterization of contaminants composition in landfill leachate under different disposing ages. *The Science of the Total Environment*, *407*(10), 3385–3391. doi:10.1016/j.scitotenv.2009.01.028 PMID:19217144

Zong, X., Kim, K., Fang, D., Ran, S., Hsiao, B. S., & Chu, B. (2002). Structure and process relationship of electrospun bioabsorbable nanofiber membranes. *Polymer*, *43*(16), 4403–4412. doi:10.1016/S0032-3861(02)00275-6

Zoronda, P., Vazquez, S., Esteban, E., Fernandez, M., & Carpena, R. (2002). Cdamium stress in nodulated white lupin: Strategies to avoid toxicity. *Plant Physiology and Biochemistry*, *40*(12), 1003–1009. doi:10.1016/S0981-9428(02)01464-X

Zouboulis, A. L., Matis, K. A., & Hancock, I. C. (1997). Biosorption of metals from dilute aqueous solutions. *Seperation and Purification Methodology*, *26*(2), 255–295. doi:10.1080/03602549708014160

About the Contributors

Satarupa Dey, PhD, is an Assistant Professor in Department of Botany, Shyampur Siddeswari Maavidyalaya, affiliated under University of Calcutta, West Bengal, India. She earned her BSc in Botany from University of Calcutta, West Bengal, India. She holds also an MSc in Botany with Microbiology specialization from the same university. She completed her PhD in the year 2012 and her field of expertise is Bioremediation of Toxic metals. She was selected as a Department of Biotechnology, Government of India research fellow and later worked as visiting scientist in Agricultural and Ecological research Unit of Indian Statistical Institute, Kolkata. She is the recipient of the Woman Scientist (WOS A) fellowship of Department of Science and Technology, Government of India. She has published in several journals of National and International repute.

Biswaranjan Acharya is an academic currently associated with Kalinga Institute of Industrial Technology Deemed to be University along with pursuing Ph.D. in computer application from Veer Surendra Sai University of Technology (VSSUT), Burla, Odisha, India. He has received MCA in 2009 MCA from IGNOU, New Delhi, India and M.Tech in Computer Science and Engineering in the year of 2012 from Biju Pattanaik University of Technology (BPUT), Odisha, India. He is also associated with various educational and research societies like IACSIT, CSI, IAENG, and ISC. He has along with 2 years industry experience as a software engineer, total 9 years experience in both academia of some repute university like Ravenshaw University and software development field. He is currently working on research area multiprocessor scheduling along with different fields like Data Analytics, Computer Vision, Machine Learning and IoT.

* * *

Jayanthi Balakrishnan, PhD, working as Assistant Professor in PG & Research Department of Biotechnology, Mahendra Arts and Science College (Autonomous), Affiliated under Periyar University, Salem, Tamil Nadu, India. She obtained her Undergraduate from Nandha Arts and Science College and Postgraduate from Vivekanandha College of Arts and Science for Women, Namakkal, India. She has been selected for Indian Academies' Summer research Fellowship during her Master'. She secured University 1st Rank and availed DST-INSPIRE fellowship for pursuing doctoral research work. She has research experience and specialized in Microbial Technology, Food Chemistry, food processing technology, traditional foods and medicines, nutritional genomics, Agricultural Biotechnology, Stress Biology and waste water management. Her current research work is concentrated on the nutritional changes during the food processing and packing, identification of metabolites used as drugs, and waste water treatment through traditional ways. For her credit, she has published several research articles in National and International reputed journals and 2 book chapters.

Saroj Deep is working as a PhD Research Scholar in School of Life sciences, Sambalpur University, Odisha.

Suchhanda Ghosh, studied at Lady Brabourne College, Kolkata, and did her Masters from Calcutta University in Botany in 2004, securing the third position. She obtained her doctoral degree from the same university in the field of Microbiology. Presently she is an Assistant Professor in the Department of Botany at Shri Shikshayatan College, Kolkata. She had been actively involved in both teaching and research for past decade and half, and has a number of publications in journals of international repute to her name. Her primary area of research interest is related to microbes and environmental sustainability.

Anjan Hazra works as a project linked personnel at Indian Statistical Institute, Kolkata. He has accomplished B.Sc. (Hons.) from University of Calcutta, M.Sc. in Botany from West Bengal State University and Ph.D. from University of Kalyani and Indian Statistical Institute, Kolkata. Heis actively involved in plant based research programsincluding agricultural and evolutionary genomics, crop improvements and molecular markers. He authors more than 12 research articles in various SCI/Scopus indexed journals ofinternational repute and two book chapters of Springer publishing house.

Nallakumar Kannan, MSc., M.Phil., PhD., SET, is Assistant Professor, PG & Research Department of Biotechnology, Mahendra Arts and Science College (Autonomous) Affiliated to Periyar University, Tamil Nadu, India. He earned his BSc in Zoology from Periyar University, Tamil Nadu, India in the year of 1997. He holds his MSc in Zoology from the University of Madras, Tamil Nadu, India and he hold his M.Phil., in Zoology from the same University in the year of 2001and he awarded his M.Phil in Periyar University in the year of 2006. He has more than 16 years teaching experience and he acted as a board member in the Autonomous colleges affiliated to Periyar University and Bharathiyar University. He received 6 students projects from TNSCST, Chennai, Tamil Nadu, India.

Junaid Ahmad Malik has completed his Ph.D in Zoology (Wildlife) in 2015. He is working as a Lecturer in Zoology at Govt. Degree College, Bijbehara, Anantnag. Dr. Malik has published many research articles and technical papers in many International and National reputed Journals like Springer, Elsevier, Taylor and Francis etc. He has also authored several books and book chapters of reputed publishers. He is in the editorial board and a regular reviewer of several journals of reasonable reputation. He is a lifetime member of several organizations and is in the editorial board of various journals. Dr. Malik has participated in various national and international conferences/seminars/symposiums/workshops as well. He is the life member of SBBS (Society for Bioinformatics and Biological Sciences) with membership id LMJ-243.

Arup Kumar Mitra is an Associate Professor, Department Of Microbiology, St. Xavier's College, Kolkata, India with teaching experience of UG -26Years and PG- 17 Years. He has more than 20 years of research experience. He has completed his Ph.D. in the year 1995 and the title of his thesis was Studies on the uptake of heavy metal pollutants and its effect on their growth, productivity and mammalian system. He has supervised total of 11 projects (Recently One DBT and One WB DST projects are running) and has 6 students presently registered for their PhD degree in the Department of Marine Science and Botany, University of Calcutta and Department of Microbiology, St.Xavier's College. He as a total of 86 publication in journal of national and international repute and has written 36 books.

Govarthanan Muthusamy, PhD, is a Research Professor in the Department of Environmental Engineering, Kyungpook National University, South Korea. He earned his B.Sc., M.Sc, and M.Phil from Periyar University, Salem, Tamil Nadu (India). He completed his PhD in Chonbuk National University, South Korea in the year of 2015 and his field of expertise is Bioremediation of toxic metals. He worked as a Postdoc in South Korea and Japan. He has published more than 70 papers in several journals with good impact factor.

Joan Mwihaki Nyika is a PhD Researcher and Tutorial Fellow at the University of South Africa and the Technical University of Kenya, respectively. She obtained her Msc. (Land and Water Management) and Bsc. (Biochemistry) from the University of Nairobi and the Jomo Kenyatta University of Agriculture and Technology in Kenya in 2017 and 2011, respectively. She has conducted vast research in solid waste management, heavy metal pollution and remediation in land and water and hydrological modelling. She has more than 10 publications in peer reviewed articles and book chapters. She has more than 5 years experience in mentoring undergraduate students.

Amit Kumar Pal is working as an assistant teacher of life sciences in West Bengal Schools Education Service. He completed his B.Sc. (Hons.) and B.Ed. from University of Burdwan and M.Sc. from University of Kalyani. He was awarded his Ph.D. degree in Botany from University of Kalyani in 2018. Dr. Pal also served as a guest lecturer in Kanchrapara College, West Bengal. His research interest includes – isolation of heavy metal tolerant microorganisms from soil and their exploitation in modern agriculture. He authors about 11 publication in reputed national and international journals.

Raikamal Pal obtain her Master's degree in Botany from University of Calcutta in the year 2003 and PhD from University of Calcutta on 2013 under the supervision of Professor Rita Kundu. Her PhD thesis title was " **Heavy metal accumulation and their effects in some plants from contaminated areas with special reference to lead ".**She got her Bachelor's degree in Botany from Lady Brabourne College under University of Calcutta. She is now working as Lecturer in Shri Shikshayatan College from 2010 in the Department of Botany. She has excellent academic record throughout her carrier.

Suparna Pal, is an Assistant Professor, PG Depertment of Botany, Lady Brabourne College, affiliated under University of Calcutta, West Bengal, India. She had obtained her B.Sc in Botany from University of Calcutta, WB, India. She pursued her M.Sc in Botany with Plant physiology, Biochemistry and molecular biology specialization from University of Calcutta. She qualified UGC-NET on 2004 and was the recipient of UGC-RFSMS fellowship on 2005. She was awarded PhD in the year of 2013 and her field of expertise is heavy metal toxicity with the special emphasis on cadmium, chromium induced genotoxicity assessment and exploration of phytoremedial potential of local weeds. She has few publications in national and international journals. She had successfully completed UGC-Minor research project on 2011. Now she has a registered PhD candidate under Calcutta University. In 2020 she is awarded with West Bengal DST (Department of Science and Technology) - Major research project on nannoarticle mediated amelioration of Cd toxicity in rice plant.

Surya Narayan Pradhan is currently working as an Assistant Professor in School of Life Sciences, Sambalpur University Odisha.

Yuvarajan Ragumathan, M.Sc, M.Phil, PhD, is Assistant Professor, PG & Research Department of Biotechnology, Mahendra Arts and Science College (Autonomous), Affiliated to Periyar University, Tamil Nadu, India. He earned his BSc Biotechnology & M.Sc Bioinformatics from Periyar University, Tamil Nadu, India. He completed his PhD in the Year 2018 and his field of expertise in Nano-Biotechnology. He was selected as adjunct Professor in Saveetha Dental College Hospital, Chennai, Tamil Nadu, India in the year of 2019. He has published in several journals of national and international repute.

Karthika Rajamanickam, PhD, is an Assistant professor in PG & Research Department of Biotechnology, Mahendra arts and science college (Autonomous), Affiliated under Periyar University, Salem, India. She obtained her B.Sc & M.Sc in Biotechnology in Vivekandha College of Arts and Science for Women, Namakkal, India. She has carried out her research programme M.Phil-Biotechnology in the field of Hydrocarbon Degradation. She has completed Ph. D in the year of 2018 and expertise in the field of Bioremediation of hydrocarbon pollutants. She received a Research Fellowship for Research Scholar (RFRS) from Tamil Nadu State Council for Science and Technology (TNSCST), Government of Tamil Nadu. She has published several papers in National and International reputed journals.

Vivek Rana is currently working as a Research Associate in the field of water quality management at Central Pollution Control Board, Ministry of Environment, Forest and Climate Change, Delhi (India). He has completed his Ph.D. on "Assessment of natural wetlands and its evaluation for the treatment of wastewater using *Typha latifolia* and *Colocasia esculenta* – A sustainable approach" from the Indian Institute of Technology (ISM) Dhanbad (India). He has a working experience of 6 years in the field of phytoremediation, wastewater treatment, constructed wetlands, and metal pollution. He has published several papers in journals of high international repute.

Thirumalaisamy Rathinavel, PhD, is an Assistant Professor in the PG & Research Department of Biotechnology, Mahendra Arts and Science College (Autonomous), Namakkal (Dt.), Affiliated to Periyar University, Tamil Nadu, India. He completed Graduate and Postgraduate degrees in Biochemistry discipline during the year 2001-2006 in Vysya College, Salem (Dt.), Affiliated to Periyar University, Salem (Dt.), Tamil Nadu, India. He also completed M.Phil Biochemistry in Bharathidasan University, Tiruchirappalli (Dt.), Tamil Nadu, India under Directorate of Distance Education (DDE) mode in February 2008. He started his Teaching career as Lecturer in the year 2007 in Biotechnology Department at PGP College of Arts and Science, Namakkal (Dt.), Tamil Nadu, India. He also completed Post Graduate Diploma in Biotechnology (PGDBT) and Master of Business Administration (MBA) in Finance under the Directorate of Distance Education of Periyar University, Salem (Dt.), Tamil Nadu, India during the year 2008-2011. He has proven his teaching excellence through clearing Lectureship for Life Sciences subject in Tamil Nadu State Eligibility Test for Lectureship (TNSET 2016) conducted by the Nodal Agency Mother Teresa Women's University, Kodaikanal, Tamil Nadu, India. He successfully completed Doctoral degree (PhD) in Biochemistry at Rajah Serfoji Government College (Autonomous), Thanjavur (Dt.), Affiliated to Bharathidasan University, Tiruchirappalli (Dt.), Tamil Nadu, India. His research finding in the fields of Anti-inflammatory Phytocompounds and it's In Silico Docking works published in reputed national and international journals and conference proceedings.

Dipankar Roy, is a PhD Research Scholar under the guidance of Dr. Arup Kumar Mitra, Department of Microbiology, St. Xavier's College, Kolkata. He has 2 years of research experience and his research interest includes bioremediation.

Chandan Sengupta is a retired professor of Microbiology at Department of Botany, University of Kalyani, India. He completed B.Sc. (Hons.), M.Sc. and Ph.D. in Botany from University of Kalyani. He served as a lecturer at Raigunj University College during 1982 to 1985 and later on joined Kalyani University. Dr.Sengupta has more than 40 years of research and teaching experience in various arena of microbiology. His active research area includes – plant growth promoting rhizobacteria, heavy metal tolerant microorganisms, fishery microbiology, mangrove ecosystem, and crop improvements. He has authored 110 research articles, 23 conference papers and 10 book chapters so far. Throughout his academic career, he has supervised 29 Ph.D. scholars as well as 92 M.Sc. dissertations. He acted as Chancellor's nominee and expert in theselection committees for appointment of Assistant Professors, Associate Professors and Professors at various Universities in West Bengal time to time. He also involved in various recruitment boards for selection of research scholarsand acted as chairperson of several conferences and symposia. He is serving as an external peer reviewer and editorial board member of some reputed journals like - BMC Plant Biology, International Journal of Biological Macromolecules, Current Plant Biology, Plant Science Today, and Frontiers in Environmental Microbiology

Selvankumar Thangaswamy, PhD, is Associate Professor and Head, PG and Research Department of Biotechnology, Mahendra Arts & Science College (Autonomous), Namakkal, Tamil Nadu, India. He obtained his PhD in Environmental Biotechnology, Bharathidasan University, Tiruchirappalli, Tamil Nadu, India. He has honoured with several awards including Best Teacher Award -2019 of Periyar University. He has 21 years of teaching, research experience in Microbiology, Immunology, Bioprocess Technology and Nanotechnology Biotechnology. His current research work is concentrated on the identification of peptide drugs, Nanoparticle-based drug delivery and metagenomics. For his credit, he has published 76 research articles in National and International Journals and submitted 42 microbial nucleotide sequence data bank submitted in NCBI. He has carried out research projects from UGC, DST-FIST and DBT.

Index

16srRNA 98

A

Adsorption 2-3, 7, 13-15, 24, 26, 29, 34, 40, 50, 58, 74, 99, 102, 171, 181, 186, 208-211, 217-218, 223-227, 231, 233, 235, 244, 247, 257, 260-261, 263-267, 269, 271-275, 278-280, 283, 286-287
Alum 6, 13
Antagonistic 98-99, 122-123, 177
Antibiotics 45, 48, 50, 62, 98, 106, 194, 281
Aqueous 4, 10-11, 13, 16, 33-34, 56, 90, 105, 200, 207-208, 211-213, 215-218, 220, 225, 227, 230, 233-234, 257, 259, 261, 271-272, 275, 279, 283, 285-286

B

Bacteria 3-5, 12-13, 20, 24, 26-29, 31-35, 39-40, 43-45, 47-49, 52-53, 56-57, 60, 62-64, 69-70, 74-81, 83-84, 86, 89-90, 92-95, 97-100, 102-103, 122, 124-125, 181, 193-195, 197-202, 204-206, 208, 212-213, 216, 224, 228-229, 231-232, 242, 251, 259, 272, 275, 278-281, 283, 286
bioaccumulation 29, 31-34, 39, 99, 106, 116, 127, 137, 144, 154, 164, 181, 190, 195, 207, 209, 216-218, 229-230, 251, 258, 269, 285
Biochemical Oxygen Demand 239, 244, 256
Biofilms 29, 31, 95, 274-275, 278-281, 284-288
Biogenic nanoparticle 221, 229-230
Bioremediation 1-3, 10, 13-14, 20-22, 24-26, 29-31, 33-35, 37, 39-42, 45-47, 49, 51, 53-60, 62-72, 74, 86, 89-90, 99-100, 102-103, 123-127, 166, 181, 183, 190, 193, 197-198, 200, 217, 219, 221-223, 229, 232-233, 235, 245, 256, 258, 274, 277-278, 282-283, 286
biosorption 3, 31, 33-35, 124-126, 181, 194-195, 207-213, 215-220, 260, 265, 269, 272, 278-279, 283-288

C

Cadmium 20, 34-35, 98-99, 101, 124, 127-129, 139-147, 186, 190-195, 197-205, 207, 213, 217-220, 224, 229-231, 233, 271, 275-276, 279-280, 282-283, 287
Cadmium and lead tolerance 190
Chemical Oxygen Demand 239, 244, 256
chemicals 7, 21, 38-39, 44, 50, 55, 59, 63, 66, 147, 177, 213, 228, 243-244, 256, 259, 281-282
Chromate Efflux 73
Chromate Reductase 73-75, 78-83, 86-87, 90, 92, 94-96, 100, 124
Chromium 22, 32-33, 73-77, 79-81, 83, 86, 88-101, 103-104, 106-109, 116-125, 127, 129, 139-147, 197, 203, 205, 212, 216-219, 225, 230-231, 254, 271-272, 275-276, 280, 283, 287-288
Chromium Bioremediation 73, 123
Chromium Toxicity 73-74, 80, 99, 122, 145
Consortium 29, 91, 98-99, 122-123, 215, 229, 242
Constructed Wetland 238-239, 246-247, 249-256
Constructed wetlands 238-239, 246-256
contaminants 1-4, 6-7, 9-10, 20-21, 23-24, 26-27, 31, 40, 43, 47, 49, 52, 55, 57-60, 62, 64-67, 70, 74, 92, 125, 128, 131-132, 137, 148, 152, 166-167, 181-182, 190, 193, 221-224, 226-227, 229-232, 235, 238-239, 242, 252, 255-256, 274, 276, 278
contamination 1-2, 20, 38, 49, 52, 55, 57, 70, 73, 92-93, 95, 99, 125, 127, 129, 131-132, 135, 146, 148-152, 164, 166-167, 181, 183, 185, 188, 190-193, 198-199, 203, 223-225, 230-231, 233, 239, 251, 253, 275-276, 280-281, 283, 285-286
Cow Dung 1, 5, 10, 15
Cytotoxicity 257-258

D

desorption 208, 213, 279
detoxification 26, 31-32, 37, 43-44, 46, 51, 53, 69, 71-

72, 75-77, 86, 90-91, 96, 102-103, 124, 127, 133, 135, 137-138, 143, 182, 197, 206, 218, 225, 227
Drinking water 5, 149, 166-168, 174-178, 183-184, 187-189, 232, 258, 271, 276-277, 280-281, 284, 286, 288

E

Emergent plants 238, 247
environment 9, 14, 20-21, 23-25, 29, 31-32, 34, 37-40, 45, 47, 49, 51-52, 55-57, 63-64, 68, 70, 72-74, 79, 95, 98-99, 101-102, 123-125, 127-131, 139-141, 143-147, 149, 151, 160, 164-168, 180, 182, 184, 187, 191, 193-194, 197-198, 200-201, 204, 206-209, 213, 215, 222-223, 225-226, 230-231, 238-239, 242, 249-251, 254-256, 258, 275-276, 278-279, 283-285, 288
enzymes 22, 24, 37, 39, 41-52, 55-63, 65-71, 73-76, 80-83, 86-87, 99-101, 133, 136, 138, 140, 143, 183, 191-194, 197, 202, 228-230, 242, 246, 258, 282
EPS 31, 272-274, 278-280, 284-286
Exopolymer 95, 257, 259, 265, 269

F

Fabrications 221
Fluoride 12-13, 18, 166-174, 176-181, 183-188
fungal biosorption 206, 211
fungal cell wall 197, 206, 209-210

G

genotoxicity 131, 134, 143

H

Health effect 166
heavy metal 10, 20-30, 32-36, 39, 50, 59, 72-73, 91, 98, 101-103, 124-130, 133-138, 140-146, 148-149, 151, 155, 164-165, 190-195, 197-202, 204, 206-209, 215-216, 219, 221-227, 229, 232-233, 235, 258, 269-271, 273, 275-276, 278-280, 282-284, 286-288
Heavy metal pollution 20, 32, 148, 206-207, 216
heavy metals 6-7, 10, 14-15, 18, 20-29, 31-36, 38-39, 57-58, 60-61, 66-67, 71, 80, 89, 96, 98-99, 101-103, 124-129, 131-136, 138-140, 143, 147-151, 153-154, 165, 187, 191-194, 196-202, 205-208, 213, 215-219, 221-227, 231-232, 234, 239, 243, 249-251, 258, 261, 269, 271, 274-279, 282-286, 288
Hexavalent Chromium 73-75, 79, 86, 88-91, 93-97, 106, 142-143, 147, 203, 217, 287
Hyper accumulator 148, 161, 193

I

Iron binding 77, 257-258, 260, 263, 267, 272

L

Landfill 12, 91, 222-223, 238-244, 246-247, 249-256, 275
LD50 98-99, 104, 106, 108-109, 123, 281
Leachate 12, 238-256
Lead 23, 28, 35, 45, 98-99, 101, 125, 139-141, 144-146, 148, 150-151, 153-154, 160, 164-165, 174, 190-195, 197, 199-201, 203, 205-207, 219, 224-225, 230, 233, 238, 242, 275-276, 280, 282-283, 286

M

mechanism of biosorption 206
Metabolism 22, 39, 43, 45, 54, 59, 64, 75, 79, 98-101, 128-130, 138, 141, 144-145, 166, 176, 180, 191, 193, 205, 209
metal detoxification 26, 46, 72, 133, 197, 206
metal transporter 137
Microorganism 24, 27, 39, 64, 75, 92, 102, 193, 209, 258, 274, 282
microorganisms 1-2, 20-33, 36, 40, 43, 45, 54-57, 60, 62-65, 67, 86, 89-91, 98-99, 101-103, 124-125, 181-182, 193, 200, 204, 206, 208-209, 216, 219, 221, 228-229, 232, 234, 246-247, 259, 261, 263, 274, 277-278, 280-282
Municipal Solid Waste 199, 238-242, 249, 251, 253, 256

N

Nano-bioremediation 221-223, 225, 232-233, 235
Nanomaterials 7, 9, 16, 52, 66, 71-72, 221, 227, 229, 231, 234, 237
Nano-Materials 1-3, 9
Nanotechnology 1-3, 5, 12, 14-15, 68, 221, 223-224, 228, 231-236
NCBI 33, 79, 82, 87, 98, 104
NCMR 98

O

oxidoreductase 37, 43-44, 58, 69, 72-73, 82

P

Palm Kernel Shell 1, 6, 10
PGPR 190, 193-195, 197, 202
Phyto-bioremediation 230
Phytoremediation 14, 43, 52, 70, 127, 131-132, 137-142, 144, 146, 148, 151-153, 163-164, 166, 181, 183, 193, 198-204, 230-231, 238-239, 245, 249, 251-256
Phytostabilization 127, 132, 139, 181, 245, 250
Phytovolatilization 132, 245-246
plants 5-6, 20-21, 32, 35, 38, 40, 43-45, 47, 49, 56-58, 60, 62, 68, 75, 89, 101, 127-147, 150-159, 161-164, 166-167, 169, 171-172, 181, 183-186, 191-197, 199-202, 204-206, 222, 228, 230-232, 238, 246-247, 250, 252, 256, 258, 275
Pollutants 1-2, 6, 12, 15, 24-25, 31, 37-40, 43, 46-47, 49, 51, 55-58, 60-62, 64-65, 69, 71, 101-103, 124, 128, 141, 152, 184, 198-200, 221-223, 225, 227, 230-231, 234, 237, 239, 244-247, 249-250, 252, 258, 274-275, 282-283
pollution 16, 18, 20-22, 29, 32-33, 35-36, 38, 40, 49, 51-53, 55-57, 67, 94, 102, 128, 131-132, 137, 140-141, 147-149, 160, 164-165, 169, 171, 176, 183-185, 188, 191, 197-199, 201-202, 204, 206-207, 216, 221-222, 225, 227, 238, 250-253, 274-276, 278, 282, 285
proline 133-134, 136-137, 140-141, 146-147
Pseudomonas sp 68, 124, 194, 198, 201, 257-259, 261-262, 267, 269, 286

R

Rhizofiltration 127, 132, 151, 200, 230
Ricinus communis 148, 153, 155, 161-163

S

safety,Efficacy 257
Sanitary landfill 238, 252
Sensor 1, 270
Sludge 2, 5-6, 10, 14, 18, 74, 98-100, 102, 107-108, 122-123, 142, 148, 165, 191, 199, 201-203, 208, 218, 244-245, 247, 251, 254, 258-259, 272, 275, 278, 280, 282, 287
soil contamination 70, 150-151, 190-191
soils 12, 24, 36, 51, 55-68, 74, 89-90, 92, 103, 125, 131-132, 139, 142, 144, 148, 151, 154-155, 164-165, 167, 169-173, 183-188, 191, 198-199, 201-205, 222, 230, 232-233, 235, 275-276, 287
Solid waste management 238, 251, 254
sorption 20, 24, 30, 34, 57, 170, 173, 209, 211-213, 215-216, 226-227, 234, 258, 260, 263-264, 270, 272, 277, 279-280
Sustainable development 67, 232

T

Tooth 166, 179
Toxic heavy metals 26, 32, 57, 98, 101-102, 124, 126, 136, 150, 202, 276, 282
transformation 18, 24, 32, 37, 40, 44, 46, 48, 50-51, 53, 57, 62, 65-66, 88, 195, 200, 229, 242, 247, 252, 274

U

Urbanization 38, 148-149, 222, 238-240, 249

W

water pollution 35, 38, 148, 274, 285
Water Treatment 1-7, 11, 14-18, 130, 227, 236, 258, 277-279

X

xenobiotic compounds 55, 59, 61-62, 65-66, 68

www.ingramcontent.com/pod-product-compliance
Lightning Source LLC
LaVergne TN
LVHW060410300926
849162LV00006B/1395